ENERGY SCIENCE, ENGINEERING AND TECHNOLOGY

BIODIESEL

BLENDS, PROPERTIES AND APPLICATIONS

ENERGY SCIENCE, ENGINEERING AND TECHNOLOGY

Additional books in this series can be found on Nova's website
under the Series tab.

Additional e-books in this series can be found on Nova's website
under the e-books tab.

RENEWABLE ENERGY: RESEARCH, DEVELOPMENT AND POLICIES

Additional books in this series can be found on Nova's website
under the Series tab.

Additional e-books in this series can be found on Nova's website
under the e-books tab.

ENERGY SCIENCE, ENGINEERING AND TECHNOLOGY

BIODIESEL

BLENDS, PROPERTIES AND APPLICATIONS

JORGE MARIO MARCHETTI

AND

ZHEN FANG

EDITORS

New York

For permission to use material from this book please contact us:
Telephone 631-231-7269; Fax 631-231-8175
Web Site: http://www.novapublishers.com

NOTICE TO THE READER

The Publisher has taken reasonable care in the preparation of this book, but makes no expressed or implied warranty of any kind and assumes no responsibility for any errors or omissions. No liability is assumed for incidental or consequential damages in connection with or arising out of information contained in this book. The Publisher shall not be liable for any special, consequential, or exemplary damages resulting, in whole or in part, from the readers' use of, or reliance upon, this material. Any parts of this book based on government reports are so indicated and copyright is claimed for those parts to the extent applicable to compilations of such works.

Independent verification should be sought for any data, advice or recommendations contained in this book. In addition, no responsibility is assumed by the publisher for any injury and/or damage to persons or property arising from any methods, products, instructions, ideas or otherwise contained in this publication.

This publication is designed to provide accurate and authoritative information with regard to the subject matter covered herein. It is sold with the clear understanding that the Publisher is not engaged in rendering legal or any other professional services. If legal or any other expert assistance is required, the services of a competent person should be sought. FROM A DECLARATION OF PARTICIPANTS JOINTLY ADOPTED BY A COMMITTEE OF THE AMERICAN BAR ASSOCIATION AND A COMMITTEE OF PUBLISHERS.

Additional color graphics may be available in the e-book version of this book.

Library of Congress Cataloging-in-Publication Data

Biodiesel : blends, properties, and applications / editors, Jorge Mario Marchetti, Zhen Fang.
   p. cm.
Includes index.
ISBN: 978-1-63117-024-9(softcover)
1. Biodiesel fuels. I. Marchetti, Jorge Mario. II. Fang, Zhen.
TP359.B46B558 2011
665.37--dc23
                2011015359

Published by Nova Science Publishers, Inc. † *New York*

CONTENTS

PREFACE

Biodiesel, which consists of long-chain fatty acid methyl esters (FAME) obtained from renewable lipids such as those in vegetable oils or animal fat can be used as both an alternative fuel and an additive for petroleum diesel. This book gathers research from across the globe in the study of biodiesel blends, properties and applications. Topics discussed include biodiesel purification methods; exhaust emissions study of biodiesel operated garbage trucks; the heterogeneous catalyst for the transesterification of triglycerides into biodiesel; valorization of wastes and by-products derived from biodiesel manufacturing and biodiesel production using cation-exchange resin as heterogeneous acid catalyst

Chapter 1 - This chapter provides an insight into the study of exhaust emissions from in-use garbage trucks operating under different idling modes. Several engine and fuel characteristics that have been widely predicted to affect the emission concentrations of pollutants were used in the evaluation of exhaust emission characteristics. The emission concentrations of carbon monoxide (CO), sulfur dioxide (SO2), oxides of nitrogen (NO, NO2, and NOX), and carbon dioxide (CO2) were examined with respect to different engine parameters such as fuel temperature, coolant temperature, and percent fuel. A Testo 350 XL portable emission monitoring instrument was used to collect the data on a one-second interval scale simultaneously with the engine parameter data that were collected using OBD software. The tail pipe emissions from ultra low sulfur diesel (ULSD) are compared with biodiesel (BD/B20) emissions. Biodiesel garbage trucks emitted higher concentrations of NO_X during both hot and cold starts. A comprehensive study of different vehicle, fuel, and operating parameters that affect the exhaust emission concentrations were conducted to find a better alternative to ULSD.

Chapter 2 - The world's air pollution problems are increasingly been related to automotive exhaust emissions. The implementation of diesel and biodiesel blends in passenger vehicle engines have gradually produced a new ecotoxicological profile of urban and rural air pollution, where nanoparticles, volatile exhaust fractions, microparticles and aerosol agglomerates dominate the spectrum of emission species. The effects of these species are increasingly associated with cardiovascular diseases, lung cancer and increase in all-cause mortality in the human population, particularly in urban and other highly trafficked areas. Also, the size of particles and agglomerates from exhaust has been related to particular diseases, risks of contracting types of pathologies and development of cardiovascular complications. Particles have therefore selectively been addressed in this literature review for adverse health effects. With particular focus has biodiesel blending been extensively reviewed for chemical species and associated adverse health effects. The reviewed data suggests that the legislatory environmental health authorities worldwide are not fully updated with all aspects of air pollution and that filtering technologies, fuel types and threshold values for particle content in the air are not up to date with the medical and patho-physiological knowledge.

Chapter 3 - Oil provides energy for 95% of transportation and the demand of transport fuel continues to rise. According to a forecast made by the International Energy Agency (IEA) global oil demand will rise by about 1.6% from 75 mb/d in the year 2000 to 120 mb/d in 2030. Almost three quarters of the increase will be from the transport sector. Oil will remain the fuel of choice in road, sea and air transportation. In developing countries, the increase in demand for oil for use in transport sector is expected to grow at a much higher rate. All countries including India are grappling with the problem of meeting the ever increasing demand of transport fuel within the constraints of international commitments, legal requirements, environmental concerns and limited resources. In this connection transport fuels of biological origin have drawn a great deal of attention during the last two decades.

Bio-fuel is a generic term that is used to refer to liquid or gaseous fuels that are produced from a biological source. The term liquid 'bio-fuel' is more commonly used to refer to specific types of bio-fuels used as fossil fuel substitutes. These are further defined by the particular type of biomass from which they are made, and the degree to which they are refined before use. The most common types of liquid bio-fuel are straight vegetable oils, ethanol and biodiesel.

Chapter 4 - In the existent process to produce biodiesel, alkali hydroxide dissolved in methanol catalyzes the transesterification of the feedstock oil consisting of triglycerides with methanol. The homogeneous base catalysis is very effective in enhancing the reaction efficiency, but it is necessary to wash the homogeneous base catalyst off the crude biodiesel with water after the reaction. A massive amount of wastewater originating in the purification mentioned above is one of the problems causing costly production of biodiesel. Also, the purification allows stable emulsion to be formed, which not only disturbs the process operation but also decreases the productivity. Therefore, many researchers took their interests in utilizing heterogeneous catalytic process for biodiesel production. In the present work, research papers studying the heterogeneous catalyst for the transesterification of triglycerides are reviewed.

Chapter 5 - Expansion of non-traditional oilseed crops is expected to have great possibilities at worldwide level, while in other countries is already considerably increasing. These crops do not compete for lands with those destined to food production.

It was necessary to identify the requirements, limits and bio-meteorological tolerance and conditions for each of species, taking into account the climatological characteristics of native areas and the regions of successful cultivation around the world.

Afterwards, the resulting bio-climatological indicators had to be extrapolated to the Argentine lands. It was mandatory to consider growth aspects such as development and death chances by excess or deficiency.

In parallel, an agro-climatological inventory was performed based on: information of available climatological statistics, agro-climatological derived values and estimated indexes. With all the information gathered, the data was evaluated. As from the available database, geographical limits were mapped for the different variables that define aptitude classes: optimal zone, suitable, marginal and non-suitable.

Among the species investigated by our working group, this the review includes: Perennial species: *Jatropha curcas; Ricinus communis; Moringa oleifera; Pongamia pinnata* and *Cynara cardunculus* and Annual species: *Sesamus indicum; Guitzotia abyssinica; Sinapis alba; Thlaspi arvense* and *Crambe abyssinica*); identifying their agro-climatological aptitude for their cultivation in Argentina.

Chapter 6 - Renewable fuels have now come to play an important role in meeting the world's energy requirements. Due to the economic, environmental and supply problems derived from the use of fossil fuel, liquid biofuels are a growing and interesting complement to petroleum-based fuel. Biodiesel, which consists of long-chain fatty acid methyl esters (FAME) obtained from renewable lipids such as those in vegetable oils or animal fat can be used as both an alternative fuel and an additive for petroleum diesel.

After transesterification reaction, the obtained biodiesel must be purified before being used as diesel fuel in compliance with the EN 14214 Standard. The main disadvantages of using biodiesel are the high cost of the raw material and the need to purify the product, resulting in a very expensive final product in comparison to petrol-diesel. The first drawback to using biodiesel could be overcome by employing a cheap raw material such as used cooking oils or by using microalgae oils. On the other hand, improving product purification is one of the aims of this chapter with a view to fulfilling EN 14214 specifications.

Because FAME cannot be classified as biodiesel until the EN 14214 Standard specifications are met, the purification stage is essential. Depending on the raw material (used cooking oils, crude vegetable oils or refined vegetable oils) and the catalyst used, the purification stage will be more or less difficult. The purity level of the biodiesel has a strong effect on fuel properties and on engine life.

Several studies have examined the purification of biodiesel using different methods: liquid-liquid extraction (with water, acids or organic solvents), adsorption (silica gel, activated carbon, rice hull ash, bentonite, and magnesium silicate), ion exchange (cation or anion resins), molecular distillation and membrane separation.

The aims of this chapter are to revise the different purification methods described in the literature and examine the efficiency of three methods for removing methanol, glycerol and soaps under conditions that have been kept as close to commercial operating practice as possible: (a) adsorption (magnesium silicate and bentonite), (b) liquid-liquid extraction (distilled water, tap water, glycerol), and (c) ion exchange (cation resin). Further objectives are to study water and free fatty acid removal and the effect of these processes on final density, viscosity, glycerides and FAME.

Chapter 7 - Biodiesel fuels have recently drawn much attention due to the various advantages they have over petroleum-based fuels. However, the recent growth of the biodiesel industry has created a surplus of impure glycerol. This has resulted in a dramatic decrease in crude glycerol prices over the last few years and given rise to environmental concerns regarding the disposal of contaminated glycerol. On the other hand, the biodiesel obtained by alkaline transesterification requires purification. Water washing is frequently used; a process that generates large amounts of highly polluted wastewater. Consequently, the biodiesel industry has been forced to implement effective treatment technologies in order to fulfill the severe quality standards related to environmental protection that are currently being developed and to recover valuable compounds contained in wastes and by-products.

Anaerobic digestion is an interesting way to treat and revalorize abundant and low-priced glycerol and wastewater streams. This chapter examines the performance and the stability of the anaerobic digestion process of glycerol-containing waste derived from biodiesel manufacturing. The glycerol was previously acidified with phosphoric acid and centrifuged in order to recover the catalyst used in the transesterification reaction (KOH) as agricultural fertilizer (potassium phosphates). The results reveal that pre-treated glycerol has a high level of anaerobic biodegradability (around 100% COD) and that a substantial quantity of methane can be obtained in this way (292 mL CH_4/g COD removed at 1atm, 25°C). In addition, the treatment of wastewater derived from biodiesel manufacturing (428 g COD/L) was studied. Firstly, wastewater was acidified to recover free fatty acids. After centrifuging, the aqueous phase underwent two different physical-chemical pre-treatments to demulsify the remnant organic matter: coagulation-flocculation and electrocoagulation. The effluents of both pre-treatments were subjected to biomethanization. Of the two physical-chemical pre-treatments performed, electrocoagulation was found to be the most suitable. Furthermore, this chapter describes the anaerobic co-digestion of acidified glycerol and electrocoagulated wastewater after mixing at a proportion that allows the different glycerol and wastewater flows generated in the biodiesel industry to be simultaneously absorbed. Biodegradability was found to be around 100%, while the methane yield coefficient reached a mean value of 310 mL CH_4/g COD removed (1atm, 25°C), thus showing the convenience of treating both wastes simultaneously. This research forms part of what is known as the biorefinery concept; a relatively new term referring to the conversion of feedstock into a host of valuable chemicals and energy with the consequent economic and environmental benefits.

Chapter 8 - This study examined the effects of pressure, temperature and solvent to solid ratio (SSR) on extraction efficiency of triglycerides from powdered *Jatropha curcas* seeds and *Aquilaria crassna* seeds by using supercritical carbon dioxide (SC-CO_2) extraction. According to the results of response surface methodology, pressure presented more effective in increasing the recovery of triglycerides, but both temperature and the SSR value are important in obtaining high concentration of triglycerides from two plant seeds that are useful for biodiesel materials. The experimental equilibrium solubilities of these two oils dissolved in SC-CO_2 at temperatures of 318–338 K and pressures of 200–350 bar were also investigated. The results indicated that the solubility of the oils linearly increased with the density of SC-CO_2 and the experimental data were successfully correlated by the Chrastil equation. Finally, this study demonstrated that subcritical hydrolysis followed by supercritical methylation of the SC-CO_2 extracted oil is a feasible two-step process in producing a 98.5% purity level of biodiesel from *Jatropha curcas* kernels.

Chapter 9 - This chapter presents a summary of the results obtained from a field program using the Toledo Area Regional Transit Authority (TARTA) public transport buses, running on biodiesel blend and ultra-low sulfur diesel (ULSD) fuels. This TARTA research program is divided into two studies: exhaust emissions and indoor air quality.

A comprehensive analysis of pollutant levels released from the tailpipe exhaust emissions resulting from the use of different biodiesel blends under different operating modes is included. Indoor air quality (IAQ) in buses operating on B20 grade biodiesel and ULSD has been studied for more than two years. Study of vehicular pollutant exposure resulted in concluding that it is safe for the passengers as well as drivers, with the indoor pollutant levels being well below the available health guidelines. Particulate matter filters collected between 2007 and 2009 were analyzed using computer controlled environmental scanning electron microscopy and energy dispersive X-ray spectrometry (SEM/EDS) to characterize shape and aspect ratio of the fine particulate matter.

Chapter 10 - Biodiesel production through free fatty acids (FFA) esterification with methanol using cation-exchange resin as heterogeneous acid catalyst is becoming a hotspot. In this work, three types of cation-exchange resins (NKC-9, 001×7 and D61) were employed to prepare biodiesel via the batch esterification of FFA from acidified oils generated from waste frying oils. According to this study, the catalytic activity of NKC-9 was higher than that of 001×7 and D61. The FFA conversion by NKC-9 increased with increasing in the amount of catalyst, reaction temperature and time and methanol/oil molar ratio. The maximal conversion of reaction was approximately 90.0%. Further, NKC-9 resin exhibited good reusability in batch esterification. Gas chromatography-mass spectrometry analysis revealed that the production was simplex and mainly composed of C16:0 (palmitic), C18:2 (linoleic), and C18:1 (oleic) acids of methyl ester, respectively. Furthermore, continuous esterification of acidified oil with methanol was carried out with NKC-9 resin in a fixed bed reactor with an internal diameter of 25 mm and a height of 450 mm to produce biodiesel. The results showed that the FFA conversion increased with increases in methanol/oil mass ratio, reaction temperature and catalyst bed height, whereas decreased with increases in initial water content in feedstock and feed flow rate. The FFA conversion kept over 98.0% during 500 h of continuous esterification processes under 2.8:1 methanol to oleic acid mass ratio, 44.0 cm catalyst bed height (the amount of packed resin 87.5 g), 0.62 ml/min feed flow rate and 65 °C reaction temperature, showing a much high conversion and operational stability as compared with the batch esterification. Furthermore, the loss of sulfonic acid groups from NKC-9 resin into the production was not found during continuous esterification. In sum, NKC-9 resin as heterogeneous acid catalyst shows the potential commercial applications to esterification of FFA.

Chapter 11 - The esterification of lauric acid and methanol is explored using thermally coupled distillation sequences with side columns and with minimum number of reboilers. The product of the esterification can be used as biodiesel. This is a major step forward since thermally coupled reactive distillation sequences with side columns and with minimum number of reboilers offer significant benefits, such as: reductions on both capital investment and operating costs due to the absence of the reboilers, higher conversion and selectivity as no products are recycled in the form of reflux or boil-up vapors, as well as no occurrence of thermal degradation of the products due to a lower temperature profile in the column. In the traditional process, lauric acid is fed as saturated liquid, while the methanol is fed as saturated vapor stream to the main column, then we can take heat from vapor stream to deliver it to the system without a reboiler. Rigorous simulations were performed in AspenTech AspenONE engineering suite to design this novel reactive distillation process and evaluate the technical and economical feasibility. The results indicate that the energy consumption of the complex distillation sequences with a side column can be reduced significantly by varying operational conditions, in comparison with conventional reactive distillation sequences.

Chapter 12 - Among the different possible sources, the methyl esters of vegetable oils, known as biodiesel is receiving increasing interest because of their low environmental impact. In addition biodiesel fuel may not require any significant modification of existing diesel engines. The present investigation focuses on the study on simultaneous reduction of NOx and smoke by water emulsion of Karanji oil methyl ester in a single cylinder, direct injection diesel engine with exhaust gas recirculation. The quantity of EGR was varied from 5 % (by vol.) to 15 % (by vol.) in steps of 5 % (by vol.). The quantity of water was varied from 5 % (by vol.) to 15 % (by vol.) in steps of 5% (by vol.). Water-emulsified diesel fuel with EGR technology has been proven to reduce nitrogen oxides (NOx) and smoke simultaneously at relatively low cost compared to other pollution-reducing strategies. From the test results, it is observed that water emulsion and EGR technology results in a substantial reduction of NO_x by 66 % and smoke emissions by 50 % with a compromise of 2.5 % on thermal efficiency.

Chapter 13 - At a time when the use of fossil fuel means increasing energy scarcity and an environmental crisis in the world in which we live, we need green innovations now more than ever. Growing attention has been drawn to the use of biofuels, such as bioethanol and biodiesel, which have gradually come to make up part of the total energy supply. Despite its rapid development, uncertainties about the environmental and ecological aspects of the production and consumption of biofuel still exist. A life cycle analysis (LCA), with deliberate system boundaries and life cycle inventories, has been applied to evaluate two principal functional parameters, the energy efficiency and GHG balance, from the cradle to the grave of different biofuel feedstocks. However, the results of the entire lifetime estimates varied dramatically in production chains, which make it difficult to take a holistic view about energy intake and yields, economic costs and values, environmental impacts and their benefits. Apart from the diversity in system boundaries and life cycle inventories, a variance in terminologies and the limitations of interdisciplinary communication are the main factors that affect the quality of the results. In this paper, various life cycle analysis models are reviewed and evaluated with a special emphasis on two important biofuels, i.e., biodiesel and ethanol.

Chapter 14 - In the last one and a half decades efforts for applying biofuels have increased all over the world. The main reasons of this are the rising demands for gas oil and the bio regulations in the European Union. Nowadays approx. 95% of the used bio blending components in diesel fuels is FAME (biodiesel). The use of bio gas oils (mixture of n-and i-paraffins) is gradually increasing. The biodiesels are produced by catalytic transesterification of different triglyceride-containing feedstocks (e.g., vegetable oils, used frying oils, animal fats, algae oils, brown grease, etc.) mainly with methanol. Many factors affect the quality and performance properties of these first generation biofuel; e.g., the type of the feedstock, and the alcohol; the transesterification technology and the catalytic system. The quality of the feedstock has the strongest effect on final FAME quality.

The aim of our systematic research and development activity was to produce high quality biodiesels. The feedstocks from various sources were transesterified in the presence of *Candida antarctica* immobilized lipase in a solvent free medium. The effect of the purity and the fatty acid composition of different feedstocks on the quality and performance properties of FAME were investigated.

Based on the experimental results the authors concluded that the physical and/or chemical refining of the raw materials is definitely necessary to produce high quality biodiesel. The significantly different fatty acid composition of the feedstock does not affect the enzymatic transesterification. It can be stated that biodiesels with adequate oxidation stability and CFPP value can be produced only from vegetable oils with high oleic acid content.

Chapter 15 - In recent years, the development of alternative fuels from renewable resources, like biomass, has received considerable attention. Biodiesel is defined as fatty acid methyl or ethyl esters from vegetable oils, when it is used as fuel in diesel engines and heating systems.

In this context, the use of used frying oil (UFO) as a raw material for biodiesel production has gained special interest since it has provided a potential added value to what has traditionally been treated as a residue.

Methyl or ethyl esters are the product of transesterification of vegetable oils with alcohol (methanol/ethanol) using an alkaline catalyst. In addition, the process yields glycerol, which has large applications in the pharmaceutical, food and plastics industries.

In the present work, the process of biodiesel production for pilot plant using sunflower oil as raw material with methanol and using potassium hydroxide as catalyst has been studied providing a mass and energy balance of the process.

The biodiesel quality meets European specifications defined by pr EN 14214:2003(E). The obtained results have been used for industrial scale up of the process.

Chapter 16 - Biodiesel production is gaining more and more relevance due to its environmental benefits such as its biodegradability, it has no CO or sulfur emissions, its produce a reduction in the CO_2 emissions, prolongs the engines lives, could be blend with regular diesel and it is produce from renewable and sustainable sources.

The name "biodiesel" is normally associated to any blend of regular diesel and biodiesel; however, the ASTM defines as biodiesel to the B100 (100% pure biodiesel) while any other blend is label in relation to the percentage of biodiesel in the mixture. The most common blend is B20, which has a 20% of biodiesel.

Based on the percentage of the biodiesel, the amount of regular diesel and the raw material from where the biofuel is produce, different properties, emissions and engine performance can be seen. A comparison of the biofuel based on these three main parameters will be presented with the aim of pointing out the differences of each blend, the influence on the emissions based on the quantity of biodiesel in the fuel and how this percentage affects the combustion of the internal engine of a car.

Even more, it will be shown international policies regarding the type of blend that should be used and when this should be reached.

In: Biodiesel: Blends, Properties and Applications
Editors: Jorge M. Marchetti and Zhen Fang

ISBN: 978-1-63117-024-9
© 2014 Nova Science Publishers, Inc.

Chapter 1

ANALYSIS OF EMISSIONS FROM ULTRA LOW SULFUR DIESEL AND BIODIESEL OPERATED GARBAGE TRUCKS

Venkata Naga Ravikanth Garimella and Ashok Kumar
Department of Civil Engineering, University of Toledo,
Toledo, Ohio, U.S.A.

ABSTRACT

This chapter provides an insight into the study of exhaust emissions from in-use garbage trucks operating under different idling modes. Several engine and fuel characteristics that have been widely predicted to affect the emission concentrations of pollutants were used in the evaluation of exhaust emission characteristics. The emission concentrations of carbon monoxide (CO), sulfur dioxide (SO2), oxides of nitrogen (NO, NO2, and NOX), and carbon dioxide (CO2) were examined with respect to different engine parameters such as fuel temperature, coolant temperature, and percent fuel. A Testo 350 XL portable emission monitoring instrument was used to collect the data on a one-second interval scale simultaneously with the engine parameter data that were collected using OBD software. The tail pipe emissions from ultra low sulfur diesel (ULSD) are compared with biodiesel (BD/B20) emissions. Biodiesel garbage trucks emitted higher concentrations of NO_X during both hot and cold starts. A comprehensive study of different vehicle, fuel, and operating parameters that affect the exhaust emission concentrations were conducted to find a better alternative to ULSD.

1. INTRODUCTION

1.1. Overview of Vehicular Emissions

Motor vehicle emissions are a result of combustion and evaporation of fuel. The most common types of transport fuels are gasoline and diesel. When the fuel in a vehicle is burned, pollutants such as carbon monoxide (CO), carbon dioxide (CO_2), total nitrogen oxides (NO_x), sulfur oxides (SO_x), hydrocarbons (HC) and particulate matter (PM) are emitted. According to the third national report on commuting trends, there was a 33% increase in population during the years 1990-2000 (Transportation Research Board, 2006). The report concluded that 65% of the commuters live in urban areas and spend almost 7% of their daily time in commuting. Also, there was a 2 minute increase observed in the overall commute time from the previous decade. This rapid increase in population, vehicle usage, and commute time has led to a striking increase in vehicular pollution. The increasing concern on cleanliness of the environment led to a demand in significant reduction of these toxic emissions from vehicles such as transportation buses, trucks (heavy-duty vehicle), and cars (light-duty vehicle). With increase in population, there is an increase in the number of heavy duty garbage collection trucks that are being deployed to collect the garbage from the curb sides. These waste collection services in the United States (U.S.) are catered to 75 million homes, 7 million businesses, and 100,000 government enterprises (Wright, 2001). Though a majority of the public concerns are related to the growing waste quantities and dangers posed by landfills, one element of the waste management industry that is being neglected is the health and environment hazard caused by the vast garbage truck fleets. Majority of the garbage trucks weigh more than 33,000 lbs and are classified as "heavy – heavy duty vehicles" (HHDV) (US Census Bureau, 1997). Some of the facts established by INFORM Inc. are listed below (INFORM, 2003).

- An estimated 136,000 garbage trucks, 12,000 transfer vehicles, and 31,000 dedicated recycling vehicles haul away America's garbage (179,000 vehicles in total which is twice the size of municipal transit sector (Beck, 2001)).
- An average garbage truck travels 25,000 miles annually, gets less than 3 miles per gallon, and uses approximately 8,600 gallons of fuel each year. Although garbage trucks constitute only a small percentage of vehicles on roads, they log an industry wide total of approximately 3.4 billion miles per year when compared to just 2 billion miles logged by the transit buses. Yet, not many studies are found in the literature that evaluates exhaust emissions from a large fleet of garbage disposal trucks.
- The garbage truck sector alone is responsible for consuming approximately 1 billion gallons of diesel fuel annually, representing nearly 3% of total diesel fuel consumed in the U.S.
- The average diesel-powered garbage truck costs over $170,000 and is not retired for 12 years. However, over 40% of garbage trucks are over 10 years old, making it the oldest among all vehicle fleets in the US.
- 82% of collection services are carried out by private companies and 18% by public entities, and the industry revenues, including waste hauling and recycling, exceed $43 billion per year.

Most garbage trucks are rear loader trucks that are equipped with a compactor to compact the garbage collected. Solid waste collection from household as well as commercial complexes accounts to a lot of idling time when loading the garbage and operating the compactor. Unlike commercial heavy duty trucks, these garbage trucks are operated on a constant stop-and-go driving style with speeds of 10 mph that involves instant acceleration and constant braking. This driving behavior results in heavy emission of air pollutants mentioned above and low fuel mileage (approximately 2.8 miles per gallon). Emission of air pollutants and fuel consumption are the key issues associated with truck idling. Garbage truck idling, in particular, is a major concern since it idles for nearly 70% of its operating time. Driving modes such as acceleration, cruise, and deceleration are also significant factors influencing the emissions from diesel engines.

A National scale air toxic assessment (NATA) was carried out by EPA in 2002 to identify and prioritize the sources of emissions and air toxics. According to the results of these assessments, mobile emissions accounted for 30% of the overall average cancer risk in 2002 compared to 50% during 1999.

NATA estimated the average cancer risk for 2002 to be 36 in a million. NATA also reported that the "background" pollutants, pollutants whose source is not currently known but exist in the environment and can cause health effects, accounted for nearly 45% of the overall average cancer risk. A national multi-pollutant emissions comparison conducted by EPA in 2005 concluded that on-road vehicles were major contributors of CO, NO_X and VOCs as shown in Figure 1. Figure 2 illustrates the multi-pollutant emissions comparison for Lucas County, Ohio in 2005 which also confirms the same.

Gasoline fueled vehicles tend to be the main source of CO emissions, substantial amount of volatile organic compounds (VOC), NO_x and modest amounts of PM_{10}, $PM_{2.5}$ (Schauer et al., 2002; Huang et al., 1999). On the other hand, diesel fueled vehicles are major contributors of PM_{10}, $PM_{2.5}$, NO_x, hydrocarbons to ambient air pollutant inventories (EPA, 2003; Perrot et al., 2004).

The notations PM10 and PM 2.5 are used to express particles having an aerodynamic diameter of 10μm or less and 2.5μm or less respectively. NOx and reactive hydrocarbons from these diesel vehicle emissions form low-level ozone (O_3) which causes severe respiratory hazards such as coughing, throat irritation, chest tightness, wheezing, inflammation of airways etc.

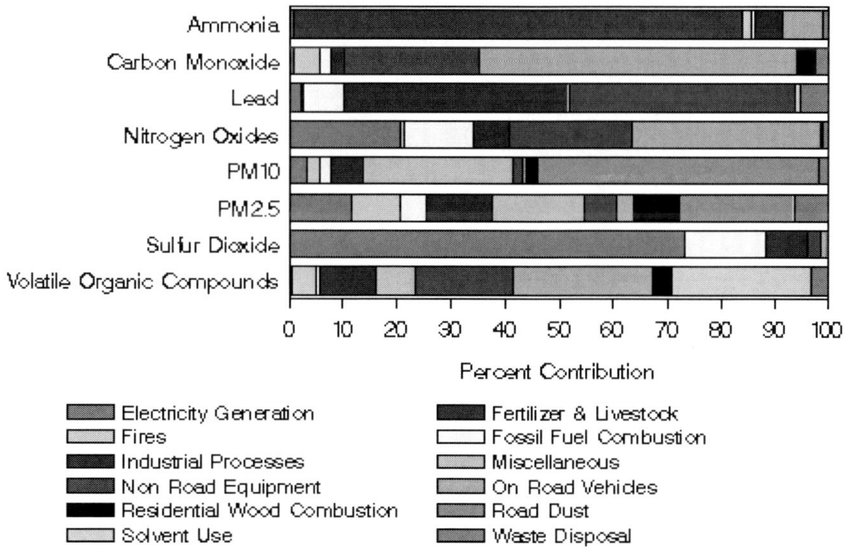

Figure 1. National multi-pollutant emissions comparison by source sector, 2005 (EPA, 2010).

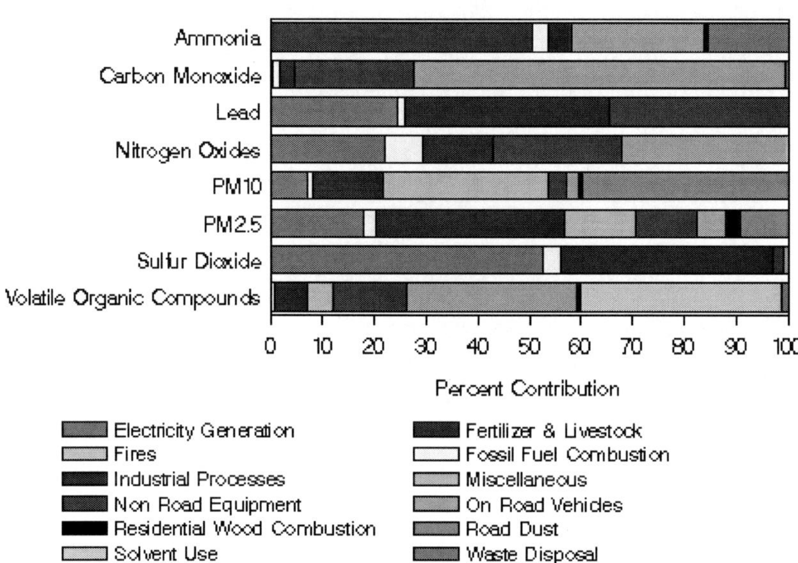

Figure 2. Lucas County, Ohio: Multi-pollutant emissions comparison by source in 2005 (EPA, 2010).

Since NO_x and reactive hydrocarbons form low level ozone (O_3), study of diesel fueled vehicles has been the focus area in developing various diesel fuel blends to reduce these emissions. More than 90% of the garbage collection trucks are fuelled by regular diesel fuel and most of these operate in and around residential areas. Exhaust from diesel fuels has long been recognized as a carcinogenic agent (Gierczak et.al, 1994). In order to curb the increase in vehicular pollution and thereby decrease the possible health effects due to its exposure, various federal source emission standards have been established by the United States Environmental Protection Agency (USEPA) under the Clean Air Act (CAA), 1990 which is a comprehensive federal law that regulates air emissions from stationary and mobile sources. According to the national air trends provided by the EPA, these pollutants have been at very high concentrations during the early 80's and have come down reasonably well due to the availability of efficient engine technologies in the recent past and also due to the stringent emission standards being maintained (Fernandez,1997; EPA, 2010). From 1990 to 2008, emissions of air toxins from on-road vehicles decreased by 35% and also, emission of six principal air pollutants decreased by 54% as shown in Table 1.

Table 1. Percent change in emissions of air pollutants (from USEPA)

Pollutants	1980 vs. 2008	1990 vs. 2008
Carbon Monoxide (CO)	-56	-46
Lead (Pb)	-99	-79
Nitrogen Oxides (NO_x)	-40	-35
Volatile Organic Compounds (VOC)	-47	-31
Direct $PM_{10.0}$	-68	-39
Direct $PM_{2.5}$	---	-58
Sulfur Dioxide (SO_2)	-56	-51

[*]Trend data not available, Direct PM10 emissions for 1980 are based on data since 1985; Negative numbers indicate reductions in emissions.

Very few studies have been carried out that have determined the indoor air quality in public transport systems operating on alternative fuels with an emphasis on the health effects associated with exposure to vehicular pollutants that are predominantly found as a consequence of infiltration of lead vehicular traffic emissions and self-exhaust infiltration. Kadiyala et al. (2010) have studied the in-vehicle exposure in association with the temporal behavior of the common indoor air pollutants that include CO_2, CO, NO, NO_2, SO_2, and $PM_{2.5}$ in both BD and ULSD buses.

Several studies have estimated the emissions from vehicles through real time exhaust emission testing as well as through the use of driving cycle based emission models such as EPA's MOBILE6 and CARB's EMFAC (Emission Factor Model). Several studies using these models also demonstrated that vehicular emissions are episodic in nature and average emissions for a trip are influenced by short-term events. However, in reality, these models are not the exact representatives of the real world conditions in which the vehicle operates and lack the spatial and temporal resolution to accurately characterize the episodic nature of vehicle emissions (Gierczak et al., 1994; Tong et al., 2000; Rouphail et al., 2001). Furthermore, the standard driving cycles adopted in studies conducted on characterizing exhaust emissions may not properly represent the real world driving patterns of a particular route or region. Hence, the emission estimates thus obtained can only be used on a conservative (rough) scale. As such on-board emission measurement using portable emission measuring systems (PEMS) during actual vehicle use has been widely identified as the most accepted way to study tail pipe emissions (Cicero-Fernandez and Long, 1997; Frey et al., 2001, Frey et al., 2002a; Frey et al., 2002b)

The current focus of EPA is to reduce the sulfur content from diesel fuels. Effective 2007, EPA has mandated the use of 15 parts per million (ppm) of sulfur content in all diesel fuels. A variety of fuels and fuel blends have been proposed as alternatives to the existing fossil fuels to meet these strict emission and fuel standards. Several state and private fleet authorities have begun testing various fuels blends and alternatives such as ethanol, compressed natural gas (CNG), hydrogen, biodiesel (BD/Bxx) blends and ultra-low sulfur diesel (ULSD, sulfur content 15ppm). Very few studies have reported tail pipe emissions comparison for truck fleets operating on biodiesel and ULSD. The major drawback in most of these studies has been the small size of the vehicle fleet considered and also the need to collect engine and emission data simultaneously to study the effect of engine parameters on emissions (Gierczak et al., 1994, Kelly et al., 1993 Fernandez et al. 1997, Denis et al., 1994, Leblank et al., 1994). A detailed literature review shows the lack of information on tail pipe emissions from in-use garbage disposal trucks running on different fuel types.

1.2. Research Approach and Objectives

To fill the knowledge gaps regarding the exhaust emissions from garbage trucks and to create awareness about the environmental and health affects due to the oldest, least fuel efficient and most polluting fleet which operates in closer proximity to residential areas than any other type of heavy duty diesel vehicles, an idle emission protocol was developed to provide real world assessment of the emissions from the city of Toledo garbage disposal trucks (heavy duty diesel vehicles). The operation of garbage trucks poses certain health risks to personnel who are exposed to exhaust emissions while working around the vehicle during garbage collection. This calls for the need to identify lesser polluting unconventional fuels that can be domestically grown. To achieve these objectives, two fleets consisting of garbage trucks manufactured by Cummins and Mack Inc., make were considered. Major engine parameters that affect tail pipe emissions were identified from literature and considered as a part of the protocol in this study. Field tests were conducted on these trucks using the standard Testo'350XL' portable emission measuring system (PEMS). The emission data were collected on a one second scale and the coolant temperature were monitored to classify the condition of the vehicle as cold or hot. The alternative fuel considered in this study is BD which is derived from soy bean and can be locally grown and harvested in the United States. Basic vehicular information such as fleet type, year of manufacture, vehicle age, model, mileage, maintenance, and type of fuel is obtained from fleet records. The effect of domestically grown alternate fuels on idle exhaust emissions from these truck fleets is presented. Reducing greenhouse gas (GHG) emissions from diesel engines by identifying alternative transportation fuels, improving the fuel economy and implementing proper idle reduction strategies can help address issues related to climate change, air pollution, energy security thereby improving the nation's economy and protect the health of the people.

2. LITERATURE REVIEW

This chapter summarizes the regulations related to ambient concentration of pollutants, vehicle emissions standards and fuel standards set forth by the EPA. A detailed literature review is carried out on previous research works related to vehicle tail pipe emission characterization using various test procedures such as dynamometer testing based on federal test protocols (FTP) and real world test cycles to identify the technology gaps and health effects of diesel exhaust.

2.1. Federal Standards

2.1.1. Ambient Air Quality Standards

Under the Clean Air Act Amendments of 1970, the Environmental Protection Agency (EPA) Office of Air Quality Planning and Standards (OAQPS) has set the National Ambient Air Quality Standards (NAAQS) for six principal air pollutants (EPA, 2010), called the criteria pollutants, to protect public health and the environment. Table 2 illustrates the two types of standards set by EPA under the Clean Air Act.

NAAQ sets a *Primary standard* on the ambient concentration of criteria pollutants to protect public health and a *Secondary standard* to protect public welfare, including protection against decreased visibility, damage to animals, crops, vegetation, and buildings. The Units of measure for both the standards are parts per million (PPM) by volume, milligrams per cubic meter of air (mg/m^3), and micrograms per cubic meter of air (μg/m^3). Geographical location where the ambient concentration of a criteria pollutant exceeds the NAAQS is said to be in non-attainment for that pollutant. As such, severe restrictions are applied to permit any new source of pollution in that area and the regional/state authorities are required to come up with plans to reduce the concentration of pollutants to acceptable levels. According to EPA report on National Air Quality and emission trends, over 121 million people lived in counties that did not attain NAAQ standards in the year 1999 (EPA 2001).

Table 2. NAAQ standards for criteria pollutants

	Primary Standards		Secondary Standards	
Pollutant	Level	Averaging Time	Level	Averaging Time
Carbon Monoxide	9 ppm (10 mg/m^3)	8-hour	None	
	35 ppm (40mg/m^3)	1-hour		
Lead	0.15 µg/m^3	Rolling 3-Month Average	Same as Primary	
	1.5 µg/m^3	Quarterly Average	Same as Primary	
Nitrogen Dioxide	53 ppb	Annual (Arithmetic Average)	Same as Primary	
	100 ppb	1-hour	None	
Particulate Matter (PM$_{10}$)	150 µg/m^3	24-hour	Same as Primary	
Particulate Matter (PM$_{2.5}$)	15.0 µg/m^3	Annual (Arithmetic Average)	Same as Primary	
	35 µg/m^3	24-hour	Same as Primary	
Ozone	0.075 ppm (2008 std)	8-hour	Same as Primary	
	0.08 ppm (1997 std)	8-hour	Same as Primary	
	0.12 ppm	1-hour	Same as Primary	
Sulfur Dioxide	0.03 ppm	Annual (Arithmetic Average)	Same as Primary	
	0.14 ppm	24-hour	0.5 ppm	3-hour
	75 ppb	1-hour	None	

Source: (USEPA- http://www.epa.gov/air/criteria.html)

2.1.2. Fuel Standards

With increased concerns over the toxic mobile emissions, the U.S. Environmental Protection Agency (EPA) mandated the use of ULSD starting in 2006. The diesel fuel regulation limits the sulfur content in on-highway diesel fuel to 15 ppm from the previous 500 ppm weight (Shida et al., 2007; Laroo et al., 2005). The American Society for Testing and Materials (ASTM) defines the consensus on fuels to help manufacturers design engines for a fuel that has certain characteristics. With stricter emission standards being set and fossil fuel reserves running low, various fuel alternatives have been considered and tested in vehicles to assess the concentrations of toxic pollutants released. As a result, in December 2001, ASTM approved the full standard of biodiesel in US with a designation of D-6751 which covers pure biodiesel for blending with diesel in levels up to 20% by volume. Biodiesel has been registered with the U.S. EPA as a fuel and a fuel additive under Section 211(b) of the Clean Air Act. The letter "B" followed by a numeral denotes the biodiesel/ULSD blend fuels according to the volume percent biodiesel in the fuel. For example, B20 is a 20% biodiesel, 80% ULSD blend where as B100 is pure biodiesel.

2.1.3. Exhaust Emission Standards

Emission standards are limits set to the amount of pollutant that can be emitted from a vehicle. EPA has set different emission standards for cars, light trucks when compared to heavy trucks, transit buses and motor vehicles running on different types of engines (spark ignition (SI) and compression ignition (CI)). Heavy-duty vehicles are defined as vehicles of GVWR (gross vehicle weight rating) above 8,500 lbs in the federal jurisdiction and above 14,000 lbs in California (model year 1995 and later). Diesel engines used in heavy-duty vehicles are further divided into service classes by GVWR, as follows:

- Light heavy-duty diesel engines: 8,500 < LHDDE < 19,500 (14,000 < LHDDE < 19,500 in California, 1995+)
- Medium heavy-duty diesel engines: $19,500 \leq MHDDE \leq 33,000$
- Heavy heavy-duty diesel engines (including urban bus): HHDDE > 33,000

The basic standards are expressed in g/bhp-hr and require emission testing over the Transient FTP engine dynamometer cycle. The standards set for heavy-duty diesel vehicles presented in Table 3. The state of California is the only state in U.S. with the authority to set its own emission standards.

Table 3. EPA emission standards for heavy-duty diesel engines, g/bhp-hr

Year	CO (g/bhp-hr)	NO$_x$(g/bhp-hr)	THC (g/bhp-hr)	PM (g/bhp-hr)
1988	15.5	10.7	1.3	0.60
1990	15.5	6.0	1.3	0.60
1991	15.5	5.0	1.3	0.25
1994	15.5	5.0	1.3	0.10
1998	15.5	4.0	1.3	0.10
2004+	15.5	2.4	1.3	0.10
2007+	15.5	0.2	1.3	0.01

Source: Compiled from Emission Standards Reference Guide, EPA420-F-97-014.

2.2. Background on Emission Studies

According to the test procedures defined by the EPA in the Code of Federal Regulations in the United States, diesel engines are regulated for CO, NO$_X$, THC and total particulate matter less than 10μm ($PM_{10.0}$ or PM). In order to reduce these emissions from diesel engines and decrease the dependence on foreign sources for oil, considerable efforts are being undertaken to develop alternative fuels and diesel blends in United States. Some of the domestic sources that have been recognized as potential alternate fuels based on reduction in exhaust emission concentrations include 1) biodiesel, 2) compressed natural gas, 3) liquefied petroleum gas, 4) alcohol fuels (methanol, ethanol), and 5) hydrogen.

The source and environmental impacts of these alternative fuels are given in the Table 4. Each of the fuel type mentioned below has reductions in emissions of one or more regulated pollutants and even the cost comparison might be useful in assessing the most suitable fuel to develop possible scenarios of operating as an alternative fuel. Biodiesel, having similar properties as that of conventional diesel, gained popularity as an alternative fuel because it can be derived from renewable sources such as vegetable oil or animal fat. Biodiesel is basically mono-alkyl esters of medium to long-chain fatty acids produced by transesterifying organic oils to a viscosity close to that of petroleum diesels (Knothe et al., 2004; Tat et al., 2007). Biodiesel fuel is biodegradable and non-toxic and is mixed in various amounts with regular diesel.

Biodiesel has an advantage of being able to be easily substituted for regular diesel with little or no engine modifications (Sheehan et al., 1998). Also, the usage of biodiesel is expected to increase due to the legislation that was passed by the congress as a part of the EPA act, which allows federal and state fleet managers to use 20% or higher blends of biodiesel mixed with regular diesel in vehicles.

Due to its characteristic to clean the carbon deposition inside the combustion chamber, it appears that biodiesel fuelled vehicles provide complete combustion than regular fuels. Biodiesel is available anywhere in the country now and can be easily transported. It requires no new refueling infrastructure and works in heavy duty applications as well. In order to assess any emission benefits or liabilities due to biodiesel and its blends, it is important to conduct well planned field studies.

Table 4. Environmental impacts of alternative fuels

Fuel	Biodiesel	Ethanol (E85)	Compressed Natural Gas (CNG)	Liquefied Petroleum Gas (LPG)	Hydrogen
Source	Sun flowers, soybeans, animal fats or recycled cooking oil.	Starch crops, corn, wheat, agricultural waste	Underground reserves	A by-product of petroleum refining or natural gas processing	Natural gas, electrolysis and other energy sources (gasoline, diesel, methanol etc)
Environmental impacts of burning fuel (in comparison to gasoline and diesel)	Reduces PM and GHG emissions. NOx emissions may be increased.	1) Reduces GHG emissions by 20-80%. 2)Reduces ozone-forming emissions (CO and NOx) by 25%	1) Reduces CO2 emissions by 25% and PM by 95%, compared to gasoline. 2) Reduces NOx emissions by 80% and PM by 95% compared to Diesel. HC emissions may be increased.	1) Reduces CO2 emissions by 10-15%, NOx emissions may be increased, compared to gasoline. 2) Causes 80% reduction in NOx emissions and 95% reduction in PM, compared to diesel.	Zero regulated emissions for fuel cells operating on pure H2 CO_2 and NOx emissions equivalent to gasoline, for internal combustion engines

Source: http://www.eere.energy.gov.

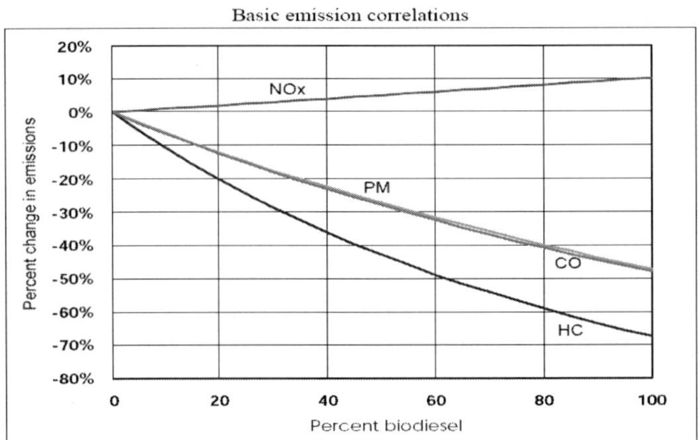

Figure 3. Change in emissions due to change in % biodiesel.

The US EPA (EPA, 2002) has reviewed the published biodiesel emissions data for heavy duty engines and the results for NO_X, PM, CO, HC are summarized in Figure 3. It has been concluded that the emission concentrations of these pollutants decreased with increase in biodiesel percentage in the base fuel (ULSD).

As it can be observed, a substantial reduction in emissions of PM, CO and HC were found due to the use of biodiesel. On the other hand, the emissions of NO_x have been found to be on the higher side. The average emission changes found by EPA for B20 have been summarized in Table 5.

Studies conducted by Schumacher (1996), Smith (1998), Spataru (1995), Graboski et al. (1996) have concluded that biodiesel has greater emission reduction potential compared to regular diesel. Schumacher (1996) and Marshall et al. (1996) also observed that an increase in the percentage of biodiesel blend in diesel fuel resulted in the reduction of HC, CO emissions and increase in NO_x emissions. Most of the above mentioned studies were conducted on Detroit Diesel Corporation (DDC) engines which are widely used in most of the heavy trucks and transit buses. In a study conducted on a 4 cylinder turbo charged engine to evaluate BD emissions compared to low sulfur diesel fuel called the no.2 diesel, significant emission reductions of CO and HC were observed as compared to a 12% increase in the NO_x emissions. No significant power losses were expected due to the use of BD blends and studies conducted by Marshall et al. (1995) using the Federal test procedures on Cummins engines also confirmed the same.

Table 5. Average heavy-duty emission impact of 20% biodiesel relative to average conventional diesel fuel (EPA, 2002)

Air Pollutant	% Change B20
NOx	2
PM	-10.1
CO	-11
Hydrocarbon	-21.1

Although there is a consensus on the overall reduction of emissions in using BD, the specific percentage of reduction varied from study to study. McCormick et al. (2000), observed that the particulate matter (PM) decreased 25% and NO_x increased 3% in all engines using biodiesel. F.I. Hiller, chief in the equipment department for Arlington County, VA reported a 22% reduction in PM, 20% reduction in carbon monoxide (CO), and 30% reduction of hydrocarbons (Laroo, 2005). However, Knothe et al. (2004) reported a 10% reduction in PM, 21% in hydrocarbons, 11% in CO, and a 2% increase in NO_x. Although 100% biodiesel, or neat biodiesel, emits much less pollutants, it produces more NO_x and causes frequent filter clogging. According to the literature review conducted by Shandilya and Kumar (2009), majority of the studies conducted by previous researchers on exhaust emissions from biodiesel found an increase in NO_x, CO, and THC (Table 6).

Table 6. Percentage share of literature reporting increase, same or decrease in emissions over regular diesel due to the use of biodiesel (Shandilya and Kumar, 2009)

Emissions	Percentage of Total Studies		
	Increase	Same	Decrease
NOx	87	9	4
PM	2	2	96
THC	2	2	96
CO	2	3	95

A number of vehicles, fuel, operating and environmental variables that are found to influence the tail pipe emissions are shown in Table 7 (Faiz et al., 1997). Emission values depend on road conditions, driving behavior (speed, acceleration or speed time acceleration) and on driving habits, which is also, an important factor affecting the exhaust emissions (Ergeneman et al., 1999).

The increase in NO_x emissions in B20 trucks observed in the present study complies with the results of previous studies. Southwest Research Institute contributed the increase of NO_x to the percentage of biodiesel blend in the fuel. It was concluded that larger the percentage of biodiesel higher the NO_x emissions (Laroo et al., 2005). Ban-Weiss et al. presented another theory using the chemical structure of biodiesel that has more double bonds than petroleum diesel, to explain the increase in NO_x emissions (Ban-Weiss et al., 2007). Since soybean is the organic product used to make 90% of biodiesel in the U.S. market, Tat et al., experimented with genetically modified high-oleic soybeans and found a significant reduction in NO_x emission (Tat et al., 2007). The differences found between Mack and Cummins vehicles running on the same fuel and the variation between the hot and cold starts indicate that fuel, though important, is not the only influential factor in emissions (Wilson et al., 2006). Different characteristics of trucks and temperatures also affect the concentration of the exhaust gases. Weilenmann et al., (2005) reported that exhaust emissions were found to be considerably higher for cold start. This behavior was found to be very significant during the phase in which the engine tries to attain optimum operating temperature called the warm – up phase.

Table 7. Factors influencing vehicle emissions

Vehicle Parameters	Fuel Parameters	Operating conditions	Environmental parameters
Vehicle Manufacturing year, type, technology and class	Type of Fuel	Average Vehicle speed	Altitude
Fuel delivery system	Fuel properties like oxygen content, sulfur content, volatility, benzene content, lead and metal content , etc.	Load on the engine due to heavy load on the vehicle, accessory loading due to the usage of air conditioners etc.	Humidity
Emission control system		Cold or hot start	Ambient temperature
Air conditioning and other vehicle appurtenances.	Fuel Quality	Traffic patterns and Driver behavior	Diurnal temperature changes
Vehicle miles travelled (VMT)		Emissions standards in effect	
Inspection/maintenance and anti-tampering (ATP) program		Type of road and road grade	

The emissions produced during this warm phase and the total time taken by the engine to attain optimum temperature were found to be dependent on the ambient temperature as well as the initial temperature of the engine (Weilenmann et al., 2009). In a study conducted by Lyons et al., (2000) on the emissions from nine heavy trucks fueled by diesel and biodiesel blends without engine modifications, the heavy trucks fueled by B35 emitted lower PM, CO, HC compared to trucks fuelled by no. 2 diesel. The NOx emissions in both the cases were found to be the same. In a study by Wang et al. (1997) the emission data from the use of alternative fuels (Natural Gas, Methanol100, Ethanol) and diesel powered heavy duty vehicles (buses) were measured and collected from 32 transit agencies in 17 states using the two West Virginia University (WVU) 34 transportable heavy duty vehicle emission testing laboratories (THDVETLs). A special driving cycle called the Central Business district (CBD) cycle was selected and buses were tested using this new cycle. The emissions measured were influenced by the engine technology used, the test cycle employed, the presence or absence of after treatment devices, atmospheric conditions and fuel types. They have also concluded that particulate matter formation depends on fuel composition and molecular structure, while the significant factor for NOx formation is combustion temperature.

Motor vehicles emit air pollutants from three distinct sources: combustion process, evaporation of fuel and wear of tires and brakes. Combustion emissions are a function of ambient temperature, power output of the engine, emission control systems, fuel characteristics and air fuel ratio (Delucchi, 1996). Evaporative emissions are HC vapors lost into the atmosphere mainly from fuel tank and carburetor. Loss from fuel tank is mainly due to diurnal and temperature changes whereas loss from carburetor called the hot soak emissions occur mainly when a hot engine is stopped. Since, most of the older engines did not have the technology to maintain the fuel under pressure to curb hot soak emissions, they have been found to release more HC vapors into atmosphere than the new engines. Refueling of the vehicle can also lead to evaporative losses in some cases.

Although biodiesel has many benefits, it also has some drawbacks. Biodiesel offers advantages such as less engine wear, better lubricity and needs little or no engine and fuel system changes. However, it emits higher NO_x, freezes at higher temperature, frequent clogging of filters and has a shorter shelf life than petroleum diesels. As alternative fuels are relatively new to the public, there are still many questions and health issues not addressed or even known. Unlike biodiesel, ULSD fuel has low lubricity (Knothe et al., 2005). When the sulfur was taken out of ULSD, the fuel has virtually no capacity for lubrication and would therefore need additives to enhance its lubricity in order to meet product specifications and reduce engine wear (Wilson et al., 2006). On the other hand, biodiesel has a potential to clean carbon cleanups from inside the carburetors due to its chemical structure. One major drawback due to this characteristic is the necessity to frequently change the filters in the vehicle.

Among the most common, yet most ignored vehicles on the roads are garbage trucks. According to INFORM, Inc., a New York based environmental research firm, heavy-duty diesel-powered vehicles, including garbage trucks, make up only 7 percent of vehicles on the road, but they produce 69 percent of on-road fine particulate pollution and 40 percent of the NO_x emissions (INFORM, 2010; Kilcarr et al., 2007). In 2007, U.S. residents, businesses, and institutions produced more than 254 million tons of municipal solid waste (MSW), which is approximately 4.6 pounds of waste per person per day. Between 1960 and 2007, the amount of waste generated by each person has almost doubled from 2.7 to 4.6 pounds per day. The waste management sector represents 4% (i.e., 260 out of 6750 tetra grams of CO_2 equivalents) of total U.S. anthropogenic GHG emissions (Methane and CO_2) of which collection and transportation of MSW and recyclables accounted for ~ 0.5 and 1 million metric tons carbon equivalents in 1974 and 1997 respectively. This increase in the contribution of collection and transportation sector was mainly due to doubling of MSW generated since 1974. Since 1974, a significant increase in the release of non-GHG pollutants such as SO_x, NO_x, CO, O_3 was also observed. Despite the comparisons presented in the literature on biodiesels and regular diesels, the data, analysis, and comparison of the realistic exhaust emissions between the recently mandated ULSD and biodiesel are lacking for garbage trucks. As of today, very vague information is available on emission characteristics from in-use garbage disposal trucks running on biodiesel blends and Ultra Low Sulfur Diesel (ULSD). This study deals with the determination and evaluation of idle emission rates from the city of Toledo garbage disposal trucks running on different biodiesel blends and ULSD. Field tests were conducted on garbage disposal trucks of different makes using the standard Testo350XL emission testing instrument. The emission data pertaining to six gases was collected on a one second scale. A detailed analysis is presented using these emission rates collected from field tests conducted.

2.3. Diesel Exhaust Health Effects

Diesel engines are a major source of pollution emitting particulate matter (soot), nitrogen oxides which contribute to the production of ground-level ozone (smog), acid rains, hydrocarbons and air toxics. These emissions can damage plants, animals, crops, and water resources. Emissions from diesel exhaust can lead to serious health conditions, such as asthma and allergies. They can also worsen heart and lung diseases, especially for vulnerable populations such as children and older individuals. EPA estimates that every $1 spent on clean diesel projects produces up to $13 of public health benefits. Studies conducted in the past have established the fact that a huge percentage of pollutant emissions in ambient air are emitted by vehicles (EPA's National-Scale Air Toxics Assessment, 2009). They have numerous effects on human health as well as environment. In fact, a study in the Journal of the American Medical Association cited that people who live in the most heavily polluted areas have a 12% higher risk of getting lung cancer than people in the least polluted areas (Burnett et al., 2002). A summary of health effects is given in Table 8.

Table 8. A summary of health concerns due to different diesel pollutants

Pollutant	Health Concern
Nitrogen	Lung irritation, respiratory illness, and premature death
Carbon monoxide	Headaches and reduces mental alertness
Particulate matter	Increased respiratory disease, lung damage, cancer and premature death
Sulfur dioxide	Increase in existing heart disease, breathing difficulties and respiratory illness
Ozone	Breathing difficulties, respiratory infections and lung and tissue damage

3. METHODOLOGY

3.1. Fleet Characteristics

The city of Toledo solid waste division has two fleets with around 40 trucks being dispatched for garbage collection and disposal duties every day. The typical operation times of these trucks are from 7 am to 2:30 pm. Each truck has a different route each day and by the end of the week these trucks full fill the task of collecting and disposing curb side residential as well as commercial garbage at the city of Toledo landfill. Most of these trucks are fuelled with ULSD and a few of them operate on B20 grade biodiesel (20% methyl ester bio-fuel + 80% ULSD). Fuel was completely drained off from trucks chosen to run on biodiesel and were then operated on the required fuel blend for at least a day before the testing commenced. This method was adopted to make sure the fuel system is rinsed with the required fuel and no amount of residual diesel fuel can influence the emission characterization. Since tail pipe emissions depend on the vehicle characteristics like vehicle make, model, model year, engine type and size, proper attention has been paid while picking the sample trucks that would be fuelled with biodiesel blends for testing purposes. Trucks manufactured during the year 1997 were not included in the emission reduction analysis carried out for biodiesel and ULSD fuels. Since each of these trucks weighed more than 62000 lbs, they were classified as Heavy heavy-duty diesel vehicles. The characteristics of trucks in each fleet running on different fuels are presented in Table 9.

Table 9. Fleet characteristics

Engine Make	Model	Year	Engine Capacity	Horse power	No. of Cylinders	OBD II Compatible	Vehicle Class
Mack	MR688S	1997	12L	300	6	NO	Heavy
Mack	MR688S	2003	12L	300	6	Yes	Heavy
Cummins	LET-2	2001	8.3L	275	6	Yes	Heavy

3.2. Fuel Characteristics

The city of Toledo solid waste division has adopted using biodiesel blends to fuel garbage trucks since August, 2006 to help reduce the pollution caused by diesel trucks that operate in the city. Pure biodiesel i.e., 99.9% biodiesel is supplied by Peter Cremer which is mixed using the splash blend technique with ULSD supplied by Amoco fuels to fuel the trucks. For example, B20 fuel blend contains 20% of 99.9% pure biodiesel and 80% ULSD. Similar acronyms such as B5, B10, and B0 were given to biodiesel blends that were developed by mixing 99.9% biodiesel with ULSD as base fuel in appropriate proportions. Both biodiesel and ULSD supplied by the providers meet EPA's requirements and conform to the ASTM D-6751 (EPA-4627) and ASTM D-975 fuel specifications respectively. The biodiesel used in this study is derived from Soybean oil. A summary of the fuel properties is provided in Table 10. Currently the garbage trucks run on ULSD and B20 and the details of the test fleet are provided in Table 11.

Biodiesel contains very low amount of sulfur as compared to ULSD. This can help reduce SO_2 emissions that lead to acid rain. Since biodiesel has a very high weight percentage of oxygen, it reduces emission of products related to incomplete combustion. On the otherhand the Cetane number of biodiesel is higher compared to ULSD which may lead to higher NO_x emissions. Higher cloud point of biodiesel might also create handling and storage issues during winter. However, if mixed in appropriate quantities the resulting mixture is predicted to find a perfect balance in all the mentioned properties.

Table 10. Properties of biodiesel and ULSD

Fuel Property	Fuel Type	
	Biodiesel (B99.9%)	ULSD
Cetane number	47	40
Cloud point (summer) (°F)	40	20
Cloud point (winter) (°F)	42.8	15
Flash point (°F)	>260	125
Sulfur (ppm)	<1	15
Water and sediment (moisture) (Vol. %)	<0.005	0.05
Kinematic Viscosity, 40°C (mm^2/sec)	4	1.9-3.4
Oxygen, wt-%	10.87	0.0

Table 11. Test fleet details

Engine Make	Year	No. of Trucks fuelled with ULSD	No. of Trucks fuelled with B20
Mack	1997	7	0
Mack	2003	28	2
Cummins	2001	3	1

3.3. Experimental Study Design

Preparation for field data collection involved four major components: (1) coordination with the division of solid waste regarding scheduling of the test and access to the truck; (2) verification of the status of the PEMS and the availability of all necessary accessories and consumables; (3) zeroing the PEMS by purging it with natural air; (4) complete data collection. The most crucial factor in this study was to successfully communicate with the officials at the solid waste division specifying which vehicles are to be tested, when they are to be tested, what fuel will be used, what type of duty cycle will be performed, and where should the truck be parked inside the garage to enable a comfortable data collection.

Tail pipe emission data was collected during cold idle and hot idle. Hot idle emission testing was carried out when the trucks returned from their regular scheduled routes in the afternoons. Cold idle emission testing was carried out early in the morning between 5am to 7 am. The trucks have to leave the garage by 7 am for their daily routine and the window to schedule emission testing was limited to just 2 hours during mornings. A particular truck tested in the morning for cold idle emissions was tested in the afternoon on the same day for hot idle emissions in order to nullify any external parameters that may change the status of the vehicle thereby affecting the exhaust emission values. The TESTO 350XL is portable and can be installed in approximately 15 minutes. The probe attached to the emission testing instrument is plugged into the exhaust of the truck and the OBD was connected to gather information about the engine performance variables. The engine performance data and emission data were collected simultaneously without any time lag. The OBD used for obtaining engine performance data from trucks were supplied by the Mack trucks Inc., and Cummins Inc. The truck was started in the idling mode and emission data along with the engine performance data was collected for 15 minutes. This test protocol was repeated for each truck in the entire fleet. Since trucks were tested on the same day for both hot and cold idle, not more than two trucks were tested on a single day due to the time constraints. Since the solid waste division had a limited fleet of vehicles that were needed to run constantly to ensure timely garbage collection and cater to the needs of a large city, it was difficult to schedule trucks for emission testing. Such constraints contributed to the delay in the experimental part of the project.

3.4. Instrumentation

The PEMS used in this study to collect tail pipe emissions is the TESTO'350XL' system manufactured by Testo Inc. It consists of non-dispersive infrared sensors that can continuously measure up to six gases: oxygen (O_2), carbon dioxide (CO_2), carbon monoxide (CO), sulfur dioxide (SO_x), nitric oxide (NO) and nitrogen dioxide (NO_2) with calculated nitrogen oxides (NO_x) and a temperature sensor. The unit of measurement for all the above gases except CO_2 is parts per million of volume (PPMV). CO_2 is measured in % volume which is 10^4 PPMV. Only two vehicle modes were considered i.e., hot idle and cold idle. A laptop containing the "EasyEmission" software was connected to the instrument to export data in Microsoft Excel format. Figure 3 shows the instrument connected to the laptop for downloading tail pipe concentrations of the pollutants. Simultaneously the engine operating parameters such as idle rpm, power, fuel ratio, ambient temperature, exhaust temperature, coolant temperature, engine load % are gathered during vehicle operation using the On board diagnostic (OBD) unit which was connected to the engine computer module (ECM) port located at the driver side foot well of the truck. The engine scanner was interfaced with the computer using the software purchased from Cummins and Mack as shown in Figure 3.

Figure 4. TESTO 350XL unit interfaced with a computer and collecting OBD data simultaneously.

Figure 5. Sampling probe and heat shield clamped to the exhaust pipe.

An aluminum attachment, clamped to the exhaust pipe, was also designed to shield the instrument analyzer probe from the heat and to ensure uniform insertion depth of the probe into the vehicle exhaust (see Figure 5). Continuous power supply was ensured to both the laptops as well as the instrument to avoid interruptions in data collection. The emission data obtained for all series/vehicle/fuel/load combinations are compared for both the fuels i.e., B20 and ULSD and a comparative analysis is carried out.

3.5. Database

The emission monitoring for the trucks considered in the study was carried out in the month of April '08. For cold idle monitoring the data was collected between 5am and 7am during weekdays. The Trucks left for garbage collection route at 7am on weekdays and return to the garage in the afternoon. For the hot idle monitoring the data was collected during the afternoons when the trucks return back from their regular route. The instrument was purged with fresh air between each idle test so as to clear the electrochemical sensors for any gas residues and all the particulate filters in the instrument were replaced in accordance with recommendation from the manufacturers. This zeroing is to sample fresh air which is considered to be free from significant levels of CO, NO, SO_2. The average ambient values for O_2 and CO_2 are assumed to be at 20.9 Vol % and 0.03 Vol % respectively.

Figure 6 shows the firmware on the instrument being updated to make sure it is running on the latest version and no software glitches are encountered during the emission testing. Engine parameters were collected using dedicated OBD software's provided by Mack Inc., and Cummins Inc. The OBD software for Cummins trucks was customized to log data at one second interval along with the emission data as shown in Figure 7. The OBD software for Mack trucks could not be customized to log second by second values of the desired engine parameters. The software logs data whenever there is a change in the value of any of the desired parameters. Since the instrument was sent for calibration just before the data collection started, the data thus obtained is assumed to be free from errors and all the data has been used for statistical analyses.

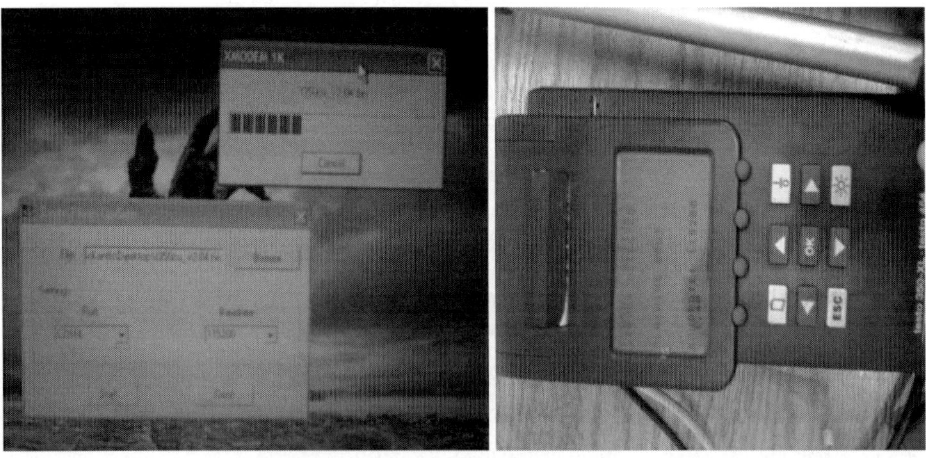

Figure 6. Instrument firmware update.

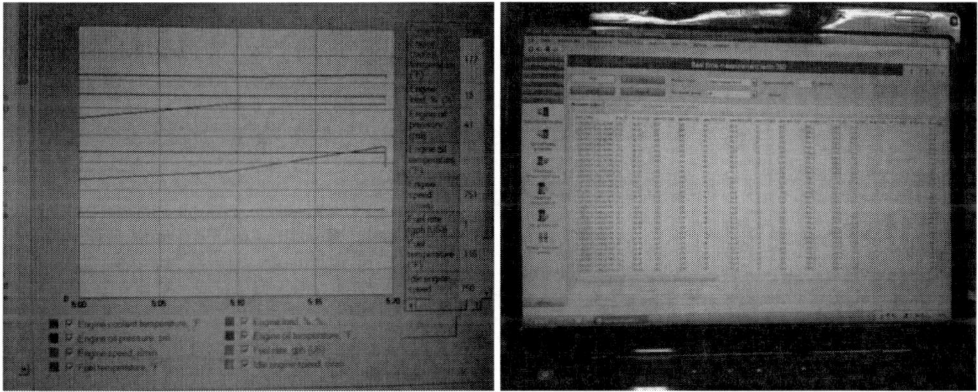

Figure 7. OBD and emission data being logged simultaneously at one second interval.

3.6. Quality Assurance and Quality Control

To ensure data quality and control, the combined (emission and engine) data set for a vehicle test is screened for errors. If any errors were found the data set is not used for analysis. Whenever the instrument "freezes" where in a value that is expected to change dynamically stays constant for a longer interval and the instrument fails to update the value, the test is considered a fail and the data was not used for analysis. The same approach is applied to OBD data too. Special attention was paid to the engine RPM values to make sure no abnormal values are being reported since this was the leading cause of loss of data. Overall, 97% of the raw data from idle testing was found to be valid in estimating exhaust emissions from garbage trucks. However, more than 90% of the on-road emission tests failed due to the bad suspension of the garbage trucks which resulted in a very bumpy ride and the instrument frequently froze or turned off, unable to handle the jarring. As a result the testing had to be stopped to preserve the instrument for the ongoing research.

3.7. Analysis Methodology

The following pointers were followed during data analysis.

- Idling => speed and acceleration both are zero.
- About 63750 observations after data preparation.
- Disregard data sets containing negatives/null/spikes in values.
- 15 minute average emissions.
- Data used -- Idling with RPM values 700 ± 50.
- Trucks were parked inside the garage which is maintained at room temperature, so the effect of ambient temperature on fuel temperature was assumed to be nullified.

4. RESULTS AND DISCUSSIONS

This study presents the idle emissions collected from garbage disposal trucks running on biodiesel and ULSD. The effect of changing fuels on exhaust emission concentrations of criteria pollutants in different truck fleets will be evaluated. The above described methodology was followed and emission data for a 15 minute time interval along with the engine performance data was collected using the Testo350XL instrument and the Onboard Diagnostic System (OBD) available. The Testo350XL emission testing instrument measures the emission data on a 1 sec time scale. These are then averaged out to a 1 min scale. The one minute data is then averaged out for 15 minutes and a conservative observation is made. In order to understand the variations in the measured values, sensitivity analysis is also carried out. A consistency analysis is carried out to present the uniformity in the readings obtained by the instrument from the same test sample.

4.1. Consistency Analysis

The TESTO 350XL instrument was sent to calibration in the month of May, 2008 and was received in June, 2008. The NO and SO_2 sensors were found to be out of tolerance after conducting the calibration tests and new replacements were ordered. New O_2 and CO_2 sensors were also added to the instrument.

Since this study involves a large number of tests, it is important to assess the consistency of the instrument readings. In order to do so, three sets of tests were conducted on three different trucks in cold idle mode fuelled with ULSD.

The 15 minute average test to test variations are presented in the Figures 8 and 9. Error bars are also shown which give an indication of the standard deviation of the sample sets. It can be observed that the average readings have been uniform in the case of all the three trucks for all the pollutants except for NO (truck 2) and CO_2 (truck 3).

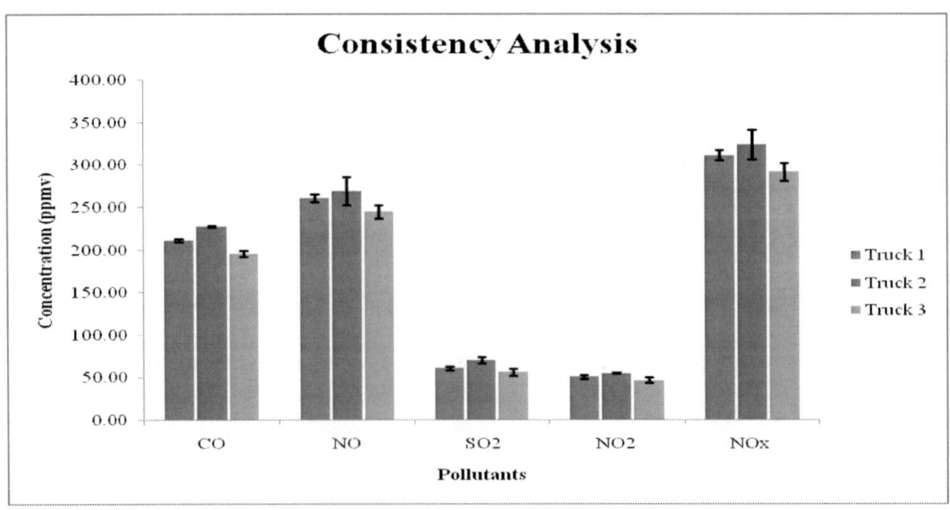

Figure 8. Consistency analysis for Testo 350 XL.

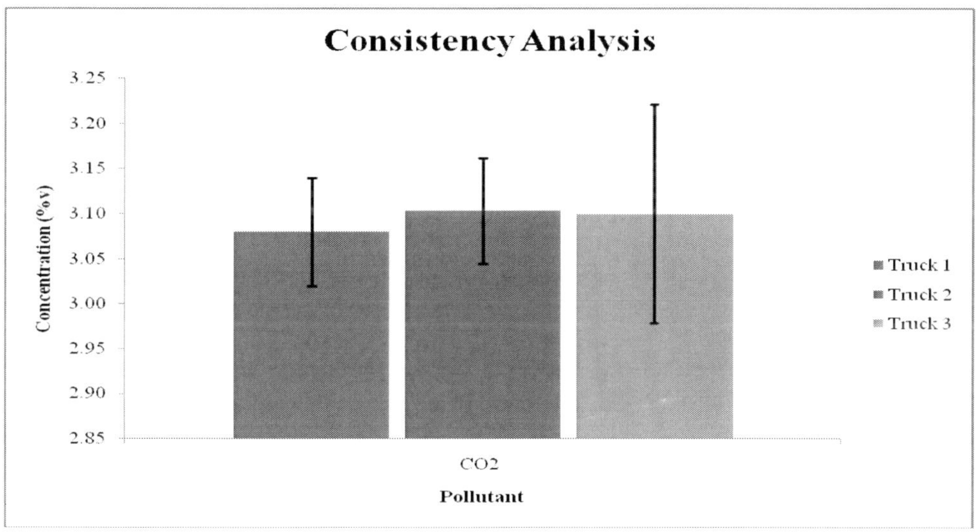

Figure 9. Consistency analysis for Testo 350 XL – CO_2 Emissions.

Similar trend has also been observed in the second to second data collected during the testing. This consistency analysis demonstrates the uniformity in the data collected by the instrument and the accuracy of the sensors.

4.2. Idle Emission Characterization

Since garbage trucks idle for a longer time than other conventional diesel fuelled heavy duty trucks, it is necessary to understand the emissions characterization of pollutants from the exhaust. Extensive real time measurement of exhaust emissions is carried out as a part of this study. Different vehicle, fuel characteristics and their effects on exhaust emissions will be presented in the following sections. The test fleet presented in chapter 3 (Table 11) is used to study the effects of various vehicle and fuel characteristics.

4.2.1. Effect of Engine Parameters

Garbage trucks spend a lot of time in idling during garbage collection. So, it is very important for us to understand the exhaust emission profile of these trucks. Also since garbage collection personnel work around the truck during the daily routine, it is a key factor to evaluate in order to understand the kind of emissions they will be exposed to. A quick overview of the literature suggests that most studies found that vehicles emit higher concentrations of pollutants during cold starts as compared to hot starts. In fact, one of the most discussed topics regarding vehicular emissions is the influence of engine temperature. The cold start emissions are usually higher compared to hot start emissions due to the formation of a rich fuel mixture inside the ignition chamber. A rich fuel mixture is required to facilitate fuel vaporization during cold starts. To enable proper vaporization, extra fuel needs to be pushed in to the combustion chamber. This extra fuel pushed creates improper combustion and leads to the emission of CO and HC. In order to study this behavior, cold start idle and hot start idle tests were conducted for both ULSD and B20. The second by second exhaust emissions were collected along with other engine parameters such as fuel percent (FP), coolant temperature (CP) and fuel temperature (FT). Coolant is a fluid which flows through the engine to prevent it from overheating. Coolant temperature refers to the temperature of the coolant running inside the engine. Generally, coolant temperature is a representative of the engine temperature since it absorbs the heat and transfers it to other devices in the engine that dissipate it. The term fuel temperature refers to the temperature of the fuel inside the fuel tank of the vehicle.

Fuel percentage is the percentage of fuel the vehicle is commanding while idling or on other operational modes. Ideal fuel percentage is derived from the torque curve. Torque curve is the curve that shows the range of torque the vehicle is being used at and consequently indicates the percentage of fuel the vehicle is commanding to work at a given amount of torque. For a vehicle in idling mode, since no amount of torque is applied on the engine, the fuel percentage is expected to stay constant at an X amount of value depending on the vehicle specifications. To examine the predicted fact that extra fuel is required during the warming phase of an engine, minute by minute averages of fuel rate and emissions of various pollutants from a truck running on ULSD were plotted and presented in Figure 10 and Figure 11 during cold idling mode. Emissions have been found to be really high during the first few minutes and a rapid fall in concentrations before stabilizing has been found to be the most common behavior of all pollutants. It can be observed from the graphs below that this huge spike in the emissions concentration is due to higher percentages of fuel rate and thus it is a major factor contributing to episodes of high emission rates during cold idling. Similar behavior was also seen in the case of B20 for cold idling modes.

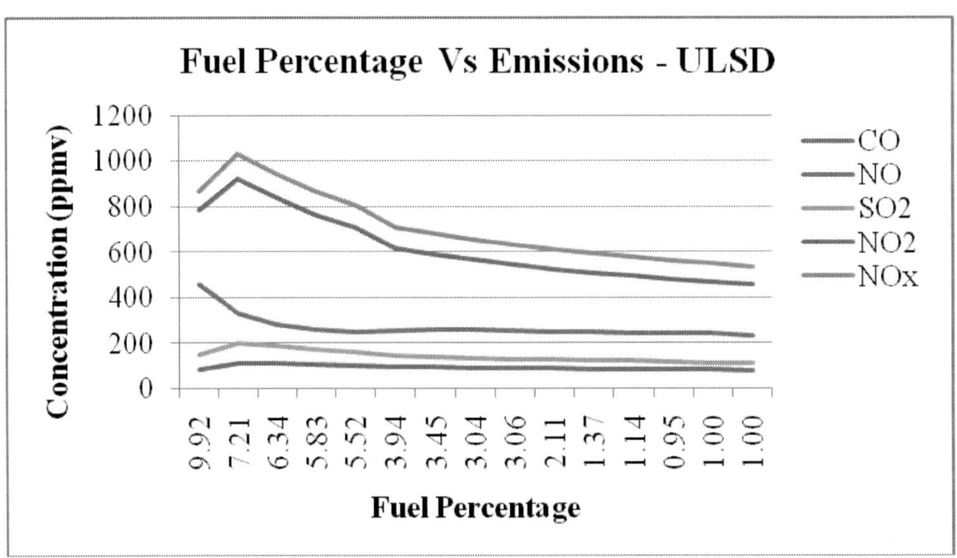

Figure 10. Effect of fuel percentage on exhaust emissions from ULSD – Cummins.

A very huge spike was observed in the concentrations of all the pollutants during the first few minutes (transient emissions) of the test during cold idling. This spike in transient emissions can be due to the high percentage of fuel being used at that point of time. This behavior was found to be uniform across all fuel and mode combinations. In order to present a thorough and detailed picture of this behavior the second by second data for all the six pollutants along with engine parameters is plotted and shown in Figure 11.

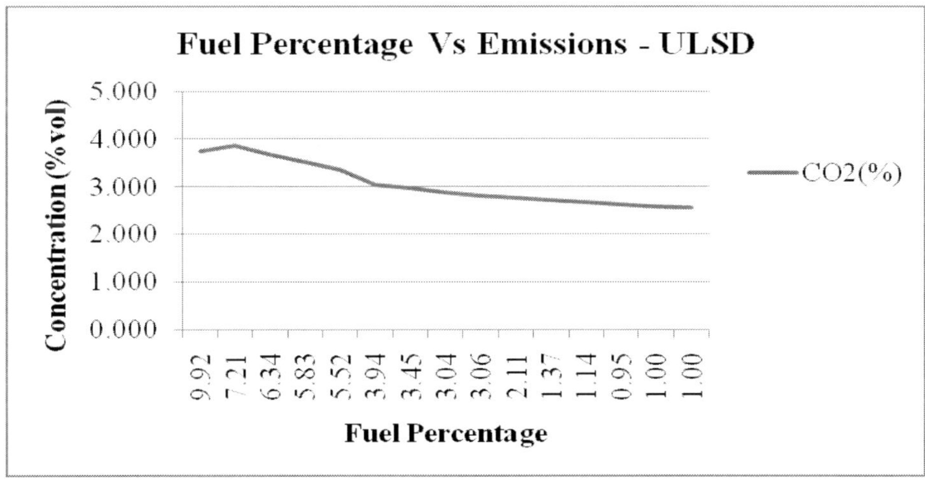

Figure 11. Effect of fuel percentage on CO_2 emissions from ULSD – Cummins.

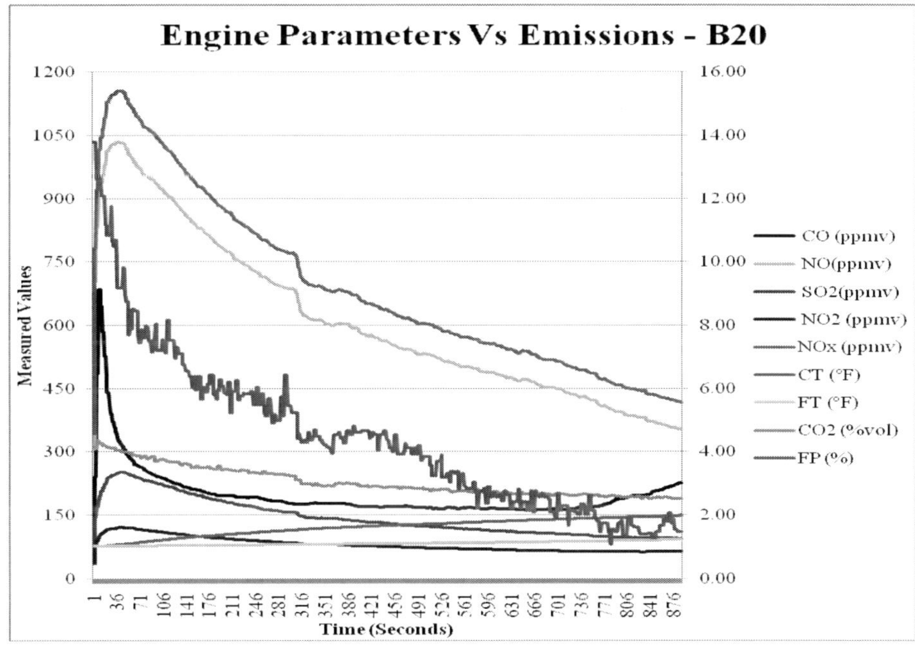

Figure 12. Effect of engine parameters on exhaust emissions for B20.

Since the readings for CO_2 and fuel percentage are small compared to other variables, they have been plotted on a secondary axis for the ease of understanding. It can be observed that the fuel temperature and coolant temperature have increased gradually while fuel percent has dropped significantly with increase in the idle time. The amount of fuel commanded by the engine has decreased over time due to the increase in CT and FT. This phenomenon helps the engine attain an ideal operating temperature. Since higher temperatures facilitate fuel vaporization inside the combustion chamber, rich fuel mixture is not formed in the combustion chamber. As the FP dropped and CT increased the concentrations of all pollutants except CO_2 decreased for B20 whereas the behavior was found to be uniform across all the pollutants for ULSD.

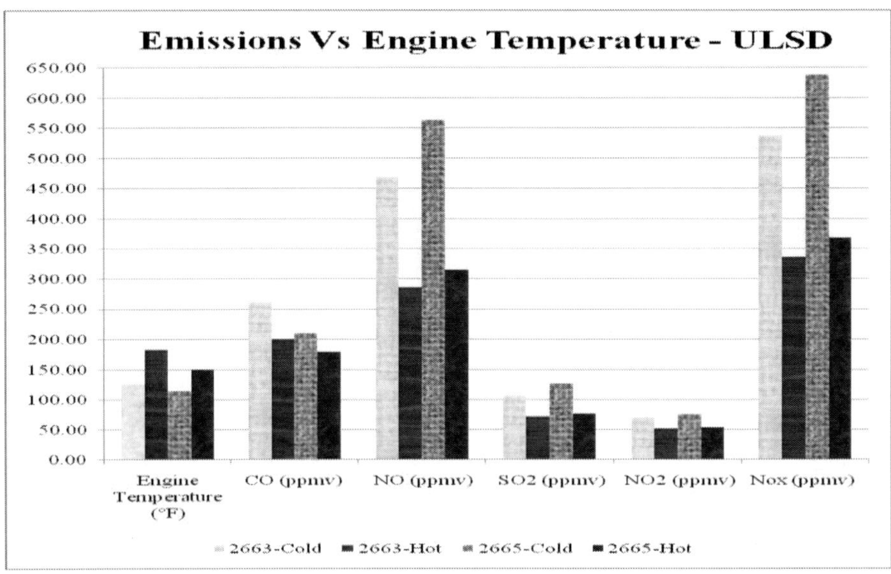

Figure 13. Pollutant concentration comparison for ULSD during hot and cold starts – Cummins.

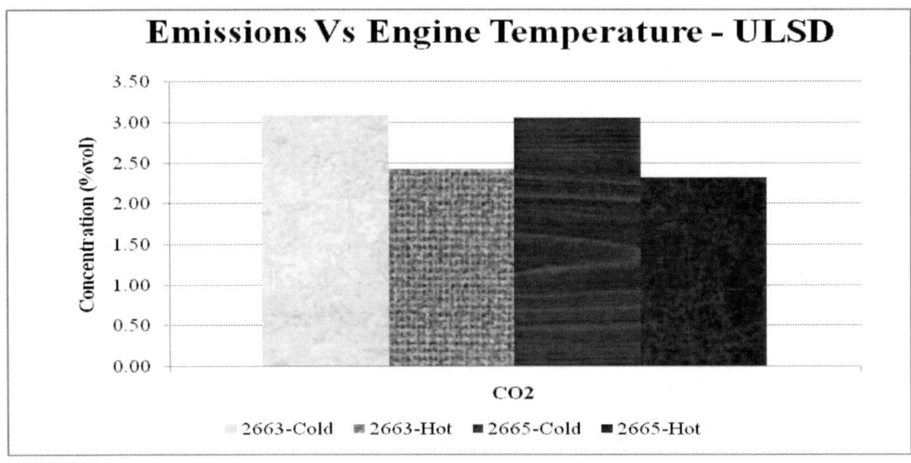

Figure 14. Pollutant concentration comparison for ULSD during hot and cold starts – Cummins.

Two cummins trucks were tested both for hot and cold start running on ULSD and one truck was tested for the same modes while running on B20. Figures 13 and 14 present the variations in 15 minute average pollutant concentrations with change in engine temperature, while fuelled with ULSD. Figures 13 and 14 show the effect of engine temperature on 15 minute average pollutant concentration for B20. For both the fuels, the concentrations of pollutants during cold idling are more than those during hot idling. This behavior of higher emission concentrations during cold starts regardless of the type of fuel further strengthens the findings documented by previous studies (Kumar and Vinay, 2009; Vijayan et al., 2008).

For both the fuels during cold and hot starts NO emission concentrations were found to be the highest among all the pollutants.

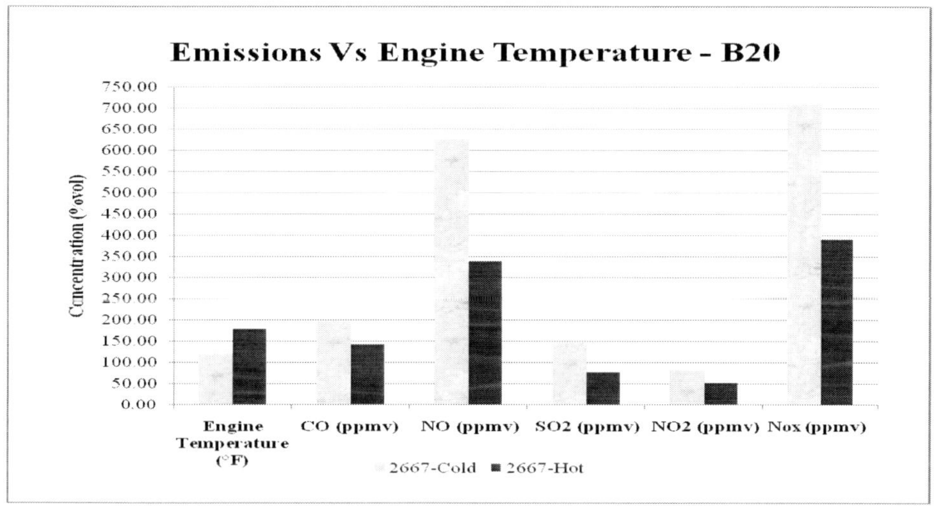

Figure 15. Pollutant concentration comparison for B20 during hot and cold starts – Cummins.

The concentrations of CO_2 showed similar behavior as other pollutants for both the fuels in different idling modes (figures 14 and 16).

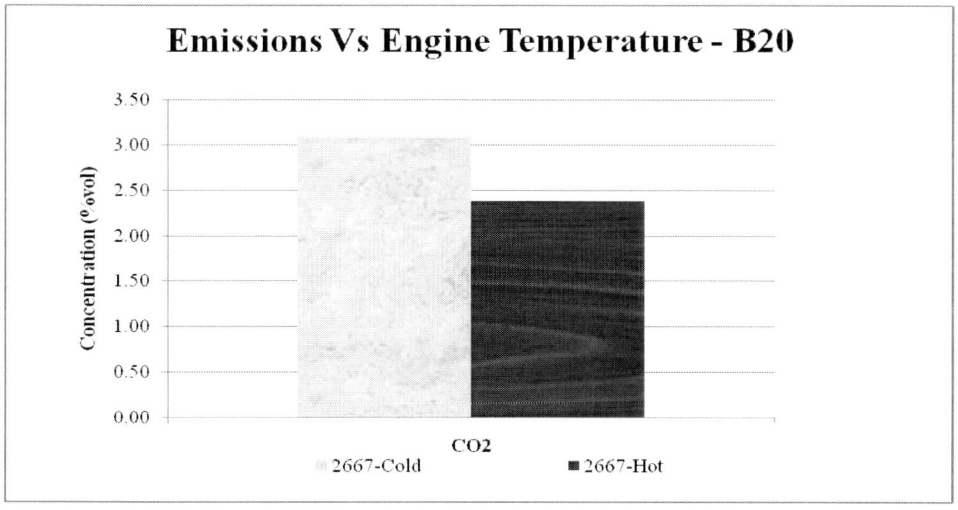

Figure 16. Pollutant concentration comparison for B20 during hot and cold starts – Cummins.

Substantial reduction in concentrations were found in NO (41% and 46%) and SO_2 (37% and 48%) emissions between hot and cold starts for both ULSD and B20 respectively. The difference in engine temperatures during cold idle for multiple tests conducted was found to be approximately 1% and as such the data sets are deemed comparable.

4.2.3. Effect of Fuel – Mack

One of the major purposes behind conducting this research is to find out a suitable alternative to ULSD. To assess the potential of biodiesel as an alternate fuel, the exhaust emissions of Mack engine are measured and analyzed. The fuels used in this study are ULSD, B5, B10 and B20. The second by second cold idling emission behavior of each pollutant from various fuel blends is presented in Figure 17. Table 12 shows the percentage change in the concentrations of each pollutant with respect to ULSD for each biodiesel blend.

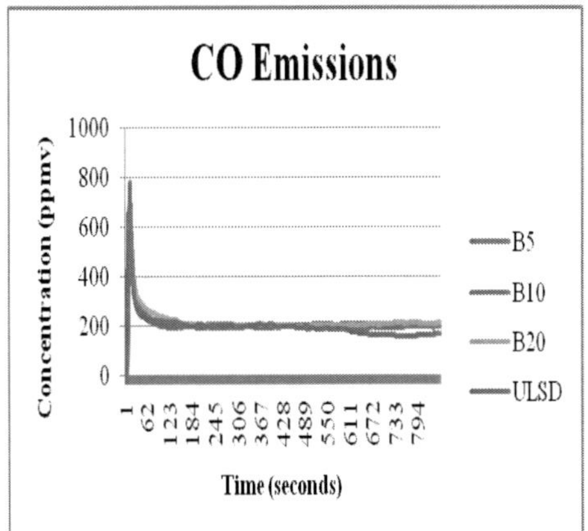
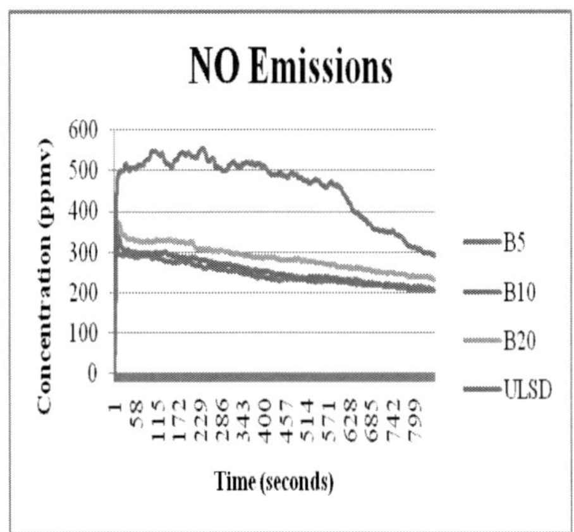
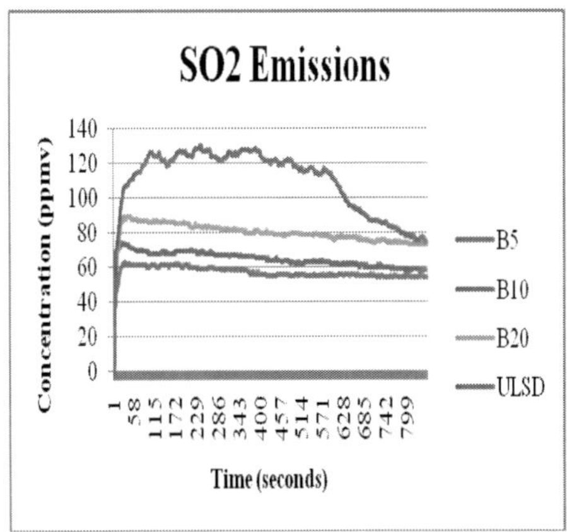
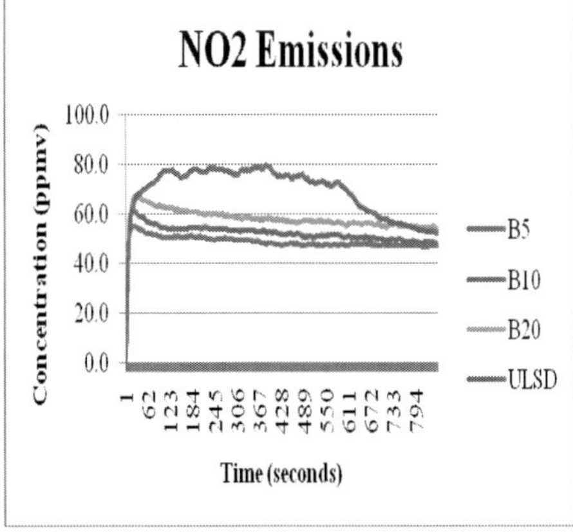

Figure 17a. Behavior of pollutant concentrations during cold starts – Mack.

It can be observed from the graphs that cold idle concentrations of NO from ULSD were almost double to those from other biodiesel blends. The concentrations of SO_2 and NO_2 were also found to be significantly higher than that of other biodiesel blends. The percentage change in emissions of NO, SO_2, NO_2 were found to decrease with an increase in the percentage of biodiesel concentration, where as for CO the percentage change in emissions has increased with an increase in the biodiesel concentration.

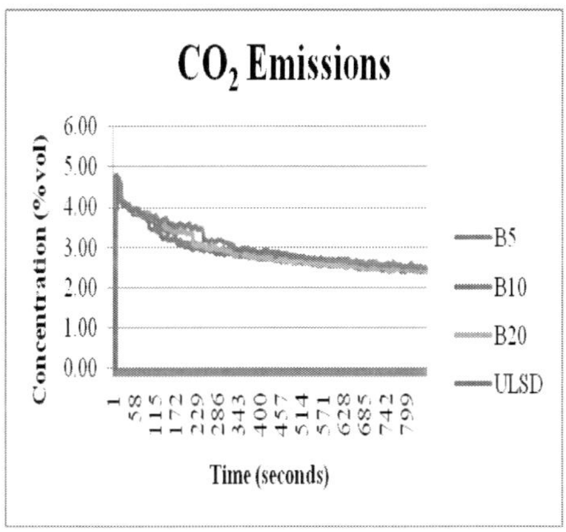

Figure 17b. Comparision of cold idle emissions for various pollutants between ULSD and biodiesel blends.

From the percentage changes presented in Table 12, biodiesel is found to be a clear winner with respect to the emission concentrations of different pollutants with the exception of CO.B10 is found to have greater reduction potential for CO_2 emissions among all the fuel blends tested while B5 is found to be burning cleaner with respect to NO, SO_2 and NO_2 concentrations. Similar results have been found for hot idling mode too.

Table 12. Percentage change in cold idle emissions for different biodiesel fuel blends

% Change	CO	NO	SO2	NO2	NOx	CO2
B5	2.80	-47.05	-48.18	-29.93	-44.81	-4.37
B10	6.23	-45.02	-41.40	-24.97	-42.39	-6.39
B20	10.64	-37.95	-27.34	-16.48	-35.13	-4.16

4.2.4. Effect of Fuel – Cummins

The effect of biodiesel blends on exhaust emissions could not carried out as comprehensively as it was for Mack engines as the trucks equipped with Cummins engines were less in number and were prone to constant mechanical breakdowns or engaged in a very busy garbage collection route. As such, the exhaust emission concentrations were studied just for B20 fuel blend and compared with ULSD for both hot and cold idling modes. Table 13 shows the 15 minute average emissions of pollutants during hot and cold idling for both ULSD and B20. Figure 18 illustrates the percentage change in emissions due to the use of B20 for both cold and hot idling modes. B20 yielded lower CO concentrations during both hot and cold idling modes. The concentrations of all other pollutants were higher for B20 over ULSD during both the idling modes.

As compared to ULSD, B20 yielded approximately 16% lower CO emissions and a very negligible decrease in CO_2 emissions during cold idling. It has also been found that B20 emitted 7% higher NO_2 emissions, 25% higher SO_2 and NO emissions. The average results from hot idling mode identified a 23% decrease in CO emissions and a negligible decrease in

NO_2 emissions whereas an increase of approximately 28%, 15% and 7% in emissions were recorded for NO, SO_2 and CO_2 respectively.

Table 13. Average emissions for B20 and ULSD during cold and hot idle

Pollutant (PPMV)	Cold Idle		Hot Idle	
	ULSD ('01)	B20('01)	ULSD ('01)	B20('01)
CO	252.58	195.28	177.6	149.35
NO	524.6	659.02	312.12	399.12
SO2	120.57	151.9	72.93	84.04
NO2	76.46	82.2	50.74	54.61
NOx	601.07	741.22	362.84	453.71
CO_2 (%)	3.14	3.13	2.44	2.61

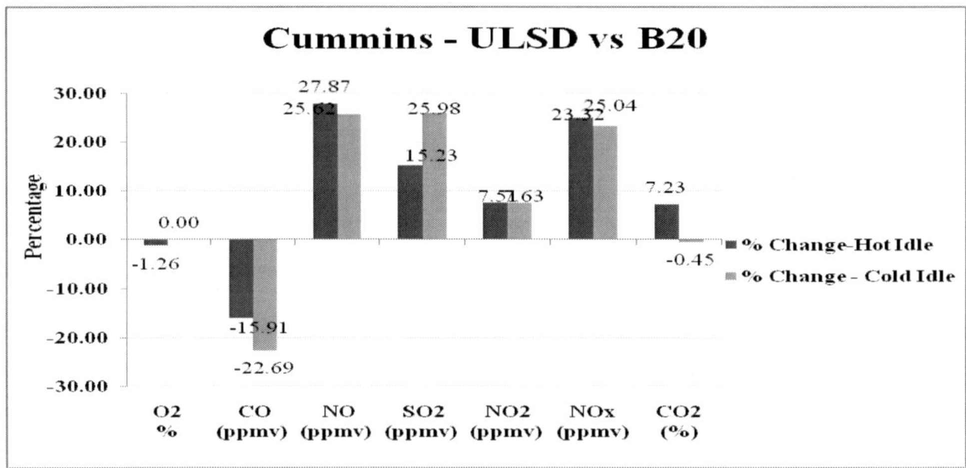

Figure 18. Percentage change in emissions.

4.2.5. Effect of Vehicle Age

The age and technology level of a vehicle also affects the exhaust emissions (Table 3). Experimental emission data from in-use vehicles collected by previous researchers also indicates the effect of vehicle age on the emissions. This behavior is mainly due to their deteriorating catalytic converters and emission control technology systems (Zachariadis et al., 2001). The function of a catalytic converter is to reduce the toxicity of the emissions produced and it requires time after starting the vehicle to reach optimum working temperature and work efficiently. Since the older vehicles have aging catalytic converters, it takes even more time to reach the operating temperature and thus becomes ineffective in reducing emissions. The test fleet considered to study this behavior is presented in Table 14.

Table 14. Mack truck test fleet for studying the effect of age on emissions

Truck Number	Make	Year	Fuel type	Mileage
2656	MACK	1997	ULSD	100948
2658	MACK	1997	ULSD	100047
2673	MACK	2003	ULSD	45693
2678	MACK	2003	ULSD	44703
2681	MACK	2003	B20	44976
2694	MACK	2003	B20	43304

Two sets of Mack trucks manufactured in 1997 (truck # 2656 and #2658) and 2003 (truck #2673 and #2678) running on ULSD were considered. Another set containing two more Mack trucks running on B20 (truck #2681 and #2694) were also considered. These sets were chosen in such a way that each truck in the set has the same year of manufacture and identical mileage on them. This will negate the effect of vehicle miles travelled, which is also an important factor that effects emission concentrations. The effect of age on the emissions from these trucks across two different fuels is studied. The trucks manufactured in 1997 were not OBD compatible. Hence, exhaust temperatures at the start of the test were considered as an indication of the engine temperature. Figure 19 to show the 15 minute average concentrations of each set of trucks across different fuels for both cold and hot engine temperatures.

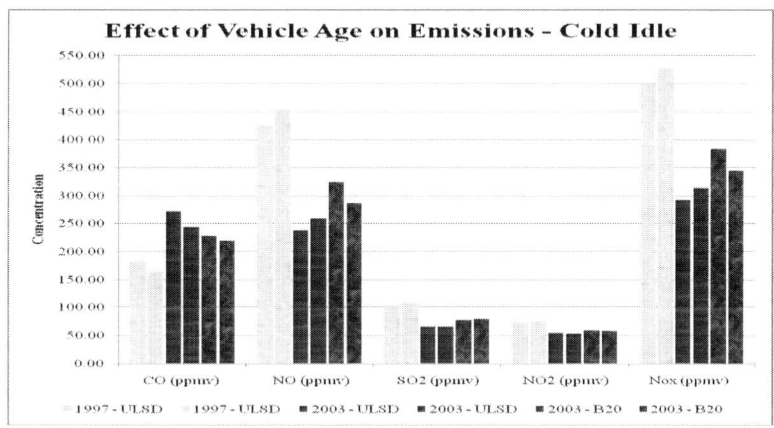

Figure 19. Influence of vehicle age on cold idle exhaust emissions.

The trucks manufactured in 1997 emitted lower concentrations of CO and CO_2 and higher concentrations of NO, SO_2 and NO_2 than those emitted by the trucks manufactured in 2003 during cold starts. Similar trends were observed during hot starts too which provides no clear indication that older vehicles generally emit higher emissions as compared to the newer vehicles (Figures 21 and 22). It is also interesting to note that the concentration of CO during hot starts from newer vehicles is almost twice as much as it is from the older vehicles (Figures 21). The number of vehicle miles travelled also had no precise affect on the exhaust emissions. Similar results were reported by Vijayan et al. (2008) and Tao et al. (2006).

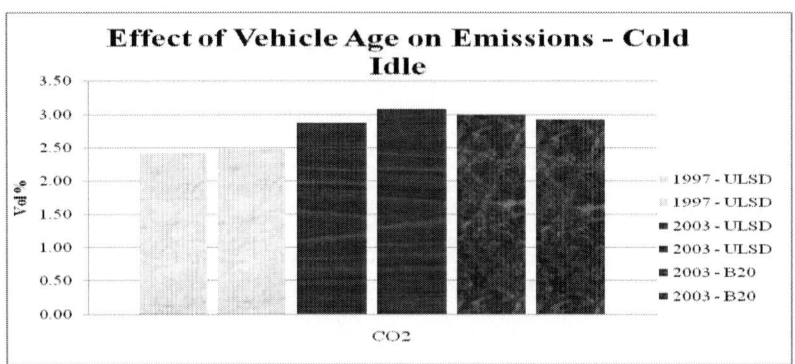

Figure 20. Influence of vehicle age on cold idle CO_2 emissions.

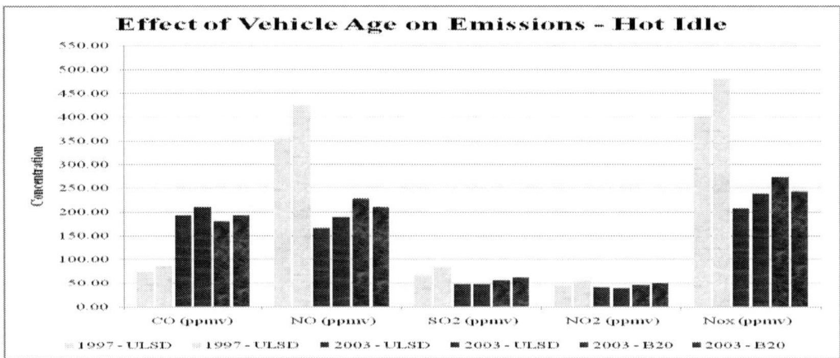

Figure 21. Influence of vehicle age on hot idle exhaust emissions.

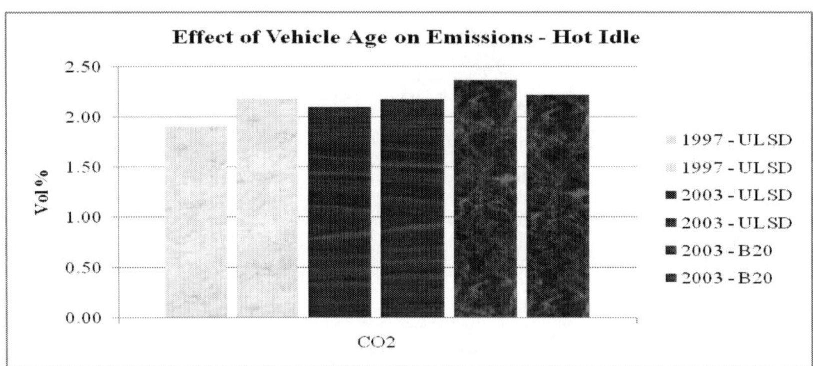

Figure 22. Influence of vehicle age on hot idle CO_2 emissions.

4.2.5. Preventive Maintenance Inspection

Preventive maintenance is the most common vehicle maintaining protocol for every vehicle. It generally includes fuel filter change, air filter change, engine oil change and spark plug inspection. The Fuel filter is a device that stops unwanted debris from entering the combustion chamber and is a very important component to ensure free flow of fuel inside the fuel lines of the vehicle. A clogged fuel filter might lead to a decrease in the amount of fuel flowing to the combustion chamber which leads to an improper air fuel ratio. This condition is generally referred to as a lean mixture which produces hotter gases and might melt the pistons inside the engine. A clogged fuel filter also results in loss of vehicle fuel mileage. An air filter does almost the same thing a fuel filter does but to the air entering the chamber. A clogged air filter will again lead to the formation of a lean mixture which might result in similar issues as mentioned above. The spark plugs provide the necessary spark required for fuel combustion inside the gas chamber. Faulty spark plugs might lead to improper cranking, rough idling and hence affect the combustion process.

Engine oil is the fluid flowing through all the parts of the engine keeping them lubricated and also absorbing some engine heat to pass it to the coolant. Old engine oil will lead to less lubrication of engine parts, thereby affecting the efficiency of the engine and hence the emissions. Poorly maintained vehicles may consume the engine oil leading to the release of heavy smoke from the exhaust. Since all the above four directly affect the exhaust emissions, proper care should be taken to make sure they are maintained in a timely manner. According to a report published by EPA (2007) on low cost ways to reduce emissions from diesel vehicles, it was found that there is a reduction in PM, NOx, CO and HC emissions when preventive maintenance is performed on diesel engines. Vijayan et al. (2008) also observed a reduction of 15% to 20% in overall emissions concentration due to the preventive and maintenance programs. For this analysis, truck#2657 was tested. Cold Idle emissions were collected from this truck before the day of PMI and after the day of PMI. The second by second emissions concentrations for each pollutant are presented in Figure 23.

The initial spike in concentrations of CO and the overall concentrations of SO_2 were found to be higher after PMI. Also, the overall concentrations of NO, NO_2 and CO_2 were found to be higher before PMI. An in depth study comprising more test trucks will be required in order to understand this unusual behavior of CO and SO_2 concentrations.

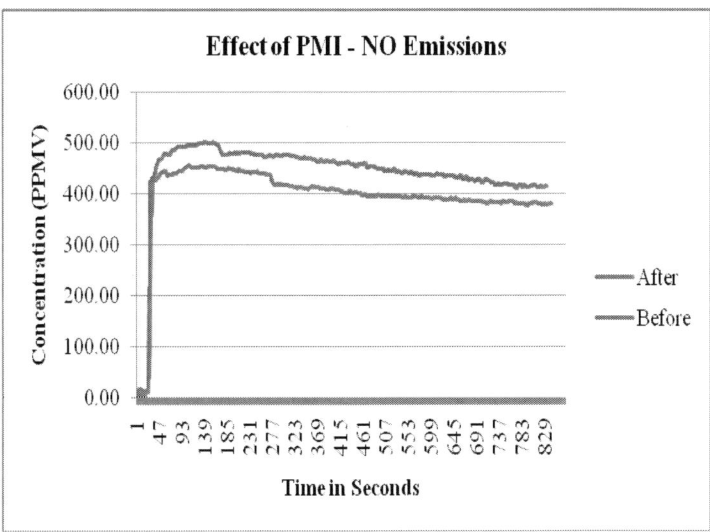

Figure 23. (Continued).

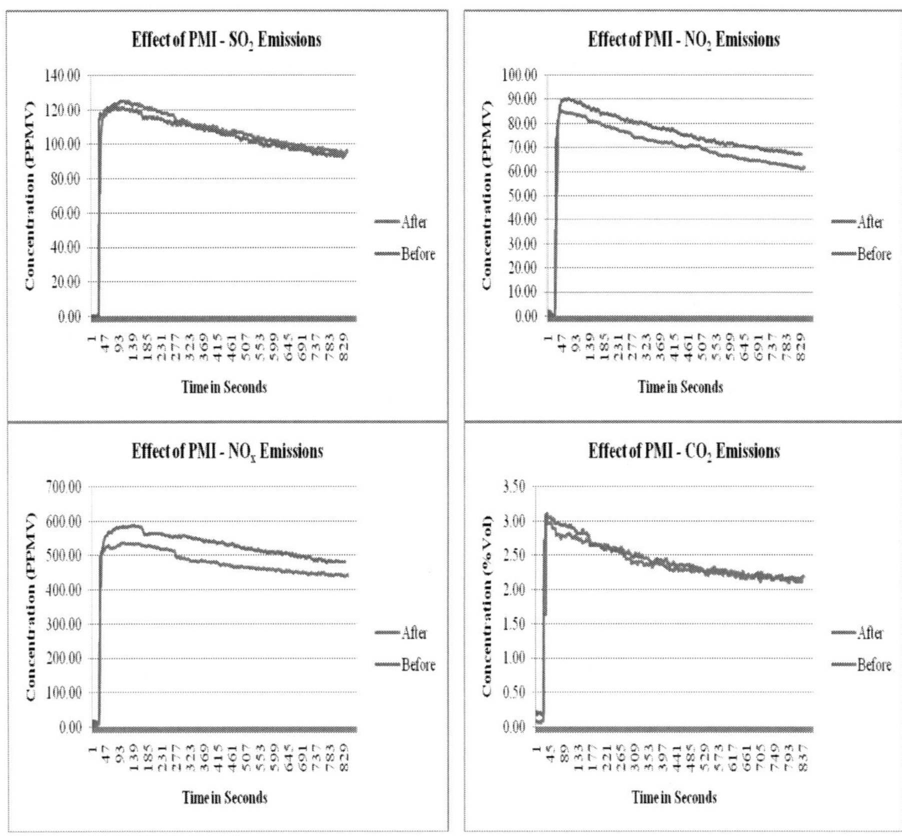

Figure 23. Comparison of preventive maintenance for various pollutants.

4.3. Behavioral Changes in Emissions due to the Use of Biodiesel

One of the major advantages of using biodiesel as a substitute to ULSD is the fact that no additional engine modifications are necessary. Most of the engine manufactures have approved the usage of B20 fuel blends in their vehicles. Since biodiesel has a higher Cetane ignition rating, there is less engine noise pollution when compared to other conventional diesel fuels. Most researchers have documented the clogging of fuel filters as a major disadvantage to the use of biodiesel (biodiesel.org, 2010). This is due to the fact that biodiesel has solvent characteristics which dissolve the sediment build up on the walls of the fuel tank and fuel lines due to the usage of regular diesel fuel. It also is known to dissolve carbon build up in the combustion chamber. But this behavior is found to have eased over time and has actually been found to help engines run cleaner and increase their overall life. Since it is found that biodiesel has such cleansing characteristics, it will be interesting to note the behavior of the exhaust emissions during the initial days of transformation from ULSD to biodiesel. As such, a sample study has been conducted to examine the behavioral changes in idle emissions due to the use of biodiesel blends during the first two days. Three Mack trucks are chosen for this study and the fuel blend used to study the behavioral changes is B10. The characteristics of the test fleet are shown in Table 15.

Table 15. Fleet characteristics of trucks

Trucks Number	Mileage	Manufacturing Year	IandM
Truck#1	128829	1997	No
Truck#2	57203	2003	No
Truck#3	52033	2003	Yes

The three Mack trucks that were chosen had different characteristics as shown in the above table. This was done to study the effect of biodiesel's solvent characteristics on the idle emissions of vehicles having different age and maintenance programs. The inspection and maintenance procedure includes replacing fuel filters, oil filters, engine oil and inspecting spark plugs and ignition wires. Truck#1 was manufactured in the year 1997 whereas trucks #2 and #3 were manufactured in 2003. Trucks #1 and #2 had no inspection and maintenance (IandM) performed on them where as truck#3 was inspected and maintained just before testing. All the test trucks were fuelled with B10 after returning from the daily routes. Emission testing started from the subsequent day. Emissions were measured using the Testo350XL instrument on a second by second basis for a period of 30 minutes. Quality analysis was carried out to make sure no missing or null values were used in the analysis. The 30 minute average values are presented in the Figures 24 and 25.

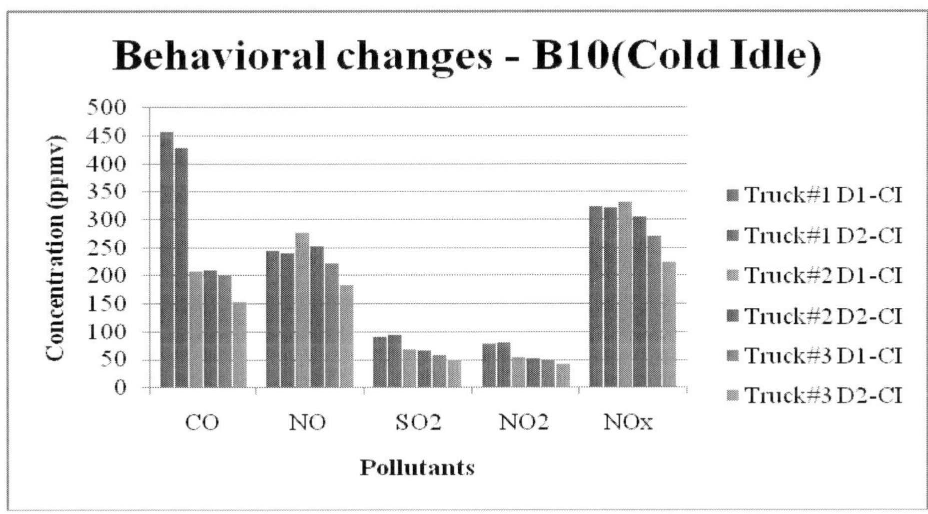

Figure 24. Pollutant concentrations for B10 fuel blend during cold idling.

After testing the truck in the morning for cold idle emissions, it is assigned to its daily route and was tested on the following afternoon for hot idle emissions. A similar approach was followed for the second day and for all the other test trucks. The term "Truck#1 D1-CI" in the legend denotes the cold idle emissions on day one from truck#1. Similar acronyms were used to denote the other trucks representing the day of testing and the mode of testing. The pollutant concentrations for B10 fuel blend during cold idle for all the three trucks are shown in Figure 24 and Figure 25. Other than CO_2 emissions from truck1, none of the pollutants showed significant increase in concentration on the second day as compared to the first. The CO concentrations of truck 1 manufactured in 1997 were found to be very high when compared to those from the newly manufactured trucks.

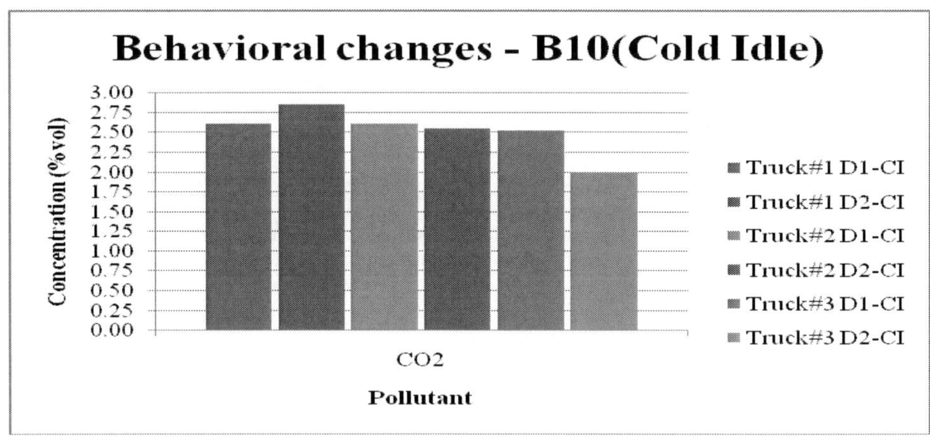

Figure 25. CO_2 concentrations for B10 fuel blend during cold idling.

A reasonable decrease in concentrations of CO was observed over the two days for truck 1 manufactured in 1997. The concentration of NO decreased over the two day period for truck manufactured in 2003 with an IandM program performed. When compared to both truck 1 and truck 2, truck 3 showed uniform trend in the decrease of pollutant concentrations over the two day testing period. This might be attributed to the fact that truck 3 had an IandM performed to it before the old fuel was substituted with the newer fuel blend.

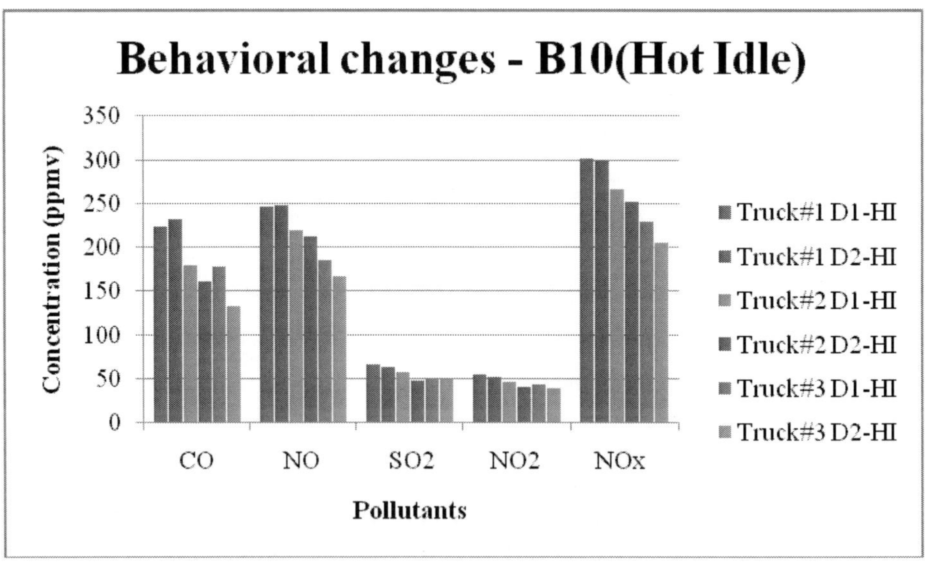

Figure 26. Pollutant concentrations for B10 fuel blend during hot idling.

Figure 26 and 27 present the behavioral changes in pollutant concentrations for B10 fuel during hot idle. A drastic decrease in the concentration of CO can be observed during hot idling from Truck 1 as compared to cold idling. A slight increase in the emissions of CO and

CO_2 were observed over the two days of testing and trends similar to those observed during cold idling were also observed during hot idling.

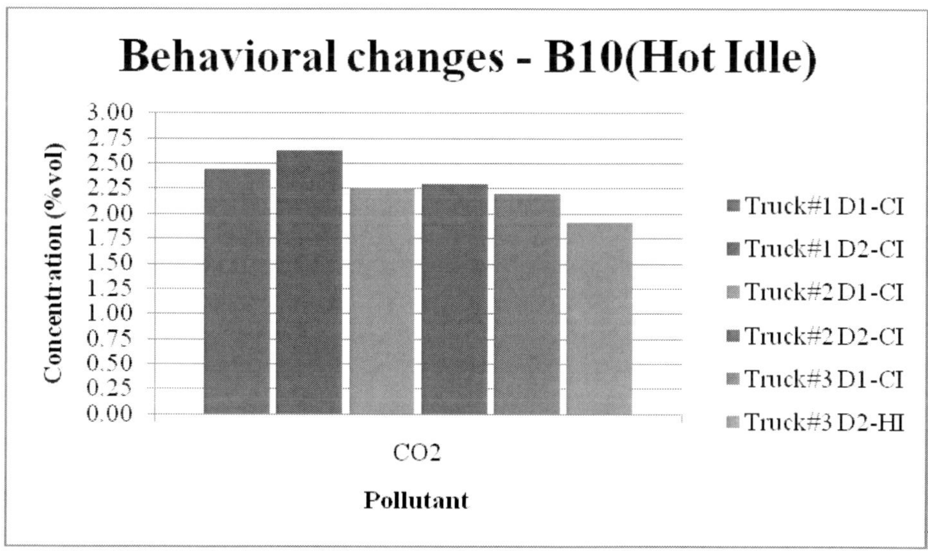

Figure 27. CO_2 concentrations for B10 fuel blend during hot idling.

Since biodiesel has the property to dissolve the sediment and carbon build up thereby clogging filters and leading to improper combustion, higher concentrations of emissions were expected to be observed during the initial days of transition from diesel fuel to the biodiesel blends. From the figures and observations discussed above, no such significant behavior was observed as expected. It might be due to the small percentage of biodiesel present in the fuel (10%). Further in depth study for a larger duration with higher concentrations of biodiesel might be needed to examine the emission profile and behavioral changes of emissions during this transition.

CONCLUSIONS

A quick glance through the literature regarding the tail pipe emissions from vehicles reveals numerous studies that have been carried out on a variety of vehicle and fuel combinations. Study of emission behavior from heavy duty diesel engines seems to be the most sought after field of research but very limited studies were conducted on garbage trucks. The garbage truck fleet in US constitutes some of the oldest trucks operating in the residential areas. Since garbage trucks spend a majority of their time idling and have very poor fuel economy, there is a necessity to evaluate the tail pipe emissions from these vehicles and find an alternative that is cleaner and safer for the environment. A comprehensive field study has been designed to evaluate and analyze the emissions from the City of Toledo garbage trucks and different idling modes for a variety of fuel lends along with engine parameters affecting the exhaust emissions was carried out. Some of the key findings from this study are detailed below.

- Idle engine emissions are directly related to engine temperature. Hotter engines produced lesser emissions compared to colder engines. This behavior was found to be uniform among all different fuel combinations.
- Vehicle age and miles travelled greatly affected the concentrations of emissions produced from garbage trucks. Higher concentrations of NO, NO_2, and SO_2 were released from the older trucks as compared to the newer ones. But this behavior was not found to be consistent across all pollutants since the older vehicles emitted lower concentrations of CO and CO_2 compared to the newer trucks
- Preventive maintenance has a definite effect on the pollutant concentrations in the tail pipe emissions. NO_X emissions were greatly influenced and the emissions of NO decreased by 10% due to the inspections and maintenance programs. A detailed study involving a larger test sample needs to be planned to appropriately characterize the pollutant behavior.
- A study was designed to examine the behavior of exhaust emissions during the transition phase from a base fuel to a fuel blend. No significant observations could be made from the results obtained. This could be due to the low percentage of biodiesel mix in the base fuel.
- Vehicle operating parameters like coolant temperature, fuel temperature, percent fuel, engine speed are the most important variables affecting exhaust emissions. A 90% decrease in the amount of fuel commanded as the idling time increase is a definite indication of the need for new technologies to avoid such high fuel usage during the initial period of vehicle idling.
- The use of biodiesel in garbage trucks yielded mixed results. Biodiesel was found to be a clear winner with the exception of CO in Mack trucks while there was no clear winner among the two fuels in Cummins trucks. This might be due to the difference in engine technologies and engine characteristics between the two fleets.
- Truck drivers have been educated about the importance of reducing the idling time. The IandM program statistics for both Cummins and Mack trucks revealed that there was a significant delay in scheduling the programs. A stricter protocol is being developed to make sure all the vehicles are serviced on the due date. Also, any unwanted idling is being strictly discouraged.
- Although, the behavior of pollutants was not constant with respect to the age of the vehicle, the technician's service records indicated a higher failure rate of mechanical parts and equipment on the older vehicles. The need for new trucks has been identified and active decisions are being taken towards procurement of new or leased trucks to replace the older trucks thereby cutting down on repair costs and fuel usage.

ACKNOWLEDGMENTS

The authors would like to thank the United States Department of Transportation (US DOT) and Toledo Area Regional Transit Authority (TARTA) for a grant awarded to the Intermodal Transportation Institute (ITI) at The University of Toledo. The authors would also like to express their sincere gratitude to the management and employees of the TARTA and the City of Toledo for their continued interest and involvement in this work. The views expressed in this paper are those of the authors and do not represent the views of the funding organizations.

REFERENCES

Ban-Weiss, G.; Chen, J.; Buchholz, B.; Dibble, R. "A Numerical Investigation into the Anomalous Slight NOx Increase When Burning Biodiesel", *Fuel Processing Technology*, 2007, Vol. 88, pp: 659-667.

Beck R.W. (April 2001). "Size of the U.S. Solid Waste Industry", Environmental Research and Education Foundation, Chartwell Information Publishers, (http://www. environmentalisteveryday.org/publications-solid-waste-industry-research/no-cost/size-of-the-industry-study/index.php), accessed September, 2010.

Delucchi, M. "Emissions of criteria pollutants, toxic air pollutants and Green house gases, from the use of Alternative Transportation Modes and Fuels", University of California, Davis, 1996.

Denis, M.J.St.; Cicero-Fernandez, P.; Winer, A. M.; Butler, J.W.; Jesion, G. "Effects of In-UseDriving Conditions and Vehicle/Engine Operating Parameters on "Off-Cycle" Events: Comparison with Federal Test Procedure Conditions,"*J. Air Waste Manage. Assoc.*, 1994, Vol. 44, issue 1, pp 31-38.

Ergeneman, M., Sorusbay, C. and Goktan, A. G. (1999) "Exhaust Emission and Fuel Consumption of CNG Diesel Fueled City Buses Calculated Using a Sample Driving Cycle", Energy Sources, Part A: Recovery, Utilization, and Environmental Effects, Vol. 21, issue 3, pp. 257 – 268.

Faiz, A.; Christopher S.; Walsh, M.P. "Air Pollution from Motor Vehicles: Standards and Technologies for Controlling Emissions", ISBN 0-8213-3444-1, World Bank publications, 1997.

Fernandez, P.; Long, J.R. "Effects of Grades and Other Loads on On-Road Emissions of Hydrocarbons and Carbon Monoxide," *J. Air Waste Manage. Assoc.* 1997, Vol. 47, issue 8, pp. 898-904.

Frey, H.C.; Rouphail, N.M.; Unal, A.; Colyar, J.D. "Emission Reductions Through Better Traffic Management: An Empirical Evaluation Based Upon On-Road Measurements", Department of Civil Engineering, North Carolina State University for North Carolina Department of Transportation, Raleigh, NC, 2001.

Frey, H.C.; Unal, A; Rouphail, N.M.; Colyar, J. D. "On-Road Measurement of Vehicle Tailpipe Emissions Using a Portable Instrument". *J. Air and Waste Manage. Assoc*, 2002a.

Frey, H.C.; Unal, A.; Chen, J. "Recommended Strategy for On-Board Emission Data Analysis and Collection for the New Generation Model", Department of Civil Engineering, North Carolina State University for Office of Transportation and Air Quality, U.S. EPA. 2002b.

Frey H.C.; Unal, A (2003). *"Emissions Measurement Program"*. North Carolina State University, Department of Civil (http://www.epa.gov/ttn/chief/conference/ei11/mobile/frey unal.pdf), accessed Sep 2010.

Gierczak, C. A.; Jesion, G.; Piatak, J.W.; Butler, J.W. "On-Board Vehicle Emissions Measurement Program: CRC VE-11-1", *Coordinating Research Council*, 1994.

Gordon, D.; Burdelski, J.; Cannon, J.S.; (2003). "Greening Garbage Trucks: New Technologies for Cleaner Air", Society of Automotive Engineers, http://www. informinc.org/reportpdfs/st/GreeningGarbageTrucks.pdf, accessed October, 2010.

Graboski, M. S.; Ross, J. D.; McCormick, R.L. "Transient emissions from no. 2 diesel and biodiesel blends in a DDC series 60 engine", *Society of Automotive Engineers Paper* 961166, 1996.

Health Effects Institute (2001). "Research Directions to Improve Estimates of Human Exposure and Risk from Diesel Exhaust", http://pubs.healtheffects.org/view.php?id=5, accessed August, 2010.

INFORM, Inc. (2007), "Facts on Greening Garbage Trucks: New Technologies for Cleaner Air", http://informinc.org/fact_ggt.php, accessed September 2010.

Kadiyala, A.; Kumar, A.; Vijayan, A. Study of occupant exposure of drivers and commuters with temporal variation of in-vehicle pollutant concentrations in public transport buses operating on alternative diesel fuels. *The Open Environmental Engineering Journal*, 3, 55-70, 2010.

Kelly, N. A.; Groblicki, P.J. "Real-World Emissions from a Modern Production Vehicle Driven in Los Angeles," *J. Air Waste Manage. Assoc.*, 1993, Vol. 43, issue 10, pp.1351-1357.

Kilcarr, S. (2007). "The Adoption of Technologies, Policies, and Systems for Improving Fleet Fuel Performance". http://www.intrans.iastate.edu/mtc/reports/fleet-fuel.pdf, accessed 2009.

Knothe, G.; Sharp, C.; Ryan, T. "Exhaust Emissions of Biodiesel, Petro-diesel, Neat methyl esters, and Alkanes in a New Technology Engine, Energy and Fuels", 2004, Vol. 20, issue 1, pp. 403-408.

Knothe, G.; Steidley, K. "Lubricity of Components of Biodiesel and Petrodiesel" The Origin of Biodiesel Lubricity, *Energy Fuels,* 2005., Vol.19, pp.1192-1200.

Kumar, A.; Nerella, V. V. K. "Experimental analysis of exhaust emissions from transit buses fuelled with biodiesel", *The Open Environmental Engineering Journal*, 2009, Vol. 2, pp. 81-96.

Laroo, C.; Mason, R.; Buckingham, J. "Ultra low sulfur diesel (ULSD) Sulfur Test Method Variability: a Statistical Analysis of Reproducibility", US EPA ULSD Round-Robin Test Program, SAE International, 2005.

LeBlanc, D.C.; Meyer, M.D.; Saunders, F.M.; Mulholland, J.A. "Carbon Monoxide Emissions from Road Driving: Evidence of Emissions Due to Power Enrichment," *Transportation Research Record* 1994, Vol.1444, pp. 126-134.

Marshall, W.; Schumacher, L.G.; Howell, S. "Engine exhaust emissions evaluation of a Cummins L10E when fuelled with a biodiesel blend." 1995 SAE paper, 952363.

Mc Cormick, R. L.; Graboski, M. S.; Alleman, T. L.; Yanowitz, J. "Idling emissions from heavy-duty diesel and natural gas vehicles at high altitude", *Journal of Air and Waste Management Association*, 2000, Vol. 50, issue 11, pp. 1992-1998.

Perrot, T. L.; Constantino, M. S.; Kim, J. C.; Tario, J. D.; Hutton, D.B.; Hagan, C. Truck Stop Electrification as a Long-Haul Tractor Idling Alternative, Transportation Research Board, 2004.

Pisarski, A (2006). "Commuting in America. Third national report on commuting trends". Transportation Research Board, Cooperative Research Programs, http://onlinepubs. trb.org/onlinepubs/nchrp/CIAIII.pdf, accessed November, 2010.

Pope, C.A,III.; Burnett, R.T.; Thun, M.J.; Calle, E.E.; Krewski, D.; Ito,K.; Thurston, G.D. "Lung Cancer, Cardiopulmonary Mortality, and Long-term Exposure to Fine Particulate Air Pollution", *JAMA*, 2002, Vol. 287, pp. 1132-1141.

Rouphail, N.M.; Frey, H.C.; Colyar, J.D.; Unal, A., "Vehicle Emissions and Traffic Measures: Exploratory Analysis of Field Observations at Signalized Arterials", *Transportation Research Board*, 2001.

Schafer, J.J.; Kleeman, M.J.; Cass, G.R.; Simoneit, B.R.; "Measurement of Emissions from Air Pollution Sources. 5. C1−C32 Organic Compounds from Gasoline-Powered Motor Vehicles" *Environ. Sci. Technol.*, 2002, Vol. 36, issue 6, pp 1169–1180.

Schumacher, L.; Chandler, K. "Alternate Fuel Transit Bus Evaluation Program results", *Society for automotive engineers*, 1996.

Shandilya, K. K. and Kumar, A. "Analysis of research studies on exhaust emission from the heavy duty diesel engine fueled by biodiesel", Chapter 2, Handbook of Environmental Research, ISBN: 978-1-60741-492-6, Nova Science Publishers, Inc., 2009.

Sheehan, J.; Duffield, J.; Graboski, M.; Shapouri, H. "Life Cycle Inventory of Biodiesel and Petroleum Diesel for Use in an Urban Bus.", National Renewable Energy Laboratory for U.S. Department of Energy's Office of Fuels Development and U.S. Department of Agriculture's Office of Energy, 1998, Rep. No. NREL/SR-580-24089.

Shida, T.; Frank, B,; Lanni, T.; Rideout, G.; Meyer, N.; Beregszaszy, C. "Unregulated Emissions from a Heavy-Duty Diesel Engine with Various Fuels and Emission Control Systems", *Environment Science and Technology*, 2007, Vol. 41, issue 14, pp. 5037 − 5043,.

Smith, J.A.; Endicott, D.L.; Graze, R.R. "Biodiesel Engine Performance and Emissions Testing," *Caterpillar Technical Center*, May 1998.

Spataru, A.; Roming, C, "Emissions and Engine Performance from Blends of Soya and Canola Methyl Esters with ARB #2 Diesel in a DCC 6V92TA MUI Engine". *SAE Tech. Pap. Ser. 1995, No. 952388*.

Tao, H.; Shah, S. D.; Wayne, M. J.; Younglove, T.; Chernich, T. J.; Ayala, A. "Analysis of Heavy-Duty Diesel Truck Activity and Emissions Data", *Atmospheric Environment*, 2006, Vol. 40, issue 13, pp. 2333-2344.

Tat, M.; Wang, P.; Van Gerpen, J.; Clemente, T. "Exhaust Emissions from an Engine Fueled with Biodiesel from High-Oleic Soybeans", *Journal of the American Oil Chemists' Society*, 2007, Vol. 80, pp. 865-869.

Tong, H.Y.; Hung, W.T.; Cheung C.S.J, "On-Road Motor Vehicle Emissions and Fuel Consumption in Urban Driving Conditions," *J. Air Waste Manage. Assoc.*, 2000, Vol. 50, issue 4, pp. 543-554.

U.S Environmental Protection Agency (2007). "California Greenhouse Gas Emission Inventory". (http://www.arb.ca.gov/cc/inventory/inventory.html), accessed December, 2009.

U.S Environmental Protection Agency. "Air and Radiation". (http://www.epa.gov/air/airtrends/index.html), accessed December, 2010.

U.S Environmental Protection Agency. "Basic information". (http://www.epa. gov/airtrends/sixpoll.html), accessed November, 2010.

U.S Environmental Protection Agency "National-Scale Air Toxics Assessment for 2002", June 24, 2009. (http://www.epa.gov/ttn/atw/nata2002/factsheet.html, accessed October, 2010).

U.S Environmental Protection Agency (September, 1997). "Emission Standards Reference Guide for Heavy-Duty and Non road Engines", (http://www.epa.gov/otaq/ cert/hd-cert/stds-eng.pdf), accessed August, 2010.

U.S. Environmental Protection Agency. National Air Quality and Emissions Trends Report; 2001, USEPA 454/K-01-004.

U.S. Environmental Protection Agency, National Air Quality and Emissions Trends Report, 2003, USEPA 454/R-03-005.

U.S Environmental Protection Agency. A Comprehensive Analysis of Biodiesel Impacts on Exhaust Emissions, Draft Technical Report, 2002, EPA420-P-02-001, (http://www.epa.gov/oms/models/biodsl.htm), accessed November, 2010.

"Vehicle Inventory and Use Survey", (1999). U.S Department of Commerce, Economics and Statistics Administration, Bureau of Census, (http://www.census.gov/prod/ec97/97tv-tn.pdf), accessed October, 2010.

Vijayan, A.; Kumar, A.; Abraham, M. A.; Experimental Analysis of Vehicle Operation Parameters Affecting Emission Behavior of Public Transport Buses with Alternative Diesel Fuels, *Journal of the Transportation Research Board*, 2008, No. 2058, 68-78.

Wang, W.; Clark, N. N.; Lyons, D. W.; Yang, R. M.; Gautam, M.; Bata, R. M.; Loth, J. L. "Emissions comparisons from alternative fuel uses and diesel buses with a chassis dynamometer testing facility", *Environmental Science and Technology*, 1997, Vol. 31, issue 11, 3132-3137.

Wang, W.; Clark, N. N.; Lyons, D. W.; Gautam, M. "Emissions from nine heavy trucks fueled by diesel and biodiesel blend without engine modification", *Environmental Science and Technology*, 2000, Vol. 34, issue 6, 933-939.

Weilenmann, M.; Soltic, P.; Saxer, C.; Forss, A.; Heeb, N. "Regulated and non-regulated diesel and gasoline cold start emissions at different temperatures", *Atmospheric environment*, 2005, Vol. 39, issue 13, pp. 2433-2441.

Weilenmann, M.; Favez, J. Y.; Alvarez, R. "Cold-start emissions of modern passenger cars at different low ambient temperatures and their evolution over vehicle legislation categories", *Atmospheric environment*, 2009, Vol. 43, issue 15, pp. 2419-2429.

Wilson, E.C. "Biodiesel Promise ", Bulk Transporter Magazine, Prism Business Media, 2006, pp. 31-34.

Wright, J.W. "2001 New York Times Almanac". New York, NY: Penguin Putnam Inc., accessed Nov 2010.

www.Biodiesel.org (2005). "Biodiesel Myths and Facts", (www.biodiesel.org/pdf_files/ fuelfactsheets/Myths_Facts.pdf) accessed October, 2010.

www.1eere.energy.gov (2003), "Biodiesel, Just the basics", (http://www1.eere. energy.gov/vehiclesandfuels/pdfs/basics/jtb_biodiesel.pdf) accessed October, 2010.

Zachariadis, T.; Ntziachristos, L.; Samaras, Z. "The effect of age and technological change on motor vehicle emissions", *Transportation Research Part D, Transport and Environment*, 2001, Vol. 6, issue 3, 221-227.

In: Biodiesel: Blends, Properties and Applications   ISBN: 978-1-63117-024-9
Editors: Jorge M. Marchetti and Zhen Fang   © 2014 Nova Science Publishers, Inc.

Chapter 2

BIODIESEL, FOSSIL DIESEL AND THEIR BLENDS: CHEMICAL AND TOXICOLOGICAL PROPERTIES

Sergio Manzetti[1], Otto Andersen[1] and Jan Czerwinski[2]*
[1]Nanotoxicology Unit, Western Norway Research Institute,
Fosshaugane Campus, Sogndal, Norway
[2]Abgasprüfstelle und Motorenlabors (AFHB),
Berner Fachhochschule für Technik und Informatik TI, Biel,
Gwerdtstrasse, Nidau, Switzerland

ABSTRACT

The world's air pollution problems are increasingly been related to automotive exhaust emissions. The implementation of diesel and biodiesel blends in passenger vehicle engines have gradually produced a new ecotoxicological profile of urban and rural air pollution, where nanoparticles, volatile exhaust fractions, microparticles and aerosol agglomerates dominate the spectrum of emission species. The effects of these species are increasingly associated with cardiovascular diseases, lung cancer and increase in all-cause mortality in the human population, particularly in urban and other highly trafficked areas. Also, the size of particles and agglomerates from exhaust has been related to particular diseases, risks of contracting types of pathologies and development of cardiovascular complications. Particles have therefore selectively been addressed in this literature review for adverse health effects. With particular focus has biodiesel blending been extensively reviewed for chemical species and associated adverse health effects. The reviewed data suggests that the legislatory environmental health authorities worldwide are not fully updated with all aspects of air pollution and that filtering technologies, fuel types and threshold values for particle content in the air are not up to date with the medical and patho-physiological knowledge.

---

*Email: oan@vestforsk.no, Ph: +47 977 10 928

INTRODUCTION

The majority of motorized road vehicles in Europe run on diesel. Almost all heavy duty lorries, buses, and an increasing number of passenger cars use this fuel. The advantage of diesel lies in its high fuel efficiency, high power output and its lower price compared with gasoline. Given the increased use of diesel, the pollution and exhaust toxicological aspects have changed considerably since the mid 90s, particularly for passenger cars. A vast number of studies has been published on diesel's toxic potential and indentified it as the major source of pollution in metropolitan and trafficked areas [1-8]. The deposition of combustion particles in the human respiratory tract and the inflammatory responses including potential carcinogenic responses have been documented [2-4, 6] and the toxicological characteristics of the combustion particles from the diesel engines are still under ongoing research. Having a great variety in chemical composition, the particles from diesel and biodiesel fuels present toxic aspects which may vary from being an inflammatory threat to carcinogenic factor. This variation depends on dose-dependent exposure, time-dependent exposure and frequency of exposure [5-8].

Durbin and colleagues [9] cited that elemental and organic carbons species (EC and OC) are the primary constituents of diesel particulate matter (DPM), consisting of approximately 73-83% of total mass. Soot, a collective terminology for OC and EC including other pollution components from diesel combustion, has been shown in Taiwan and at the Karolinska Institute in Sweden to cause cancer in children [10, 11], while EC and OC alone have been correlated to respiratory and cardiovascular diseases, and even carcinoma [12, 13]. In addition to these factors, fossil diesel has a high content of polycyclic aromatic hydrocarbons (PAH) and the incomplete combustion of these compounds generate exotic species in the exhaust, in addition to uncombusted PAHs [14-16]. These findings have made several organizations, including the US National Institute of Occupational Safety and Health [17], the International Agency for Research on Cancer (IARC) [18], the World Health organization [19], the US Environmental Protection Agency [20] and the US National Toxicology Program [21] to classify diesel exhaust as a potential human carcinogen. Additional studies characterize the composition of DPM to be highly complex and composed of hundreds of components in particulate, liquid and gaseous form [22].

The gaseous compounds of DPM are nitrogen oxide species, carbon dioxide and monoxide, sulphur compounds and several low molecular weight hydrocarbons such as aldehydes, benzene, PAHs and nitro-PAHs. The DPM core is mainly made of elemental carbon and adsorbed organic compounds, in addition to trace amounts of sulfate, nitrate, metals, and other trace elements. DPM consists of coarse and fine particles, with diameters lower than 10 and 2.5µm respectively (PM_{10} & $PM_{2.5}$), including a number of ultrafine particles of diameters less than 0.1µm, so-called nanoparticles [22]. The nanoparticles have surface that easily adsorb other aerosol compounds and are able to reach deep into the lungs and cross the blood barrier [23-25]. The organic part, encompassing the mentioned compound classes can reach up to 49% of the DPM weight and contain 5% metals (including heavy metals) and 4% sulfates and nitrates [22].

These components have been shown to affect respiration and promote cancer [26-28], through mechanisms of inflammation after deposition in the alveoli.

PUBLIC HEALTH STUDIES

Several studies have been carried out to determine the effects on public health of long term exposure to air pollution with particular emphasis on diesel exhaust [29-46]. In all these, a common theme of increased mortality and morbidity was found in relationship with nanoparticles from traffic exhaust. The effect of the smallest particles was determined to be less known than coarser particles [29]. Given the poor availability of studies on nanoparticles, a series of policy options were proposed to be reviewed and modified in accordance with public safety [29]. The expert panel that evaluated the various toxicological scenarios relating to nanoparticles (NP) was consistent in determining a medium to high risk of short-term exposure's cause to increase all-cause mortality. The correlation between NP and mortality was centered on the thrombotic and respiratory inflammatory effects of NP on human health. This proved the main theme to introduce improvements in the current public health regulations against NP.

The sensitivity of human health to various health pollutants has also been thoroughly discussed by Brunekreef and Holgate [30] who showed that reactions to pollution both at small and large doses and during short term as well as long term exposure, are of equal importance and must be evaluated consistently for better public health regulations to be implemented. The probability that we are dealing with new types of particles continuously, due to the rising number of motorized vehicles and the photochemical reactions between sun-exposure and particle clouds, gives reason to underline that toxicological studies of traffic pollution must be continuously updated, and new experiments in relation with temperature, weather conditions and city densities have to be considered [30]. The traffic pollution condition is in accord with Brunekreef and Holgate [30] evolving with society, infrastructures and not at least with the new types of combustible fuels. In relation to the high levels of pollutants in cities, life-expectancy of the population exposed has been assessed to be reduced by 1-2 years, which is considerably high compared to other life-style or environmental risk factors [30, 31]. Most of the pathophysiological complications related to this shortened life expectancy reside in the pulmonary system, where the conditions start from asthma, allergies and worsen in late stages of life to pneumonia, lung cancer and other pulmonary disorders. The risk of developing lung cancer in Danish cohorts was found to be present and to request novel strategies to reduce exposure of pollution to the population [32]. Furthermore, in another study, the association of ambient residential exposure to PM_{10}, $PM_{2.5}$, NO_2, SO_2 and mortality was examined in 53,814 men in the US trucking industry [33]. The various ranges of particle diameter, likely indicating various degrees of pathological responses, were assessed in order to provide data to the existing knowledge on the more grave danger of finer particles. In accord with expectations, the risk for lung cancer was substantially higher for finer particles, but not for PM_{10}. The overall increased risk for cardio-pulmonary complications was increased with all particle sizes [33]. In another cohort study recently published in the journal *Stroke* [34], researchers discovered the reduced survival rate of stroke patients in highly trafficked areas in the UK. In this study PM_{10} was identified to increase risk of death in stroke patients by 52%. Additionally, in a study of 21 cities, all-cause mortality was found to increase by 0.6% for every $10\mu g/m^3$ increase in PM_{10} [35]. Admission for asthma and chronic obstructive pulmonary disease among people over 65 years of age was then found to increased by 1% for every $10\mu g/m^3$ increase in PM_{10} [36]. Also admission for cardiovascular diseases increases by 0.5 % for the same increase of PM_{10} and by 1.1 % for each increase of 10 $\mu g/m^3$ of black smoke from diesel engines [37].

The case for smaller particles was surveyed in a study from Boston, where the cases of the onset of myocardial infarction for patients suffering of arrhythmia occurred at times during higher concentrations of $PM_{2.5}$ [38]. $PM_{2.5}$ was also associated with severe cases of arrhythmia leading to therapeutic intervention by an implanted defibrillator [39].

The relationship between nanoparticle diameter and type pathology may therefore be hypothesized in a sense where 1) the finer the particle are, and the longer the duration of exposure is, more likely to cause the development of chronic pulmonary and cardiac pathologies (given the ability to cross the blood barrier) 2) at short but repetitive and intense exposures to larger particles, the more chances are to develop abrupt reactions such as pulmonary inflammation, asthma, frequent common colds and immune reactions related to throat and bronchi disorders, including difficulties in recovering from stroke and circulatory disorders and interventions [34]. These hypotheses may be followed up by a third hypothesis where the variation of particle chemistry given changes in weather, sun and urban conditions [30] may increase the risk of lung cancer. This third point is hypothesized accounting for the continuous provocation by new antigens in interaction with the immune system thereby promoting continuous and frustrated inflammatory conditions. Frustrated inflammation [47] activates a series of protein complexes, where some have been involved in the proliferation of cancer [48].

The continuous exposure to traffic pollution particles in cities and large urban areas shapes therefore the epidemiological profile of urban populations in a manner that requires tighter regulations and attention on the issue, particularly monitoring of lung-related conditions. The "frustrated" behavior of the immune system in the alveoli is also observed in the cases of asthma, where asthma is triggered from the immune system upon early exposure in early childhood [40]. This underlines that the immune system's reaction to traffic pollution has the tendency to behave in an "iterative exaggerated" manner in babies and small children, because of their more frequently maturing physiome, promoting stronger and more adverse reactions than an adult immune system. Accounting for the ever-present levels of nanoparticles in urban and trafficked areas, pediatric heath becomes quickly a central aspect with scientific demands and challenges to public health, and therefore studies on pediatric health in urban and highly trafficked areas compared to low pollution areas are needed.

In this context, PM_{10} particles, in addition to NO_x and SO_2, were found to be particularly relevant to the development of asthma in children [41]. This underlines that larger particles may trigger asthma more easily given their larger physical occupation in the alveoli, affecting respiration in a suffocating manner. Smaller particles on the other hand may affect more significantly the circulatory and immune system given penetration of the blood barrier in the alveoli, causing more complications in internal tissue such as in brain and nerve tissue [49]. Indeed, the smaller particles, once crossed the blood-barrier may impair vasomotor function, and cause vascular and circulatory complications [42], thereby exerting more stress on the cardiac system causing complications such as myocardial infarction [45]. The dependency of these reactions evolves eventually on cellular and biochemical responses, which have been thoroughly studied.

In Vivo Studies

Pathophysiological mechanisms of action from nanoparticles on organs have been elucidated by various groups. A group in particular has worked on studying the effects of nanoparticulates and PAH from diesel exhaust on cells from the pulmonary and cardiac system [50-56]. On a study on cardiac cells in particular, Totlandsdal and colleagues [50] applied ultrafine carbon black particles of increasing size on rat cardiomyocytes and cardiofibroblasts. The particle size was at a minimum size of 12 nm and agglomerating in culture, reached sizes up to 100 nm. The effects on the cells were detrimental, and the increased release of interleukins was observed in addition to cell damage. A similar effect was observed on rat lung epithelial cells, where the expression of interleukin- 6 and 1β was found to be the main response to particulate matter of ultrafine carbon black [51]. The mechanisms of release of IL-6 from the lung cells was furthermore described [52], where the expression of IL-6 occurred with the expression of IL-1α and IL-1β. The findings indicated also that the reactions occur within hours from the moment of exposure to nanoparticles, and that it is facilitated by the initial release of IL-1 after only 4 hours. The exposure to nanoparticles furthermore induced the MAPKs (Mitogen-activated protein kinases) and NF-κB (Nuclear factor kappa-light-chain-enhancer of activated B cells) and p38. These mechanisms of action indicate a significant inflammatory response in the lung tissue and show also the involvement of proliferation factors, which also participate in the mechanisms of apoptosis and anti-cancer reactions. The question on whether pollution particles affect and induce cancer is therefore also relevant and of interest to review. In further studies by the same group [53-56], the carcinogenic aspects of nitro PAHs on human bronchial epithelial cells and murine hepatoma cells were demonstrated. 1-nitropyrene (1-NP) induced DNA damage and cytotoxicity while 3-nitrofluoranthene was particularly toxic to the cells and altered their cell-cycle. The mechanisms of action were through the cytokine/chemokine pathways, activating cell cycle checkpoint factors [53, 54]. These same toxic compounds were furthermore observed to affect the caspase-pathways [56] which can in turn take part in the development of cancer if repetitively disturbed by toxic substances.

BLOOD BARRIER AND INFLAMMATORY RESPONSES

At the boundary of the blood barrier, a series of molecular reactions become relevant to the assessment of damage by nanoparticles. These mechanisms of reaction to particulate matter can become involved in "persistent and non-resolving inflammation" [57]. In this review the authors describe how the non-resolving action of inflammatory agents creates long termed diseases which affect the respiratory and cardiac tissue, contributing also to necrosis and cancer. The difficulty the immune system cells experience in decomposing and attempting the decomposition on a continuous amount of nanoparticles results in many cases to deficiencies where the inflammatory system persist either excessively or subnormally. These reactions to nanoparticles trigger therefore mechanisms that may result in harming the body, through many different cell pathways.

The research on these pathways has been conducted over several years, but given the variation of chemical species in exhaust and cellular reactions, many unanswered questions still remain. In a particular study [58], the effects of diesel exhaust showed that DPM induced serum vascular cell-adhesion molecule levels (VCAM)-1 in mice and enhanced vasoconstriction. Affecting the cardiac and respiratory system, the ability of the nanoparticles to enter the blood barrier was shown and the reactions to DPM were identified at the tissue, cellular and the molecular level [58, 59]. The observations showed also that DPM blocked transportation attempts of DPM by the cellular machinery by blocking transcription or protein synthesis. In a fashion which may induce persistency in inflammation, inhibition of NADPH oxidase also occurred. This event took place catalyzed by radical scavengers, which ameliorated the up-regulation of DEP-induced P-glycoprotein. This indicated that reactive oxygen species (ROS), arisen during exposure to nanoparticles, took part in the signaling between cells and disturbed the inflammatory response in a repetitive manner [58, 59].

Once crossed the blood barrier, nanoparticles have also been shown to cause increased expression levels of brain capillary tumor necrosis factor-alpha (TNF-alpha) and leading to P-glycoprotein down-regulation [60]. The down-regulation of P-glycoprotein affects the transport of nutrients and signaling factors to the cell, thereby leading to necrosis. Nanoparticles promote more severe reactions at the blood-barrier, where combined with persistent inflammation may lead to more serious conditions. This was observed in a study where emphysema in rat was observed upon exposure to DPM [61]. Here, the proinflammatory response in the lungs of the rats was characterized by a significant infiltration of leukocytes such as macrophages, eosinophils, lymphocytes and an increased level of IL1β in lung homogenates. Lung damage was also observed in this study, which showed characteristics of emphysema-related morphological changes including airspace enlargement and progressive destruction of alveolar wall structures [61, 62]. A group of researchers also found the increased risk of spontaneous abortion for pregnant women given exposure to DPM [63]. The findings were related to a mouse model study where the ability of the embryo to bind to a fibronectin matrix was studied and found to be affected by DPM. DPM have also been shown to amplify the cellular response in the lung tissue to invading agents, thereby promoting "frustrated inflammation" thus yielding excessive phagocytosis [64].

The mechanism of frustrated phagocytosis is of particular relevant in development of asthma and allergies as an epidemic [65-67]. A significant exposure to nanoparticles on a continuous basis may lay the foundation for the immune system to promote stronger reactions to the particulate matter such as dyspnoea, respiratory difficulties, heart palpitations; arrhythmia and chest pain on a continuous basis [68-72], making a high-alert mode of the immune system a daily and almost background reference for immune activity causing serious physiological complications in the masses exposed to nanoparticles, $PM_{2.5}$ and PM_{10}.

PHYSIOLOGICAL COMPLICATIONS AND DISEASE DEVELOPMENT FROM EXHAUST EMISSION PARTICLES

The initial signs of wheezing, coughing, chest tightness and shortness of breath have been reported among a group of 249 subjects that were included in a study to assess the reactions to diesel $PM_{2.5}$ [68]. In this study, the results indicated that the smallest particles of diesel exhaust may confer a greater health risk than only larger PM. PM overall was however enough to increase risk of respiratory symptoms [68]. Asthmatics are more prone to difficulties and graver symptoms than none-asthmatics; however a 6-8 hour daily exposure introduces substantial health risks for both groups [68]. Additional complications promoted by PM in humans are: worsening peak flow, necessity of inhaler usage, respiratory symptoms, and emergency room visits in asthmatic children and adults [67]. Gradually, these complications grow to more serious conditions such as cardiopulmonary mortality and lung cancer [73]. Furthermore, general air pollution has been correlated to lung cancer and mesothelioma [74] and the particulate matter is pinpointed as the central component for particularly the cardiovascular complications developed during a prolonged exposure in urban and trafficked areas [75]. This statement from the American Heart Association states also that life span is reduced by several months up to years by only a few years of exposure to PM; in particular $PM_{2.5}$. Other events occurring from the exposure to particulate matter are stroke, myocardial infarctions, and heart failure exacerbation [75]. The causes for this are changes in prothrombotic and coagulant content in the blood, progression of atherosclerosis and vasoconstriction [75].

Inhalation of exhaust particles is ranked as the 13 leading cause of death in the world and responsible for approx. 800.000 premature deaths pr year [76]. Even short-term exposure to $PM_{2.5}$ in particular is responsible for thousands of deaths pr year in the US alone [77, 78]. Still related to particulate matter, admission at the hospitals increased with only a 10 $\mu g/m^3$ increase, causing cerebrovascular disease increment, and heart failure [79].

In an Italian study, particulate matter has also been shown to be correlated to deep vein thrombosis [80]. Particulate matter has also been correlated cardiac arrest and arrythmia [80, 81]. Interestingly, heart failure, ischemic stroke, cardiovascular hospitalizations, systemic inflammations and heart-rate variability are reported to be more likely to be contracted during short-term exposure to particulate matter rather than long-term [77]. Long-term exposure to particulate matter is instead more associated with ischemic heart disease and cardiovascular mortality [77].

Therefore, in a general overview of the complications that are involved on exposure to particulate matter, the more grave exposure is short-term, which gives more severe reactions to various clinical conditions as mentioned here. Systemic and pulmonary arterial bypass is also caused in many cases by DPM on short-term exposure in addition to repolarization abnormalities of the heart [77].

Most of the cardiopulmonary conditions hereby mentioned are seemingly caused ultimately by the effects of DPM on unbalancing the brain and the autonomic nervous system [82]. This observation indicates the crucial aspect of DPM, its ability to cross not only the blood barrier, but most likely also the membrane barrier of cells due to its chemical properties and surface components.

CHEMICAL PROPERTIES OF NANOPARTICLES

The properties of exhaust nanoparticles vary according to the type of fuel combusted. A series of studies are here reported in the delineation of the differences between biodiesel and fossil diesel fuel nanoparticles, and also nanoparticles from the combustion of fossil/biodiesel blends.

Nanoparticles from Biodiesel/Fossil Diesel Blends and their Toxicity

Few studies have been conducted on nanoparticles in the exhaust from the combustion of biodiesel blends in fossil diesel [83-85]. Today's biodiesel consists of fatty acid methyl esters (FAME) derived from the trans-esterification of the triglycerides in plant oils, e.g., soya oil (generating soya methyl esters - SME) and rapeseed oil for the production of rapeseed methyl esters (RME). FAME generates lower levels of CO_2 in its life-cycle [86], however it generates more of the smallest particles. In addition non-combusted FAME are emitted, particularly in colder countries. The exhaust contains little PAH, aldehydes, and ketones, but more NOx , in comparison with fossil diesel [87-89].

Fossil diesel on the other hand, is quite rich in toxic combustion products. The majority of chemical compounds from the soluble organic fraction (SOF) of fossil diesel arranges in agglomerates which contain unburned hydrocarbons, semi-oxidized hydrocarbons and PAHs [90-92]. The level of PAHs in fossil diesel has been documented to be high, reaching as much as 7 times higher concentrations than in biodiesel [88]. Their carcinogenic potential of PAHs is well known [93]. Some studies have been done on the toxicological aspects of biodiesel, but very few on biodiesel in blends with fossil diesel. It has been shown that dust deriving from a biofuel plant caused strong inflammation in mice [94], and a similar response was found in tractor drivers employing biodiesel as fuel [95]. Another group proved a significant cytotoxic effect of combustion products from RME on mouse lung cells and on a mammalian microsome model [88, 96]. In the attempt to further characterize the toxicity of biodiesel and biodiesel blends, the PM from biodiesel/fossil diesel blends has been demonstrated to promote a higher cytotoxicity than regular fossil diesel. However, the reason for this observation was not fully understood [89]. Bünger and colleagues suggested in 2000 [97] that the unburned parts of biodiesel represent the main toxicity aspect of biodiesel.

Blends of biodiesel in fossil diesel introduces therefore new toxicological aspects given the mixtures of two different chemistries, giving rise to potentially "toxic partnerships" between biodiesel components and fossil diesel components. Given the indications in [87-89, 97], the component of interest from biodiesel which is particularly relevant for toxicological assessments is uncombusted FAME as a "Trojan horse" component [98]. Nanoparticles from biodiesel are therefore of different size and composition than those from fossil diesel. Lin and colleagues studied and characterized the particle size and distribution from diesel engines running on palm-biodiesel and fossil fuel blends [99]. Their results showed primarily that pure biodiesel fuel could cause incomplete combustion, thereby generating higher amount of particulate matter and gaseous FAME. Furthermore, blending palm biodiesel with fossil diesel would contribute to an increment in nanoparticle size. The range of nanoparticles diameter was below 0.31µm, indicating that the smallest fractions would be affected by blending biodiesel and fossil diesel [99]. Noting that the smallest fraction has a higher ability to penetrate the lung tissue and cross the blood barrier, the hazardous aspects of blending biodiesel and fossil diesel is therefore important. The B20 category (20 % biodiesel, 80% fossil diesel) generated a larger amount of smaller nanoparticles than the other tested blends [99]. The highest level of nanoparticles emission was generated with 100% biodiesel, then with 100% fossil diesel and ultimately with the B20 blend particularly during the first stages of combustion [99]. Paraffinic emulgation of the blends reduced particle size substantially for all stages of combustion [99].

In a study where soy bean derived biodiesel was assessed for emissions, a higher number of PM concentrations were found for B0, B10, B20 and B50 blends [100]. The highest concentrations of PM were found for B50 fuels, indicating the contribution of biodiesel in the generation of higher amounts of nanoparticles than fossil diesel solely [86, 87]. Biodiesel fuels have also recently been found to contribute with higher concentration of organic carbon and total carbon during light load-conditions [101]. Biodiesel has furthermore been found to increase NO_x emissions by 10% [102, 103], implying a more toxic contribution of blends. CO_2 emissions have also been found to increase with biodiesel blending [102], which is contradictive with earlier reports [86]. Additionally and of interest, the total emission of carbonyl compounds (ketones and aldehydes) increased with biodiesel blends, compared to regular diesel [102]. Ketone- and adehyde emissions have originally been found to be highly generated primarily by alcohol fuels [104, 105], and have not been previously mapped in biodiesel emission studies as clearly as in the study by Karavalakis et al. [102]. Interestingly, the PAH-content was found to be slightly higher in biodiesel blend B20 than in 100% fossil diesel [102]. This pattern was also observed in two other studies [106, 107] presenting the intriguing finding that biodiesel may not be as non-toxic as previously presumed [86, 87]. The mixture with fossil diesel appears to generate different chemical exhaust species depending on percentage of mixture, type of biodiesel blended (RME, SME etc.), the cetane number, outside temperature, engine type and age of the biodiesel.

Studies on nanoparticles and chemical compositions from the combustion of blends are therefore imminent and require greater attention, given their increased implementation in society [83]. Blends of fossil diesel and used cooking oil methyl ester (UCOME) were regarded to be 3-4 times more toxic than the unblended fuels [102]. The level of PAH was found to be 4 times higher in UCOME fuels than in fossil diesel [102]. Karavalakis and co [103] also presented interesting findings on that formaldehyde and acetaldehydes were the dominant aldehydes emitted from rapeseed and palm fuels. Lower emissions of total PAH and nitro-PAH have however been reported for biodiesel [103] but also increases of the low-molecular weight PAHs (anthracene and phenanthrene) and oxy-PAH [108]. Oxy-PAHs and anthracene are known to be toxic and carcinogenic [109, 110].

Blending biodiesel into fossil diesel has also been reported to increase toxicity of the semi-volatile fraction (SVF) in the exhaust [111]. This includes carbonyls, semi-combusted hydrocarbons and emulgates arising with the moisture in the exhaust system. The knowledge of the generation of additional bi-products from the cooling process and interaction with moisture in the exhaust pipe is therefore a topic of interest; however it has not been mentioned in the literature hereby reviewed. Particularly in cold climates this can be a specific problem [86]. This is related to that the higher viscosity of biodiesel contributes to the generation of more viscous combustion emissions, particularly based on the presence of uncombusted FAME particles. This is caused by the known drop in combustion efficiency, particularly in certain blends [111]. Blending biodiesel with fossil diesel can therefore be expected to generate novel exhaust compounds and also more water soluble nanoparticles: hydrosol particles which can be a more hazardous part of pollution than dry particles. In this context, the uncombusted part of biodiesel may be of central importance: uncombusted FAME may play a pivotal role in giving novel toxicochemical properties and pollution-challenges in bioblended diesel. This is stated because uncombusted FAME is reported to be a central aspect for the increase in emissions of the smallest nanoparticles from combustion of biodiesel blends and pure biodiesel [112-114].

In this context, another point is of interest. The difference between the types of methyl esters (e.g., palm oil, soy etc.) may be of importance to asses which biodiesel generate the highest nano-agglomerating potential (the most poorly combusted at various conditions), SVF and toxicity in combination with fossil diesel. Because of the effect mentioned in the previous paragraph where outside temperatures play a role [86], and cold start emissions representing 10% of total emissions [115], more analyses on the agglomeration profiles based on different types of methyl esters, temperatures, and engine types, cetane numbers is substantial for new projects. This is particularly explained by contrasting results among studies [86, 102, 111-117], where two of these studies [111, 116] report inverse results of the levels of SVF in emissions from combustion of B100 and blends compared to the other groups [86, 102, 117]. Therefore, there is a need to make affirmations on whether biodiesel contributes to increase or decrease in SVF and PM and in accord with which blends, cetane number, iodine number and FAME type.

Therefore, to what extent blends increase toxicity is still not properly determined. It has been stated that "Biodiesel was more toxic than diesel because it promoted cardiovascular alterations as well as pulmonary and systemic inflammation" [115]. This is a statement which is contradicting with toxicological expectations based on that fossil diesel contains a higher level of PAH than biodiesel and generates a higher level of hydrocarbon species, while biodiesel reduces these emission components [117-122]. The level of various PM generated from bio-blending may also provide information to explain the toxic aspects of biodiesel such as stated [102, 106-108, 111, 115, 123].

Interestingly, the importance of studying blends and differences within blends has been studied by Lin and colleagues amongst others [123]. In their study, all blends below B15 (B10, B5 etc) were concluded to contribute to increase the PAH content in the nanoparticles with diameters between 0.056 μm and 0.31 μm. The PAH content, being the prime cytotoxic, carcinogenic and toxic part in fossil fuels, is therefore a key component to be evaluated for aerosol-potential, hydrosol-potential so the toxicity of nanoparticles generated in general can be better assessed. Its ability to attach to a specific particle type may also explain the differences in results in toxicological assessments of blends. Knowledge of PAH-FAME-PM interactions may therefore aid in developing better exhaust filters.

FAME-PAH: Potential Toxic Partnerships?

The type of FAME applied in the biodiesel fraction may prove crucial for the interaction with PAH, and potentially a higher toxicity. This is of particular interest because the higher toxicity may be correlated to an *increase of delivery* of PAH molecules to the cells in the alveolar passages. This point of toxicology research promotes the importance of assessing whether FAME (and which types of FAME) actively carries, or aids in carrying PAHs to the cell nucleus, a theory of a so-called "Trojan-horse mechanism" where micelles are created by merging aromatic (PAH) and semi-polar compounds (FAME). This theory is currently being explored through molecular simulations [124] and promotes for further studies to be carried out on the basis of investigating whether the PAH increase is the prime suspect for increase toxicity of blends, or that PAH is more efficiently delivered. Also, the important points of that toxicity increases in biodiesel blends [102, 106-108, 111, 115, 123] has not been investigated using molecular biology methods, where levels of various cellular factors are monitored while exposed to various concentrations of PAH in combinations with FAME. The significance of relating FAME and PAH to molecular biology studies is founded on that PAH's toxic effects are partly known, but not in context with the chief component in uncombusted biodiesel, FAME, which may be of significant importance given the high levels of uncombusted fuel in biodiesel exhaust [112-114], and the relatively high part of total exhaust emissions of 10% of cold-start phase [87, 111, 115, 125].

Naturally, hydrocarbons, carbonyls and other combustion products may be of equal importance, however these have been tested for toxicity, particularly carbonyls [102]. Uncombusted FAMEs are more viscous than carbonyls, and may act in the formation of nanoparticles making the emissions hydrosolic and increase their size and ability to bind more pollution components and also more efficiently contaminate water-sources. The action of uncombusted FAMEs in a nanoagglomerate context may therefore be participation in the formation of micelles with PAH and un-/semi-combusted hydrocarbons, and substantiate the formation of "toxic partnerships" between these two molecules, increasing toxicity and carcinogenicity [124] when blending.

Are Blends More Toxic than Pure Biodiesel / Fossil Diesel?

A recent study from Finland has shown that there is a significant reduction of PM_{10} in the emissions when comparing B100 RME to EN590 diesel, without DOC/POC (Dissolved and Particulate Organic Carbon) filter [126]. However, when using DOC/POC filter the emissions were virtually the same, something that implies the need to improve the understanding of the generation of nanoparticles in blends, and their properties. The same study reported that the levels of PAH were reduced by nearly a third, when applying B100 RME compared to EN590 fossil diesel [126], however the cytotoxicity of B100 RME and fossil diesel were still nearly the same. Furthermore, the release of TNF-alpha by macrophages was far lower in RME-combustion exposed cultures, including ROS production. These findings may suggest that the reduction of PAH content may reduce the inflammatory reactions observed in macrophages, but not affect the cytotoxicity. Cytotoxicity may therefore be dependent on the emission species from blends, either carbonyls, hydrocarbons or other species primarily. Genotoxicity was also lower for RME-emission exposed cells, which supports the potential in that fossil diesel does cause more damage to DNA than biodiesel, but blends were not studied in this study [126]. Controversially to these findings, Kado and Kuzmicky [127] report higher mutagenic activity pr. particle from biodiesel/fossil diesel blends, but given reductions in total mutagenic emission rates finds blends to be less toxic than fossil diesel. Not changing the subject on that particles from blends may be more toxic than particles from fossil diesel, the increase of PM in blends [86] debates these findings, and instigates that blends may increase toxicity [102, 106-108, 111, 115, 123] and not decrease it [126]. Biodiesel PM extracts have also been found to be more inflammatory potent than fossil diesel PM extract [128], which correlates well with the observed findings on that biodiesel may irritate the mucous membranes and cause irritation, dizziness, and nausea [128, 130]. A review by McCormick [131] showed that the generation of emission species, ranging from PAH, nPAH, aldehydes and other emission species differ extensively when applying different engines. This review also suggested B20 blends to be the ideal blend to reduce PM, HC, carbonyls and toxic compounds including volatile C1-C12 carbon species.

However, in a report by Krahl and colleagues [132] the mutagenicity of the exhaust from biodiesel and biodiesel blends was measured and gave different results. The results showed that biodiesel blends shows a peak in mutagenicity at B20 and that this mutagenic effect most likely derives from species of molecules and nanoparticles generated in its interaction with fossil diesel. The results by Krahl and colleagues also showed that the fossil diesel condensates from the tested fuels where the most mutagenic, and that these condensates are most likely carried more efficiently in nanoparticle formation when blending at B20. 100% RME showed interestingly 4-5 times lower mutagenicity than B20 which delineates that the potential of mutagenicity derives from blending. Similar findings were found earlier by Fang and McCormick [133]. Published recently from the conference "Euro Oil and Fuel" in Krakow [134] utterly sustaining data to these findings was introduced. The findings by Mayer and colleagues [134] showed that nanoparticle formation in the diameter range from 10-100nm in standard fossil diesel is 5-10 times lower than of B10, B20 and B30 RME biodiesel blend. These results showed also that blends increasing from 0 to 100% biodiesel behave parabolically in the generation of carcinogenic PAHs (highest generations of PAHs at medium concentrations than low and high concentrations). Additionally, the findings by Mayer et al. [134] showed that 100% RME had 5-7 fold higher generation of PAHs than regular fossil diesel during the first point of engine operation, indicating poorer combustion efficiency and higher generation of heavy aromatic compounds. The results were similar for particle-bound PAHs and for particle-bound carcinogenic PAHs [134].

Of equal interest, Munack and colleagues presented in 2010 at the XVIIth World Congress of the International Commission of Agricultural and Biosystems Engineering in Canada [132] evidence on that PM content in blends increases from B30 toward B100 RME compared to fossil diesel. Fossil diesel presented higher HC content than blends and B100 however NOx content was higher in blends surpassing B30. In the same presentation [132], Munack and colleagues concluded that particles from B20 blends are the most mutagenic, and that mutagenicity increases with increasing addition of biodiesel in fossil diesel (while fossil diesel particles has the lowest mutagenicity). The reason for these findings is based on the tendency of blends to form sediments, particularly if aged (partly oxidized) biodiesel is used [136]. The mutagenicity is therefore hypothesized to be caused when carotenoids in the aged biofuel lose their anti-oxidative effect in the biodiesel fraction forming oligomers and thereby sediments [136]. These reactions may therefore account for the 9.7 − 59 fold increment in mutagenicity of RME compared to fossil diesel as reported by Krahl and co [137, 138], and additionally be co-involved with the higher generation of NOx by RME combustion [139]. RME produces also more particles below 10nm in diameter than fossil diesel and gas-to-liquid fuels [139, 140], which is consistent with the particle formation observed by Manzetti and colleagues using molecular simulations [124]. The results from the simulations [124] showed that FAME gas agglomerates to stable nanoparticles with a mean diameter of 10nm at a pressure of 100bar, as found in the combustion chamber of diesel engines (J. Czerwinski, personal communication).

The evaluation of applying biodiesel in blends should therefore evolve around its toxicological aspects, unanswered questions, and potential increase of toxic and ecotoxic effects, rather before implementation is applied in traffic vehicles. The application of a small percentage of biodiesel in fossil diesel has taken place in Norway, accordingly with EU-regulations, and data show that even low percentages of biodiesel addition in blends increases PAH generation and nanoparticle formation [134]. The effects of blending on public health are therefore important to explore based on the generation of new emission compounds, which also may pass through DPF-systems and not to mention on DPF-devoid vehicles. DPF technologies may also have to be revised accordingly with blending implementations. After all, biodiesel and diesel generate a higher amount of nanoparticles than gasoline [86, 139, 140], and however detrimental for the environment and health, all three options should face more stringent evaluation as sustainable energy options (fossil diesel, biodiesel and biodiesel blends), particularly in large cities. The application of fossil fuels in general does represent such hazardous long term effects, as mentioned particularly in the previous paragraphs, that research focus should be intensified on alternative fuel sources and follow the precautionary principle before implementation of new alternatives is granted.

Diesel Nanoparticles

Diesel nanoparticles have been studied for the past 20 years in particular, with the increasing implementation of diesel engines in cars during the 1990's. The implementation of diesel cars in city traffic occurred primarily due to lower prices of diesel compared to gasoline in Europe. Secondary, lower CO_2-emissions from diesel engines was also used as an argument to use diesel instead of gasoline. However diesel engines generate up to 800 times more organic and elementary carbon than gasoline [136] due to the lower grade of refining from crude oil. Diesel engines generate also more than 30 times more particulate matter than gasoline engines alone, and include a wide range of toxic components such as PAH, HC and sulphur. Ultra-low sulphur diesel was proposed in the USA in 2006 and in Europe in 2005 with EURO IV standard in order to reduce emissions. DPF filters have been applied in many diesel vehicles during the late 2000's so to reduce PM. However diesel emissions still encompass a large fraction of air pollution [141] and display a higher sum added potential carcinogenicity than gasoline cars [142].

The majority of the carcinogenic components in diesel are PAHs. The levels of PAH emissions in diesel exhaust are to an extent reduced through DPF filtering technologies, urea de-nitration and other purification technologies and outside temperatures [143, 144]. However, the emissions from purified exhaust are still highly toxic and present a threat to urban public health in particular.

Bergvall and Westerholm [142] demonstrated the presence of two isomers of the highly carcinogenic PAH dibenzopyrene in nanoparticles from two diesel engines which were tested on ARTEMIS rural road running cycles, urban- and highway cycles. The results show that diesel engines emitted a net sum of ~1μg of PAHs pr kilometer in the urban driving cycles, which encompassed 5-10 times more the equivalent value for rural driving cycles and 5-10 times more than motorway driving. The majority of PAH types in the nanoparticles from the diesel engines used in the tests was represented by benzo(b)fluroanthene, benzo(e)pyrene, benzo(a)pyrene and benzo(ghi)perylene. The National Institute for Occupational Safety and Health [145] considers the first three substances as human carcinogens, whereof the latter, has been involved in studies to measure for carcinogenicity in humans [146]. The apparent reason for the lower PAH emission in rural and motorway driving than urban driving with diesel engines was hypothesized to be linked to the more reduced speeds in urban driving and more frequent starts and stops [142].

Diesel exhaust is a central cause to asthma in large cities and dense traffic areas also [147-149]. The content of diesel particles is akin to tobacco smoke, with a high content of organic and elemental carbon and also sulphates and nitrates. Other particularly health threatening substances are PAHs and related compounds such as quinones which have a high inflammatory potential. The inflammation is mediated through the penetration of diesel particles deep in the lung tissue [150] which are defined into sizes as mentioned in the previous paragraphs. The inflammatory mechanism of diesel particles is at its most potent level when quinones in the polar fraction interact with the immune system. The main cause of the triggering of asthma lies in the enhanced IgE antibody production and airway hyper-responsiveness and oxidative stress as demonstrated in mice and rats [151, 152], increased expression of bronchial adhesion molecule, and the novel finding of bronchoalveolar easinophilia cells [153], which explain the high correlation between asthma and diesel and [146].

Diesel exhaust particles has high content of aldehydes, aliphatic ketones and aliphatics [154]. In the study by Jakober and colleagues [154], a collection of more than 60 chemical compounds were identified in diesel exhaust during a five phase analysis of a heavy duty diesel truck engine. The highest concentrations were of the aldehydes butanal and heptanal amongs other aldehydes, both in the gas phase as well as in the particle phase. The concentrations of these two compounds were 27.000 µg/L in the gas phase and 2.100 µg/L in the particle phase for butanal, and 13.000 µg/L and 2.300 µg/L in the respective phases for heptanal. These concentrations were measured during a 17 min creep phase. Other compounds emitted at relative high concentrations were the aliphatic dicarbonyl methyl glyoxal, 2-3 hexanedione and the aromatic aldehyd benzaldehyde. All species of compounds were found at much higher concentrations in the diesel measurements than the gasoline measurements, indicating the significant difference in toxicity between gasoline and diesel fuels.

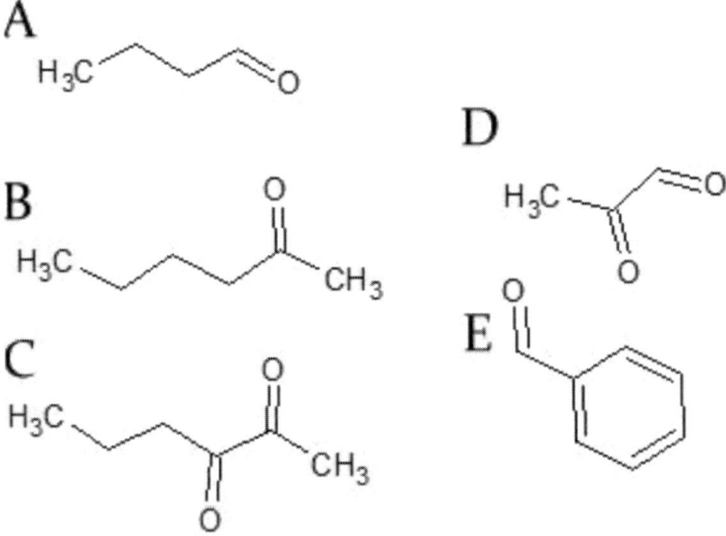

Figure 1. Aldehydes, ketones and dicarbonyls found in diesel exhaust emissions [148]. Alphabetically, A: Butanal; B: Heptanal; C:2-3 hexanedione; D: dicarbonyl methyl glyoxal; E: Benzaldehyde.

Further studies on nanoparticles from diesel exhaust have been conducted by Giordano and colleagues, with particular focus on the content of emissions from diesel engines with diesel particulate filter [155]. The content of the nanoparticles from traffic pollution (nowadays represented mostly by diesel cars in urban areas) was found to be composed of a significant amount of the heavy metals of Cu, Ni and Zn based on the biomonitors moss and lichen used in the study, and of Al and Cr in moss, that was significantly increased. Of particular interest, found high levels of silica fibers was also found, deriving from the diesel particulate filters from modern diesel cars, and the group exerted alertness towards the potential danger of this material deriving from the catalytic muffler and posing as a possible source very hazardous material to human health. This finding particularly shows how even the filtering solutions themselves may be a source of pollution in these otherwise intricate pollution aspects of diesel combustion. Therefore the aspect of heavy metals in nanoparticles is an equally, if not even more important aspect than carbonaceous compounds from diesel emissions, given the stronger toxic aspect of heavy metals to cells and particularly the nerve system [156]. The application of DPF and in diesel fuel catalysts in the diesel combustion system in modern cars leads to other ecotoxicological aspects, where the application of cerium oxide and silica carbide generates nanoparticles composed of CeO and fibrous compounds of silica, which are found to generate cytoxocity, blood-barrier damage and phospholipidosis with enlarged alveolar macrophages [157]. This aspect of nanoparticles from diesel-cars promotes another important view on the nanoparticle composition from modern diesel cars, where the technologies applied contribute to diversifying the chemical composition of diesel-combustion generated nanoparticles. The size of these ranges is still in the area of PM definitions, PM_{10} and $PM_{2.5}$, however the CeO/CeO_2 particles may also be substantially smaller depending on the fabric of the filter, however laboratory studies applied particles ranging 20nm in diameter for experiments [157].

The health effects of CeO_2 are still ubiquitous. The US National Health Effects Institute [158] reports that inhaled CeO_2 induces enlargement of lymph nodes, increased lung weight, and dose-dependent increases in segmental blood neutrophils. Also, the same institute reports that both pulmonary and systemic toxicity in rats has been provoked by inhaled CeO [158]. Studies were cerium is the major component has also shown to induce rare earth pneumoconiosis with pathologic features of granulomas and interstitial fibrosis [159-161]. The direct linkage between pulmonary inflammation in rats and CeO particles has also been recently demonstrated [157] and the inclusion of these durable but toxic materials in DPF filters and catalysts has therefore been questioned [157, 158].

Diesel nanoparticles are otherwise, except from being substantially represented by a solid elemental carbon core, covered by a multitude of known and unknown organic compounds. A study performed in 2001 by Tobias and colleagues [90], revealed the composition of such particles to be mostly made of branched alkanes and alkyl-substituted cycloalkanes. The size of the smallest nanoparticles analyzed was in the range of 50-80nm and a second group in the range between 30-70nm. Within these classes, the unknown carbon compounds were estimated to be composed of chains of a minimum of 17 carbons, and reach up to 25 carbon long. The correlation between length of carbon chains and engine load seem to be poor, showing that there were no particular patterns of chain length of alkanes and engine load. The same accounts for the diameter of the nucleus of the nanoparticles, which was largely 7-13 nm overall. Fuel additives and oil content alternated the composition of the particles to reach up to 45 carbon atoms in length, still encompassing structurally and chemically unknown molecular arrangements. Also in this study, the vast number of PAHs have been discussed and partly mapped, and of particular importance, the PAHs have been reported to exist mostly separated from the particulate material as found in earlier studies [162, 163].

CONCLUSIONS

Particulate matter and exhaust emissions present a significant environmental threat and hazardous aspects to health. The classification the exhaust species and their detrimental effects need to be furthermore emphasized in the toxicological studies and assessments of fuel types. Biodiesel and fossil diesel introduce both serious aspects to health and the environment, and their chemical aspects reveal indeed the serious threat to the body's immune system, respiratory system and cardiovascular system. Ultimately, blends seem not to reduce toxicity, and may in many case worsen the toxic aspect of emissions. The nanoparticles formed from biodiesel, and the nanoparticles deriving from fossil diesel may be significantly altered during the combustion of blends, thereby producing novel toxicological aspects. In order to limit these problems, exploration of novel filtering technologies, and the search for alternative fuels must continue. This to avoid disastrous conditions for the increasing population of the world, resulting from the continuous combustion of fuels.

REFERENCES

[1]   Zielinska B, Samy S, McDonald JD, Seagrave J; HEI Health Review Committee. (2010). Atmospheric transformation of diesel emissions. *Res. Rep. Health Eff. Inst.* 147:5-60.

[2]   Mar TF, Koenig JQ, Primomo J. (2010). Associations between asthma emergency visits and particulate matter sources, including diesel emissions from stationary generators in Tacoma, Washington. *Inhal. Toxicol.* 22:445-8.

[3]   Li R, Ning Z, Majumdar R, Cui J, Takabe W, Jen N, Sioutas C, Hsiai T. (2010). Nano particles from diesel vehicle emissions at different driving cycles induce differential vascular pro-inflammatory responses: implication of chemical components and NF-κB signaling. *Part Fibre Toxicol.* 22;7:6.

[4]   Hofmann W, Morawska L, Winkler-Heil R, Moustafa M. (2009). Deposition of combustion aerosols in the human respiratory tract: comparison of theoretical predictions with experimental data considering nonspherical shape. *Inhal. Toxicol.* 21:1154-64.

[5]   Braun A, Bewersdorff M, Lintelmann J, Matuschek G, Jakob T, Göttlicher M, Schober W, Buters JT, Behrendt H, Mempel M. (2009). Differential impact of diesel particle composition on pro-allergic dendritic cell function. *Toxicol. Sci.* 113:85-94.

[6]   Roller M. (2009). Carcinogenicity of inhaled nanoparticles. *Inhal. Toxicol.* 21 Suppl 1:144-57.

[7]   Seagrave J. (2008). Mechanisms and implications of air pollution particle associations with chemokines. *Toxicol Appl Pharmacol.* 232:469-77.

[8] Yokota S, Ohara N, Kobayashi T. (2008). The effects of organic extract of diesel exhaust particles on ischemia/reperfusion-related arrhythmia and on pulmonary inflammation. *J. Toxicol. Sci.* 33(1):1-10.

[9] Durbin TD, Norbeck JM. (2002). Effects of biodiesel blends and Arco EC-diesel on emissions from light heavy-duty diesel vehicles. *Environ. Sci. Technol.* 36:1686-91.

[10] Weng HH, Tsai SS, Chen CC, Chiu HF, Wu TN, Yang CY. (2008). Childhood leukemia development and correlation with traffic air pollution in Taiwan using nitrogen dioxide as an air pollutant marker. *J. Toxicol. Environ. Health A.* 71:434-8.

[11] Feychting M, Svensson D, Ahlbom A. Exposure to motor vehicle exhaust and childhood cancer. (1998). *Scand. J. Work Environ. Health.* 24:8-11.

[12] Krieger, R. K. Report to the Air Resources Board on the Proposed Identification of Diesel Exhaust as a Toxic Air Contaminant, 1998.

[13] Truex TJ, Norbeck JM, Smith MR. (1998). Evaluation of Factors that Affect Diesel Exhaust Toxicity - Other Documents and Presentations, Final Report to the California Air Resources Board under Contract 94-312. June 25. 98-VE-RT83-002-FR.

[14] Yang H.-H., Chiang C.-F., Lee W.-J., Hwang K.-P., Ming-Yang Wu E. (1999). Size distribution and dry deposition of road dust PAHs. *Environment International* 25:585-597

[15] Fromme H, Oddoy A, Piloty M, Krause M, Lahrz T. (1998). Polycyclic aromatic hydrocarbons (PAH) and diesel engine emission (elemental carbon) inside a car and a subway train. *Sci. Total Environ.* 217:165-73.

[16] Tsai JH, Chen SJ, Huang KL, Lin YC, Lee WJ, Lin CC, Lin WY. (2010). PM, carbon, and PAH emissions from a diesel generator fuelled with soy-biodiesel blends. *J. Hazard Mater.* 179:237-43.

[17] National Institute for Occupational Safety and Health (NIOSH). Current Intelligence Bulletin No 50: Carcinogenic effects of exposure to Diesel exhaust, National Institute for Occupational Safety and Health (NIOSH), Cincinnati, OH, USA, 1998.

[18] International Agency for Research on Cancer (IARC). Monographs on the Evaluation of Carcinogenic Risks to Humans: Diesel and Gasoline Engine Exhaust and some Nitroarenes. International Agency for Research on Cancer (IARC), Lyon, France, 1989.

[19] World Health Organization (WHO). Diesel Fuel and Exhaust emissions; Environmental Health Criteria, vol 171. World Health Organization (WHO), Geneva, Swtzerland, 1996.

[20] US EPA, Health Assessment Document for Diesel Emission, Review Draft, EPA//8-90/057E, Washington, DC, 2000.

[21] US National Toxicology Program, Ninth Report on Carcinogens, Research Triangle Park, NC, 2000.

[22] Wichmann HE. (2007). Diesel exhaust particles. *Inhal. Toxicol.* 19 Suppl 1:241-4.

[23] Heinrich, J., and Wichmann, H. E. (2004). Traffic related pollutants in Europe and their effects on allergic disease. *Curr. Opin. Allergy Clin. Immunol.* 4:341–348.

[24] Pope, C. A., Burnett, R. T., Thun, M. J., Calle, E. E., Krewski, D., Ito, K., Thurston, G. D. (2002). Lung cancer, cardiopulmonary mortality,and long-term exposure to fine particulate air pollution. *J. Am. Med.Assoc.* 287:1132–1141.

[25] Riedl, M., and Diaz-Sanchez, D. (2005). Biology of diesel exhaust effects on respiratory function. *J. Allergy Clin. Immunol.* 115(2):221–228

[26] Bhatia, R., Lopipero, P., and Smith, A. H. (1998). Diesel exhaust exposure and lung cancer. *Epidemiology* 9:84–91.

[27] Brüske-Hohlfeld, I.,M"ohner, M., Ahrens,W., Pohlabeln, H., Heinreich,J., Kreuzer, M., J"ockel, K. H., Wichmann, H. E. (1999). Lung cancer risk in male workers occupationally exposed to diesel motor emissions in Germany. *Am. J. Ind. Med.* 36:405–414.

[28] Cohen, A. J. (2000). Outdoor air pollution and lung cancer. *Environ. Health Perspect.* 108:743–750. 2000.

[29] Knol AB, de Hartog JJ, Boogaard H, Slottje P, van der Sluijs JP, Lebret E, Cassee FR, Wardekker JA, Ayres JG, Borm PJ, Brunekreef B, Donaldson K, Forastiere F, Holgate ST, Kreyling WG, Nemery B, Pekkanen J, Stone V, Wichmann HE, Hoek G. (2009). Expert elicitation on nano particles: likelihood of health effects and causal pathways. *Part Fibre Toxicol.* ;6:19.

[30] Brunekreef B, Holgate ST. (2002). Air pollution and health. *Lancet* 360:1233-42.

[31] Brunekreef B. (1997). Air pollution and life expectancy: is there a relation? *Occup Environ Med* 54:781–84.

[32] Raaschou-Nielsen O, Bak H, Sørensen M, Jensen SS, Ketzel M, Hvidberg M, Schnohr P, Tjønneland A, Overvad K, Loft S. (2010). Air pollution from traffic and risk for lung cancer in three Danish cohorts. *Cancer Epidemiol Biomarkers Prev.* 19:1284-91.

[33] Hart JE, Garshick E, Dockery DW, Smith TJ, Ryan L, Laden F. (2010). Long-term Ambient Multi-pollutant Exposures and Mortality. *Am. J. Respir. Crit. Care Med.* doi:10.1164/rccm.200912-1903OC.

[34] Maheswaran R, Pearson T, Smeeton NC, Beevers SD, Campbell MJ, Wolfe CD. (2010). Impact of outdoor air pollution on survival after stroke: population-based cohort study. *Stroke* 41:869-77.

[35] Katsouyanni K, Touloumi G, Samoli E, Gryparis A, Le Tertre A, Monopolis Y, Rossi G, Zmirou D, Ballester F, Boumghar A, Anderson HR, Wojtyniak B, Paldy A, Braunstein R, Pekkanen J, Schindler C, Schwartz J. (2001). Confounding and effect modification in the short-term effects of ambient particles on total mortality: results from 29 European cities within the APHEA2 project. *Epidemiology* 12: 521–31.

[36] Atkinson RW, Anderson HR, Sunyer J, Ayres J, Baccini M, Vonk JM, Boumghar A, Forastiere F, Forsberg B, Touloumi G, Schwartz J, Katsouyanni K. (2001). Acute effects of particulate air pollution on respiratory admissions: results from APHEA 2 project. Air Pollution and Health: a European Approach. *Am. J. Respir. Crit. Care Med.* 164: 1860–66.

[37] Le Tertre A, Medina S, Samoli E, Forsberg B, Michelozzi P, Boumghar A, Vonk JM, Bellini A, Atkinson R, Ayres JG, Sunyer J, Schwartz J, Katsouyanni K. (2002). Short-term effects of particulate air pollution on cardiovascular diseases in eight European cities. *J. Epidemiol. Community Health.* 56:773-9.

[38] Peters A, Dockery DW, Muller JE, Mittleman MA. (2001). Increased particulate air pollution and the triggering of myocardial infarction. *Circulation* 103: 2810–15.

[39] Peters A, Liu E, Verrier RL, Schwartz J, Gold DR, Mittleman M, Baliff J, Oh JA, Allen G, Monahan K, Dockery DW. (2000). Air pollution and incidence of cardiac arrhythmia. *Epidemiology* 11: 11–17.

[40] Goldsmith CA, Kobzik L. (1999). Particulate air pollution and asthma: a review of epidemiological and biological studies. *Rev. Environ. Health* 14: 121–34.

[41] Clark NA, Demers PA, Karr CJ, Koehoorn M, Lencar C, Tamburic L, Brauer M. (2010). Effect of early life exposure to air pollution on development of childhood asthma. *Environ. Health Perspect.* 118:284-90.

[42] Barath S, Mills NL, Lundbäck M, Törnqvist H, Lucking AJ, Langrish JP, Söderberg S, Boman C, Westerholm R, Löndahl J, Donaldson K, Mudway IS, Sandström T, Newby DE, Blomberg A. (2010). Impaired vascular function after exposure to diesel exhaust generated at urban transient running conditions. *Part Fibre Toxicol.* 23;7:19.

[43] Cesaroni G, Badaloni C, Romano V, Donato E, Perucci CA, Forastiere F. (2010). Socioeconomic position and health status of people who live near busy roads: the Rome Longitudinal Study (RoLS). *Environ. Health.* 21;9:41.

[44] Künzli N, Kaiser R, Medina S, Studnicka M, Chanel O, Filliger P, Herry M, Horak F Jr, Puybonnieux-Texier V, Quénel P, Schneider J, Seethaler R, Vergnaud JC, Sommer H. (2000). Public-health impact of outdoor and traffic-related air pollution: a European assessment. *Lancet.* 356:795-801.

[45] Peters A, Dockery DW, Muller JE, Mittleman MA. (2001). Increased particulate air pollution and the triggering of myocardial infarction. *Circulation* 103:2810-5.

[46] Pope CA 3rd, Burnett RT, Thun MJ, Calle EE, Krewski D, Ito K, Thurston GD. (2002). Lung cancer, cardiopulmonary mortality, and long-term exposure to fine particulate air pollution. *JAMA.* 287:1132-41.

[47] O'Neill LA. Immunology. How frustration leads to inflammation. (2008). *Science.* 320:619-20.

[48] Okamoto M, Liu W, Luo Y, Tanaka A, Cai X, Norris DA, Dinarello CA, Fujita M. (2010). Constitutively active inflammasome in human melanoma cells mediating autoinflammation via caspase-1 processing and secretion of interleukin-1beta. *J. Biol. Chem.* 285:6477-88.

[49] van Berlo D, Albrecht C, Knaapen AM, Cassee FR, Gerlofs-Nijland ME, Kooter IM, Palomero-Gallagher N, Bidmon HJ, van Schooten FJ, Krutmann J, Schins RP. (2010). Comparative evaluation of the effects of short-term inhalation exposure to diesel engine exhaust on rat lung and brain. *Arch. Toxicol.* 84:553-62.

[50] Totlandsdal AI, Skomedal T, Låg M, Osnes JB, Refsnes M. (2008). Pro-inflammatory potential of nano particles in mono- and co-cultures of primary cardiac cells. *Toxicology* 247:23-32.

[51] Totlandsdal AI, Refsnes M, Låg M. (2010). Mechanisms involved in nano carbon black-induced release of IL-6 from primary rat epithelial lung cells. *Toxicol In Vitro.* 24:10-20.

[52] Totlandsdal AI, Refsnes M, Skomedal T, Osnes JB, Schwarze PE, Låg M. (2008). Particle-induced cytokine responses in cardiac cell cultures--the effect of particles versus soluble mediators released by particle-exposed lung cells. *Toxicol. Sci.* 106:233-41.

[53] Ovrevik J, Arlt VM, Oya E, Nagy E, Mollerup S, Phillips DH, Låg M, Holme JA. (2010). Differential effects of nitro-PAHs and amino-PAHs on cytokine and chemokine responses in human bronchial epithelial BEAS-2B cells. *Toxicol. Appl. Pharmacol.* 242:270-80.

[54] Asare N, Tekpli X, Rissel M, Solhaug A, Landvik N, Lecureur V, Podechard N, Brunborg G, Låg M, Lagadic-Gossmann D, Holme JA. (2009). Signalling pathways

involved in 1-nitropyrene (1-NP)-induced and 3-nitrofluoranthene (3-NF)-induced cell death in Hepa1c1c7 cells. *Mutagenesis*. 24:481-93.

[55]	Ovrevik J, Låg M, Holme JA, Schwarze PE, Refsnes M. (2009). Cytokine and chemokine expression patterns in lung epithelial cells exposed to components characteristic of particulate air pollution. *Toxicology* 259:46-53.

[56]	Asare N, Låg M, Lagadic-Gossmann D, Rissel M, Schwarze P, Holme JA. (2009). 8.3-Nitrofluoranthene (3-NF) but not 3-aminofluoranthene (3-AF) elicits apoptosis as well as programmed necrosis in Hepa1c1c7 cells. *Toxicology* 255:140-50.

[57]	Nathan C, Ding A. (2010). Nonresolving inflammation. *Cell* 140:871-82.

[58]	Quan C, Sun Q, Lippmann M, Chen LC. (2010). Comparative effects of inhaled diesel exhaust and ambient fine particles on inflammation, atherosclerosis, and vascular dysfunction. *Inhal. Toxicol.* 22:738-53.

[59]	Miller MR, Borthwick SJ, Shaw CA, McLean SG, McClure D, Mills NL, Duffin R, Donaldson K, Megson IL, Hadoke PW, Newby DE. (2009). Direct impairment of vascular function by diesel exhaust particulate through reduced bioavailability of endothelium-derived nitric oxide induced by superoxide free radicals. *Environ. Health Perspect.* 117:611-6.

[60]	Hartz AM, Bauer B, Block ML, Hong JS, Miller DS. (2008). Diesel exhaust particles induce oxidative stress, proinflammatory signaling, and P-glycoprotein up-regulation at the blood-brain barrier. *FASEB J.* 22:2723-33.

[61]	Inoue KI, Koike E, Takano H. (2010). Comprehensive analysis of elastase-induced pulmonary emphysema in mice: Effects of ambient existing particulate matters. *Int Immunopharmacol.* 10:1380-9.

[62]	Nemmar A, Inuwa IM. (2008). Diesel exhaust particles in blood trigger systemic and pulmonary morphological alterations. *Toxicol. Lett.* 176:20-30.

[63]	Januário DA, Perin PM, Maluf M, Lichtenfels AJ, Nascimento Saldiva PH. (2010). Biological effects and dose-response assessment of diesel exhaust particles on in vitro early embryo development in mice. *Toxicol. Sci.* 117:200-8.

[64]	Chaudhuri N, Paiva C, Donaldson K, Duffin R, Parker LC, Sabroe I. (2010). Diesel exhaust particles override natural injury-limiting pathways in the lung. *Am. J. Physiol. Lung Cell Mol. Physiol.* 299:L263-71.

[65]	McCreanor J, Cullinan P, Nieuwenhuijsen MJ, Stewart-Evans J, Malliarou E, Jarup L, Harrington R, Svartengren M, Han IK, Ohman-Strickland P, Chung KF, Zhang J. (2007). Respiratory effects of exposure to diesel traffic in persons with asthma. *N. Engl. J. Med.* 357:2348-58.

[66]	Morgenstern V, Zutavern A, Cyrys J, Brockow I, Koletzko S, Krämer U, Behrendt H, Herbarth O, von Berg A, Bauer CP, Wichmann HE, Heinrich J; GINI Study Group; LISA Study Group. (2008). Atopic diseases, allergic sensitization, and exposure to traffic-related air pollution in children. *Am. J. Respir. Crit. Care Med.* 177:1331–1337.

[67]	Devalia JL, Bayram H, Rusznak C, Calderón M, Sapsford RJ, Abdelaziz MA, Wang J, Davies RJ. (1997). Mechanisms of pollution-induced airway disease: in vitro studies in the upper and lower airways. *Allergy.* 52:45-51.

[68]	Patel MM, Chillrud SN, Correa JC, Hazi Y, Feinberg M, Kc D, Prakash S, Ross JM, Levy D, Kinney PL. (2010). Traffic-related particulate matter and acute respiratory symptoms among New York City area adolescents. *Environ. Health Perspect.* 118:1338-43.

[69] Wold LE, Simkhovich BZ, Kleinman MT, Nordlie MA, Dow JS, Sioutas C, Kloner RA. (2006). In vivo and in vitro models to test the hypothesis of particle-induced effects on cardiac function and arrhythmias. *Cardiovasc Toxicol.* 6:69-78.

[70] Reed MD, Gigliotti AP, McDonald JD, Seagrave JC, Seilkop SK, Mauderly JL. (2004). Health effects of subchronic exposure to environmental levels of diesel exhaust. *Inhal. Toxicol.*16:177-93.

[71] Grella E, Paciocco G, Caterino U, Mazzarella G. (2002). Respiratory function and atmospheric pollution. *Monaldi Arch. Chest Dis.* 57:196-9.

[72] Vincent R, Kumarathasan P, Goegan P, Bjarnason SG, Guénette J, Bérubé D, Adamson IY, Desjardins S, Burnett RT, Miller FJ, Battistini B. (2001). Inhalation toxicology of urban ambient particulate matter: acute cardiovascular effects in rats. *Res. Rep. Health Eff. Inst.* 104:5-54.

[73] Pope CA 3rd, Thun MJ, Namboodiri MM, Dockery DW, Evans JS, Speizer FE, Heath CW Jr. (1995). Particulate air pollution as a predictor of mortality in a prospective study of U.S. adults. *Am. J. Respir. Crit. Care Med.* 151(3 Pt 1):669-74.

[74] Cohen AJ, Higgins MWP. (1995). Health Effects of Diesel Exhaust: Epidemiology. Diesel Exhaust: A Critical Analysis of Emissions, Exposure, and Health Effects. A Special Report of the Institute's Working Group. Cambridge, Health Effects Institute pp. 251–292.

[75] Brook RD, Rajagopalan S, Pope CA 3rd, Brook JR, Bhatnagar A, Diez-Roux AV, Holguin F, Hong Y, Luepker RV, Mittleman MA, Peters A, Siscovick D, Smith SC Jr, Whitsel L, Kaufman JD; American Heart Association Council on Epidemiology and Prevention, Council on the Kidney in Cardiovascular Disease, and Council on Nutrition, Physical Activity and Metabolism. (2010). Particulate matter air pollution and cardiovascular disease: An update to the scientific statement from the American Heart Association. *Circulation.* 121:2331-78.

[76] World Health Organization. World Health Report 2002. Geneva, Switzerland: World Health Organization; 2002.

[77] Brook RD, Franklin B, Cascio W, Hong Y, Howard G, Lipsett M, Luepker R, Mittleman M, Samet J, Smith SC Jr, Tager I; Expert Panel on Population and Prevention Science of the American Heart Association. Air pollution and cardiovascular disease: a statement for healthcare professionals from the Expert Panel on Population and Prevention Science of the American Heart Association. (2004). *Circulation.* 109:2655–2671.

[78] Pope CA 3rd, Dockery DW. (2006). Health effects of fine particulate air pollution: lines that connect. *J. Air Waste Manag Assoc.*56:709 –742.

[79] Dominici F, Peng RD, Bell ML, Pham L, McDermott A, Zeger SL, Samet JM. (2006). Fine particulate air pollution and hospital admission for cardiovascular and respiratory diseases. *JAMA.* 295:1127–1134.

[80] Forastiere F, Stafoggia M, Picciotto S, Bellander T, D'Ippoliti D, Lanki T, von Klot S, Nyberg F, Paatero P, Peters A, Pekkanen J, Sunyer J, Perucci CA. (2005). A case-crossover analysis of out-of-hospital coronary deaths and air pollution in Rome, Italy. *Am J Respir Crit Care Med.* 172:1549 –1555.

[81] Santos UP, Terra-Filho M, Lin CA, Pereira LA, Vieira TC, Saldiva PH, Braga AL. Cardiac arrhythmia emergency room visits and environmental air pollution in Sao Paulo, Brazil. (2008). *J. Epidemiol. Community Health.* 62:267–272.

[82] Dvonch JT, Kannan S, Schulz AJ, Keeler GJ, Mentz G, House J, Benjamin A, Max P, Bard RL, Brook RD. (2009). Acute effects of ambient particulate matter on blood pressure: differential effects across urban communities. Hypertension. 53:853–859.

[83] Hellmann F, Verburg PH. (2010). Impact assessment of the European biofuel directive on land use and biodiversity. *J. Environ. Manage.* 91:1389-96.

[84] Neeft J, van Thujil, Wismeijer R, Mabee WE. (2007). Biofuel implementation agendas. IEA Task Report; T39-P5, 52pp.

[85] Hill, J., Nelson, E., Tilman, D., Polasky, S., Tiffany, D. (2006). Environmental, economic, and energetic costs and benefits of biodiesel and ethanol biofuels. *PNAS* 103,11206–11210.

[86] Office of Weatherization and Intergovernmental Programs. Clean Cities. Energy Efficiency and Renewable Energy. DOE/GO-102005-2029. April 2005.

[87] Heikkilä J, Virtanen A, Rönkkö T, Keskinen J, Aakko-Saksa P, Murtonen T. (2009). Nanoparticle emissions from a heavy-duty engine running on alternative diesel fuels. *Environ. Sci. Technol.* 43:9501-6.

[88] Bünger J, Krahl J, Franke HU, Munack A, Hallier E. (1998). Mutagenic and cytotoxic effects of exhaust particulate matter of biodiesel compared to fossil diesel fuel. *Mutat. Res.* 415(1-2):13-23.

[89] Biodiesel, the Comprehensive Handbook. Mittelbach M and Remschmidt C. 1st Ed. 2004. Boersedruck G.m.b.h. Vienna. Pp.35, 41, 231

[90] Tobias HJ, Beving DE, Ziemann PJ, Sakurai H, Zuk M, McMurry PH, Zarling D, Waytulonis R, Kittelson DB. (2001). Chemical analysis of diesel engine nanoparticles using a nano-DMA/thermal desorption particle beam mass spectrometer. *Environ. Sci. Technol.* 35:2233-43.

[91] Lee J, Patel R, Schnborn A, Ladommatos N, Bae C. (2009). Effect of biofuels on nanoparticles emission from spark and compression-ignited single cylinder engines with same exhaust displacement volume. *Energy Fuels* 23:4363-9.

[92] Guilloteau A, Bedjanian Y, Nguyen ML, Tomas A. (2010). Desorption of polycyclic aromatic hydrocarbons from a soot surface: three- to five-ring PAHs. *J Phys Chem A.* 114:942-8.

[93] Scheepers PT, Bos RP. (1992). Combustion of diesel fuel from a toxicological perspective. II. Toxicity. *Int. Arch. Occup. Environ Health.* 64:163-77. Review.

[94] Madsen AM, Saber AT, Nordly P, Sharma AK, Wallin H, Vogel U. (2008). Inflammation but no DNA (deoxyribonucleic acid) damage in mice exposed to airborne dust from a biofuel plant. *Scand. J. Work Environ Health.* 34:278-7.

[95] Krahl, J., G. Vellguth, M. Graef, A. Munack, K.-H. Stalder, and M. Bahadir, "Utilization of Rape Seed Oil and Rape Seed Oil Methylester as Fuels – Exhaust Gas Emissions and their Effects on Environment and Human Health," Proceedings of the 8th European Conference on Biomass for Energy, Environment, Agriculture, and Industry, Vienne, Austria, Oct. 3-5, 1994.

[96] Bünger J, Krahl J, Munack A, Ruschel Y, Schröder O, Emmert B, Westphal G, Müller M, Hallier E, Brüning T. (2007). Strong mutagenic effects of diesel engine emissions using vegetable oil as fuel. *Arch. Toxicol.* 81:599-603.

[97] Bünger J, Krahl J, Baum K, Schröder O, Müller M, Westphal G, Ruhnau P, Schulz TG, Hallier E. (2000). Cytotoxic and mutagenic effects, particle size and concentration

analysis of diesel engine emissions using biodiesel and petrol diesel as fuel. *Arch. Toxicol.* 7:490-8.

[98]   Auffan M, Rose J, Benameur L, Hotze EM, Masion A, Bottero J, Proux O. (2010). Nanoparticles - contaminant interactions: sorption mechanisms and biological effects (Trojan horse effect). Platform Presentation Abstracts, Nano2010. Clemson Univ.

[99]   Lin YC, Lee CF, Fang T. (2008). Characterization of particle size distribution from diesel engines fueled with palm-biodiesel blends and paraffinic fuel blends. *Atmospheric Environment,* 42:1133-1143

[100] Tsai JH, Chen SJ, Huang KL, Lin YC, Lee WJ, Lin CC, Lin WY. (2010). PM, carbon, and PAH emissions from a diesel generator fuelled with soy-biodiesel blends. *J. Hazard Mater.* 179(1-3):237-43.

[101] Bugarski AD, Cauda EG, Janisko SJ, Hummer JA, Patts LD. (2010). Aerosols emitted in underground mine air by diesel engine fueled with biodiesel. *J. Air Waste Manag. Assoc.* 60:237-44.

[102] Karavalakis G, Bakeas E, Stournas S. (2010). Influence of oxidized biodiesel blends on regulated and unregulated emissions from a diesel passenger car. *Environ. Sci. Technol.* 44:5306-12.

[103] Karavalakis G, Stournas S, Bakeas E. (2009). Light vehicle regulated and unregulated emissions from different biodiesels. *Sci. Total Environ.* 407:3338-46.

[104] Song C, Zhao Z, Lv G, Song J, Liu L, Zhao R. (2010). Carbonyl compound emissions from a heavy-duty diesel engine fueled with diesel fuel and ethanol-diesel blend. *Chemosphere.* 79:1033-9.

[105] Hansen AC, Zhang Q, Lyne PW. (2005). Ethanol-diesel fuel blends -- a review. *Bioresour Technol.* 96(3):277-85.

[106] Zhou, L., Atkinson, S., (2003). Characterising vehicle emissions from the burning of biodiesel made from vegetable oil. *Environmental Technology* 24, 1253–1260.

[107] Turrio-Baldassarri, L., Battistelli, C.L., Conti, L., Crebelli, R., Berardis, B., Iamicelli, A.L., Gambino, M., Iannaccone, S., (2004). Emission comparison of urban bus engine fueled with diesel oil and biodiesel blend. *Science of the Total Environment* 327, 147–162.

[108] Karavalakis G, Fontaras G, Ampatzoglou D, Kousoulidou M, Stournas S, Samaras Z, Bakeas E. (2010). Effects of low concentration biodiesel blends application on modern passenger cars. Part 3: impact on PAH, nitro-PAH, and oxy-PAH emissions. *Environ. Pollut.* 158:1584-94.

[109] Schuetzle, D., (1983). Sampling of vehicle emissions for chemical analysis and biological testing. *Environmental Health Perspectives* 47, 65–80.

[110] Stelzer MK, Pitot HC, Liem A, Schweizer J, Mahoney C, Lambert PF. (2010). A Mouse Model for Human Anal Cancer. *Cancer Prev. Res. (Phila).* doi: 10.1158/1940-6207.CAPR-10-0086

[111] Liu, Y. Y., Lin, T. C., Wang, Y. J., and Ho, W. L. (2008). Biological toxicities of emissions from an unmodified engine fueled with diesel and biodiesel blend. *J. Environ. Sci. Health A Tox. Hazard. Subst. Environ. Eng.* 43, 1735–1743.

[112] Krahl J, Bünger A, Munack A. Biodiesel exhaust emissions and determination of their environmental and health effects. In: *Plant oils as fuels: present state of science and future developments.* Springer, Berlin. 1998, 104-122.

[113] Bagley S, Gratz LD, Johnson JH, McDonald JF. (1998). Effects of an Oxidation Catalytic Converter and a Biodiesel Fuel on the Chemical, Mutagenic, and Particle Size Characteristics of Emissions from a Diesel Engine. *Environ. Sci. Technol.* 32:1183–1191.

[114] Bünger J, Krahl J, Franke HU, Munack A, Hallier E. (1998). Mutagenic and cytotoxic effects of exhaust particulate matter of biodiesel compared to fossil diesel fuel. *Mutat Res.* 415:13-23.

[115] Schifter I, Díaz L, Rodríguez R. (2010). Cold-start and chemical characterization of emissions from mobile sources in Mexico. *Environ. Technol.* 31:1241-53.

[116] Brito JM, Belotti L, Toledo AC, Antonangelo L, Silva FS, Alvim DS, Andre PA, Saldiva PH, Rivero DH. (2010). Acute cardiovascular and inflammatory toxicity induced by inhalation of diesel and biodiesel exhaust particles. *Toxicol. Sci.* 116:67-78.

[117] Tsai JH, Chen SJ, Huang KL, Lin YC, Lee WJ, Lin CC, Lin WY. (2010). PM, carbon, and PAH emissions from a diesel generator fuelled with soy-biodiesel blends. *J. Hazard Mater.* 179:237-43.

[118] Tsolakis, A. (2006). Effects on particle size distribution from the diesel engine operating on RME-biodiesel with EGR. *Energy Fuels* 20:1418–1424.

[119] McCormick, R.; Graboski, M.; Alleman, T.; Herring, A. (2001). Impact of biodiesel source material and chemical structure on emissions of criteria pollutants from a heavy-duty engine. *Environ. Sci. Technol.* 35:1742–1747.

[120] Sharp, C. A.; Howell, S. A.; Jobe, J. (2000). The effect of biodiesel fuels on transient emissions from modern diesel engines, part I regulated emissions and performance. *SAE Tech. Pap. Ser.* 2000-01-1967.

[121] Chang, D.; Gerpen, J. (1997). Fuel properties and engine performance for biodiesel prepared from modified feedstocks. *SAE Tech. Pap. Ser.* 971684.

[122] Mi HH, Lee WJ, Chen CB, Yang HH, Wu SJ. (2000). Effect of fuel aromatic content on PAH emission from a heavy-duty diesel engine. *Chemosphere.* 41:1783-90.

[123] Lin YC, Tsai CH, Yang CR, Wu CHJ, Wu TY, Chang-Chien GP. (2008). Effects on aerosol size distribution of polycyclic aromatic hydrocarbons from the heavy-duty diesel generator fueled with feedstock palm-biodiesel blends. *Atmospheric Environment* 42:6679-6688

[124] Manzetti S, Andersen O, Czerwinski J, van der Spoel D. (2010). Molecular Simulations of a fatty acid methyl ester (oleic methyl ester) and a polycyclic aromatic hydrocarbon (phenanthrene): new toxicological aspects for biodiesel/fossil diesel blends. Nano2010: Environmental Effects of Nanoparticles and Nanomaterials: 2010.

[125] Lin YC, Lee WJ, Hou HC. (2006). PAH emissions and energy efficiency of palm-biodiesel blends fueled on diesel generator. *Atmospheric Environment* 40:3930-3940.

[126] Jalava PI, Tapanainen M, Kuuspalo K, Markkanen A, Hakulinen P, Happo MS, Pennanen AS, Ihalainen M, Yli-Pirilä P, Makkonen U, Teinilä K, Mäki-Paakkanen J, Salonen RO, Jokiniemi J, Hirvonen MR. (2010). Toxicological effects of emission particles from fossil- and biodiesel-fueled diesel engine with and without DOC/POC catalytic converter. *Inhal. Toxicol.* doi:10.3109/08958378.2010.519009.

[127] Kado, N.Y., and Kuzmicky, P. A. (2003). Bioassay analyses of particulate matter from a diesel bus engine using various biodiesel feedstock fuels. National Renewable Energy Laboratory, NREL/SR-510-31463. Available at http://www.nrel.gov/publications.

[128] Swanson, K., Kado, N., Madden, M. C., and Ghio, A. J. (2007). Release of IL-8 and IL-6 by BEAS-2B cells following in vitro exposure to biodiesel PM extracts. Toxicologist 96:813(A).

[129] The Department of Energy. United States of America. (2001). Biodiesel: Handling and use guidelines. NREL/TP-580-30004.

[130] Penn State University. College of Agricultural Sciences. (2008). Biodiesel Safety and Best Management Practices for Small-Scale Noncommercial Use and Production. CODE# AGRS-103.

[131] McCormick RL (2007). The Impact of Biodiesel on Pollutant Emissions and Public Health. *Inhalation Toxicology* 19:1033–1039.

[132] Krahl J, Munack A, Ruschel Y, Schröder O and Bünger J. (2008). Exhaust gas emissions and mutagenic effects of diesel fuel, biodiesel and biodiesel blends. *SAE Technical Paper Series*. SAE International. 2008-01-2508.

[133] Fang, H.; McCormick, R. L. (2006). Spectroscopic Study of Biodiesel Degradation Pathways. *SAE Technical Paper Series*. SAE International. 2006-01-3300.

[134] Mayer A, Czerwinski J, Wyser M, Heitzer A. (2010). Particle Emissions of Diesel Busses with RME/Diesel blends. Euro Oil and Fuel. Bio-components in Diesel fuels – impact on emission and ageing of engine oil. In Ed. Michal Krasodomski. pp. 37-59.

[135] Munack A, Krahl J, Schröder O, Bünger J. (2010). Potentials of biofuels. 3 XVIIth World Congress of the International Commission of Agricultural and Biosystems Engineering (CIGR). CSBE100337 – Presented at Section IV: Rural Electricity and Alternative Energy Sources Conference.

[136] Krahl J, Munack A, Schmindt L, Urban B, Petchatnikov M, Schröder O. Wechselswirknungen zwischen Biodiesel und modernen Dieselkraftstoffen. (2010). 8. FAD Konferenz. 3.11-4.11.2010 in Dresden. Pp. 209-223.

[137] Krahl J, Munack A, Ruschel Y, Schröder O, Bünger J. (2007). Comparison of emissions and mutagenicity from biodiesel, vegetable oil, GTL and Diesel Fuel. *SAE Technical Paper Series*. SAE International, 2007-01-4042.

[138] Krahl J, Knothe G, Munack A, Ruschel Y, Schröder O, Hallier E, Westphal G, Bünger J. (2009). Comparison of exhaust emissions and their mutagenicity from the combustion of biodiesel, vegetable oil, gas-to-liquid and petrodiesel fuels. *Fuel* 88:1064-1069.

[139] Cheung KL, Polidori A, Ntziachristos L, Tzamkiozis T, Samaras Z, Cassee FR, Gerlofs M, Sioutas C. (2009). Chemical characteristics and oxidative potential of particulate matter emissions from gasoline, diesel, and biodiesel cars. *Environ. Sci. Technol.* 43:6334-40.

[140] Krahl J, Munack A, Ruschel Y, Schröder O, Bünger J. (2009). Ultrafine particles from a heavy duty diesel engine running on rapeseed oil methyl ester. *SAE Technical. Paper Series*. SAE International, 2009-01-2691.

[141] Ayres J, Maynard R, Richards R. (2006). Air pollution and Health. In: Air Pollution Review, vol. 3. Imperial College Press, London, ISBN 1-86094-191-5.

[142] Bergvall C, Westerholm R. (2009). Determination of highly carcinogenic dibenzopyrene isomers in particulate emissions from two diesel- and two gasoline-fuelled light duty vehicles. *Atmospheric Environment* 43:3883-3890.

[143] Westerholm R, Christensen A, Törnqvist M, Ehrenberg L, Rannug U, Sjögren M, Rafter J, Soontjens C, Almén J, Grägg K. (2001). Comparison of exhaust emissions

from Swedish environmental classified diesel fuel (MK1) and European Program on Emissions, Fuels and Engine Technologies (EPEFE) reference fuel: a chemical and biological characterization, with viewpoints on cancer risk. *Environ. Sci. Technol.* 35:1748-54.

[144] Ludykar D., Westerholm R., Almén J. (1999). Cold start emissions at +22, −7 and −20°C ambient temperatures from a three-way catalyst (TWC) car: regulated and unregulated exhaust components. *The Science of The Total Environment* 235:65-69.

[145] Occupational Safety and Health Administration, 200 Constitution Avenue, NW, Washington, DC 20210 . http://www.osha.gov/SLTC/healthguidelines.

[146] Alomirah H, Al-Zenki S, Husain A, Sawaya W, Ahmed N, Gevao B, Kannan K. (2010). Benzo[a]pyrene and total polycyclic aromatic hydrocarbons (PAHs) levels in vegetable oils and fats do not reflect the occurrence of the eight genotoxic PAHs. *Food Addit Contam Part A Chem Anal Control Expo Risk Assess.* 27:869-78.

[147] Balmes JR. (2011). How does diesel exhaust impact asthma? Thorax. ;66:4-6.

[148] McCreanor J, Cullinan P, Nieuwenhuijsen MJ, Stewart-Evans J, Malliarou E, Jarup L, Harrington R, Svartengren M, Han IK, Ohman-Strickland P, Chung KF, Zhang J. (2007). Respiratory effects of exposure to diesel traffic in persons with asthma. *N. Engl. J. Med.* 357:2348-58.

[149] Salam MT, Islam T, Gilliland FD. (2008). Recent evidence for adverse effects of residential proximity to traffic sources on asthma. *Curr. Opin. Pulm. Med.* 14:3-8.

[150] US Environmental Protection Agency (EPA). Health effects assessment document for diesel engine exhaust. EPA/600/8-90/057F. National Center for Environmental Assessment, Office of Research and Development. Washington, DC: US EPA, 2002.

[151] Whitekus MJ, Li N, Zhang M, Wang M, Horwitz MA, Nelson SK, Horwitz LD, Brechun N, Diaz-Sanchez D, Nel AE. (2002). Thiol antioxidants inhibit the adjuvant effects of aerosolized diesel exhaust particles in a murine model for ovalbumin sensitization. *J. Immunol.* 168(5):2560-7.

[152] Kooter IM, Gerlofs-Nijland ME, Boere AJ, Leseman DL, Fokkens PH, Spronk HM, Frederix K, Ten Cate H, Knaapen AM, Vreman HJ, Cassee FR. (2010). Diesel engine exhaust initiates a sequence of pulmonary and cardiovascular effects in rats. *J. Toxicol.* 2010;2010:206057.

[153] Sehlstedt M, Behndig AF, Boman C, Blomberg A, Sandström T, Pourazar J. (2010). Airway inflammatory response to diesel exhaust generated at urban cycle running conditions. *Inhal. Toxicol.* 22:1144-50.

[154] Jakober CA, Robert MA, Riddle SG, Destaillats H, Charles MJ, Green PG, Kleeman MJ. (2008). Carbonyl emissions from gasoline and diesel motor vehicles. *Environ. Sci. Technol.* 42:4697-703.

[155] Giordano S, Adamo P, Spagnuolo V, Vaglieco BM. (2010). Instrumental and bio-monitoring of heavy metal and nanoparticle emissions from diesel engine exhaust in controlled environment. *J. Environ. Sci. (China).*22:1357-63.

[156] Dorne JLCM., Kass GEN., Bordajandi LR, *et al.* Metal Ions in Toxicology. Vol. 8, pp. 27-60. In: Metal Ions in Toxicology: Effects, Interactions, Interdependencies (Metal Ions in Life Sciences). Ed. Astrid Sigel, Helmut Sigel, Roland K. O. Sigel. Royal Society of Chemistry; 1st Edition. edition (December 17, 2010).

[157] Ma JY, Zhao H, Mercer RR, Barger M, Rao M, Meighan T, Schwegler-Berry D, Castranova V, Ma JK. Cerium oxide nanoparticle-induced pulmonary inflammation and

alveolar macrophage functional change in rats. *Nanotoxicology.* 2010 Oct 6. [Epub ahead of print].

[158] Health Effects Institute (HEI). 2001. Evaluation of human health risk from cerium added to diesel fuel. In: Hibbs JB Jr, editor. HEI Communication 9, Boston, MA, USA. Integrated Cerium oxide-induced lung injury 13.

[159] Sabbioni E, Pietra R, Gaglione P, Vocaturo G, Colombo F, Zanoni M, Rodi F. 1982. Long-term occupational risk of rare-earth pneumoconiosis. A case report as investigated by neutron activation analysis. *Sci. Total Environ.* 26:19–32.

[160] Waring PM, Watling RJ. 1990. Rare earth deposits in a deceased movie projectionist. A new case of rare earth pneumoconiosis? *Med. J. Aust.* 153:726–730.

[161] McDonald JW, Ghio AJ, Sheehan CE, Bernhardt PF, Roggli VL. 1995. Rare earth (cerium oxide) pneumoconiosis: Analytical scanning electron microscopy and literature review. *Mod. Pathol.* 8:859–865.

[162] Rogge WF., Hildemann LM., Mazurek MA., Cass GR., Simoneit BRT. (1993). Sources of Fine Organic Aerosol. 2. Non-Catalyst and Catalyst-Equipped Automobiles and Heavy Duty Diesel Trucks. *Environ. Sci. Technol.* 27, 636-651.

[163] Schauer JJ; Kleeman MJ; Cass GR; Simoneit BRT. (1999). Measurement of emissions from air pollution sources 2 C1 through C30 organic compounds from medium duty diesel trucks. *Environ. Sci. Technol,* 33: 1578-1587.

In: Biodiesel: Blends, Properties and Applications   ISBN: 978-1-63117-024-9
Editors: Jorge M. Marchetti and Zhen Fang   © 2014 Nova Science Publishers, Inc.

Chapter 3

BIODIESEL PREPARATION FROM NON-EDIBLE OILS

S. Sivanesan, K. V. Thiruvengadaravi and V. Sathya Selva Bala

Department of Chemical Engineering,
Anna University, Tamilnadu, India

ABSTRACT

Oil provides energy for 95% of transportation and the demand of transport fuel continues to rise. According to a forecast made by the International Energy Agency (IEA) global oil demand will rise by about 1.6% from 75 mb/d in the year 2000 to 120 mb/d in 2030. Almost three quarters of the increase will be from the transport sector. Oil will remain the fuel of choice in road, sea and air transportation. In developing countries, the increase in demand for oil for use in transport sector is expected to grow at a much higher rate. All countries including India are grappling with the problem of meeting the ever increasing demand of transport fuel within the constraints of international commitments, legal requirements, environmental concerns and limited resources. In this connection transport fuels of biological origin have drawn a great deal of attention during the last two decades.

Bio-fuel is a generic term that is used to refer to liquid or gaseous fuels that are produced from a biological source. The term liquid 'bio-fuel' is more commonly used to refer to specific types of bio-fuels used as fossil fuel substitutes. These are further defined by the particular type of biomass from which they are made, and the degree to which they are refined before use. The most common types of liquid bio-fuel are straight vegetable oils, ethanol and biodiesel.

STRAIGHT VEGETABLE OILS

The simplest form of bio-fuel is pure vegetable oil, such as the oil from peanuts, olives or sesame seeds. This oil has similar energy content and some similar physical characteristics to diesel fuel. In fact the inventor of the diesel engine, Mr. Rudolph diesel originally designed his engine to be run on peanut oil (Shay 1993).

ETHANOL

Ethanol is used as a fuel or as oxygenate for gasoline. Generally ethanol is mixed with gasoline in varying concentrations. This can be produced from a wide range of biomass using a relatively complex process. Raw material used for producing ethanol varies from sugar in Brazil, cereals in United States of America, sugar beet in Europe to molasses in India. Brazil uses ethanol as 100% fuel in about 20% vehicles and 25% blend with gasoline in the rest of the vehicles. United States of America uses 10% ethanol-gasoline blends whereas 5% blend in Sweden. Australia uses 10% ethanol gasoline blend. Use of 5% ethanol-gasoline is already approved by Bureau of Indian Standards (BIS) and is in progressive implementation in India.

BIODIESEL

Biodiesel is defined as a fuel comprised of mono alkyl esters of long chain fatty acids derived from vegetables oils or animals fats, very similar to fossil based fuel and can be used in almost any type of diesel engine.

BIODIESEL AND ITS PREPARATION

The alternative diesel fuels must be technically and environmentally acceptable and also economically competitive. From the view point of these requirements, triglycerides and their derivatives may be considered as viable alternatives for diesel fuels.

Biodiesel is the name for a variety of ester based fuels generally defined as the mono-alkyl esters made from vegetable oils or animal fats through a simple transesterification process. Biodiesel is a clear amber coloured liquid with viscosity similar to petro-diesel, the industry term for diesel produced from petroleum. Biodiesel has physical properties very similar to petro-diesel but its emission properties are superior.

Biodiesel Scenario in India

India ranks sixth in the world in terms of energy demand accounting for almost 4.8% of world commercial energy demand. The growth in energy demand in all forms is expected to continue unabated owing to increasing urbanization, standard of living and expanding population. The increase in gap between demand and domestically produced petroleum is a matter of serious concern. Transport sector remains the most problematic sector as no alternative to petro-diesel has been successful so far.

While the country is in dearth of petroleum reserve, it has large extractable land as well as good climatic conditions with adequate rainfall in large parts of the area to account for large biomass production each year. For the reason of edible oil demand being higher than its domestic production, there is no possibility of diverting this oil for production of biodiesel. Fortunately there is a large junk of degraded forest land and un–utilized public land, field boundaries and follow lands of farmers where non-edible oil seeds can be grown. There are many tree species which bear seeds rich in oil. The main commodity sources for biodiesel in India can be non-edible oils obtained from plant species such as *Jatropha curcas* (Ratanjyot), *Pongamia pinnata* (Karanja) *Calophyllum inophyllum* (Nagchampa), *Madhuca indica* (Mahua), *Herca Brasiliensis* (Rubber) etc.

Cost of Biodiesel

In India the estimated cost of biodiesel produced by transesterification of non-edible oils shall be approximately same as that of petro diesel. The cost components of biodiesel includes price of seed, oil extraction, oil transesterification, transport of seed and oil. Cost recovery will be through the sale of oil cake and of glycerol. The use of biodiesel is thus economically feasible. Also, the cost of biodiesel can be further reduced with the full utilization of wastelands for the production of non-edible oils.

Advantages of Biodiesel

Biodiesel can be considered as an alternate fuel owing to the following advantages:

- Biodiesel is environmentally friendly and ideal for heavy polluted cities.
- Biodiesel is non toxic and biodegradable.
- It conserves natural reserves.
- It helps mitigate possible negative impacts of global climate change by lowering net CO_2 emissions from transportation sector.
- Biodiesel has improved lubricity and ignition quality relative to diesel fuel.
- Biodiesel is sulphur free.
- Biodiesel is renewable i.e., it is not a fixed resource like fossil fuels that could be completely consumed.

Disadvantages of Biodiesel

There are some disadvantages involved with the production, storage and use of biodiesel:

- Biodiesel can create problems in cold weather conditions, because certain of its constituent compounds can form crystals in the fuel. These crystals cause undesired effects like plugging of fuel filters.
- Biodiesel is often susceptible to oxidative and biological instabilities than conventional diesel fuel.
- Impurities such as unreacted fatty acids or alcohol, as well as glycerine or catalyst left over from the production process can lead to accelerated ulcer or corrosion of engine components.

Addition of properly chosen stabilizers can help avoiding the above mentioned disadvantages.

Methods of Preparation

The problems with substituting triglycerides for diesel fuels are mostly associated with their high viscosities, low volatilities and poly unsaturated character. These problems have been mitigated by developing vegetable oil derivatives and making them comparable with the hydrocarbon – based diesel fuels through:

Pyrolysis

Pyrolysis refers to a chemical change caused by the application of thermal energy in the absence of air or oxygen. The liquid fractions of the thermally decomposed vegetable oil are likely to approach diesel fuels. Pyrolyzed soybean oil was found to contain 79% carbon and 11.88% hydrogen (Schwab et al. 1988). The pyrolyzate has lower viscosity, flash point and pour point than diesel fuel and equivalent calorific values. The cetane number of the pyrolyzate is low. The pyrolyzed vegetable oils contain acceptable amounts of sulphur, water and sediment and acceptable corrosion values but unacceptable ash, carbon residues and pour point.

Emulsification

The formation of micro emulsions (co-solvency) is one of the potential solutions for solving the problem of vegetable oil viscosity. Micro emulsions are defined as transparent and thermodynamically stable colloidal dispersions. The droplet diameters in micro emulsions range from 100 to 1000 Å.

A micro emulsion can be made of vegetable oils with an ester and dispersant (co-solvent) or of vegetable oils, an alcohol and a surfactant and a cetane improver, with or without diesel fuels. The use of 2-octanol as a surfactant for emulsification of soybean oil in the presence of methanol has been demonstrated by Bagby (1987).

Dilution

Dilution of vegetable oils can be accomplished with such materials as diesel fuels, solvent or ethanol. The dilution of sunflower oil with diesel fuels in the ratio of 1:3 by volume has been studied and engine tests were carried out by Ziejewski et al. (1983). They concluded that the blend could not be recommended for long term use in the direct injection diesel engines because of severe injector nozzle choking and sticking.

Transesterification

Transesterification also called alcoholysis, is the displacement of alcohol from an ester by another alcohol in a process similar to hydrolysis. This process has been widely used to reduce the viscosity of triglycerides.

Among the above mentioned processes, transesterification is considered to be a better process owing to the simplicity of the process involved and with respect to the usage of the product formed.

Chemistry of Transesterification Process

The transesterification reaction is represented by the general equation.

$$RCOOR' + R''OH \longrightarrow RCOOR'' + R'OH$$

In the transesterification of vegetable oils, a triglyceride reacts with an alcohol in the presence of a strong acid or base, producing a mixture of fatty acid alkyl esters and glycerol (Wright et al. 1944, Freedman et al. 1986). The reaction of vegetable oils is represented by the general equation

$$
\begin{array}{c}
\text{H}_2\text{C}-\text{OCOR}' \\
| \\
\text{HC}-\text{OCOR}'' \quad + \quad 3 \;\; \text{ROH} \quad \overset{catalyst}{\rightleftharpoons} \\
| \\
\text{H}_2\text{C}-\text{OCOR}'''
\end{array}
\qquad
\begin{array}{c}
\text{ROCOR}' \\
+ \\
\text{ROCOR}'' \\
+ \\
\text{ROCOR}'''
\end{array}
\quad + \quad
\begin{array}{c}
\text{H}_2\text{C}-\text{OH} \\
| \\
\text{HC}-\text{OH} \\
| \\
\text{H}_2\text{C}-\text{OH}
\end{array}
$$

$$\text{triglyceride} \qquad \text{alcohol} \qquad\qquad \text{mixture of alkyl esters} \qquad \text{glycerol}$$

Transesterification of Vegetable Oils

The overall process is a sequence of three consecutive and reversible reactions, in which di and monoglycerides are formed as intermediates. The stoichiometric equation is 1 mole of a triglyceride requires 3 moles of the alcohol. However, an excess alcohol is used to increase the yields of the alkyl esters and to allow its phase separation from glycerol formed. Several aspects, including the type of catalyst (alkaline or acid), alcohol/vegetable oil molar ratio, temperature, purity of reactants (mainly water content) and free fatty acid content have an influence on the course of transesterification (Schuchardt et al. 1998).

Factors Affecting Transesterification

The important factors that affect the transesterification process are discussed below.

Reaction Temperature

The rate of reaction is strongly influenced by the reaction temperature. However, given enough time the reaction will proceed to near completion even at room temperature. Generally, the reaction is conducted close to the boiling point of methanol (60 to 70°C) at atmospheric pressure. These mild conditions, however, require the removal of free fatty acids from the oil by refining or pre-esterification. The pre-treatment is not required if the reaction is carried out under high pressure (9000 kPa) and high temperature (240°C). Under these conditions simultaneous esterification and transesterification take place (Hui 1996).

Molar Ratio of Alcohol to Oil

Another important variable affecting the yield of ester is the molar ratio of alcohol to vegetable oil. The stoichiometry of the transesterification reaction requires 3 moles of alcohol per mole of triglyceride to yield 3 moles of fatty esters and 1 mole of glycerol. To shift the transesterification reaction to the forward direction, it is necessary to use either a large excess of alcohol or to remove one of the products from the reaction mixture. The second option is preferred wherever feasible, since in this way, the reaction can be driven to completion. Sharma and Singh (2008a) employed a two step transesterification process and took 8:1 molar ratio of alcohol to oil for acid esterification and 9:1 for alkaline esterification, for biodiesel production from karanja oil.

Nature of Catalysts

Alkali metal alkoxides are the most effective transesterification catalyst compared to the acidic catalyst. Sodium alkoxides are among the most efficient catalysts used for this purpose, although KOH and NaOH can also be used. Transmethylations occur approximately 4000 times faster in the presence of an alkaline catalyst than those catalyzed by the same amount of acidic catalyst (Formo 1954). Most commercial transesterifications are conducted with alkaline catalysts because they are not only less corrosive but also faster. The alkaline catalyst concentration in the range of 0.5 to 1% by weight yields 94 to 99% conversion of vegetable oil into esters (Feuge and Gross 1949, Krisnangkura and Simamaharnnop 1992). Further, increase in catalyst concentration does not increase the conversion and it adds to extra costs because it is necessary to remove it from the reaction medium after the completion of the transesterification process.

Mixing Intensity

In the transesterification reaction, the reactants initially form a two-phase liquid system. The reaction is diffusion-controlled and poor diffusion between the phases results in a slow rate. As methyl esters are formed, they act as mutual solvent for the reactants and a single-phase system is formed. The mixing effect is more significant during the slow rate region of the reaction. As the single phase is established, mixing becomes insignificant. A better yield with mechanical stirring has been reported by Sharma and Singh (2008b).

Purity of Reactants

Impurities present in the oil also affect conversion of vegetable oils to biodiesel. Under the same conditions, 67 to 84% conversion into esters using crude vegetable oils can be obtained, compared with 94 to 97% when using refined oils (Srivastava and Prasad 2000). The free fatty acids (FFA) in the original oils interfere with the catalyst. However, under conditions of high temperature and pressure this problem can be overcome.

ACID CATALYSTS

Homogeneous acid catalysts are effective replacements to base catalysts as they perform well even in the presence of high FFA in oils. The idea is to use the triglycerides with alcohol and an acid catalyst instead of a base. The most commonly used acid is sulphuric acid (Freedman et al. 1984, Freedman et al. 1986, Harrington and Darcy-Evans 1985) and some others prefer sulfonic acid (Stern et al. 1990). This type of catalyst gives high yield in esters but slow in its action, requiring almost more than one day (Marcheti et al. 2007). The disadvantage with the use of homogeneous acid catalysts are high reaction temperature, high alcohol to oil molar ratio, separation of catalyst, serious environmental issues and corrosion related problems.

HETEROGENEOUS ACID CATALYSTS

Heterogeneous catalysis is often called surface catalysis as it mainly occurs between a solid surface and a gas. There are basically three stages in this process:

a)   adsorption of the reactant on the catalyst surface,
b)   chemical reaction on the surface,
c)   desorption of product from the catalyst surface.

The following points are to be noted about the action of catalysts. First, catalysts do not alter the thermodynamics of the reactions. No catalyst favours a thermodynamically unfeasible reaction. Consequently, the reaction would proceed even without the catalyst, though perhaps too slowly to be observed or be of use in a given context. Further more, the use of catalyst does not change the equilibrium composition. Heterogeneous catalysts have several intrinsic advantages over their homogeneous counterparts. The following Table gives a comparison of homogeneously and heterogeneously catalyzed transesterification reaction (Viswanathan and Ramaswamy 2008).

Comparison of Homogeneous and Heterogeneously Catalyzed Transesterification Reaction

S.No.	Factors	Homogeneous catalysis	Heterogeneous catalysis
1.	Reaction rate	Fast and high conversion	Moderate conversion
2.	After treatment	Catalyst cannot be recovered, must be neutralized leading to waste chemical production	Can be recovered
3.	Processing methodology	Limited use of continuous methodology	Continuous fixed bed operation possible
4.	Presence of water / free fatty acids	Sensitive	Not sensitive
5.	Catalyst reuse	Not possible	Possible
6.	Cost	Comparatively costly	Potentially cheaper

Solid acid catalysts can replace liquid acids thereby eliminating corrosion problems associated with their use and consequent environmental hazards posed by them. Though environmental benefits have been the major reasons for the introduction of solid acid catalysts in many commercial processes, these catalysts have also proved to be more economical and often produce better quality products. However, the efforts in exploiting solid acid catalysts for esterification of FFA in non-edible oils are limited due to pessimistic expectations on the possible reaction rates and possible side reactions. This situation does exist, since better understanding of the factors that govern the reactivity of solid catalysts has not yet been reached. For example, simple correlations between acid strength and activity of the catalyst have not yet been formulated. Secondly, due to diffusion restrictions one has to have a porous system with interconnecting pores, so that the entire surface of the solid is available for promoting the esterification reaction. Even though, it is possible to generate such textural characteristics in solids, it is not yet routinely possible to obtain uniform pore architecture with absolute control over the size or radius or geometry of the pores as well as the stability of the solid. It is obvious that the surface should possess some hydrophobic character for promoting the preferential adsorption for oily hydrophobic species on the catalyst surface and to avoid deactivation of catalytic sites by strong adsorption of polar by products like glycerol or water. It is essential that a reliable quantitative measure of the hydrophobicity of solids is evolved so that one can appropriately correlate this function with the observed activity.

Among the different types of inorganic solids that have been employed as catalysts for the production of biodiesel such as ion exchange resins, hydrotalcites, sulphated zirconia, mesoporous silica etc., the most popular ones are the zeolites.

USE OF NON-EDIBLE OILS

The raw materials being exploited commercially by countries working on biodiesel constitute the edible fatty oils derived from rapeseed, soybean, palm, sunflower, coconut, linseed, etc. (Korbitz 1999). Use of such edible oil to produce biodiesel in India is not feasible in view of a big energy gap in demand and supply of such oils in the country. Increased pressure to augment production of edible oil has also put limitation on the use of these oils for production of biodiesel. Under Indian conditions only plants which produce non-edible oil in appreciable quantity and can be grown in large scale on non-cropped marginal lands and wastelands (Azam et al. 2005) can be considered for biodiesel production.

The technology for converting edible oil to biodiesel has been well established, however, the main concern for converting non-edible oil into biodiesel is always associated with the high free fatty acid content. Use of conventional base catalysts with such oils results in soap formation which hinders the separation of biodiesel from glycerine (Demirbas 2003).

A two-step method is reported by several researchers (Karme and Chadha 2005, Saka 2005, Berchmans and Hirata 2008) as the best method to produce biodiesel from non-edible oils. At the initial step, FFA content of oil is reduced by acid-catalyzed esterification process as shown in the following scheme; meanwhile, in the second step an alkali-catalyzed process is used to convert oil and methanol, to methyl esters and glycerol. This two step process was found to be very effective with the yield of biodiesel in the overall process reaching up to above 90%.

$$RCOOH + R'OH \longrightarrow RCOOR' + H_2O$$
Esterification of free fatty acids

LITERATURE REVIEW

The use of vegetable oils as alternative fuels for diesel engines dates back almost nine decades (Srivastava and Prasad 2000) e.g., palm, soyabean, sunflower, peanut and olive oils. Due to the rapid decline in crude oil reserves, the use of vegetable oils as diesel fuels is again promoted in different countries. Depending upon climate and soil conditions different nations are looking into different vegetable oils for diesel fuels. Saydut et al. (2008) and Ahmad et al. (2010) have studied the direct transesterification of sesame oil with NaOH catalyst for biodiesel production. Biodiesel has been synthesized from tucum oil with homogeneous alkali catalyst using methanol and ethanol (Lima et al. 2008). Transesterification of Terminalia catappa L. oil has been carried out using homogeneous acid and base catalysts (dos Santos et al. 2008). Cerbera odollam (sea mango) oil has been transesterified using homogeneous and heterogeneous catalysts by Kansedo et al. (2009a).

Heterogeneous transesterification method has been proved to be more superior as compared to homogeneous transesterification method especially on the separation and purification of the product. Transesterification of crude palm kernel oil and crude coconut oil by different solid catalysts have been studied by Jitputti et al. (2006) and Benjapornkulaphong et al. (2009). Soybean oil transesterification using sulfonic acid functionalised polymeric membranes (Guerreiro et al. 2006), SrO as a solid base catalyst (Liu et al. 2007), sulphated zirconia (Garcia et al. 2008) and basic solids (Antunes et al. 2008) has also been carried out. Vanadyl phosphate catalyzed transesterification of soybean oil has been performed by Di Serio et al. (2007). Transesterification of triolein with various anionic ion-exchange resins has been done (Shibasaki-Kitakawa 2007). Lopez et al. (2007) have studied the transesterification of triacetin over nafion acid resins. A study of palm kernel oil transesterification with modified dolomites and Calcium and Zinc mixed oxides has been performed (Ngamcharussrivichai et al. 2007 and 2008). Montmorrilonite KSF has been used as a heterogeneous catalyst for biodiesel production from palm oil (Kansedo et al. 2009b). Transesterification of rapeseed oil with ethanol over heterogeneous heteropolyacids have been studied by Hamad et al. (2008). Biodiesel production from soybean oil via an enzymatic/acid-catalyzed hybrid process has been carried out (Ting et al. 2008).

The use of vegetable oils and animal fats for biodiesel production is of great concern because they compete with food materials. As the demand for vegetable oils for food has increased tremendously in recent years, it is impossible to justify the use of these oils for fuel use purposes such as biodiesel production. Moreover, these oils could be more expensive to use as fuel (Chhetri et al. 2008). Hence, the contribution of non-edible oils will be significant for biodiesel production. Rathore et al. (2007) have studied the synthesis of biodiesel from edible and non-edible oils in supercritical alcohols and enzymatic synthesis in supercritical carbon dioxide. Biodiesel production from high FFA rubber seed oil has been carried out by Ramadhas et al. (2005). Production of biodiesel from crude tobacco seed oil has been investigated (Veljkovic et al. 2006). Sahoo et al. (2007) have developed biodiesel from high acid value polanga seed oil and studied its performance evaluation in compression ignition engine. Esterification of Zanthoxylum bungeanum seed oil with high free fatty acids for biodiesel production has been investigated (Zhang et al. 2008). Moringa oleifera oil has been pre-treated and transesterified by Rashid et al. (2008). Biodiesel production from pomace oil and further studies has been carried out by Caynak et al. (2009).

Mahua, Madhuca longifolia of the family Sapotaceae, is a medium to large tree with a wide round canopy. Mahua is a slow-growing species, attaining a mean height of 0.9 -1.2 m at the end of the fourth year but may attain a height of up to 20 m. The variety latifolia is common throughout the Indian sub-continent, including Bangladesh. It is of deciduous nature and thrives in dry tropical and sub-tropical climates. As a plantation tree, mahua is an important plant having a socio-economic value. This species can be planted along the roadside and canal banks on a commercial scale and in social forestry programs, particularly in tribal areas. The seed kernel contains about 50% of oil. The oil yield by screw pressing is 34-37% and the fresh oil from properly stored seed is yellow in colour (Bringi 1987). Mahua oil is contaminated with high amount of free fatty acids (FFA) which depends upon the moisture content in the seed during collection as well as oil expression. Hence its difficulty in processing by transesterification with conventional base catalysts. A study of the preparation and emission characteristics of mahua oil methyl ester (Puhan et al. 2005a) and mahua oil ethyl ester (Puhan et al. 2005b) has been made. Ghadge et al. (2005 and 2006) have studied the production of biodiesel from mahua oil and the process optimization for biodiesel production using response surface methodology. Evaluation of methyl esters of mahua oil as diesel fuel has been carried out by Kapilan and Reddy (2008). Kapilan et al. (2009) have investigated the effect of using mahua oil biodiesel in compression ignition engines.

Pongamia pinnata (Leguminoceae, subfamily Papilionoideae) a native plant to India produce seeds containing 30 – 40% of oil. It is a medium sized tree that generally attains a height of about 8 m and a trunk diameter of more than 50 cm. Pongamia (Karanja) is often planted as an ornamental and shade tree. A thick yellow-orange to brown oil is extracted from seeds and a yield of 25% of volume is possible using a mechanical expeller. The oil has a bitter taste and a disagreeable aroma and is used as a fuel for cooking and lamps. Process optimization of direct transesterification of low FFA Karanja oil has been studied using a base catalyst (Agarwal and Bajaj 2009). Biodiesel production from high FFA pongamia pinnata oil has been carried out using sulfuric acid and KOH catalysts (Sharma and Singh 2008a, Naik et al. 2008). Methyl ester of karanja oil has been investigated as an alternative renewable fuel by Srivastava and Verma (2008). A study of the performance and exhaust emission characteristics of a compression ignition engine fueled with pongamia pinnata methyl ester and its blends has been carried out (Sureshkumar et al. 2008).

Esterifications are industrially important reactions where an organic acid and alcohol combine, usually in the presence of an acid catalyst, to give ester plus water. Strong bronsted acids, such as sulphuric acid, hydrochloric acid or orthophosphoric acid, are commonly used as catalyst. However, these mineral acids are corrosive and they have to be usually neutralized after reaction. These problems can be overcome by the use of solid acid catalysts. Heterogeneous catalysts are preferable, offering easy separation from the reactants and products by filtration (Corma et al. 1989, Gimenez et al. 1987). Design of multifunctionalized mesoporous silica for esterification of fatty acid has been investigated by Mbaraka et al. (2005). Heterogeneous esterification of oil with high amount of free fatty acids has been studied (Marchetti et al. 2007). Ni and Meunier (2007) have studied the esterification of free fatty acids in sunflower oil over solid acid catalysts using batch and fixed bed reactors. Simultaneous esterification and transesterification with HT2 hydrotalcite catalyst (Barakos et al. 2008), zinc hydroxide nitrate (Cordeiro et al. 2008) and zinc stearate immobilized on silica gel (Jacobson et al. 2008) have been investigated. Continuous conversion of free fatty acids in used vegetable oils for production of biodiesel with WO_3/ZrO_2 has been carried out by Park et al. (2008). Polyoxometalate-tantalum pentoxide composite has been used for efficient esterification of fatty acid (Xu et al. 2008). Liquid phase esterification and transesterification have been studied with heteropoly acids as catalysts (Alsalme et al. 2008). Caetano et al. (2008) have investigated the use of heteropolyacids immobilized on silica for esterification of free fatty acids. Esterification of free fatty acids in waste cooking oils with acidic ion-exchange resins (Ozbay et al. 2008) and carbohydrate-derived solid acid catalysts (Lou et al. 2008) have been investigated.

Zeolites are finding applications in many areas of catalysis, generating intense interest in these materials in industrial and academic laboratories. As catalysts, zeolites exhibit appreciable acid activity with shape selective features not available in the compositionally equivalent amorphous catalysts. Transesterification of soybean oil with zeolite and metal catalysts has been investigated (Suppes et al. 2004). NaX zeolites loaded with KOH has been used as a heterogeneous catalysts for the preparation of soybean methyl esters (Xie et al. 2007). Zeolite beta modified with La^{3+} has been employed for the synthesis of biodiesel from soybean oil (Shu et al. 2007). A study of the transesterification of sunflower oil over zeolite using different metal loading has been made by Ramos et al. (2008). Preparation of biodiesel from low free fatty acid containing pongamia pinnata has been carried out by Karmee and Chadha (2005). Kirumakki et al. (2006) have performed the esterification of alcohols with acetic acid over three different zeolites. Removal of free fatty acid in waste frying oil by esterification with methanol on zeolite catalysts has been studied (Chung et al. 2008). A comparison of different heterogeneous catalysts including zeolites for the esterification of oleic acid has been investigated by Marchetti and Errazu (2008).

Kinetics of the transesterification of palm oil has been investigated by Darnoko and Cheryan (2000). The reaction kinetics of acid-catalyzed transesterification of waste cooking oil has been studied (Zheng et al. 2006). Berrios et al. (2007) have made an elaborate study on the kinetics of the esterification of free fatty acids in sunflower oil. Kinetic measurements of the esterification of oleic acid catalyzed by $SnCl_2$ (Cardoso et al. 2008) has been investigated.

Carmo et al. (2009) have studied the kinetics of the esterification of palmitic acid over mesoporous aluminosilicate Al-MCM-41.

SCOPE OF THE PRESENT STUDY

Zeolites such as NaY, Hβ, HY, HZSM 5, mordenite, faujasite and silicalite have been used as catalysts for the esterification of free fatty acids in non edible oils, waste cooking oils, oleic acid and acetic acid. In view of the rising demand for fuels, several edible and non-edible oils are being investigated for the production of biodiesel, to be used an alternative diesel fuel. Very less work has been carried out with respect to the production of biodiesel via a two step catalyzed process from non-edible oils using solid acid catalysts. Suitable modification of zeolites using phosphoric acid has to be carried out to use them as solid acid catalysts in the first, esterification step of the two step method. The modification of the zeolites leads to an increase in the acidity, pore size, hydrophobicity and density of the zeolites and hence their enhanced activity towards esterification reaction. Hardly any literature is found available on the use of phosphoric acid modified HY and HZSM 5 zeolites for the esterification of free fatty acids in mahua and karanja oils.

MODIFICATION OF ZEOLITES

Modification of the zeolites Y and Z were carried out according to an established method. The resultant Y zeolite was abbreviated as MY and Z zeolite as MZ.

CHARACTERIZATION

Characterization of Zeolites and Modified Zeolites

Analysis of the zeolites before and after modification was performed to understand the changes in their catalytic properties due to phosphoric acid modification.

X-Ray Diffraction Analysis

In the powder diffraction characterization of materials, the diffraction pattern is the fingerprint of any crystalline phase. Powder diffraction is used extensively to identify the mixture of phases. The modified zeolites were ground with the help of a mortar and pestle and the fine powder was then packed on the surface of the sample holder. The powder X-ray diffraction (XRD) patterns of all the zeolites were recorded on an X-ray diffractometer (Rigaku D/Max Ultima III) operating at 40 kV and 40 mA with the scan rate of 2°/ min. Cu K α radiation was nickel filtered ($\lambda = 1.5406$ Å).

Most of the patterns of unmodified Y are retained in the spectrum of MY illustrating that there is no crystallization loss as a result of phosphoric acid modification. There is a decrease in the intensity of the patterns but no complete disappearance of peaks. Hence, as a result of phosphoric acid modification, there might not be any structural degradation. The main pattern is same in both Z and MZ. Similar changes in the intensity of peaks were observed, confirming the absence of crystallization loss due to phosphoric acid modification.

Scanning Electron Microscope (SEM) Analysis

The SEM measurements were carried out using a JEOL JSM-6360. The images were taken with an emission current = 100 µA by the Tungsten filament and an accelerator voltage = 12 kV. The samples were secured onto brass stub with carbon conductive tape, sputter coated with gold and viewed in JEOL JSM-6360 microscope. The pretreatment of the samples consisted of coating with an evaporated Au film in a polaron SC 500 sputter Coater metallizator to increase the catalyst electric conductivity.

From the SEM image of Y, the presence of aggregation of small sponge like morphology is evidenced. All of them appear nearly similar without any impure phase. Similar result was also reported in the literature (Sato et al. 1999). Aggregation of sponge like morphology is also seen in the image of MY similar to that of the Y. Compared to Y, the aggregates appear larger in size in the modified image. It is also supported by XRD results as discussed above. Hence phosphoric acid modification might have some role in the agglomeration of particles altogether. In other words, phosphoric acid in each particle, instead of lying exclusively inside the pore, might also expose outwards to interact with other particles in order to facilitate agglomeration. From the SEM image of Z also, the presence of aggregation of small sponge like morphology is evidenced. Aggregation of sponge like morphology is also seen in the image of MZ similar to that of the Z.

Fourier Transform Infra-Red Analysis

Infra-red (IR) spectra of the zeolites were recorded on a Nicolet (AVATAR 360) instrument using the KBr pellet technique The changes in the spectra of MY as compared to Y, confirm phosphoric acid modification. Similarly, the changes in the spectra of modified zeolite evidences phosphoric acid modification.

NH_3-Temperature Programmed Desorption Studies

Temperature programmed desorption (TPD) technique is commonly used to determine the concentration and strength of acid sites (Cvetanovic and Amenomiya 1972, Miyamoto et al. 2000) in solid catalysts. The probe molecules most frequently used are ammonia, pyridine and n-butyl amine. NH_3-TPD profiles of the zeolites were obtained using a chemisorption analyzer. The amount of acid sites of the catalysts can be determined from desorption peak of ammonia. The acid strength can be evaluated by maximum peak temperature (T_{max}) of desorption peak relating to the activation energy for desorption of ammonia. The zeolite samples were activated in a helium flow at 550°C for 1 h, then cooled to 150°C. Ammonia pulses were added to the zeolite samples till they were saturated with it. The samples were purged with a helium flow at 150°C for 1 h to remove physically adsorbed ammonia. The temperature of the zeolite sample was increased to 600°C with the ramping rate of 10°C/min. The desorption peaks observed at 350 - 550°C are attributed to strong acid sites, whereas those at 150-250°C to weak sites or physically adsorbed ammonia (Lonyi and Vaylon 2001).

It is clear from TPD studies, that there is an increase in the acidity of MY compared to Y catalyst due to phosphoric acid modification.

It is evident from the TPD studies, that the strong acid sites in Z are converted to weak acid sites in MZ due to phosphoric acid modification and also that there is an increase in acidity.

Thermo Gravimetric Analysis

Variation in hydrophobic property, which is an important criteria in the esterification of non-edible oils, due to phosphoric acid modification was studied by recording the thermogram of zeolites.

Thermograms were recorded using a Thermo Gravimetric Analyzer (TGA Q 50 V 20.6 Build 31 instrument). MY is confirmed to be more hydrophobic than Y from the thermograms recorded.

This thermograms confirm that the MZ zeolite is more hydrophobic than Z, which is due to phosphoric acid modification.

Characterization of Non-Edible Oils

The non-edible oils selected for esterification studies were mahua and karanja.

Gas Chromatographic Analysis

The physical, chemical and fuel properties of oils depend on the different types of fatty acids present in oils. The fatty acid profiles of both the selected oils were determined by gas chromatography and are listed in the following tables.

Fatty Acid Profile of Mahua Oil

Fatty acid	Chemical name	Formula	Structure	Wt %
Palmitic	Hexadecanoic	$C_{16}H_{32}O_2$	16:0	22.9
Stearic	Octadecanoic	$C_{18}H_{36}O_2$	18:0	21.9
Oleic	Cis-9-Octadecenoic	$C_{18}H_{34}O_2$	18:1	37.1
Linoleic	Cis-9,cis-12-Octadecadienoic	$C_{18}H_{32}O_2$	18:2	15.6
Arachidic	Eicosanoic	$C_{20}H_{40}O_2$	20:0	1.1

Fatty Acid Profile of Karanja Oil

Fatty acid	Chemical name	Formula	Structure	Wt %
Palmitic	Hexadecanoic	$C_{16}H_{32}O_2$	16:0	10.4
Stearic	Octadecanoic	$C_{18}H_{36}O_2$	18:0	7.2
Oleic	Cis-9-Octadecenoic	$C_{18}H_{34}O_2$	18:1	53.7
Linoleic	cis-9,cis-12-Octadecadienoic	$C_{18}H_{32}O_2$	18:2	16.6
Arachidic	Eicosanoic	$C_{20}H_{40}O_2$	20:0	1.6
Behenic	Docosanoic	$C_{22}H_{44}O_2$	22:0	4.4
Lignoceric	Tetracosanoic	$C_{24}H_{48}O_2$	24:0	1.3
Linolenic	cis-9,cis-12,cis-15-Octadecatrienoic	$C_{18}H_{30}O_2$	18:3	3.7

Physical and Chemical Properties of Oils

The physical and chemical properties of mahua and karanja oil were determined by ASTM methods and are given in the following Table. It is found that the calorific value is high in both the oils, hence with respect to this property the oils match with petroleum based oils. The viscosity, density and acid value were found to be high, causing difficulty with their direct use in combustion engines.

Physical and Chemical Properties of Mahua and Karanja Oil

Property	Mahua oil	Karanja oil
Density at 15°C, kg/m^3	960	913
Viscosity at 40°C, mm^2/s	24.58	27.84
Calorific Value, kJ/kg	36,100	37,304
Flash Point, °C	232	205
Pour Point, °C	15	6.4
Iodine Value	58-70	80-96
Specific Gravity	0.8560	0.9090
Ash Content, %	0.900	0.014
Carbon Residue, %	3.70	1.51
Acid Value, mg KOH/g of oil	23.10	15.96
Cetane Value	45	55

ESTERIFICATION STUDIES

Esterification reaction was carried out in a three necked round bottomed flask fitted with a reflux condenser. Stirring of the reaction contents was done with the help of a mechanical stirrer. The rate of stirring was maintained at 200 rpm. Reaction temperature was maintained at different values using a silicon oil bath connected to a dimmer stat. Thermometer was inserted into the third neck to monitor the temperature of the flask. The reaction time was fixed as 1 h after a preliminary study of the reaction with time. Equilibrium reached within a span of 50 to 60 min and hence 1 h was found to be optimum. A study of the effect of reaction temperature, methanol to oil ratio and catalyst loading on the esterification of FFA in oils were carried out. Kinetic studies were undertaken for the esterification reaction. Samples were withdrawn from the reaction vessel at regular intervals and the acid value was monitored. Acid value of the oil was an indicator of the progress of the reaction. The acid value was also determined for the oils before the commencement of the reaction and after the completion of the reaction. The acid value was found out by the standard titrimetry procedure (Pasias et al. 2006). After the reaction, excess methanol was removed from the reaction mixture by centrifuging at 8000 rpm for 5 min. The solid acid catalyst was separated by decantation and the esterified oil was obtained.

A known weight of the oil sample was dissolved in ethanol and a few drops of phenolphthalein was added as indicator. 0.02N alkaline solution of KOH was used to perform the titration. The amount of KOH consumed was noted and the acid value, A was calculated using the equation,

$$A = \frac{V1000M_wC}{W}$$

where   W   -   weight of the sample taken for analysis, g
          V   -   Volume of KOH consumed, mL
          M_w -   Molecular weight of the solution, g/g mol
          C   -   Concentration of the solution, mol/L

Esterification of Mahua and Karanja with Y and MY

The esterification reaction is strongly influenced by the reaction temperature. With Y and MY as catalyst, esterification of mahua oil was carried out separately at different temperatures in the range of 30 to 70°C to evaluate the optimum temperature. Esterification of Karanja oil was also done with Y and MY in the same temperature range.

The molar ratio of methanol to oil was studied as it is one of the most important factors affecting the ester yield. The stoichiometric ratio for esterification requires one mole of alcohol and one mole of FFA to yield one mole of fatty acid ester and one mole of water. The molar ratio is associated with the type of catalyst used (Ma and Hanna 1999). The methanol to oil ratio studied was in the range of 3:1 to 15:1. Mahua oil was taken in a round bottomed flask to which the calculated amount of methanol was added followed by 1% weight of the Y zeolite as catalyst. The esterification was carried out at 60°C for a period of 1 h. The experiment was repeated with mahua oil by taking MY as catalyst. A similar study was performed with Y and MY as catalysts with g of karanja oil.

To optimize the amount of catalyst required for esterification, reactions were carried out with varying amount of the catalyst from 0.5 to 1.5% by weight. To mahua oil containing a known quantity of methanol, the Y zeolite was added in varying amounts and the experiments were carried out. Similar study with varying amounts of MY was also carried out for mahua oil. Catalyst quantity optimization studies were also carried out for karanja oil, with both Y and MY.

Kinetic studies for the esterification of mahua oil with Y zeolite were done. During the esterification reaction of FFA in mahua oil with varying methanol to oil ratios and with different catalyst loadings at 60°C, samples were withdrawn at an interval of 10 min and analyzed for their acid values. Monitoring of the acid value with time during the esterification of FFA in mahua with MY was also carried out. Similar studies were done by taking karanja oil with both Y and MY.

To evaluate the effect of phosphoric acid modification on zeolites, esterification reactions were carried with both Y and MY under a set of optimized conditions by taking mahua oil. The influence of the catalysts on the conversion of FFA was evaluated and a comparison was made. A similar comparison of the performance of Y and MY on the esterification of FFA in karanja oil under optimized conditions was done. Esterification of oleic acid was also carried out with both Y and MY, for comparison at optimum conditions.

Esterification of Mahua and Karanja with Z and MZ

Temperature is an important factor which controls the esterification process. Esterification of mahua oil was done in the temperature range from 30 to 70°C with Z and MZ in separate experiments. From the results an optimum temperature was evaluated. A similar study was also done with karanja oil in the presence of Z and MZ.

The molar ratio of methanol to oil taken for esterification reaction is associated with the type of catalyst used. Methanol to oil ratio in the range of 3:1 to 15:1 was taken for the studies. Mahua oil was taken in a round bottomed flask to which methanol was added in different quantities and the reaction was carried out with Z as catalyst. A similar study was done with mahua oil with MZ as catalyst. The experiments were repeated with karanja oil by taking both Z and MZ as catalysts.

Optimization of the catalyst quantity was done by taking 0.5 to 1.5% by weight of Z with mahua oil under a set of standard reaction conditions. The experiments were repeated with MZ as catalyst for mahua oil. A similar study with varying amounts of Z and MZ was also carried out for karanja oil.

Kinetic studies were performed for all the runs with different methanol to oil ratios and different catalyst loadings for mahua oil esterification with Z. Acid values were monitored during the study by taking samples at a regular interval of 10 min. Similar studies were also undertaken for mahua oil with MZ as catalyst. Kinetics of esterification of FFA in karanja oil in the presence of Z and MZ as catalyst was also performed and studied.

The esterification of FFA in mahua oil was done under the optimum conditions with both Z and MZ in different experiments to study the effect of phosphoric acid modification on zeolites. The influence of modified zeolite over the esterification of karanja was also studied by carrying out the reaction with MZ and a comparison was made with that of the reaction with Z. A comparison of the activity of the catalysts under optimum conditions was done by carrying out the esterification of oleic acid with Z and MZ.

CALCULATION OF CONVERSION EFFICIENCY

Conversion efficiency is the percentage of FFA converted into their corresponding methyl esters during the esterification of non-edible oils.

It is calculated from the residual acid value during or after the esterification reaction under a set of reaction conditions. The acid value of non-edible oil before the start of the reaction and at any instant of time is found out by titrimetric method.

$$X_{FFA} = \frac{a_i - a_t}{a_i} \times 100$$

where X_{FFA} is the percentage conversion
       a_i is initial FFA content of oil
       a_t is the FFA content at time t.

TRANSESTERIFICATION STUDIES

Transesterification of Esterified Mahua and Karanja Oils

Several review works (Schuchardt et al. 1998, Ma and Hanna 1999, Srivastava and Prasad 2000, Barnawal and Sharma 2005, Marcheti et al. 2007, Demirbas 2007, Sharma et al. 2008b, Sharma and Singh 2009, Gui et al. 2008) have been carried out on the transesterification of oils towards production of biodiesel. The technology for converting edible oil to biodiesel has been well established: however the main concern for converting non-edible oil into biodiesel is always associated with the high FFA content. The FFA in non-edible oil will react with alkaline catalyst to produce soap that inhibits the separation of ester and glycerine. A two-step transesterification process is reported by several researchers as the best method to produce biodiesel from non-edible oil. At the initial step, the FFA content of oil is reduced by acid-catalyzed esterification process; meanwhile, at the second step, an alkaline-catalyzed process is used to convert oil and methanol to methyl esters and glycerol. This two step process was found to be very effective with the yield of biodiesel reaching above 90% (Karme and Chadha 2005, Saka 2005, Berchmans and Hirata 2007).

Transesterification was carried out with the mahua oil obtained after esterification with Y zeolite. The temperature at which the reaction was carried out was fixed as 60°C, maintained with a silicon oil bath controlled by a dimmer stat, reaction time 1 h, with KOH as catalyst. The reaction was carried out in a three necked round bottomed flask fitted with a condenser and equipped with a thermometer. Mixing of the reaction contents was done with the help of a mechanical stirrer at a constant speed of 200 rpm. The catalyst quantity was fixed at 1% by weight and a methanol to oil ratio at 6:1 as optimized by Meher et al. (2006). The same procedure was followed for the transesterification of the esterified mahua obtained with MY, Z and MZ catalysts.

A similar transesterification was carried out with esterified karanja oil obtained using Y, MY, Z, MZ catalysts to obtain the biodiesel product.

CATALYST RECOVERY AND REUSE

The catalysts separated from the reaction mixture by centrifugation were initially washed with hexane to remove polar compounds such as methyl esters on the surface. Further, the catalysts were washed with methanol to remove polar compounds such as glycerol and finally dried at 80°C overnight. The recovered catalysts were reused for esterification studies repeatedly.

SUMMARY OF WORK DONE

Biodiesel is one among the sustainable fuels and is a non-petroleum based fuel that consists of alkyl esters derived from either the transesterification of triglycerides or the esterification of free fatty acids or both with short-chained alcohols. Currently a dual step process is being used for biodiesel preparation from high FFA containing materials. The first step of the process is to reduce the FFA content in the oil by esterification with methanol to methyl esters catalyzed by a solid acid, followed by the transesterification process, in which the triglyceride portion of the oil reacts with methanol and KOH to form ester and glycerol. Among the different types of inorganic solids that have been employed as catalysts for the production of biodiesel, the most popular ones are the zeolites. Zeolites are finding applications in many areas of catalysis, generating intense interest in these materials in industries and academic laboratories.

In this study, two zeolites HY and HZSM 5 were selected as solid acid catalysts for the esterification of FFA in mahua and karanja oils. These oils have a high FFA content, and are non-edible and available to a large extent in India. In order to improve the catalytic properties of the selected zeolites, they were modified using phosphoric acid. These modified zeolites possess enhanced acidity and hydrophobicity. The entry of phosphoric acid moiety has been confirmed by XRD, SEM and FTIR analyses. The increase in acidity due to phosphoric acid modification was confirmed from the NH_3-TPD profiles of the catalysts. The enhancement of hydrophobicity was seen from the thermograms of the modified and unmodified zeolites.

Esterification of mahua and karanja oils were carried out with the zeolite catalysts. Optimization of the reaction temperature, methanol to oil ratio and the amount of catalyst were undertaken during the course of the esterification studies. A comparison of the catalytic activities and kinetic studies were done during the esterification process. Acid value was determined during the course of the reaction, to monitor the progress of the reaction. NMR analysis of the reaction products were under taken to establish the formation of products. Studies on catalyst recovery and reuse were also performed. Transesterification of the esterified oils were done by alkali catalyzed method. The results of the esterification and transesterification studies are summarized below:

- There was a considerable increase in the acidity of both Y and Z zeolites as a result of phosphoric acid modification. The strong acid sites were converted into weak acid sites and the overall acidity of the zeolites were high due to phosphoric acid modification. XRD studies showed that there is no loss in crystallinity of zeolites due to the modification. TGA analysis showed the increase in hydrophobicity of the zeolites due to phosphoric acid modification. From the SEM micrographs it is understood that there is a sponge like surface morphology and it gets enlarged with modification in both the zeolites.

- Temperature is an important factor which controls the esterification process. Rate of the esterification process was found to increase with increase in temperature till 60°C, beyond which the rate was found to decrease. The decrease in the rate of the reaction beyond 60°C was attributed to the loss of methanol, whose boiling point is 64.7°C. This was observed for both modified and unmodified zeolites.

- The molar ratio of methanol to oil taken for esterification reaction is associated with the type of catalyst used. The esterification reaction was found to occur at a higher rate with increase in the methanol to oil ratio for both mahua and karanja with modified and unmodified zeolites. Beyond a ratio of 9:1 the increase was very minimal. Hence the optimum ratio of methanol to oil was fixed as 9:1 for all further studies in all the cases.

- The role of a catalyst is very important in the esterification reaction. The amount of catalyst used for the reaction depends on the nature and type of the catalyst. Increase in the amount of catalyst increases the rate of FFA conversion with modified and unmodified zeolites for both the oils. A catalyst loading of 1% by weight was found to be optimum, whereas with higher amounts of catalysts the increase in rate was nominal.

- Kinetic studies revealed that the solid acid catalyzed esterification reaction follows pseudo first order kinetics. Kinetic constants were computed from the kinetic plots obtained making use of the kinetic equations arrived, based on a set of assumptions. From the kinetic constant values obtained for mahua and karanja with modified and unmodified zeolites, it was clear that the forward reaction, esterification of FFA occurs to a higher extent, in comparison to the reverse ester hydrolysis.

- NMR spectra were recorded for raw oils, esterified oils and the product biodiesel. Esterification of FFA in presence of the modified and unmodified zeolites were confirmed by the peak obtained in the spectra corresponding to that of the methyl ester proton. An increase in intensity of the same peak in the spectra recorded for the biodiesel product confirmed the conversion of triglycerides into methyl esters.

- The activity of Y, MY, Z and MZ were compared by carrying out the esterification process at optimum reaction conditions with both mahua and karanja oils. It was found that there was an increase in efficiency of MY and MZ in comparison to Y and Z respectively towards esterification. The increase in acidity, enhanced hydrophobicity, agglomeration of particles at the surface, an increase in density in both Y and Z zeolites due to phosphoric acid modification was attributed to the increased efficiency of modified zeolites. It was also found that between Y and Z zeolites, the activity of Y both in the unmodified and modified conditions is higher. This may be due to pore topology and surface area. Y has high surface area and pores with a wide and straight channel which facilitate easy and a high rate of FFA conversion. Z has low surface area and a narrow and bent pore topology. A comparison of the esterification was done by carrying out the esterification with oleic acid and the zeolite catalysts.

- Transesterification of the esterified oils were carried out with KOH catalyst under standard conditions of temperature 60°C, methanol to oil ratio 6:1 and catalyst loading 1% by weight. The yield of biodiesel in all the cases were around 90%.
- Catalyst recovery and reuse was studied. The zeolites were recovered and reused three times for the esterification reactions and there was no appreciable decrease in the activity of the catalyst.

Based on the above observations, it can be concluded that zeolites with suitable modifications to increase their acidity and hydrophobicity, apart from an increase in the surface area and other pore properties could be effectively used for the esterification of FFA in non-edible oils towards biodiesel production.

SCOPE FOR FUTURE WORK

The modification of zeolites with appreciable surface area and a favourable pore topology with such mildly acidic inorganic entities or more hydrophobic organic moieties make them suitable as efficient solid acid catalysts for the esterification of FFA with low cost starting materials. By a proper selection of the modifying molecule, the efficiency of the catalyst could further be improved.

There are many low cost feed stocks such as waste cooking oil, animal fats and non-edible oils which could be subjected to initial esterification with modified zeolite catalysts and then subjected to further transesterification, leading to an cost effective and enhanced production of biodiesel.

Furthermore, a detailed study of biodiesel prepared through such processes with regard to their fuel related properties can be done. The biodiesel product could be tested in engines and emission characteristics can be studied and compared with biodiesel prepared using other conventional catalysts.

In the Indian context, there are a lot of unexplored non-edible oils having a high FFA content and reasonable calorific value, which could be subjected to esterification and subsequent biodiesel production by making use of such heterogeneous acid catalysts.

REFERENCES

Agarwal A.K. and Bajaj T.P., "Process optimization of base catalyzed transesterification of karanja oil for biodiesel production", *Int. J. of Oil, Gas and Coal Technology*, Vol. 2, pp. 297-310, 2009.

Aguado J., Arsuaga J.M., Arencibia A., Lindo M. and Gascon V., "Aqueous heavy metals removal by adsorption on amine-functionalized mesoporous silica", *J. Hazard. Mater.*, Vol. 163, pp. 213-221, 2009.

Ahmad M., Khan M.A., Zafar M. and Sultana S., "Environment-friendly renewable energy from sesame biodiesel", *Energy Sources, Part A*, Vol. 32, pp. 189-196, 2010.

Ahmad M.I., Zaidi S.M.J., Rahman S.U. and Ahmed S., "Synthesis and proton conductivity of heteropolyacids loaded Y-zeolite as solid proton conductors for fuel cell applications", *Microporous Mesoporous Mater.*, Vol. 91, pp. 296-304, 2006.

Alsalme A., Kozhevnikova E.F. and Kozhevnikov I.V., "Heteropoly acids as catalysts for liquid-phase esterification and transesterification", *Appl. Catal., A*, Vol. 349, pp. 170-176, 2008.

Ames L.L., "The cation sieve properties of clinoptilolite", *Amer. Mineral.*, Vol. 45, pp. 689-700, 1960.

Antunes W.M., Veloso C.O. and Henriques C.A., "Transesterification of soybean oil with methanol catalyzed by basic solids", *Catal. Today*, Vol. 133-135, pp. 548-554, 2008.

Auroux A., Weirchowski P. and Gravelle P.C., "Micro calorimetric evidence of the non-uniform distribution of acid sites in a ZSM type zeolite", *Thermochem. Acta*, Vol. 32, pp. 165-170, 1979.

Azam M.M., Waris A. and Nahar N.M., "Prospects and potential of fatty acid methyl esters of some non-traditional seed oils for use as biodiesel in India", *Biomass Bioenergy*, Vol. 29, pp. 293-302, 2005.

Bagby M.O., "Vegetable oils for diesel fuel: opportunities for development", International winter meeting of the ASAE, Chicago, Pap. 87, pp. 1588-1589, 1987.

Barakos N., Pasias S. and Papayannakos N., "Transesterification of triglycerides in high and low quality oil feeds over an HT2 hydrotalcite catalyst", *Bioresour. Technol.*, Vol. 99, pp. 5037-5042, 2008.

Barnawal B.K. and Sharma M.P., "Prospects of biodiesel production from vegetable oils in India", *Renewable Sustainable Energy Rev.*, Vol. 9, pp. 363-378, 2005.

Barrer R.M., "Ion exchange and ion sieve process in crystalline zeolites", *J. Chem. Soc.*, Vol. 21, pp. 2342-2350, 1951.

Barthomeuf D., "A general hypothesis on zeolites physicochemical properties. Application to adsorption, acidity, catalysis and electrochemistry", *J. Phys. Chem.*, Vol. 83, pp. 249-256, 1979.

Bayer H.K., Karge H.G. and Borbely G., "Solid –state ion exchange in zeolites: Part I – Alkali metal chlorides/ZSM-5", *Zeolites*, Vol. 8, pp. 79-82, 1988.

Behrens P. and Stucky G.D., "Ordered molecular arrays as templates: a new approach to synthesis of mesoporous materials", *Angew. Chemie. Int. Ed. Engl.*, Vol. 32, pp. 696-699, 1993.

Benjapornkulaphong S., Ngamcharussrivichai C. and Bunyakiat K., "Al_2O_3-supported alkali an alkali earth metal oxides for transesterification of palm kernel oil and coconut oil", *Chem. Eng. J.*, Vol. 145, pp. 468-474, 2009.

Berchmans H.J. and Hirata S., "Biodiesel production from crude Jatropha curcas L. seed oil with a high content of free fatty acids", *Bioresour. Technol.*, Vol. 99, pp. 1716-1721, 2008.

Berrios M., Siles J., Martin M.A. and Martin A., "A kinetic study of the esterification of free fatty acids (FFA) in sunflower oil", *Fuel,* Vol. 86, pp. 2383-2388, 2007.

Biaglow A.J., Parrillo D.J., Kokotailo G.T. and Grote R.J., "A study of dealuminated Faujasites", *J. Catal.*, Vol. 148, pp. 213-223, 1994.

Bringi N.V., "Non traditional oil seed and oils of India", Oxford and IBH, New Delhi, pp. 121-124, 1987.

Caetano C.S., Fonseca I.M., Ramos A.M., Vital J. and Castanherio J.E., "Esterification of free fatty acids with methanol using heteropolyacids immobilized on silica", *Catal. Comun.*, Vol. 9, pp. 1996-1999, 2008.

Cardoso A.L., Neves S.C.G. and Da silva M.J., "Esterification of oleic acid for biodiesel production catalyzed by $SnCl_2$: A kinetic investigation", *Energies*, Vol. 1, pp. 79-92, 2008.

Carmo A.C., de Souza L.K.C., de Costa C.E.F., Longo E., Zamian J.R., and da Rocha Filho G.N., "Production of biodiesel by esterification of palmitic acid over mesoporous aluminosilicate Al-MCM-41", *Fuel*, Vol. 88, pp. 461-468, 2009.

Caynak S., Guru M., Bicer A., Keskin A. and Icingur Y., "Biodiesel production from pomace oil and improvement of its properties with synthetic manganese additive", *Fuel*, Vol. 88, pp. 534-538, 2009.

Chhetri A.B., Tango M.S., Budge S.M., Watts K.C. and Islam M.R., "Non-edible plant oils as new sources for biodiesel production", *Int. J. Mol. Sci.*, Vol. 9, pp. 169-180, 2008.

Chung K.H., Chang D.R. and Park B.G., "Removal of free fatty acid in waste frying oil by esterification with methanol on zeolite catalysts", *Bioresour. Technol.*, Vol. 99, pp. 7438-7443, 2008.

Cordeiro C.S., Arizaga G.G.C., Ramos L.P. and Wypych F., "A new zinc hydroxide nitrate heterogeneous catalyst for the esterification of free fatty acids and the transesterification of vegetable oils", *Catal. Commun.*, Vol. 9, pp. 2140-2143, 2008.

Corma A., Garcia H., Iborra S. and Primo J., "Modified faujasite zeolites as catalysts in organic reactions: Esterification of carboxylic acids in the presence of HY zeolites", *J. Catal.*, Vol. 120, pp. 78-87, 1989.

Cvetanovic R.J. and Amenomiya Y., "A temperature programmed sorption technique for investigation of practical catalysts", *Catal. Rev. Sci. Eng.*, Vol. 6, pp. 21-48, 1972.

Csicsery S.M., "Shape-selective zeolites science and technology", *Pure Appl. Chem.*, Vol. 58, pp. 841-856, 1984.

Darnoko D. and Cheryan M., "Kinetics of palm oil transesterification in a batch reactor", *J. Am. Oil Chem. Soc.*, Vol. 77, pp. 1263-1267, 2000.

Davis M.E., "Zeolites and molecular sieves: Not just ordinary catalysts", *Ind. Eng. Chem. Res.*, Vol. 30, pp. 1675-1683, 1991.

Demirbas A., "Biodiesel fuels from vegetable oils via catalytic and non catalytic supercritical alcohol transesterifications and other methods: a survey", *Energy Convers. Manage.*, Vol. 44, pp. 2093-2109, 2003.

Demirbas A., "Progress and recent trends in biofuels", *Prog. Energy Combust. Sci.*, Vol. 33, pp. 1-18, 2007.

Denayer J.F., Baron G.V., Vanbutsele G., Jacobs P.A. and Martens J.A., "Evidence for alkyl carbonium ion reaction intermediates from intrinsic reaction kinetics of C6-C9 n-alkane hydroisomerisation and hydrocracking of Pt/H-Y and Pt/USY zeolites", *J. Catal.*, Vol. 190, pp. 469-473, 2000.

Di Serio M., Cozzolino M., Tesser R., Patrono P., Pinzari F., Bonelli B. and Santacesaria E., "Vanadyl phosphate catalysts in biodiesel production", *Appl. Catal., A* Vol. 320, pp. 1-7, 2007.

dos Santos I.C.F., de Carvalho S.H.V., Solleti J.I., de La Salles W.F., da Silva de LA Salles K.T. and Meneghetti S.M.P., "Studies of Terminalia catappa L. oil: Characterization and biodiesel production", *Bioresour. Technol.*, Vol. 99, pp. 6545-6549, 2008.

Dyer A., "An introduction to zeolite molecular sieves", John Wiley and Sons, Chichester, pp. 95-98, 1988.

Echoufi N. and Gélin P., "Bronsted acidity in HY zeolites measured by IR study of pyridine adsorption: influence of steric effects", *Catal. Lett.*, Vol. 40, pp. 249-252, 1996.

El-Mashad H.M., Zhang R. and Avena-Bustillos R.J., "A two-step process for biodiesel production", *Biosystems Eng.*, Vol. 99, pp. 220-227, 2008.

Ferreira T.R., Lope C.B., Lito P.F., Otero M., Lin Z., Rocha J., Pereira E., Silva C.M. and Duarte A., "Cadmium(II) removal from aqueous solution using microporous titanosilicate ETS-4", *Chem. Eng. J.*, Vol. 147, pp. 173-179, 2009.

Feuge R.O. and Gros A.T., "Modification of vegetable oils. VII. Alkali catalyzed interesterification of peanut oil with ethanol", *J. Am. Oil Chem. Soc.*, Vol. 26, pp. 97-102, 1949.

Flanigen E.M., "Zeolite chemistry and catalysis", in Raoo J.A.(ed.), ACS Monograph 171, Washington D.C, pp. 231-241, 1976.

Formo M.W., "Ester reactions of fatty materials", *J. Am. Oil Chem. Soc.*, Vol. 31, pp. 548-559, 1954.

Freedman A., Pryde E.H. and Mounts T.L., "Variables affecting the yield of fatty esters from transesterified vegetable oil", *J. Am. Oil Chem. Soc.*, Vol. 61, pp. 1638-1643, 1984.

Freedman A., Butterfield R.O. and Pryde E.H., "Transesterification kinetics of soybean oil", *J. Am. Oil Chem. Soc.*, Vol. 63, pp. 1375-1380, 1986.

Garcia C.M., Teeixeira S., Marciniuk L.L. and Schuchardt U., "Transesterification of soybean oil catalyzed by sulfated zirconia", *Bioresour. Technol.*, Vol. 99, pp. 6608 – 6613, 2008.

Gelbard G., Bres O., Vargas R.M., Vielfaure F. and Schuchardt U. F., "^{1}H magnetic resonance determination of the yield of the transesterification of rapeseed oil with methanol", *J. Am. Oil Chem. Soc.*, Vol. 72, pp. 1239-1241, 1995.

Ghadge S.V. and Raheman H., "Biodiesel production from mahua (Madhuca indica) oil having high free fatty acids", *Biomass Bioenergy*, Vol. 28, pp. 601-605, 2005.

Ghadge S.V. and Raheman H., "Process optimization for biodiesel production from mahua (Madhuca indica) oil using response surface methodology", *Bioresour. Technol.*, Vol. 97, pp. 379-384, 2006.

Gimenez J., Costa J. and Cervera S., "Vapor-phase esterification of acetic acid with ethanol catalyzed by a macroporous sulfonated styrene-divinylbenzene(20%) resin", *Ind. Eng. Chem. Res.*, Vol. 26, pp. 198-202, 1987.

Guerreiro L., Castanheiro J.E., Fonseca I.M., Martin-Aranda R.M., Ramos A.M. and Vital J., "Transesterification of soybean oil over sulphonic acid functionalised polymeric membranes", *Cat. Today*, Vol. 118, pp. 166-171, 2006.

Gui M.M., Lee K.T. and Bhatia S., "Feasibility of edible oil vs. non-edible oil vs. waste edible oil as biodiesel feedstock", *Energy*, Vol. 33, pp. 1646-1653, 2008.

Hamad B., de Souza R.O.L., Sapaly G., Rocha M.G.C., de Oliveira P.G.P., Gonzalez W.A., Sales E.A. and Essayem N., "Transesterification of rapeseed oil with ethanol over heterogeneous heteropolyacids", *Catal. Commun.*, Vol. 10, pp. 92-97, 2008.

Harrington K.J. and Darcy-Evans C., "Transesterification insitu of sunflower seed oil", *Ind. Eng. Chem. Prod. Res. Dev.*, Vol. 24, pp. 314-318, 1985.

Holderich W.F. and Bekkum H.V., "Zeolites in organic synthesis", *Stud. Surf. Sci. Catal.*, Vol. 58, pp. 631-726, 1991.

Hui Y.H., Editors, "Bailey's Industrial Oils fats: Industrial and Consumer Non Edible Products from Oils and Fats, 5th ed., Wiley-Interscience, New York, 1996.

Jacobs P.A. and Heylen C.F., "Active sites in zeolites. III. Selective poisoning of bronsted acid sites on synthetic Y zeolites", J. Catal., Vol. 34, pp. 267-274, 1974.

Jacobs P.A. and Martens J.A., "Conceptual background for the conversion of hydrocarbons on heterogeneous acid catalysts in theoretical aspects of heterogeneous catalysis", Ed. Moffat J.B., Van Nostrand Reinhold, 1990.

Jacobson K., Gopinath R., Meher L.C. and Dalai A.K., "Solid acid catalyzed biodiesel production from waste cooking oil", *Appl. Catal., B* Vol. 85, pp. 86-91, 2008.

Jitputti J., Kitiyanan B., Rangsunvigit P., Bunyakiat K., Attanatho L. and Jenvanitpanjakul P., "Transesterification of crude palm kernel oil and crude coconut oil by different solid acid catalysts", *Chem. Eng. J.,* Vol. 116, pp. 61-66, 2006.

Kansedo J., Lee K.T. and Bhatia S., "Cerebra odollam (sea mango) oil as a promising non-edible feedstock for biodiesel production", *Fuel*, Vol. 88, pp. 1148-1150, 2009a.

Kansedo J., Lee K.T. and Bhatia S., "Biodiesel production from palm oil via heterogeneous transesterification", *Biomass Bioenergy*, Vol. 33, pp. 271-276, 2009b.

Kapilan N. and Reddy R.P., "Evaluation of methyl esters of mahua oil (Madhuca Indica) as diesel fuel", *J. Am. Oil Chem. Soc.,* Vol. 85, pp. 185-188, 2008.

Kapilan N., Ashok Babu T.P. and Reddy R.P., "Characterization and effect of using Mahua oil biodiesel as fuel in compression ignition engines", *J. Therm. Sci.* Vol. 18, pp. 382-384, 2009.

Karme S.K. and Chadha A., "Preparation of biodiesel from crude oil of Pongamia pinnata", *Bioresour. Technol.*, Vol. 96, pp. 1425-1429, 2005.

Kaushik A. and Kaushik C.P., " Perspectives in environmental studies", New Age International (P) Limited, New Delhi, pp. 38-39, 2009.

Kirumaaki S.R., Nagaraju N. and Chary K.V.R., "Esterification of alcohols with acetic acid over zeolites Hβ, HY and HZSM5", *Appl. Catal., A* Vol. 299, pp. 185-192, 2006.

Knothe G., Krahl J. and Gerpen J.V., "The biodiesel handbook", Eds., AOCS press: Champaign, IL, pp. 76-80, 2005.

Kocaoba S., Orhan Y. and Akyuz T., "Kinetic and equilibrium studies of heavy metal ions removal by use of natural zeolite", Desalination, Vol. 214, pp. 1-10, 2007.

Korbitz W., "Biodiesel production in Europe and North America, an encouraging prospect", *Renewable Energy*, Vol. 16, pp. 1078-1083, 1999.

Krishnangkura K. and Simamaharnnop R., "Continuous transmethylation of palm oil in an organic chromatography", *J. Am. Oil Chem. Soc.*, Vol. 69, pp. 166-169, 1992.

Kurniawan T.A. and Chan Y.S., "Comparisons of low-cost adsorbents for treating wastewaters laden with heavy metals", *Sci. Total Environ.*, Vol. 366, pp. 409- 426, 2006.

Kustov A.L., Hansen T.W., Kustova M. and Christensen C.H., "Selective catalytic reduction of NO by ammonia using mesoporous Fe containing H ZSM-5 and H ZSM-12 zeolite catalysts: An option for automotive applications", *Appl. Catal., B*, Vol. 76, pp. 311-319, 2007.

Lima J.R.O., da Silva R.B., de Moura E.M. and de Moura C.V.R., "Biodiesel of tucum oil, synthesized by methanolic and ethanolic routes", *Fuel*, Vol. 87, pp. 1718-1723, 2008.

Liu X., He H., Wang Y. and Zhu S., "Transesterification of soybean oil to biodiesel using SrO as a olid base catalyst", *Catal. Commun.*, Vol. 8, pp. 1107-1111, 2007.

Lonyi F. and Vaylon J., "On the interpretation of the NH$_3$-TPD patterns of H-ZSM-5 and H-mordenite", *Micropor. Mesopor. Mater.*, Vol. 47, pp. 293-301, 2001.

Lopez D.E., Goodwin Jr. J.G. and Bruce D.A., "Transesterification of triacetin with methanol on Nafion acid resins", *J. Catal.*, Vol. 245, pp. 381- 391, 2007.

Lou W.Y., Zong M.H. and Duan Z.Q., "Efficient production of biodiesel from high free fatty acid-containing waste oils using various carbohydrate-derived solid acid catalysts", *Bioresour. Technol.*, Vol. 99, pp. 8752-8758, 2008.

Ma F. and Hanna M.A., "Biodiesel production: A review", *Bioresour. Technol.*, Vol. 70, pp. 1-15, 1999.

Marchetti J.M., Miguel V.U. and Errazu A.F., "Heterogeneous esterification of oil with high amount of free fatty acids", *Fuel*, Vol. 86, pp. 906-910, 2007.

Marchetti J.M. and Errazu A.F., "Comparison of different heterogeneous catalysts and different alcohols for the esterification reaction of oleic acid", *Fuel*, Vol. 87, pp. 3477-3480, 2008.

Mbaraka I.K. and Shanks B.H., "Design of multifunctionalized mesoporous silicas for esterification of fatty acid", *J. Catal.*, Vol. 229, pp. 365-373, 2005.

Meher L.C., Vidya S., Dharmagadda S. and Naik S.N., "Optimization of alkali-catalyzed transesterification of Pongamia pinnata oil for production of biodiesel", *Bioresour. Technol.*, Vol. 97, pp. 1392-1397, 2006.

Miyamoto Y., Katada N. and Niwa M., "Acidity of β zeolite with different Si/Al$_2$ ratio as measured by temperature programmed desorption of ammonia", *Micropor. Mesopor. Mat.*, Vol. 40, pp. 271-281, 2000.

Mortier W.J., Samer J., Lercher J.A. and Noller H., "Bridging and terminal hydroxyls. A structural chemical and quantum chemical discussion", *J. Phys. Chem.*, Vol. 88, pp. 905-912, 1984.

Naik M., Meher L.C., Naik S.N. and Das L.M., "Production of biodiesel from high free fatty acid Karanja (Pongamia pinnata) oil", *Biomass Bioenergy*, Vol. 32, pp. 354-357, 2008.

Ngamcharussrivichai A., Wiwatnimit W. and Wangnoi S., "Modified dolomites as catalysts for palm kernel oil transesterification", *J. Mol. Catal. A: Chem.*, Vol. 276, pp. 24-33, 2007.

Ngamcharussrivichai C., Totarat P. and Bunyakiat K., "Ca and Zn mixed oxide as a heterogeneous base catalyst for transesterification of palm kernel oil", *Appl. Catal., A*, Vol. 341, pp. 77-85, 2008.

Ni Y. and Meunier F.C., "Esterification of free fatty acids in sunflower oil over solid acid catalysts using batch and fixed-bed reactors", *Appl. Catal., A* Vol. 333, pp. 122-130, 2007.

Ozbay N., Oktar N. and Tapan A.N., "Esterification of free fatty acids in waste cooking oils (WCO): Role of ion –exchange resins", *Fuel*, Vol. 87, pp. 1789-1798, 2008.

Park Y.M., Lee D.W., Kim D.K., Lee J.S. and Lee K.Y., "The heterogeneous catalyst system for the continuous conversion of free fatty acids in used vegetable oils for the production of biodiesel", *Catal. Today*, Vol. 131, pp. 238-243, 2008.

Park J.Y., Kim D.K. and Lee J.S., "Esterification of free fatty acids using water tolerable Amberlyst as a heterogeneous catalyst", *Bioresour. Technol.*, Vol. 101, pp. S62- S65, 2010.

Pasias S., Barakos N., Alexopoulos C. and Papayannakos N., "Heterogeneously catalyzed esterification of FFAs in vegetable oils", *J. Chem. Eng. Technol.*, Vol. 29, pp. 165-1371, 2006.

Poutsma M.L. and Rabo J., "Zeolite chemistry and catalysis", in Rabo J.A., *Am. Chem. Soc., Symp. Ser.*, Vol. 171, pp. 332-349, 1976.

Puhan S., Vedaraman N., Ram B.V.B., Sankaranarayanan G. and Jeychandran K., "Mahua oil (Madhuca Indica seed oil) methyl ester as biodiesel-preparation and emission characteristics", *Biomass Bioenergy*, Vol. 28, pp. 87-93, 2005a.

Puhan S., Vedaraman N., Sankaranarayanan G. and Ram B.V.B., "Performance and emisión study of Mahua oil (madhuca indica oil) ethyl ester in a 4-stroke natural aspirated direct injection diesel engine", *Renewable Energy*, Vol. 30, pp. 1269-1278, 2005b.

Ramadhas A.S., Jayaraj S. and Muraleedharan C., "Biodiesel production from high FFA rubber seed oil", *Fuel*, Vol. 84, pp. 335-340, 2005.

Ramos M.J., Casas A., Rodriguez L., Romero R. and Perez A., "Transesterification of sunflower oil over zeolites using different metal loading: A case of leaching and agglomeration studies", *Appl. Catal., A* Vol. 346, pp. 79-85, 2008.

Rashid U., Anwar F., Moser B.R. and Knothe G., "Moringa oleifera oil: A possible source of biodiesel", *Bioresour. Technol.*, Vol. 99, pp. 8175-8179, 2008.

Rathore V. and Madras G., "Synthesis of biodiesek from edible oil and non-edible oils in supercritical alcohols and enzymatic synthesis in sypercritical carbon dioxide", *Fuel*, Vol. 86, pp. 2650-2659, 2007.

Rees L.V.C., "When is a zeolite not a zeolite?", *Nature*, Vol. 296, pp. 491-492, 1982.

Sahoo P.K., Das L.M., Babu M.K.G. and Naik S.N., "Biodiesel development from high acid value polanga seed oil and performance evaluation in a CI engine", *Fuel*, Vol. 86, pp. 448-454, 2007.

Saka S., "Production of biodiesel: current and future technology", In: JSPS/VCO Core University program seminar, Universiti Sains Malaysia, pp. 40-45, September 2005.

Sand L.B., "Natural and synthetic zeolite for industrial use", *Econ. Geol., Mass., Proc. Conf. Univ. Mass*, pp. 191-198, 1967.

Sato K., Nishimura Y. and Shimada H., "Preparation and activity evaluation of Y zeolites with or without meso porosity", *Catal. Lett.*, Vol. 60, pp. 83-87, 1999.

Satyarthi J.K., Srinivas D. and Ramasamy P., "Estimation of free fatty acid content in oils, fats and biodiesel by ^{1}H NMR spectroscopy", *Energy Fuels*, Vol. 23, pp. 2273-2277, 2009.

Saydut A., Duz M.Z., Kaya C., Kafadar A.B. and Hamamci C., "Transesterified sesame (*Sesamum indicum* L.) seed oil as a biodiesel fuel", *Bioresour. Technol.*, Vol. 99, pp. 6656-6660, 2008.

Sayed M.B., Kydd R.A. and Cooney R.P., "A fourier transform infrared spectral study of H-ZSM-5 surface situ and reactivity sequences in methanol conversion", *J. Catal.*, Vol. 88, pp. 137-149, 1984.

Schuchardt U., Sercheli R. and Vargas R.M., "Transesterification of Vegetable oils: a Review", *J. Braz. Chem. Soc.*, Vol. 9, pp. 199-210, 1998.

Schwab A.W., Dykstra G.J., Selke E., Sorenson S.C. and Pryde E.H., "Diesel fuel from thermal decomposition of soybean oil", *J. Am. Oil Chem. Soc.*, Vol. 65, pp. 1781-1785, 1988.

Sharma Y.C. and Singh B., "Development of biodiesel from karanja a tree found in rural India", *Fuel*, Vol. 87, pp. 1740-1742, 2008a.

Sharma Y.C., Singh B. and Upadhyay S.N., "Advancements in development and characterization of biodiesel: A review", Fuel, Vol. 87, pp. 2355-2373, 2008b.

Sharma Y.C. and Singh B., "Development of biodiesel: Curent scenario", *Renewable Sustainable Energy Rev.*, Vol. 13, pp. 1646-1651, 2009.

Shay E.G., "Diesel fuel from vegetable oil; Status and opportunities", Biomass Bioenergy, Vol. 4, pp. 227-242, 1993.

Shibasaki-Kitakawa N., Honda H., Kuribayashi H., Toda T., Fukumura T. and Yonemoto T., "Biodiesel production using anionic ion-exchange resin as heterogeneous catalyst", *Bioresour. Technol.*, Vol. 98, pp. 416-421, 2007.

Shu Q., Yang B., Yuan H., Qing S. and Zhu G., "Synthesis of biodiesel from soybean oil and methanol catalyzed by zeolite beta modified with La^{3+}", *Catal. Commun.*, Vol. 8, pp. 2159-2165, 2007.

Srivastava A. and Prasad R., "Triglycerides-based diesel fuels", *Renewable Sustainable Energy Rev.*, Vol. 4, pp. 111-133, 2000.

Srivastava P.K. and Verma M., "Methyl ester of karanja oil as an alternative renewable source energy", *Fuel*, Vol. 87, pp. 1673-1677, 2008.

Stern R. and Hillion G., Eur P. Appl EP, Vol. 356, 317[C1.C07C67/56], 1990; R. Stern and G. Hillion, *Eur. P. Chem. Abstr.*, Vol. 113, P58504k, 1990.

Suppes G.J., Dasari M.A., Doskocil E.J., Mankidy P.J. and Goff M.J., "Transesterification of soybean oil with zeolite and metal catalysts", *Appl. Catal., A* Vol. 257, pp. 213-223, 2004.

Sureshkumar K., Velraj R. and Ganesan R., "Performance and exhaust emission characteristics of a CI engine fueled with Pongamia pinnata methyl ester (PPME) and its blend with diesel", *Renewable Energy*, Vol. 33, pp. 2294-2302, 2008.

Swarup R., "An overview of DBT's energy bioscience programme", pp. 5-9, 2007.

Ting W., Huang C., Giridhar N. and Wu W., "An enzymatic/acid-catalyzed hybrid process for biodiesel production from soybean oil", *J. Chin. Inst. Chem. Eng.*, Vol. 39, pp. 203-210, 2008.

Towsend R.P., "Ion exchange in zeolites", *Stud. Surf. Sci. Catal.*, Vol. 58, pp. 359-390, 1991.

Uytterhoeven J.B., Christner L.G. and Hall W.K., "Studies of the hydrogen held by solids. VIII. The decationated zeolites", *J. Phys. Chem.*, Vol. 69, pp. 2117-2126, 1965.

Veljkovic V.B., Lakicevic S.H., Stamenkovic O.S., Todorovic Z.B. and Lazic M.L., "Biodiesel production from tobacco (Nicotiana tabacm L.) seed oil with a high content of free fatty acids", *Fuel*, Vol. 85, pp. 2671-2675, 2006.

Viswanathan B. and Ramaswamy A.V., "Selection of solid heterogeneous catalysts for transesterfication reaction", *Chemistry Industry Digest*, pp. 91-99, August 2008.

Wan G., Duan A., Zhang Y., Zhao Z., Jiang G., Zhang D. and Gao Z., "Zeolite beta synthesized with acid-treated metakaolin and application in diesel hydrodesulfurization" *Catal. Today*, Vol. 149, pp. 69-75, 2010.

Wang S. and Peng Y., "Natural zeolite as effective adsorbents in water and wastewater treatment", *Chem. Eng. J.*, Vol. 156, pp. 11-24, 2010.

Ward J.W., "Zeolite chemistry and catalysis" in J.A. Rabo (ed.), ACS Monograph 171, Washington D.C., pp. 118-306, 1976.

Wright H.J., Segur J.B., Clark H.V., Coburn S.K., Langdon E.E. and DuPuis R.N., "A report on ester interchange", *OilandSoap*, Vol. 21, pp. 145-148, 1944.

Wu C.G. and Bein T., "Conducting polyaniline filaments in a mesoporous channel host", *Science*, Vol. 264, pp. 1757-1759, 1994.

Xavier N.M., Lucas S.D. and Rauter A.R., "Zeolite as efficient catalysts for key transformations in carbohydrate chemistry", *J. Mol. Catal. A: Chem.*, Vol. 305, pp. 84-89, 2009.

Xie W., Huang X. and Li H., "Soybean oil methyl esters preparation using NaX zeolites loaded with KOH as heterogeneous catalysts", *Bioresour. Technol.*, Vol. 98, pp. 936-939, 2007.

Xu L., Yang X., Yu X., Guo Y. and Maynurkader, "Preparation of mesoporous polyoxometalate-tantalum pentoxide composite catalyst for efficient esterification of fatty acid", *Catal. Commun.*, Vol. 9, pp. 1607-1611, 2008.

Yamamoto T., Shido T., Inagaki S., Fukushima Y. and Ichikawa M., "Ship-in bottle synthesis of $[Pt(CO)_{30}]^{2-}$ cncapsulated in ordered hexagonal mesoporous channels of FSM-16 and their effective catalysis in water-gas shift reaction", *J. Am. Chem. Soc.*, Vol. 118, pp. 5810-5811, 1996.

Zhang K., Huang C., Zhang H., Xiang S., Liu S., Xu D. and Li H., "Alkylation of phenol with trt-butyl alcohol catalyzed by zeolite Hβ", *Appl. Catal., A* Vol. 166, pp. 89-95, 1998.

Zhang J. and Jiang L., "Acid catalyzed esterification of Zanthoxylum bungeanum oil with high free fatty acids for biodiesel production", *Bioresour. Technol.*, Vol. 99, pp. 8995-8998, 2008.

Zhang L., Xian M., He Y., Li L., Yang J., Yu S. and Xu X., "A Bronsted acidicionic liquid as an efficient and environmentally benign catalyst for biodiesel synthesis free fatty acids and alcohols", *Bioresour. Technol.*, Vol. 100, pp. 4368-4373, 2009.

Zheng S., Kates M., Dude M.A. and McLean D.D., "Acid catalysed production of biodiesel from waste frying oil", *Biomass Bioenergy*, Vol. 30, pp. 267-272, 2006.

Ziejewski M., Kufman K.R. and Pratt G.L., "Vegetable oil as diesel fuel", Seminar II, Northern Regional Research Centre, Peoria, Illinois, 19-20 October, 1983.

In: Biodiesel: Blends, Properties and Applications   ISBN: 978-1-63117-024-9
Editors: Jorge M. Marchetti and Zhen Fang   © 2014 Nova Science Publishers, Inc.

Chapter 4

HETEROGENEOUS CATALYTIC PROCESS TO CONVERT VEGETABLE OIL INTO BIODIESEL

Masato Kouzu[*]
Research Center for Fine Particle Science and Technology,
Doshisha University, Kizugawadai,
Kizugawa, Kyoto, Japan

ABSTRACT

In the existent process to produce biodiesel, alkali hydroxide dissolved in methanol catalyzes the transesterification of the feedstock oil consisting of triglycerides with methanol. The homogeneous base catalysis is very effective in enhancing the reaction efficiency, but it is necessary to wash the homogeneous base catalyst off the crude biodiesel with water after the reaction. A massive amount of wastewater originating in the purification mentioned above is one of the problems causing costly production of biodiesel. Also, the purification allows stable emulsion to be formed, which not only disturbs the process operation but also decreases the productivity. Therefore, many researchers took their interests in utilizing heterogeneous catalytic process for biodiesel production. In the present work, research papers studying the heterogeneous catalyst for the transesterification of triglycerides are reviewed.

INTRODUCTION

Most of the biodiesel factories employ alkali hydroxide dissolved in methanol for conversion of the feedstock oil consisting of triglycerides into fatty acid methyl esters (FAME), because the homogeneous base catalysis results in the reasonable converting efficiency [1]. According to a research paper by Freedman et al., the yield of FAME produced by transestrerifying sunflower oil with methanol in the presence of 1 % sodium hydroxide at the temperature of 333 K was above 90 % for 0.1 h [2]. However, it is necessary to wash the

---

[*] Tel: 81-774-65-7654, Fax: 81-774-73-1902, [E-mail address] mkozu@mail.doshisha.ac.jp

homogeneous catalyst off the crude biodiesel with a massive amount of water. Not only the crude biodiesel but also the by-produced glycerol is contaminated with the homogeneous catalyst, and the contaminated glycerol is difficult to recycle for chemicals. Moreover, stable emulsion is formed by adding water into the crude biodiesel for its purification, which disturbs the process operation and leads to loss in the synthesized biodiesel. Due to a massive amount of the wastewater, the contamination of glycerol and the formation of stable emulsion, the current production of biodiesel seem to be costly [3].

For the purpose of solving the problems mentioned above, many researchers make great efforts at studying the novel process to transesterify triglycerides with methanol [4]. One of the solutions is to utilize the enzymatic process in which microbial cell producing lipase serves as the biocatalyst [5-6]. Lipase such as *Candida Antarctica* is an enzyme active in hydrolysis of triglycerides into fatty acids and glycerol, while the addition of alcohol into a mixture of lipase and triglycerides brings about its transesterification [7]. Long-chain alcohol and branched one provide the high transesterifying ratio over 90 %, whereas methanol is not proper for the enzymatic transesterification [8]. Lipase is markedly deactivated in the presence of methanol. The deactivation of lipase seems to be eased by using hexane as the solvent. Also, it was reported that the stepwise addition of methanol allowed the perfect conversion of vegetable oil into FAME to be achieved [9]. But, it took 40 hours to achieve the perfect conversion. For facilitating the handling of the biocatalyst, the use of lipase-producing microbial cells immobilized in porous material has been studying for the purpose of enhancing the lipase recovering efficiency [10].

On the other hand, Saka and Kusudiana proposed the catalyst-free process in which supercritical methanol reacts with vegetable oil [10-11]. Liquid methanol is a polar solvent and molecules of methanol are formed into cluster due to their hydrogen bond. The supercritical condition allows the hydrogen bond to be weakened, which causes a decrease in polarity of methanol. The decreased polarity enhances a solubility of triglyceride in methanol. Moreover, the ionic product increases markedly under the supercritical condition. With the help of the improved chemical property of methanol, biodiesel is produced with the very fast converting rate: it took only 4 minutes to achieve the perfect conversion of rapeseed oil into its methyl ester. As high temperature (523-673 K) and high pressure (35-60 MPa) was required for using supercritical critical, improvement of the catalyst-free process has been studying for reducing the reaction temperature and pressure [12]. The modified process is characterized by the two step reactions: the primary hydrolysis to convert triglyceride into fatty acids and the following esterification with methanol into fatty acid into its methyl ester. Both the primary hydrolysis and the following esterification were operated at 543 K under 10 MPa.

Also, utilization of the heterogeneous catalytic process is the promising way to improve the biodiesel production [13]. The reaction to produce biodiesel was initiated by nucleophilic attack of methanol to a carbonyl carbon in a molecule of triglyceride, and then the nucleophilic attack decomposes triglyceride into FAME and diglyceride [14]. By the further transesterification, triglyceride is finally converted into three molecules of FAME and a molecule of glycerol. For the transesterification, solid base catalytically acts on formation of methoxide anion that is the stronger nuclephile than methanol (Figure 1). Kabashima et al. indicated that the basic site allowed proton to be abstracted from methanol for its conjugated addition to unsaturated ketone [15], and there is a research paper showing that the same action of the basic site promoted transesterification of ethylacetate with methanol [16]. Not only solid base but also solid acid is able to functions as the heterogeneous catalyst. The Bronsted acidity results in donation of proton to triglyceride, which intensify the positive charge of the carbonyl carbon (Figure 2). Due to the intensified positive charge, the nucleophilic attack of methanol to triglyceride is promoted. The solid acid catalysis for the nucleophilic reaction was reviewed in the research paper by Lotero et al. [17].

Recently, liquid hydrocarbon produced from triglycerides begins to attract notice as the alternative diesel fuel [18-19]. A procedure to convert into the liquid hydrocarbon is dehydrogenation based on the petroleum hydrotreatment technology, in which nickel-molybdenum sulfide loaded on alumina is used as the heterogeneous catalyst for hydrotreatment of the heavy distillate [20]. The liquid hydrocarbon has an advantage of the high oxidation stability and the good cold flow properties, in comparison to FAME. The conversion into the liquid hydrocarbon was operated at the high temperature (623-673 K) under the high hydrogen pressure (5MPa).

The present work focuses on the heterogeneous catalysts active in the transesterification, and the concerning research papers are reviewed. In the published articles studying solid acid catalysis for transesterification of triglyceride, sulfated metal oxides, cation-exchange resins and the other acid materials were tested. On the other hand, researchers who are interested in the solid base catalysis investigated the catalytic activities of alumina supported alkali metals, alkaline earth metal oxides, hydrotalcite and so on. For these heterogeneous catalysts active in the transesterification, the catalytic mechanism is described in the present review paper. Additionally, the catalytic performance for transesterifying triglycerides into FAME is appreciated on the basis of data shown in the reviewed research papers.

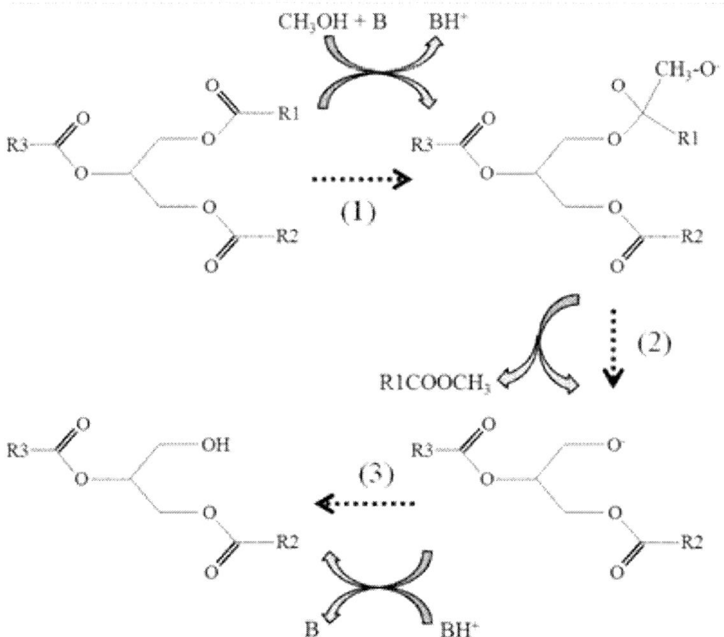

Figure 1. Mechanism on the base-catalyzed transesterification of triglyceride with methanol. (1) Attack of nucleophile generated from methanol, to triglyceride, (2) Dissociation of fatty acids methyl ester from anionic alkoxycarbonyl intermediate, (3) Addition of proton to diglyceride anion.

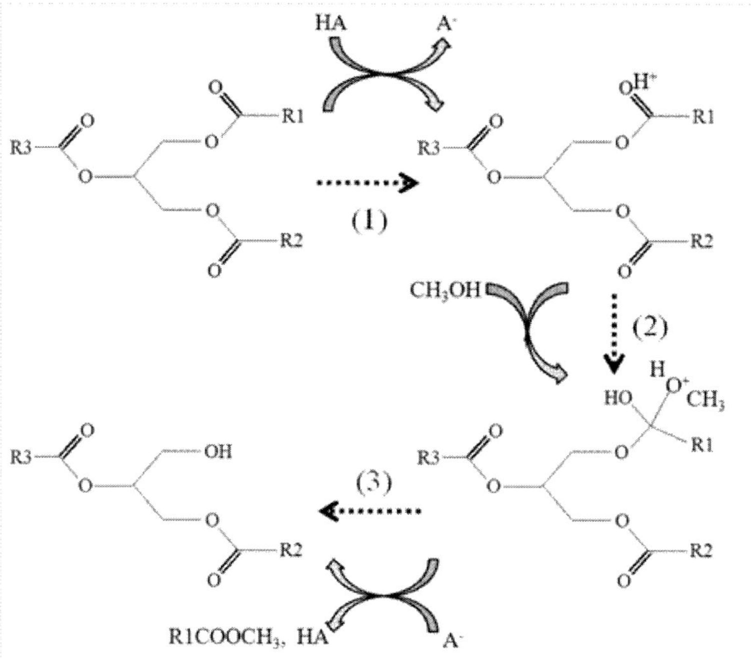

Figure 2. Mechanism on the acid-catalyzed transesterification of triglyceride with methanol. (1) Addition of proton to triglyceide, (2) Nucleophilic attack of methanol to triglyceride, (3) Dissociation of proton and fatty acids methyl ester from cationic alkoxycarbonyl intermediate.

SOLID ACID CATALYSTS

Sulfated Metal Oxides

Sulfated metal oxides, which are prepared by bonding sulfuric anion onto surface of transition metal oxides, are interpreted as "Superacidic solids", because the acidic strength of sulfated metal oxides is larger than that of pure sulfuric acid [21]. The acidity is strong enough to catalyze the isomerization of alkanes at room temperature. Among a variety of sulfated metal oxides, great attention is drawn to sulfated zirconia (SO_4/ZrO_2), due to its stronger acid property. Surface of SO_4/ZrO_2 is illustrated by giving Figure 3, and it is considered that the strong acidity originates in electron attracting force of sulfur-oxygen double bond [22-23].

Lopez et al. carried out transesterification of triacetin in the presence of SO_4/ZrO_2 for investigating its catalytic activity, on a stainless steel batch reactor [24]. At the temperature of 333 K with the methanol/triacetin molar ratio of 6, 20 % of triacetin was converted into methyl acetate after 2 hours. However, the catalytic activity of SO_4/ZrO_2 was reduced by reusing it. The catalytic deactivation resulted from leaching of the active sulfate group from the solid acid catalyst by hydrolysis [25]. Jitputti et al. used SO_4/ZrO_2 for the transesterification of coconut oil including 2.25 % fatty acid [26]. The test reaction was conducted at 473 K with the result that the yield of FAME produced after 2 hours was 86 %. Since the catalytic deactivation was appreciable, the authors regenerated the solid acid catalyst by immersing it in 0.5 M sulfuric acid solution. As a result, the catalytic activity returned to the original level. When we tested SO_4/ZrO_2 at the temperature of 353 K, 16 % of FAME was yielded from soybean oil after 12 hours. Figure 4 shows the yield of FAME produced by transesterifying soybean oil with the methanol/oil molar ratio of 12 in the presence of SO_4/ZrO_2.

For the purpose of preventing the active sulfate group from being leached away from SO_4/ZrO_2, the catalytic modification has been studied. Yadav and Murkute prepared the solid acid catalyst tolerant to the leaching by their own procedure, in which chlorosulfonic acid dissolved in an organic solvent was used as a precursor of the active sulfate group [27]. Chen et al. investigated the solid acid catalysis of SO_4/ZrO_2 loaded on mesoporous silica SBA-15 [28]. Their experimental data indicated that tolerance to the leaching was enhanced for the prepared catalyst, and it was considered that the supporting material was effective in adsorbing water causing the hydrolysis.

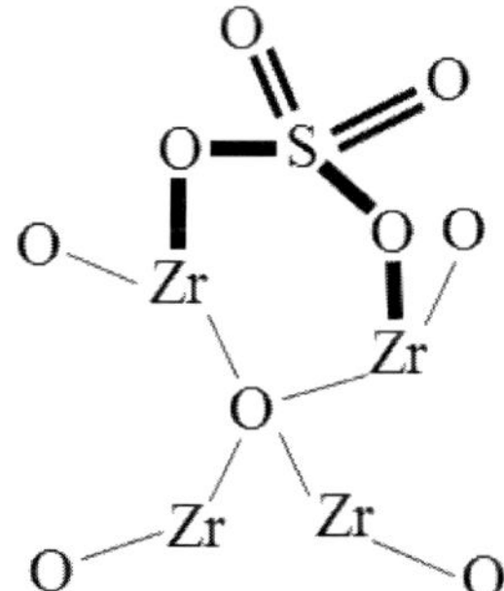

Figure 3. Surface chemical structure of sulfated Zirconia (SO_4/ZrO_2)

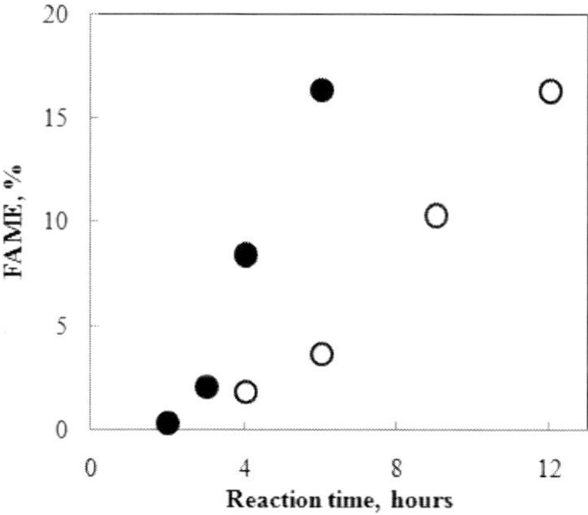

Figure 4. The yields of FAME produced by transesterifying soybean oil at 353 K in the presence of SO_4/ZrO_2 (○) and sulfonated polystyrene (●).

Furthermore, there was a research paper testing SO_4/ZrO_2 on a fixed bed continuous flow reactor unit [29]. The employed reactor unit has the advantage of simplified product separation, and is intrinsically appropriate for the heterogeneous catalytic reaction. Soybean oil and methanol were pumped into the reactor at the flow rate of 3.0 g·h⁻¹ and 4.4 g·h⁻¹, respectively, which was corresponded to the methanol/oil molar ratio of 40. The test reaction was carried out at 523 K under atmospheric pressure with WHSV of 0.54 h^{-1} as a result and oil fraction in the product effluent contained 50 % of FAME. Although sulfated tin oxide (SO_4/SnO_2) is the stronger acid than SO_4/ZrO_2, the catalytic activity was less for SO_4/SnO_2 in the research paper.

As well as SO_4/ZrO_2, tungstated zirconia (WO_3/ZrO_2) was interpreted as the superacidic solid. WO_3/ZrO_2 was more active than SO_4/ZrO_2 in the soybean oil transesterification performed on a fixed bed continuous flow reactor unit [29], while WO_3/ZrO_2 was inferior to SO_4/ZrO_2 in catalyzing the transesterification of triacetin with methanol [24]. The former test was carried out at the higher temperature with the larger methanol/oil molar ratio than the latter one. Ramu et al. conducted the esterification of palmitic acid with refluxing methanol for evaluating the catalytic activity of WO_3/ZrO_2 [30]. The solid acid catalyst calcined below 723 K was so active that 95 % of palmitic acid was converted into methyl palmitate. The crystalline structure of zirconia changed from tetragonal phase into monoclinic phase by calcining above 723 K, which was probably due to the esterifying activity of the solid acid catalyst.

Cation-Exchange Resins

Cation-exchange resin was divided into two classes: sulfonated polystyrene and perfluorinated resin sulfonic acid. They are commercially available: "Amberlyst" is one of the sulfonated polystyrenes, and "Nafion" is on the market as a perfluorinated resin sulfonic acid. For both of the classes, donation of proton from the sulfonic group is responsible for the solid acid catalysis. The solid acid catalysis of cation-exchange resin is industrially utilized for some chemical reactions: hydrolysis of ethylacetate, synthesis of bisphenol, methanolysis of methacrylic acid, alkylation of phenol and so on [31]. Therefore, cation-exchange resin is advantageous to the practical use, as compared to the other solid acids tested for the transesterifying activity.

A research work studying utilization of solid acid catalyst for biodiesel production elucidated that sulfonated polystyrene was more active than perfluorinated resin sulfonic acid [24]. The catalytic activity was measured by transesterifying triacetin with methanol at 333 K. At 2 hours of the reaction time, the yield of methyl acetate produced in the presence of sulfonated polystyrene and perfluorinated resin sulfonic acid were 35 % and 12 %, respectively. The order of the catalytic activity was correspondent with that of the acid site density. Experimental data shown by dos Reis et al. indicated that textile structure of sulfonated polystyrene affected its catalytic activity [32]. They employed various sulfonated polystyrenes differing in the textile structure for transesterification of Babassu coconut oil with methanol. When the test reaction was carried out at 333 K, the catalytic activity was larger for the macro-reticular polystyrene than for the gelatinous one. Furthermore, they found that triglyceride consisting of short-chain fatty acid brought about the large transesterifying rate. Accordingly, it was considered that penetration of the reactants into the polymer particle was crucial for the catalytic activity. The most active catalyst caused 80 % of the conversion into FAME, but the methanol/oil molar ratio was very large, above 300, in their experiment. Vicente et al. carried out transesterification of sunflower oil with the methanol/oil molar ratio of 6 at 333 K in the presence of sulfonated macroporous polystyrene

[33]. The yield of FAME did not reach 1 % even after 8 hours of the reaction. Although a great effort to control accessibility to the acid site by modifying the textile structure were made for enhancing the catalytic activity [34], it is certain that the higher temperature is required for achieving the reasonable conversion efficiency. However, at the reaction temperature above 413 K, the sulfonated polystyrene is thermally deteriorated [17].

The reasonable utilization of sulfonated polystyrene might be to catalyze esterification of the free fatty acids contaminating the feedstock oil, in advance of the base-catalyzed transesterification. Since the free fatty acids poisons active phase of the base catalyst seriously, the preliminary treatment is advantageous to the conversion of the low-quality oil into biodiesel by the base-catalyzed transesterification. Since esterification and transesterification share a common molecular pathway, it is no wonder that sulfonated polystyrene is active in the esterification of fatty acid with methanol. Ozbay et al. appreciated that sulfonated polystyrene catalyzed the esterification of the free fatty acids contained in the waste cooking oil [35]. For their experiment, the waste cooking oil containing 0.5 % of the free fatty acid was employed. The esterification was carried out at 333 K under atmospheric pressure for 3 hours as a result and the conversion ratio ranged from 35 % to 46 %. Also, it was reported that an effect of internal mass transfer on the catalytic activity was small for the macroporous polystyrene [36]. Figure 5 shows the significant data to verify that the sulfonated polystyrene could be reused without the catalytic deactivation for the esterification of the free fatty acids. The waste cooking oil containing 2.5 % of the free fatty acids was mixed with methanol in the presence of sulfonated polystyrene with the macroreticular texture, and the test reaction was repeated three times without exchanging the catalyst. The conversion into FAME was over 90 % after 4 hours for the every runs.

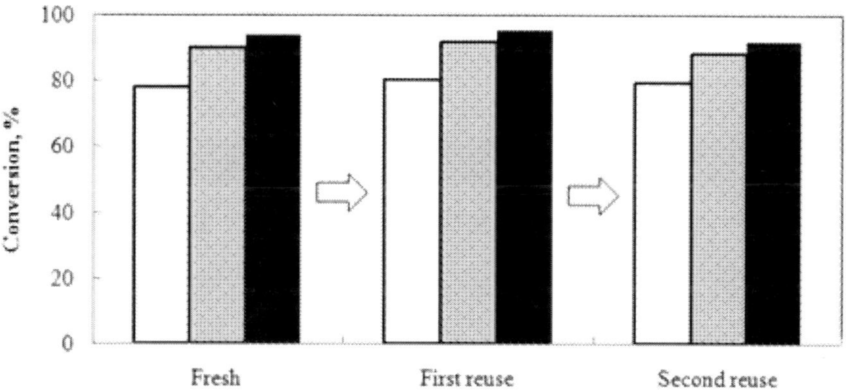

Figure 5. Conversions of the free fatty acid for three esterifying runs repeated with reusing sulfonated polystyrene catalyst. □; After 2 hours, ■; After 4 hours, ■; After 5 hours. The esterification was carried out at 333 K.

Liu et al. carried out the acid-catalyzed esterification followed by the base-catalyzed transesterification for rapeseed oil deodorizer distillate containing 49 % of the free fatty acids [37]. The acid-catalyzed esterification in the presence of sulfonated polystyrene resulted in 90 % of the conversion of the free fatty acids after 4 hours, and the residual triglycerides was transesterified into FAME in the presence of potassium hydroxide dissolved in methanol.

Talukder et al. proposed the novel process combining enzymatic hydrolysis of triglycerides with the following methanolysis of fatty acids for producing biodiesel [38]. The final product oil was measured up the standards of EN14214 and ASTM6751-03. The reaction kinetics on the esterification catalyzed by sulfonated polystyrene was investigated by Tesser et al. [39], and they showed that a pseudo-homogeneous second-order model agreed with the conversion of oleic acid mixed with soybean oil.

Perfluorinated resin sulfonic acid is characterized by the stronger acid property than sulfonated polystyrene [40]. However, the surface area is much smaller for perfluorinated resin sulfonic acid than for sulfonated polystyrene. With the view of enhancing the accessibility to the strong acid site, a new catalytic composite combining perfluorinated resin sulfonic acid with high surface silica was studied [41-42]. The solid acid composite was much more active than sulfonated polystyrene in the esterification of cyclopentadiene with acrylic acid: the product was obtained in 91 % yield for the solid acid composite and in < 5% yield for sulfonated polystyrene. Ni and Meunier studied utilization of the silica supported perfluorinated resin sulfonic acid for the preliminary treatment to catalytically convert the free fatty acid into FAME [43].

Other Solid Acids

Sulfonated carbon nano-graphite has the strong basic property combined with the enhanced durability to the thermal deterioration. Therefore, much attention has been attracted to sulfonated carbon nano-graphite as the solid acid catalyst appropriate for biodiesel production [44]. The superior catalytic properties originate in the carbon structure formed by incomplete carbonization of carbohydrates such as sugar and cellulose [45]. The small poly-cyclic aromatic carbon sheet in a three dimensional sp^3-bonded structure brings about generation of many hydroxyl and carboxyl groups. The carboxyl groups are effective in intensifying the chemical bonding between sulfonic groups and polycyclic aromatic carbon in the nano-graphite, which enhances stability of the active sites. The enhanced stability resulted in the thermal durability of the catalyst. The hydroxyl groups are useful in penetrating the reactant into the catalyst particle, which promotes the access to the catalytic sites. Okumura et al. prepared the sulfonated carbon nano-graphite with fuming sulfuric acid [45]. When the carbon nano-graphite was obtained by calcining at 673 K, the prepared catalyst was more active than Nafion perfluorinated resin sulfonic acid in esterification of acetic acid with ethanol. Also, the catalytic activity of the sulfonated carbon nano-graphite was measured by transesterification of triolein with ethanol at 353 K [46]. With the ethanol/oil molar ratio of 30, the solid base catalyst resulted in 24 % of the conversion into fatty acid ethyl esters (FAEE) after 6 hours. Under the same reacting condition, the yield of FAEE produced in the presence of Amberlyst-15, Nafion NR50 and SAC-13 were 2 %, 3% and 14 %, respectively. The similar data was shown in the research paper by Lou et al. [47]. Since the enhanced stability of the active phase led to the water tolerance in addition to the thermal durability, the sulfonated carbon nano-graphite is useful in catalyzing hydrolysis of cellulose, which is the primarily conversion to produce bioethanol and biochemicals [48].

Heteropolyacid such as 12-tungstophosphoric acid ($H_3PW_{12}O_{40}$) acts on the acid-catalyzed reaction, but is easy to dissolve in polar solvents. But, substitution of proton of $H_3PW_{12}O_{40}$ with large monovalent cation such as Cs^+ changes the soluble property to the insoluble one

[49]. Furthermore, the cation exchange results in marked increase of the surface area, which enhances the surface acidity. Since the enhancement of the surface acidity shows a maximum for $Cs_xH_{3-x}PW_{12}O_{40}$ compounds with x=2.5 [50], the heteropolyacids partially substituted with cesium-cation has been tested for the solid acid catalysis. Alsalme et al. showed on the basis of data from transesterification of short-chain FAEEs with methanol that no substituted heteropolyacid functions as the homogeneous catalyst, and that $Cs_{2.5}H_{0.5}PW_{12}O_{40}$ was tolerant to the leaching in the methanol [51]. Pesaresi et al. investigated an effect of degree of the cation exchange on the solid acid catalysis of $Cs_xH_{4-x}SiW_{12}O_{40}$ for transesterification of triglycerides [52]. The substituted heteropolyacid acted as the solid acid catalyst, above 1.3 Cs^+ loading. On the other hand, the transesterification was catalyzed in the homogeneous manner, below 0.8 Cs^+ loading. Katada et al. found that $H_4PNbW_{11}O_{40}/WO_3$-Nb_2O_5 was the novel heteropolyacid catalyst insoluble in methanol [53]. They measured the catalytic activity by transesterifying triolein with ethanol at 373 K, on an autoclave reactor. The catalytic activity was dependent of the calcination temperature: the novel heteropolyacid calcined at 773 K was most active in the transesterification. The yield of FAEE produced after 8 hours in the presence of the most active catalyst was 45 %. Also, the insoluble property originated in the combination of the heteropolyacid with WO_3-Nb_2O_5.

The solid acid catalysis of zeolite was shown in the research paper by Brito et al. [54]. They carried out transesterification of waste cooking oil with methanol at the temperature ranging from 473 K to 749 K. The conversion into FAME started above 573 K in the presence of Y-faujasite. Kiss et al. evaluated the catalytic activities of H-ZSM-5, Y and Beta by esterifying dodecanoic acid with 2-ethylhexanol at 403 K [55]. Regrettably, the conversion data indicated that all of the employed zeolites were catalytically inactive in the reaction.

Here, it should be noted that Esterfif-H^{TM}" is the process promising the utilization of solid acid catalyst for biodiesel production [13]. The promising process was developed by the French Institute of Petroleum and commercialized by Axens. Bournay et al. described a flow diagram of the "Esterfif-H^{TM}" process including two fixed bed reactor [56]. The heterogeneous catalytic process is able to convert vegetable oil into biodiesel whose properties meets the European standard, with the help of solid acid catalysis of a spinel mixed oxide of zinc and aluminium, at the temperature of 483-523 K under 3-5 MPa. Additionally, by-production of highly pure glycerol is characteristic of the heterogeneous catalytic process.

SOLID BASE CATALYSTS

Supported Alkali Metals

Alumina supported sodium was used at the opening of the research on the solid base catalyst [57]. In this research, the solid base catalysis for migration of double bond in a molecule of olefin resulted from electron donating ability of sodium. With a view of investigating the base-catalyzed reaction useful for the organic synthesis to obtain fine chemicals and pharmaceuticals, alumina supported potassium was employed for Knoevenagel condensation, aldolization and Micheal addition [58]. For these reactions, the Bronsted basicity allows proton of alpha position to be abstracted from carbonyl compounds.

In order to study the solid base catalyst for transesterification of triglycerides into biodiesel, Xie et al. prepared alumina supported potassium by impregnating potassium nitrate onto alumina. The obtained catalytic precursor was calcined at 773 K in advance of the activity measurement [59]. The catalytic activity was measured by transesterifying soybean oil at reflux of methanol. With the methanol/oil molar ratio of 15, the yield of FAME produced after 7 hours was around 90%. From IR spectrum and XRD pattern for the solid base catalyst, it was considered that potassium oxide produced by thermal decomposition of potassium nitrate and Al-O-K group were the catalytically active phase. The same catalyst was tested with jatropha oil by Vyas et al. [60]. The catalytic activity corresponded to experimental data mentioned above. As shown in Figure 6, our experimental data also verified that alumina supported potassium prepared from potassium nitrate is very active in the transesterification of soybean oil. When the soybean oil transesterification was conducted at reflux of methanol with the methanol/oil molar ratio of 12, the perfect conversion into FAME was achieved for 1 hour. Ebiura et al. impregnated potassium carbonate onto alumina in order to prepare the solid base catalyst for transesterification of triolein with methanol [61]. Their experimental data indicated that potassium carbonate was the source of the catalyst superior to potassium nitrate. Furthermore, the authors conducted double-bond isomerization of 2,3-dimethylbut-1-ene for evaluating the basic property and elucidated that the catalyst from potassium nitrate was the stronger base than that from potassium carbonate. Therefore, it was concluded that activation of carbonyl group in triolein was more important than the abstraction of proton from methanol to form nucleophile. There were research papers studying the supported potassium catalyst that was prepared with potassium iodide [62], potassium hydroxide [63] and potassium fluoride [64]. Regrettably, reuse of alumina supported potassium attended with its deactivation. According to a research paper by Alonso et al., the fresh catalyst led to the perfect conversion for 1 hour, whereas the yield of FAME decreased to 30 % by reusing the catalyst [65]. The catalytic deactivation was due to leaching of the active phase.

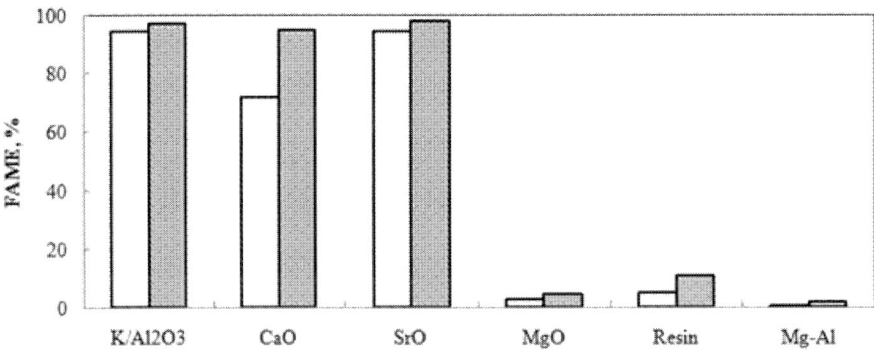

Figure 6. The yield of FAME produced by transesterifying soybean oil at reflux of methanol in the presence of alumina supported potassium (K/Al2O3), calcium oxide (CaO), strontium oxide (SrO), magnesium oxide (MgO), anion-exchange resin (Resin) and aluminum-magnesium mixed oxide from hydrotalcite (Mg-Al). □; After 0.5 hour, ■; After 1 hour.

Kim et al. prepared $Na/NaOH/Al_2O_3$ by impregnated melted sodium hydroxide and melted metal sodium onto alumina at the temperature of 593 K [66]. This catalyst is interpreted as a "superbase solid", because its basic strength is above H_- =37 [67]. The prepared catalyst was employed for transesterification of soybean oil with methanol/oil molar ratio of 9. At the temperature of 333 K, the yield of FAME produced after 2 hours reached 90 %. The authors concluded that both the sodium aluminate generated by reaction of alumina with sodium hydroxide and the ionization of sodium generated served as the stronger basic site.

Alkaline Earth Metal Oxides

When alkaline earth metal oxides were employed for transesterification of soybean oil with refluxing methanol, as shown in Figure 6, the order of the catalytic activity was in the sequence of strontium oxide > calcium oxide > magnesium oxide [68]. This corresponded to the order of the basic property determined by temperature-programmed desorption of carbon dioxide [69]. From the economical point of view, calcium oxide is a candidate for the solid base catalyst.

The solid base catalysis of alkaline earth metal oxides originate in their ionic crystal structure [70]. Since Lewis acidity of the metal cation is very weak due to its small elecronegativity, the conjugated oxygen anion is easy to attract acidic substance chemically. When the carbonyl compound comes into contact with surface of the metal oxide, the surface oxygen anion functioned as Bronsted base that abstracts proton of alpha position from the carbonyl compound. Therefore, alkaline earth metal oxides acted on the base-catalyzed reaction such as aldol reaction and Micheal addition [71-72]. In the case of transesterification of triglycerides with methanol, nucleophile attacking carbonyl carbon in a molecule of triglyceride was generated by abstracting proton from methanol. Basic property of the surface oxygen anion is dependent of its coordination number: the surface oxygen anion of low coordination number brings about the strong basic property, because ionic force bonding the oxygen anion with calcium cations becomes weak. However, the disordered surface abundant in the ion pairs of low coordination number is unstable. Therefore, calcination to prepare alkaline earth metal oxides should be carried out at as possible as low temperature. On the other hand, the high calcination temperature is required for complete elimination of carbon dioxide and water that interact with the strong basic sites.

Not only the calcination temperature but also handling after the calcination is very important for preparation of the active catalyst. Granados et al. investigated properties of calcium oxide by some instrumental methods, XRD, IR EGA-MS, and XPS [73]. The characterization data indicated that surface of calcium oxide was rapidly hydrated and carbonated by exposing it to air. The exposed catalyst resulted in the slow transesterification, while evacuation at the high temperature was effective in activating the exposed catalyst. The catalytic activity was measured by transesterifying sunflower oil at 333 K with the methanol/oil molar ratio of 13, and the yield of FAME produced after 1.5 hours increased from 60% to 95% by evacuating calcium oxide at 973 K. Additionally, they pointed out that the sunflower oil transesterification was catalyzed by not only surface of calcium oxide but also the soluble fraction leached away from calcium oxide. Also, the leaching of the catalyst was observed in our previous research paper [74]. The leached substance was concentrated in the glycerol phase after the transesterification: calcium contents of oil and glycerol phases were 139 ppm and 4602 ppm, respectively.

Reusability of calcium oxide was noted in the research papers by Liu et al. [75] and Ngamcharussrivichai et al. [76]. Both of the articles elucidated that calcium oxide could be reused without the serious deactivation for the reactions repeated several times. In our research work to evaluate the reusability, the reused catalyst consisted of the other calcium compound [77]. Since the calcium compound was identified as calcium glyceroxide by measuring XRD, it was evident that calcium oxide was combined with glycerol of the by-product. The test reaction transesterifying soybean oil was performed at reflux of methanol with the methanol/oil molar ratio of 12, and the reused catalyst resulted in around 90% of the conversion into FAME after 2 hours. Under the same reacting condition, the conversion of soybean oil in the presence of the fresh catalyst consisting of calcium oxide was approximately perfect. Interestingly, the catalytic deactivation by exposing to air was not obvious at all for calcium glyceroxide, as shown in Figure 7. Although it was deduced at that time that calcium glyceroxide functioned as the solid base catalyst, the deduction did not agree with data from our later work [78].

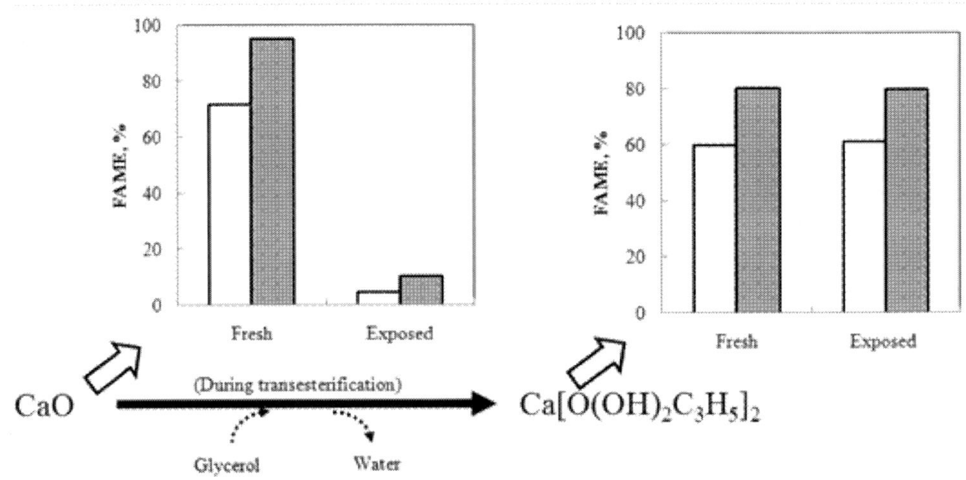

Figure 7. Enhancement of tolerance to air-exposure by transforming calcium oxide into calcium glyceroxide during the transesterification. Catalytic activity was measured by transesterifying soybean oil at reflux of methanol. □; FAME yield measured after 0.5 hour, ■; FAME yield measured after 1 hour.

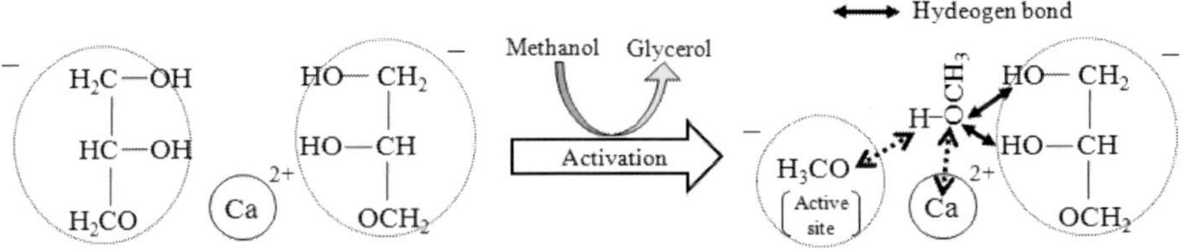

Figure 8. Catalytic mechanism of calcium glyceroxide for transesterification of triglycerides with methanol.

The later work provided data showing that calcium glyceroxide was activated by reacting with methanol under the transesterifying condition, and that chemical composition of the catalytically active phase was estimated at CH$_3$O-Ca-[O(OH)$_2$C$_3$H$_5$]: the methoxyl anion and the glyceroxyl one were bonded with calcium cation. Possibly, the glyceroxyl OH groups attracted methanol with the help of hydrogen bond, which enhanced accessibility to the basic site. This catalytic mechanism is illustrated by giving Figure 8. Taking the experimental results from a series of our works into consideration, we prepared the practical catalyst from lime stone whose size distribution was regulated within the range of 1.0 mm to 2.0 mm [79]. On a batch reactor unit consisting of a circulating loop passing through a glass column reactor packed with the practical catalyst, rapeseed oil was transesterified at 333 K with the methanol/oil molar ratio of 20. After 2 hours of the transesterifying operation, the yield of FAME reached 97%.

Since the heterogeneous catalytic process is not at all equal to the homogeneous catalytic process in the reaction rate, it is important to enhance the catalytic activity of calcium oxide. Watkins et al. doped lithium into calcium oxide for the purpose of intensifying the basic property, and reported that transesterification of glyceryl tributyrate to methyl butyrate was promoted by the intensified basic property [80]. The same doping effect was reported in the research paper by Alonso et al. [81]. The doping effect is possibly due to the electron-deficient Li^+ by the exchange of lattice Ca^{2+} for Li^+ and/or to the evolution of surface hydroxyl groups at the defect sites resulting from a reaction lithium-salt with calcium oxide [82]. Also, loading fine particle of calcium oxide onto the proper supporting material is one of the ways to enhance the catalytic activity. Umdu et al. prepared calcium oxide loaded on alumina from aluminum isopropoxide and calcium nitrate [83]. The catalytic precursor was calcined at 723 K, and then was employed for transesterification of *Nannochloropsis oculata* microalga's lipid with methanol. Albuquerque et al., who paid their attention to mesoporous materialsl, synthesized SBA-15 mesoporous silica as the supporting material [84]. Calcium oxide loaded on SBA-15 was tested by methanolysis of ethyl butyrate heated up to 333 K. In the presence of the catalyst including 14 wt% of calcium oxide, 50 % of ethyl butyrate was yielded after 2 hours. Also, increase in the amount of calcium oxide dispersed on the supporting material caused the increase in the conversion ratio. For loading a large amount of calcium oxide onto silica, Yamanaka et al. applied the industrial technology to manufacture nano-sized calcium carbonate [85]. The catalytic precursor was prepared by reaction of calcium hydroxide in its aqueous suspension with carbon dioxide, in the presence of silica gel powder. An amount of the loaded calcium compound and the calcination temperature were crucial for the catalytic activity.

Additionally, the waste matters consisting of calcium carbonate could be used as the raw material of the catalyst. In the research paper by Nakatani et al., calcium oxide prepared from oyster shell was employed for transesterification of soybean oil [86]. The catalytic transesterification was carried out at 338 K with the methanol/oil molar ratio of 6, and the yield of FAME produced after 4 hours was 95%. The similar results were shown in the research works that employed eggshell, Golden apple snail shell, Meretrix venus shell and Mud crab. [87-89]. It seemed that surface area was smaller for calcium oxides deriving from the waste matters than for that from lime stone. Therefore, a large amount of the catalyst was required for the reasonable converting efficiency.

Hydrotalcite

Hydrotalcites is a clay mineral having the unit cell formula $Mg_6Al_2(OH)_{16}CO_3 \cdot 4H_2O$. The structure of hydrotalcite is based upon the layered double hydroxides with brucite like $Mg(OH)_2$ hydroxide in which magnesium cations are octahedorally coordinated by hydroxyl anions, and the hydroxide layers are stacked with sharing their edges [90-91]. A part of magnesium cation is replaced with aluminum cation, which causes positive charge of the hydroxide layers. The positive charge is balanced with carbonate anion and water molecule, which are situated in the interlayer between the stacked brucite-like layers. By calcination at high temperature, hydrotalcite is transformed into a mixed magnesium-aluminum oxide that acts on the base-catalyzed reaction. The solid base catalysis of the calcined hydrotalcite originates in the isolated oxygen anion (strong basic site), the structural oxygen anion on the surface (medium basic site), and the surface hydroxyl group (weak basic site) [92]. Corma et al. studied the catalytic activity of the calcined hydrotalcite for the Knoevenagel condensation of benzaldehyde with activated methylenic compounds with different pKa values, and concluded that maximum strength of the basic sites was H_ up to 16.5 [93]. Moreover, the structural aluminum-oxygen pair gives moderate Lewis acidity to the calcined hydrotalcite [94]. It was reported that the acid-base property of the calcined hydrotalcite is effective in catalyzing cycloadditions of carbon dioxide to epoxides [95] and aldol condensations to form nonenal [96].

The catalytic activity of the calcined hydrotalcite for conversion of triglycerides into biodiesel was investigared in the research paper by Xie et al. [97]. They prepared the catalytic precursor by adding sodium hydroxide and carbonate into an aqueous solution of magnesium nitrate and aluminium nitrate, and then calcined the catalytic precursor at 773 K. XRD measurement appreciated that the calcined hydrotalcite consisted of magnesia, alumina and their solid solution.

When soybean oil was transesterified at reflux of methanol with the methanol/oil molar ratio of 15, 7.5% of the catalyst caused 65% of the conversion into biodiesel after 9 hours. The catalytic activity of the calcined hydrotalcite was markedly intensified by combining it with potassium compound such as potassium fluoride [98]. Additionally, Helwani et al. studied the catalytic modification by incorporating trivalent cation such as Fe^{3+} into lattice of the calcined hydrotalcite [99].

There is a research paper pointing out that hydrotalcite prepared by the conventional procedure is contaminated with alkali used as the precipitant [100]. The alkali contamination for the catalyst causes the homogeneous catalytic reaction, which disturbs the evaluation of the solid base catalysis. With a view of solving this problem, Cantrell et al. examined the alternative way to synthesis hydrotalcite [101].

In their way, hydrotalcite was precipitated with ammonium hydroxide and carbonate in the aqueous phase, and then calcined at 723K. The catalytic activity of the calcined hydrotalcite was measured by transesterifying glycerol tributyrate at 333K for 3 hours. Also, the authors investigated the effect of an amount of trivalent aluminum cation on the basic property. An increase in the amount of trivalent aluminum cation weakened the basic property of the calcined hydrotalcites. Additionally, the order of the basic properties corresponded to that of the catalytic activity.

Other Solid Bases

Some organic bases such as guanidine are catalytically active in transesterification of triglycerides with methanol [102]. Schuchardt et al. prepared alkyl guanidines grafted onto polystyrene for investigating their catalytic activity. When transesterification of soybean oil was carried out at 343K with the methanol/oil molar ratio of 27 in the presence of the prepared catalyst, the yield of FAME produced after 1 hour ranged from 60 % to 80 % [103]. However, the catalytic activity was reduced by reusing the organic solid base due to leaching of the grafted active phase. Feria et al. grafted tetramethylguanidine onto silica gel with 3-chloropropyltrithoxysilane [104]. After tetramethylguanidine was combined with 3-chloropropyltrithoxysilane, the synthesized silylant agent was attached to surface of silica gel. The catalytic activity of the organic base-silica gel composite was measure by transesterifying soybean oil with the methanol/oil molar ratio of 4 at 353K.

Trimethylamine grafted on polystyrene is a commercially available solid organic base, which is termed as "anion-exchange resin". Shibaski-Kitakawa et al. investigated the catalytic activity of the anion-exchange resin by transestrifying triolein with ethanol at 323K [105]. The anion-exchange resin characterized by a lower cross-linking density and a smaller particle size led to the fast transesterification. Although the oleic acid deriving from triolein deteriorated the catalytic site, the catalytic activity returned to the original level by washing with citric acid and alkali hydroxide solution.

Among a variety of zeolites, faujasite is transformed into a solid base catalyst by modifying with alkali metal cation. The origin of the basicity is considered to be the oxygen anion in the zeolite framework [106]. The basic strength of the alkali metal ion-exchanged zeolite increases with increasing electropositivity of the exchanged cation: $Li^+ < Na^+ < K^+ < Cs^+$. XPS data showed that the binding energy of O1s core level of the alkali metal ion-exchanged zeolite was in the sequence of $Li^+ > Na^+ > K^+ > Cs^+$ [107]. The order of the binding energy corresponds to increase in the electron density of the core level, which verified the order of the basic strength. Concerning the parent zeolite, X-faujasite provided the strong basic property than Y-faujasite. Leclercq et al. employed Cs-exchanged X-faujasites for the methanolysis of rapeseed oil [108]. The test reaction was conducted at reflux of methanol with the methanol/oil molar ratio of 275 for 22 hours with the result that the solid base catalyst gave 70% of the conversion. Suppes et al. carried out the test reaction with the methanol/oil ratio of 6 at the temperature of 333 K [109]. The yield of FAME produced in the presence of the same catalyst after 24 hours was only 7%. Under the same reacting condition, ETS-10 zeolite containing titanium resulted in 80% of the conversion into FAME, due to the strong basic property.

CONCLUSION

For the purpose of solving the problems concerning the utilization of homogeneous base catalysis for current production of biodiesel, many researchers have made great efforts to study the heterogeneous catalytic process to transesterify triglycerides composing vegetable oil into FAME. The heterogeneous catalysts tested in the concerning research papers are classified into two types: solid acids and solid bases. Solid acids are less active than solid base in catalyzing the transesterification, while the acid-catalyzed transesterification is advantageous to the conversion of the low-quality oil containing a large quantity of the free fatty acids. Since solid acids catalyzes esterification of the free fatty acids with methanol, the two step heterogeneous acid-base reactions system is one of the promising ways to convert a variety of the feedstock oil into biodiesel under the mild operating condition. However, the heterogeneous catalytic reaction is slow as compared to homogeneous one utilized for the existent process. Moreover, soluble substance is leached away from the heterogeneous catalyst under the transesterifying condition. These are the major study assignments for the practical use of the heterogeneous catalytic process, and the further research works are necessary for understanding of the catalytic mechanism. Characterization of the active sites, study on the catalytic material and theoretical quantification of the reaction kinetics will provide the significant data not only to modify the heterogeneous catalyst but also to design the appropriate process.

ACKNOWLEDGMENTS

The authors gratefully acknowledge support for this research by Kyoto Environmental Nanotechnology Cluster from Ministry of Education, Culture, Sports, Science and Technology.

REFERENCES

[1]   Knothe, G. (2010). Biodiesel: Current trends and properties. *Topics in catalysis, 53*, 714-720.

[2]   Freedman, B., Pryde, E. H. and T. L. Mounts. (1984) Variables affecting the yield of fatty esters from transesterified vegetable oils. *Journal of American Oil Chemist's Society, 61*, 1638-1643.

[3]   Helwani, Z., Othman M.R., Aziz, N., Kim, J. and Fernando, W. J. N. (2009). Solid heterogeneous catalysts for transesterification of triglycerides with methanol: a review. *Applied Catalysis A: General, 363*, 1-10.

[4]   Zabeti, M., Daud, W. M. A. W. and Aroua, M. K. (2009). Activity of solid catalyst for biodiesel production: a review. *Fuel Processing Technology, 90*, 770-777.

[5]   Mittelbach, M. (1990). Lipase catalyzed alcoholysis of sunflower oil. *Journal of the American Oil Chemists' Society, 67*, 168-170.

[6]   Kaieda, M., Samukawa, T., Matsumoto, T., Ban, K., Kondo, A., Shimada, Y., Noda, H., Nomoto, F., Ohtsuka, K., Izumoto, E. and Fukuda, H. (1999). Biodiesel fuel production from plant oil catalyzed by *Rhizopus oryzae* lipase in a water-containing system without an organic solvent. *Journal of Bioscience and Bioengineering, 88*, 627-631.

[7]   Watanabe, Y., Shimada, Y., Sugihara, A. and Tominaga, T. (2002). Conversion of degummed soybean oil to biodiesel fuel with immobilized *Candida antarctica* lipase. *Journal of Molecular Catalysis B: Enzymatic, 17*, 151-155.

[8]   Nelson, L. A., Foglia, T. A. and Marmer, W. N. (1996). Lipase-catalyzed production of biodiesel. *Journal of the American Oil Chemists' Society, 73*, 1191-1194.

[9]   Watanabe, Y., Shimada, Y., Sugihara, A. and Tominaga, Y. (2001). Enzymatic conversion of waste edible oil to biodiesel fuel in a fixed-bed bioreactor. *Journal of the American Oil Chemists' Society. 78*, 703-707.

[10]   Oda, M., Kaieda, M., Hama, S., Yamaji, H., Kondo, A., Izumoto, E. and Fukuda, H. (2005). Facilitatory effect of immobilized lipase-producing *Rhizopus oryzae* cells on acyl migration in biodiesel-fuel production. *Biochemical Engineering Journal, 23*, 45-51.

[11]   Kusudiana, D. and Saka, S. (2001). Kinetics of transesterification in rapeseed oil to biodiesel fuel as treated in supercritical methanol. *Fuel, 80*, 693-698.

[12]   Kusudiana, D. and Saka, S. (2004). Effect of water on biodiesel fuel production by supercritical methanol treatment. *Bioresource Technology, 91*, 289-295.

[13]   Kusudiana, D. and Saka, S. (2004). Two-step preparation for catalyst-free biodiesel fuel production: hydrolysis and methyl esterification. *Applied Biochemistry and Biotechnology, 115*, 781-794.

[14]   Yan, S., DiMaggio, C., Mohan, S., Kim, M., Salley, S. O. and Ng, K. Y. S. (2010). Advancement in heterogeneous catalysis for biodiesel production. *Topics in Catalysis, 53*, 721-736.

[15]   Ma, F. and Hanna, H. A. (1999). Biodiesel production: a review. *Bioresource Technology, 70*, 1-15.

[16]   Kabashima, H., Katou, T. and Hattori, H. (2001). Conjugate addition of methanol to 3-buten-2-one over solid base catalysts. *Applied Catalysis A: General, 214*, 121-124.

[17]   Hattori, H., Shima, M., and Kabashima, H. (2000). Alcoholysis of ester and epoxide catalyzed by solid bases. *Studies in Surface Science and Catalysis, 130*, 3507-3512.

[18]   Lotero, E., Liu, Y., Lopez, E., Suwannakarn, K., Bruce, D. A. and Goodwin Jr., J. G. (2005). Synthesis of biodiesel via acid catalysis. *Industrial and Engineering Chemistry Research, 44*, 5353-5363.

[19]   Stumborg, M. Wong, A. and Hogan, E. (1996). Hydroprocessed vegetable oils for diesel fuel improvement. *Bioresource Technology, 56*, 13-18.

[20]   Bezergianni, S., Voutetakis, S. and Kalogianni, A. (2009). Catalytic hydrocracking of fresh and used cooking oil. *Industrial and Engineering Chemistry Research, 48*, 8402-8406.

[21]   Huber, G. W., O'Connor, P. O. and Corma, A. (2007). Processing biomass in conventional oil refineries: production of high quality diesel by hydrotreating vegetable oil in heavy vacuum oil mixture. *Applied Catalysis A: General, 329*, 120-129.

[22]   Arata, K. and Hino, M. (1990). Preparation of superacids by metal oxides and their catalytic action. *Materials Chemistry and Physics, 26*, 213-237.

[23]   Yamaguchi, T., Jin, T. and Tanabe, T. (1986). Structure of acid sites on sulfur-promoted iron oxide. The *Journal of Physical Chemistry, 90*, 3148-3152.

[24]   Jin, T., Yamaguchi, T. and Tanabe, T. (1986). Mechanism on acidity generation on sulfur-promoted metal oxide. *The Journal of Physical Chemistry, 90*, 4794-4796.

[25]   Lopez, D. E., Goodwin Jr., J. G., Bruce, D. A. and Lotero, E. (2005). Transesterification of triacetin with methanol on solid acid and base catalysts. *Applied Catalysis A: General, 295*, 97-105.

[26]   Omota, F., Dimian, A. C. and Bliek, A. (2003). A fatty acid esterification by reactive distillation : part 2 kinetics-based design for sulphated zirconia catalyst. *Chemical Engineering Science, 58*, 3175-3185.

[27] Jitputti, J., Kitiyanan, B., Rangsunvigit, P., Bunyakiat, K., Attanatho, L. and Jenvanitpanjakul, P. (2006). Transesterification of crude palm kernel oil and crude coconut oil by different solid catalysts. *Chemical Engineering Journal, 116,* 61-66.

[28] Yadav, G. D. and Murkute, A. D. (2004). Preparation of a noble catalyst UDCaT-5: enhancement in activity of acid-treated zirconia, effect of treatment with chlorosulfonic acid vis-à-vis sulfuric acid. J*ournal of Catalysis, 224,* 218-223.

[29] Chen, X. R., Ju, Y. H. and Mou, C. Y. (2007). Direct synthesis of mesoporous sulfatedsilica-zirconia catalysts with high catalytic activity for biodiesel via esterification. *The Journal of Physical Chemistry C, 111,* 18731-18737.

[30] Furuta, S., Matsuhashi, H. and Arata, K. (2004). Biodiesel fuel production with solid superacid catalysis in fixed bed reactor under atmospheric pressure. *Catalysis Communications, 5,* 721-723.

[31] Ramu, S., Lingaiah, N., Prabhavashi Devi, B. L. A., Prasad, R. B. N., Suryanarayana, I. and Prasad, P. S. S. (2004). Esterification of palmtic acid with methanol over tungsten oxide supported on zirconia solid acid catalysts: effect of method of preparation of the catalyst on its structural stability and reactivity. *Applied Catalysis A: general, 276,* 163-168.

[32] Yadav, G. D. and Thathagar, M. B. (2002). Esterification of maleic acid with ethanol over cation-exchange resin catalysts. *Reactive and Functional Polymers, 52,* 99-110.

[33] dos Reis, S. C. M., Lacher, E. R., Nascimento, R. S. V., Rodrigues Jr., J. A. and Reid, M. G. (2005). Transesterification of Brazilian vegetable oils with methanol over ion-exchange resins. *Journal of the American Oil Chemists' Society, 82,* 661-665.

[34] Vicente, G., Coteron, A., Martinez, M. and Aracil, J. (1998). Application of the factorial design of experiments and response surface methodology to optimize biodiesel production. *Industrial Crops and Products, 8,* 29-35.

[35] de Rezende, S. M., de Castro Reis, M., Reid, M. G., Silva Jr., P. L., Coutinho, F. M. B., da Siliva San Gil and Lachter, E. R. (2008). Transesterification of vegetable oils promoted by poly (styrene-divinylbenzene) and poly (divinylbenzen). *Applied Catalysis A: General, 349,* 198-203.

[36] Ozbay, N., Oktar N. and Tapan, N. A. (2008). Esterification of free fatty acids in waste cooking oils (WCO): Role of ion-exchange resins. *Fuel, 87,* 1789-1798.

[37] Russbueldt, B. M. E. and Hoelderich, W. F. (2009). New sulfonic acid ion-exchange resins for the preesterification of different oils and fats with high content of free fatty acids. *Applied Catalysis A: General, 362,* 47-57.

[38] Liu, Y., Wang, L. and Yan, Y. (2009). Biodiesel synthesis combining pre-esterification with alkali catalyzed process from rapeseed oil deodorizer distillate. *Fuel Processing Technology, 90,* 857-862.

[39] Talukder, M. R., Wu, J. C. and Chua, L. P-L. (2010). Conversion of waste cooking oil to biodiesel via enzymatic hydrolysis followed by chemical esterification. *Energy and Fuels,24,* 2016-2019.

[40] Tesser, R., Di Serio, M., Guida, M., Nastasi, N. and Santacesaria, E. (2005). Kinetics of oleic acid esterification with methanol in the presence of triglycerides. *Industrial and Engineering Chemistry Research, 44,* 7978-7982.

[41] Okuhara, T. (2002). Water-tolerant solid acid catalysts. *Chemical Reviews, 102,* 3641-3666.

[42] Harmer, M. A., Sun, Q. and Farneth, W. E. (1996). High surface area Nafion

resin/silica nanocomposites: a new class of solid acid catalyst. *Journal of the American Chemical Society*, *118*, 7708-7715.

[43] Harmer, M. A. and Sun, Q. (2001). Solid acid catalysis using ion-exchange resins. *Applied Catalysis A: General*, *221*, 45-62.

[44] Ni, J. and Meunier, F. C. (2007). Esterification of free fatty acids in sunflower oil over solid acid catalysts using batch and fixed bed reactors. *Applied Catalysis A: General*, *333*, 122-130.

[45] Toda, M., Takagaki, A., Okumura, M., Kondo, J. N., Domen, K., Hayashi, S. and Hara, M. (2005). Green chemistry: biodiesel made with sugar catalyst. *Nature*, *438*, 178-178.

[46] Okumura, M., Takagaki, A., Toda, M., Kondo, J. N., Domen, K., Hayashi, S. and Hara, M. (2006). Acid-catalyzed reactions on flexible polycyclic aromatic carbon in amorphous carbon. *Chemistry of Materials*, *18*, 3039-3045.

[47] Hara, M. (2010). Biodiesel production by amorphous carbon bearing SO_3H, COOH and phenolic OH groups, a solid Bronsted acid catalyst. *Topics in Catalysis*, *53*, 805-810.

[48] Lou, W. Y., Zong, M. H. and Duan, Z. Q. (2008). Efficient production of biodiesel from high free fatty acid-containing waste oils using various carbohydrate-derived solid acid catalysts. *Bioresource Technology*, *99*, 8752-8756.

[49] Suganuma, S., Nakajima, K., Kitano, M., Yamaguchi, D., Kato, H., Hayashi, S. and Hara, H. (2008). Hydrolysis of cellulose by amorphous carbon bearing SO_3H, COOH and OG groups. *Journal of the American Chemical Society*, *130*, 12787-12793.

[50] Tatematsu, S., Hibi, T., Okuhara, T. and Misono, M. (1984). Preparation process and catalytic activity of $Cs_xH_{3-x}PW_{12}O_{40}$. *Chemistry Letters*, *13*, 865-868.

[51] Okuhara, T., Nishimura, T., Watanabe, H. and Misono, M. (1992). Insoluble heteropoly compounds as highly active catalysts for liquid phase reactions. *Journal of Molecular Catalysis*, *74*, 247-256.

[52] Alsalme, A., Kozhevnikova, E. F. and Kozhevnikova, I. V. (2008). Heterolpoly acids as catalysts for liquid-phase esterification and transesterification. *Applied Catalysis A: General*, *349*, 170-176.

[53] Pesaresi, L., Brown, D. R., Lee, A. F., Montero, J. M., Williams, H. and Wilson, K. (2009). Cs-doped $H_4SiW_{12}O_{40}$ catalysts for biodiesel applications. *Applied Catalysis A: General*, *360*, 50-58.

[54] Katada, N., Hatanaka, T., Ota, M., Yamada, K., Okumura, K. and Niwa, M. (2009). Biodiesel production using heteropoly acid-derived solid acid catalyst $H_4PNbW_{11}O_{40}/WO_3-Nb_2O_5$. *Applied Catalysis A: General*, *363*, 164-168.

[55] Brito, A., Borges, M. E. and Otero, N. (2007). Zeolite Y as a heterogeneous catalyst in biodiesel fuel production from used vegetable oil. *Energy and Fuels*, *21*, 3280-3283.

[56] Kiss, A. A., Dimian, A. C. and Rothenberg, G. (2006). Solid acid catalysts for biodiesel production – towards sustainable energy. *Advanced Synthesis and Catalysis, 348*, 75-81.

[57] Bournay, L., Casanava, D., Delfort, B., Hillion, G. and Chodorge, J. A. (2005). New heterogeneous process for biodiesel production: a way to improve the quality and the value of the crude glycerin produced by biodiesel plant. *Catalysis Today*, *106*, 190-192.

[58] Pines, H., Vesely, J. A. and Ipatieff, V. N. (1955). Migration of double bond in olefinic and diolefinic hydrocarnons catalyzed by sodium. Dehydration of alpha-limonene and p-cymene. *Journal of the American Chemical Society*, *77*, 347-348.

[59] Climent, M. J., Corma, A., Iborra, S. and Velty, A. (2002). Designing the adequate base

solid catalyst with Lewis or Bronsted basic sites or with acid-base pairs. *Journal of Molecular Catalysis A: Chemical, 182-183*, 327-342.

[60] Xie, W., Peng, H. and Chen, L. (2006). Transesterification of soybean oil catalyzed by potassium loaded on alumina as a solid-base catalyst. *Applied Catalysis A: General, 300*, 67-74.

[61] Vyas, A. P., Subrahmanyam, N. and Patel, P. A. (2008). Production of biodiesel through transesterification of jatropha oil using KNO_3/Al_2O_3 solid catalyst. *Fuel, 88*, 625-628.

[62] Ebiura, T., Echizen, T., Ishikawa, A., Murai, K. and Baba, T. (2005). Selective transesterification of triolein with methanol to methyl oleate and glycerol using alumina loaded with alkali metal salt as a solid-base catalyst. *Applied Catalysis A: General, 283*, 111-116.

[63] Xie, W. and Li, H. (2006). Alumina supported potassium iodide as a heterogeneous catalyst for biodiesel production from soybean oil. *Journal of Molecular Catalysis A: Chemical, 255*, 1-9.

[64] Ilgen, O. and Akin, A. N. (2009). Transesterification of canola oil to biodiesel using MgO loaded with KOH as a heterogeneous catalyst. *Energy and Fuels, 23*, 1786-1789.

[65] Xie, W. and Huang, X. (2006). Synthesis of biodiesel from soybean oil using heterogeneous KF/ZnO catalyst. *Catalysis Letters, 107*, 53-59.

[66] Alonso, D. M., Mariscal, R., Moreno-Tost, R., Poves, M. D. Z. and Granados, M. L. (2007). Potassium leaching during triglyceride transesterification using K/gamma-alumina catalysts. *Catalysis Communications, 8*, 2074-2080.

[67] Kim, H. J., Kang, B. S., Kim, M. J., Park, Y. M., Kim, D. K., Lee, J. S. and Lee, K. Y. (2004). Transesterification of vegetable oil to biodiesel using heterogeneous base catalyst. *Catalysis Today, 93-95*, 315-320.

[68] Suzukamo, G., Fukao, M. and Minobe, M. (1987). Preparation of new superbase and its catalytic activity. *Chemistry Letters, 16*, 585-588.

[69] Kouzu, M., Kasuno, T., Tajika, M., Sugimoto, Y., Yamanaka S. and Hidaka, J. (2008). Calcium oxide as a solid base catalyst for transesterification of soybean oil and its application to biodiesel production. *Fuel, 87*, 2798-2806.

[70] Zhang, G., Hattori, H. and Tanabe, K. (1988). Aldol addition of acetone, catalyzed by solid base catalysts: magnesium oxide, calcium oxide, strontium oxide, barium oxide, Lanthanum oxide and zirconium oxide. *Applied Catalysis, 36*, 189-197.

[71] Hattori, H. (2001). Solid base catalysts: generation of basic sites and application to organic synthesis. *Applied Catalysis A: General, 222*, 247-259.

[72] Matsuhashi, H. (2009). Synthesis of novel solid base of MgO covered with methanol oxide. *Topics in Catalysis, 52*, 828-833.

[73] Kabashima, H., Tsuji, H. and Hattori, H. (1997). Micheal addition of methyl crotonate over solid base catalysts. *Applied Catalysis A: General, 165*, 319-325.

[74] Granados, M. L., Poves, M. D. Z., Alonso, D. M., Mariscal, R., Galisteo, F. C., Moreno-Tost, R., Santamaria J. and Fierro, J. L. G. (2007). Biodiesel from sunflower oil by using activated calcium oxide. *Applied Catalysis B: Environment, 73*, 317-326.

[75] Kouzu, M., Yamanaka, S., Hidaka J. and Tsunomori, M. (2009). Heterogeneous catalysis of calcium oxide used for transesterification of soybean oil with refluxing methanol. *Applied Catalysis A: General, 355*, 94-99.

[76] Liu, X., He, H., Wang, Y., Zhu S. and Piao, X. (2008). Transesterification of soybean

oil to biodiesel using CaO as a solid base catalyst. *Fuel, 87*, 216-221.

[77] Ngamcharussrivichai, C., Nunthasanti, P., Tanachai, S. and Bunyakiat, K. (2010). Biodiesel production through transesterification over natural calciums. *Fuel Processing Technology, 91*, 1409-1415.

[78] Kouzu, M., Kasuno, T., Tajika, M., Yamanaka, S. and Hidaka, J. (2008). Active phase of calcium oxide used as solid base catalyst for transesterification of soybean oil with refluxing methanol. *Applied Catalysis A: General, 334*, 357-365.

[79] Kouzu, M., Hidaka, J., Wakabayashi, K. and Tsunomori, M. (2010). Solid base catalysis of calcium glyceroxide for a reaction to convert vegetable oil into its methyl esters. *Applied Catalysis A: General, 390*, 11-18.

[80] Kouzu, M., Hidaka, J., Komichi, Y., Nakano H. and Yamamoto, M. (2009). A process to transesterify vegetable oil with methanol in the presence of quick lime bit functioning as solid base catalyst. *Fuel, 88*, 1983-1990.

[81] Watkins, R. S., Lee A. F. and Wilson, K. (2004). Li-CaO catalyzed tri-glyceride transesterification for biodiesel applications. *Green Chemistry, 6*, 335-340.

[82] Alonso, D. M., Mariscal, R., Granados M. L. and Maireles-Torres, P. (2009). Biodiesel preparation using Li/CaO catalysts: Activation process and homogeneous contribution. *Catalysis Today, 143*, 167-171.

[83] Meher, L. C., Kulkarni, M. G., Dalai A. K. and Naik, S. N. (2006). Transesterification of karanja (*Pongamia pinnata*) oil by solid base catalysts. *European Journal of Lipid Science and Technology, 108*, 389-397.

[84] Umdu, E. S., Tuncer M. and Seker, E. (2009). Transesterification of *Nannochloropsis oculata* microalga's lipid to biodiesel on Al_2O_3 supported CaO and MgO catalysts. *Bioresource Technology*, 100, 2828-2831.

[85] Albuquerque, M. C. G., Jimenez-Urbistondo, I., Santamaria-Gonzalez, J., Merida-Robles, J. M., Moreno-Tost, R., Rodriguez-Castellon, E., Jimenez-Lopez, A., Azevedo, D. C. S., Cavalcante Jr., C. L. and Maireles-Torres, P. "CaO supported mesoporous silicas as basic catalysts for transesterification reactions. *Applied Catalysis A: General, 334*, 35-43.

[86] Yamanaka, S., Kouzu, M., Kadota, K., Shimosaka, A., Shirakawa Y. and Hidaka, J. (2007). Catalysis by CaO/SiO_2 composite particle for biodiesel production. *Kagaku Kougaku Ronbunsyu, 33*, 483-489.

[87] Nakatani, N., Takamori, H., Takeda K. and Sakugawa, H. (2009). Transesterification of soybean oil using combusted oyster shell waste as a catalyst. *Bioresource Technology, 100*, 1510-1513.

[88] Wei, Z., Xu, C. and Li, B. (2009). Application of waste eggshell as low-cost solid base catalyst for biodiesel production. *Bioresource Technology, 100*, 2883-2885.

[89] Boey, P. L., Maniam G. P. and Hamid, S. A. (2009). Biodiesel production via transesterification of palm olein using waste mud crab (*Scylla serrara*) shell as a heterogeneous catalyst. *Bioresource Technology, 100*, 6362-6368.

[90] Viriya-empikul, N., Krasae, P., Puttasawat, B., Yoosuk, B., Chollacoop, N. and Faungnawakij, K. (2010). Waste shells of mollusk and egg as biodiesel production catalysts. *Bioresource Technology, 101*, 3765-3767.

[91] Cavani, F., Trifiro, F. and Vaccari, A. (1991). Hydrotalcite-type anionic clay: preparation, properties and applications. *Catalysis Today, 11*, 173-301.

[92] Constantino, V. R. L. and Pinnavaia, T. J. (1995). Basic properties of $Mg^{2+}_{1-x}Al^{3+}_x$

layered double hydroxides interacted by carbonate, hydroxide, chloride, and sulfate anions. *Inorganic Chemistry, 34*, 883-892.

[93] DiCosimo, J, J., Diez, V. K., Xu, M., Iglesia, E. and Apesteguia, C. R. (1998). Structure and surface and catalytic properties of Mg-Al basic oxides. *Journal of Catalysis, 178*, 499-510.

[94] Corma, A., Fornes, V., Martin-Aranda, R. M. and Rey, F. (1992). Determination of base properties of hydrotalcites: condensation of benzaldehyde with ethyl acetoacetate. *Journal of Catalysis, 134*, 58-65.

[95] Diez, V. K., Apesteguia, C. R. and DiCosimo, J. J. (2003). Effect of chemical composition on the catalytic performance of Mg_yAlO_x catalysts for alcohol elimination reactions. *Journal of Catalysis, 215*, 220-223.

[96] Yamaguchi, K., Ebitani, K., Yoshida, T., Yoshida, H. and Kaneda, K. (1999). Mg-Al mixed oxides as highly active acid-base catalysts for cycloaddition of carbon dioxide to epoxides. *Journal of the American Chemical Society, 121*, 4526-4527.

[97] Tichit, D., Lutic, D., Coq, B., Durand, R. and Teissier, R. (2003). The aldol condensation of acetaldehyde and heptanal on hydrotalcite-type catalysts. *Journal of Catalysis, 219*, 167-175.

[98] Xie, W., Peng, H. and Chen, L. (2006). Calcined Mg-Al hydrotalcites as solid base catalysts for methanolysis of soybean oil. *Journal of Molecular Catalysis A: Chemical, 246*, 24-32.

[99] Gao, L., Xu, B., Xiao, G. and Lv, J. (2008). Transesterification of palm oil with methanol to biodiesel over a KF/hydrotalcite solid catalyst. *Energy and Fuels, 22*, 3531-3535.

[100] Macala, G. S., Robertson, A. W., Day, Z. B., Lewis, R. S., Iretskii, A. V. and Ford, P. C. (2008). Transesterification catalyst from iron doped hydrotalcite-like precursors: solid bases for biodiesel production. *Catalysis Letter, 122*, 205-209.

[101] Constantino, U., Marmottini, F., Nocchetti, M. and Vivani, R. (1998). New synthetic routes to hydrotalcite-like compounds −characterization and properties of the obtained materials. *European Journal of Inorganic Chemistry*, 1439-1446.

[102] Cantrell, D. G., Gillie, L. J., Lee, A. F. and Wilson, K. (2005). Structure-reactivity correlations in MgAl hydrotalcite catalysts for biodiesel synthesis. *Applied Catalysis A: General, 287*, 183-190.

[103] Schuchardt, U., Vargas, R. M. and Gelbard, G. (1995). Alkylguanidines as catalyst for the transesterification of rapeseed oil. *Journal of Molecular Catalysis A: Chemical, 99*, 65-70.

[104] Schuchardt, U., Vargas, R. M. and Gelbard, G.. (1996). Transesterification of soybean oil catalyzed by alkylguanidines heterogenized on different substituted styrene. *Journal of Molecular Catalysis A: Chemical, 109*, pp. 37-44.

[105] Feria, E. A., Ramalho, H. F., Marques, J. S., Suarez, P. A. Z. and Prado, A. G. S. (2008). Tetramethylguanidine covalently bonded onto silica gel surface as an efficient and reusable catalyst for transesterification of vegetable oil. *Applied Catalysis A: General, 338*, 72-78.

[106] Shibasaki-Kitakawa, N., Honda, H., Kuribayashi, H., Toda, T., Fukumura, T. and Yonemoto, T. (2007). Biodiesel production using anionic ion-exchange resin as heterogeneous catalyst. *Bioresource Technology, 98*, 416-421.

[107] Ono, Y. and Baba, T. (1997). Selective reactions over solid base catalysts. *Catalysis*

Today, *38*, 321-327.

[108] Okamoto, Y.,Ogawa, M., Maezawa, A. and Imanaka, T. (1988). Electron structure of zeolites studied by X-ray photoelectron spectroscopy. *Journal of Catalysis*, *112*, 427-436.

[109] Leclercq, E., Finiels, A. and Moreau, C. (2001). Transesterification of rapeseed oil in the presence of zeolites and related solid catalysts. *Journal of the American Oil Chemists' Society*, *78*, 1161-1165.

[110] Suppes, G. J., Dasari, M. A., Doskocil, E. J., Mankidy, P. J. and Goff, M. J. (2004). Transesterification of soybean oil with zeolite and metal catalysts. *Applied Catalysis A: General*, *257*, 213-223.

In: Biodiesel: Blends, Properties and Applications
Editors: Jorge M. Marchetti and Zhen Fang

ISBN: 978-1-63117-024-9
© 2014 Nova Science Publishers, Inc.

Chapter 5

ARGENTINA'S SEMIARID LANDS APTITUDE TO CULTIVATE NON-TRADITIONAL SPECIES FOR BIODIESEL PRODUCTION

Silvia Liliana Falasca[*1 2] **and Ana Ulberich**[2]
[1]CONICET researcher. Climate and Water Institute. INTA. Buenos Aires. Argentina.
[2]CINEA. School of Humanities. National University Center for the Province of Buenos Aires. Tandil, Buenos Aires, Argentina

ABSTRACT

Expansion of non-traditional oilseed crops is expected to have great possibilities at worldwide level, while in other countries is already considerably increasing. These crops do not compete for lands with those destined to food production.

It was necessary to identify the requirements, limits and bio-meteorological tolerance and conditions for each of species, taking into account the climatological characteristics of native areas and the regions of successful cultivation around the world.

Afterwards, the resulting bio-climatological indicators had to be extrapolated to the Argentine lands. It was mandatory to consider growth aspects such as development and death chances by excess or deficiency.

In parallel, an agro-climatological inventory was performed based on: information of available climatological statistics, agro-climatological derived values and estimated indexes. With all the information gathered, the data was evaluated. As from the available database, geographical limits were mapped for the different variables that define aptitude classes: optimal zone, suitable, marginal and non-suitable.

Among the species investigated by our working group, this the review includes: Perennial species: *Jatropha curcas; Ricinus communis; Moringa oleifera; Pongamia pinnata* and *Cynara cardunculus* and Annual species: *Sesamus indicum; Guitzotia abyssinica; Sinapis alba; Thlaspi arvense* and *Crambe abyssinica*); identifying their agro-climatological aptitude for their cultivation in Argentina.

---

[*] Email: sfalasca@conicet.gov.ar

INTRODUCTION

Being a source of renewable energy, biofuel production as from plant derived oils or biomass has great opportunities provided it indirectly converts solar energy into fuel energy derived from vegetable oils through photosynthesis. However, it may be impaired by the negative effects of climatological changes in agro-ecosystems that depend on normal weather cycles and rainfall.

Vegetable oils and biomass exploitation for biofuel production is limited owing to the competition of biomass as food/feed or fodder resource and its economic conditions. The best chance for biofuel relies on non-edible crops, either as sole block plantations or as companion crops with food crops in agricultural crop-systems.

Crops destined to biodiesel and bioethanol production to replace conventional fuels, currently deliver greenhouse gas reductions of around 55%, while future technologies are expected to offer even greater carbon savings. Blends, with conventional fossil fuels, of up to 5% can be used without engine modification or significant changes to the supply chain. Existing technologies for biofuel production are well established in many parts of the world using esterification of oil crops or fermentation.

Renewable plant materials may contribute to reduce greenhouse gases in many ways. Provided they substitute fossil-based materials. The use of physical, chemical and biochemical processes convert renewable materials into a wide range of products for industrial manufacture including polymers, lubricants, solvents, surfactants and speciality chemicals as well as healthcare products.

Due to the energy crisis of 2008 consequence of the increase of the crude oil barrel price (almost reaching US$ 150), ethanol and biodiesel production increased in the United States of America (first corn producer in the word) and thus increasing this cereal's price which affected mainly Latin America, where corn is part of the basic alimentary diet of great part of its population.

Besides, the strong pressure to extract water for intense watering of energetic crops may affect the availability of this resource, especially in semiarid o arid regions. Growing trends indicate that it is quite likely that energy and food markets be more firmly related in the future, with energy price fluctuations resulting in concomitant changes in food prices.

For the above, it is important that international strategies regarding bioenergy may not only be focused on energy generation opportunities, but also on the exhaustive appraisal of its effects on food safety and social and environmental benefits, food prices increase, deforestation and competition for the use of water and land.

The basic premise for energetic crop production should be oriented towards the use of low productivity lands so as not to compete with the food market. Water is a scarce resource and, therefore, its use in agriculture should be destined to the most productive usage: food.

For that reason, it was considered what Mc Laughlin [41] defined as the series of characteristics that energy crops must have in order to be commercially successful in arid and semiarid environments. They must have low water requirements; tolerate extreme environmental conditions; produce some commodity compound (such as resins, sub-products with properties for industrial use, etc.) with higher price than food products and not to be cultivated in areas with higher water availabilities. Furthermore, in order to maintain the ecosystem sustainability, energy crops should: a) produce high amounts of biodiesel at low cost; b) generate local employment and c) have a positive energy balance: i.e., net energy contained in biofuels must be higher than energy used by crops and in the transformation process.

The election of the suitable oil species for cultivation in semiarid climates and marginal lands is extremely important, bearing in mind the socioeconomic vulnerability of these regions farmers.

CURRENT SITUATION IN ARGENTINA

Nowadays, gas and petroleum are the most important energy sources in the Argentine Energy System. Due its institutional organization (open market), current dependence on hydrocarbons seems to be a result of the strategies of private agents [36].

Argentine Law N° 26.190 sets forth 8% participation of renewable energy for 2015. The National Law of Biofuels N° 26.093 established the obligatory proportion of 5% of bioethanol and biodiesel in petrol and diesel respectively, as from the 1st of January of 2010. This percentage was raised to 7% by mid 2010 (Resolution 554/10) while local authorities are planning to raise it to 10% in 2011.

Argentina is the 4th biodiesel producer country, after Germany, France and Brazil. In 2010, Argentina's production reached 1,900.000 tons, a 51% above 2009, with a total product value of 1,900 million dollars. According to estimations, by the end of 2011 Argentine biodiesel demand will surpass the 3 millions tons, a situation that may foster new investments in order to increase the offer.

Beyond the discussion of the ethics of burning raw materials that could be destined to human feed, it is well known that the growth of human population will increase the energy and food demand.

To face this paradigm, a challenge is presented to the world: how to produce biofuels in quantities enough to satisfy the whole world and each internal country demands without sacrificing raw materials destined to human feeding while assuring the sustainability of the productive system. It should be necessary to increase the cultivated area with oil and alcoholigen crops. But... Which crops?

Argentina is going through the so called "soybean-ization", a phenomenon that has being taking place since 2002 to date, which has been increasing the cultivated area with soybean (*Glycine max*). This is the raw material for biodiesel.

More than the 85 % of the Argentine soybean production comes from direct sowing. The highest part of soybean production is used to elaborate flour, oil and biodiesel. Oil and biodiesel elaboration sites are generally located at less than 187 miles from the seed production area.

Transformation to biodiesel is mostly carried out in Argentina, and the final product is exported. In 2009, biodiesel exports to the European Union reached up to 1,15 Mt.

The introduction of the biofuels to the national energy structure constituted a fundamental decision because of their environmental (reduction of carbon emission) and economical effects; the potentiality of the agricultural area to offer a part of its production as energy source; the generation of employment alternatives and diversification for agricultural companies, (both in the Humid Pampa and in the so called "regional" economies).

The challenge in Argentina consists in developing materials that may be used as biofuels in areas which are currently outside the traditional cultivation areas. In this way, several hundreds of thousands of acres of agricultural lands, which are nowadays abandoned, could enter into production. It would foster rural development and would let rural people obtain income from those lands as a consequence of their own productive work, apart from the assistance they might get from the government.

The target of this work was to define the agro-climatological aptitude of the Argentine semiarid and arid zones to produce biodiesel from non traditional oilseed crops.

The Agro-Climatic Zoning

In order to start working on this chapter, we evaluated the expansion of non-traditional oilseed crops which are expected to have great possibilities at worldwide level, while in other countries they are already considerably increasing. These crops do not compete for lands with those destined to food production.

It was necessary to identify the requirements, limits and bio-meteorological tolerance and conditions for each of specie, taking into account the climatological characteristics of native areas and the regions of successful cultivation around the world.

Afterwards, the resulting bio-climatological indicators had to be extrapolated to the Argentine lands. It was mandatory to consider growth aspects such as development and death chances by excess or deficiency. The agro-climatic zoning permits to identify areas with different potential yields, as per to their environmental conditions. Every plant is sensible to weather conditions. They need a minimum as well as a maximum offer from the climate in order to satisfy its physiological needs, beyond such limits they are negatively affected. The range between these two values represents the energetic level that plants need for their physiological complex to work efficiently. This range is called "ideal temperature" [49].

In order to define the agro-climatic fitness of each culture en Argentina, we worked with the weather data corresponding to the period 1961-2000 of all meteorological stations.

In parallel, an agro-climatological inventory was performed based on: information of available climatological statistics, agro-climatological derived values and estimated indexes. With all the information gathered, the data was evaluated. As from the available database, geographical limits were mapped for the different variables that define aptitude classes: optimal zone, suitable, marginal and non-suitable. In order to construe the classified areas with different grades of agro-climatic fitness, we included as Figure 1 Argentina's political map, with the toponymy of the provinces.

SELECTED SPECIES

1. *Jatropha curcas*

This plant has received trivial names like *Pinhao manso* (Brazil), *tempate* (Mexico), *Physic nut* (English), *curi-i- vai* (Paraguay), *piñón de leche o piñón botija* (Cuba), *fagiola de la India* (Italy), *purgueira* (Portugal), etc. [37].

Figure 1. Argentina's political map.

Although original from Central America and Mexico, this bush would have arrived to Africa with the Portuguese galleons that traded slaves to Brazil. This bush was only used as live fence provided its poisonous fruits kept away the cattle. According to [37], its spreading area in South America includes Bolivia, Brazil, Columbia, Ecuador, Galapagos, Paraguay, Peru and Venezuela and finally arriving in Argentina, where it was found in many provinces.

Physic nut is most likely to grow in dryer regions, but its productivity and ecosystem function delivery will be limited. As an example, production in plantation sites with 900-1200 mm rainfall is up to twice as high (5 t dry seed/ha) as in semi-arid regions (2-3 t dry seed/ha) [2]; [33].

With its life average of 30-50 years and having adapted to sub-humid and semiarid condition, the physic nut appears as an alternative culture for these regions, fostering local development through biodiesel production [26].

In Argentina, there is a strong pressure from national and foreign investment groups to impel huge enterprises related to *Jatropha curcas*. New plant nurseries in different provinces offer their services, without any previous research about a possible failure due to the agro-climatic frost regime.

Although Argentina is extended towards the Tropic of Capricorn (23°27´S), it lacks from tropical climates and, therefore, the exploitation of tropical species like this one should be very careful in order to avoid damage by frost.

In order to define the Argentine agroclimatic aptitude, bearing in mind the tolerance limit to low temperatures, the specie was considered the isotherm corresponding to the winter minimum temperature expected, with a probability of 20% (once every 5 years) for young plants, standing out the isotherm of −4°C (24.8°F) [14].

Through the agroclimatic zoning, areas could be classified with homogeneous characteristics responding to bioclimatic requirements for crops, resulting in optimum, suitable, marginal and not-suitable zones.

This figure shows "optimum" aptitude for the north western region of Corrientes and eastern regions of the provinces of Chaco and Formosa. Misiones and part of the provinces of Jujuy, Salta, Formosa, Corrientes, Entre Ríos, Chaco and Santa Fe appear with "suitable" aptitude; while some areas in Jujuy, Salta, Formosa, Chaco, Santa Fe, Corrientes and Entre Ríos can be defined as "suitable with constraints" provided they exceed the intensity of frosts (-4°C).

At the south of the latter, there is "marginal" area with shorter frost periods and winter frosts beyong -4° C, comprising the S of Salta, Tucumán, Chaco, Santiago del Estero, part of Santa Fe and Entre Ríos. The eastern region of the provinces of La Rioja and Catamarca are in the "marginal area" with frost and watering.

2. *Pongamia pinnata*

Pongamia pinnata is a medium-sized, nitrogen-fixing tree original from India, Myanmar, Malaysia and Indonesia, which is known by several names including: *Pongamia, Panigrahi, Indian beech, Honge, algarrobo aceitero* and *Karanja*. It probably ranges from Tropical Dry to Moist through Subtropical Dry to Moist Forest Life Zones [9].

Karanja grows on most soil types: from stony to sandy or clay, including Vertisols. It is very tolerant to salinity. It can be found along water courses with its roots in fresh or salty waters. It is also cultivated in canal banks, roadside avenues, and along bunds. Well-drained soils with assured moisture offer a better yield for its cultivation.

Karanja is a hardy tree that takes water from 10 m depths without competing with other crops. Although *Karanja* is most resistant to frosts than *physic nut* [19] it suffers when exposed to severe frost. It has rich leathery evergreen foliage. Poor seed yield, long gestation period (4–7 years), and alternate bearing limits its promotion. It is an excellent coppicer that can also produce root suckers readily, and can withstand pollarding.

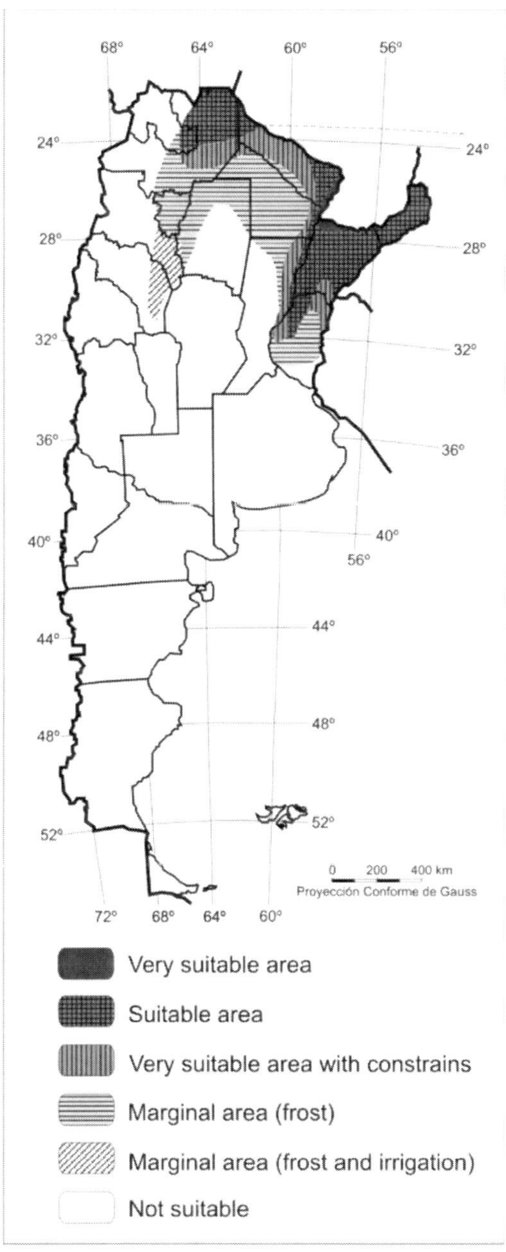

Figure 2. Agroclimatic aptitude for the cultivation of *Jatropha curcas (physic nut)*.

It takes atmospheric carbons while growing and through its roots (fostered by bacterian nodulation that fixes atmospheric nitrogen). The amount of carbons taken depends on the age of the plants [45].

Its oil has a huge volume of triglycerides and its unpleasant taste and smell are due to bitter furan flavones like "pongamol" and "karanjin". This oil is not fit for human consumption; it is thick, with an orange yellowish colour, bitter and not drying. It is suitable for biodiesel [3]; [54] and with insecticide properties [43].

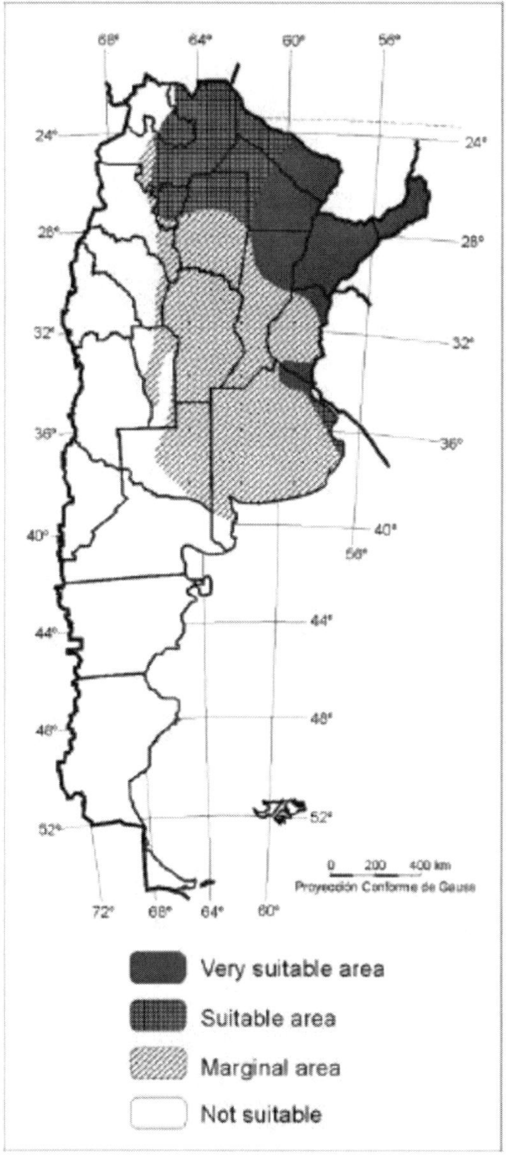

Figure 3. Agroclimatic aptitude for the cultivation of *Pongamia pinnata (karanja)*.

This tree has not been introduced in Argentina but the harvesting period would extend from June to October. Australia has all the technology for mechanized harvest, at an average of 2 trees per minute [20].

Due to its adaptation to flooding and salty soils and its resistance to droughts, we recommend it for all marginal lands in Latin America, erosionated soils or with risk of becoming deserted due to a bad management; as it would improve the soil fertility at the same time that produces raw material for biodiesel and qualifies for carbon credit [19].

Figure 3 provides a panoramic view of the expansion potentials of this specie: "optimum and suitable areas" are found in the NE and N of the country. Besides, there are three other regions, to the south of the latter: one at the northern area of Buenos Aires and the southern area of Entre Ríos; and the other two at the Buenos Aires coastal region.

These areas, due to its proximity to the river or sea respectively, have milder temperatures that may provide protection against frost. The "marginal region" is defined as such due to frost and a minimum absolute temperature beyond -5°C (23°F), with a recurrence period each 30 years.

3. *Moringa oleifera*

Moringa oleifera is an evergreen tree original of South Himalaya, from Pakistan NE until Bengala's NW, in India [46]; [52]; [56].

Ranging from Subtropical Dry to Moist through Tropical Very Dry to Moist Forest Life Zones, *moringa* is reported to tolerate annual temperature of 18.7 to 28.5°C (65.6 to 83.3°F) and an annual precipitation of 480 to 4030 mm [9].

It is a perennial tree of very quick growth, with short life -it may live up to 20 years but not more. In India, they have annual crops with mechanized cultivation [32].

The *moringa* can be used as raw material to produce biodiesel from seeds and bioethanol from their leaves. Argentina had isolated specimen in the provinces of Misiones, Chaco and Catamarca and, since 2003, in Anta Muerta, Department of Oran in the province of Salta. It is the only exotic tree specie that was declared of national interest by the Nation's Chamber of Representatives in 2002 [16].

Yield per hectare is 3000 kg of seeds, equivalent to 900 kg of oil per hectare, comparable to soybean which also yields 3000 kg of seeds/hectare but only 20% oil. The seed has 31-47% oil content, with high level of oleic acid. The oil may be used for human consumption, soap and cosmetics manufacturing, lubricant for clocks and biodiesel production among other uses [44].

Its soft leaves and flowers may be eaten; raw or cooked, provided they are rich in proteins, minerals, beta carotene, rivoflavine and vitamin C [48]. However, the tree is mainly valued by its tender and eatable sheaths that taste like asparagus. In India, they are exported fresh, refrigerated and canned to Hindu communities [21]. Moringa forestation is part of many social inclusion projects at international level.

Figure 4 shows "optimum areas" at N Corrientes, E Chaco and E Formosa. "Suitable regions" comprise almost the whole territory of Formosa, the eastern region of Salta, and NW Chaco. "Suitable areas with frost constraints" are distributed in the NE of Santiago del Estero, N and center of Chaco and Misiones. "Suitable areas with winter temperature constraints" (lower than 18°C = 64.4°F) can be found at the NE of Santa Fe, the whole province of Corrientes; S of Misiones; NE, N and NW of Entre Ríos, E of Jujuy and eastern-center of Salta. The "marginal area", with frost and watering problems, surrounds the latter towards the south and west. Non-suitable areas for its cultivation were left uncolored [18].

Its implantation may help to recover erosionated soils that have been deforested for agricultural purposes, as an alimentary source to fight malnutrition in Argentina. It shall also help to improve the quality of cattle in the northern region and as raw material for biodiesel and bioethanol production; although the cultivation areas should be carefully chosen provided this tree is affected by frost [17].

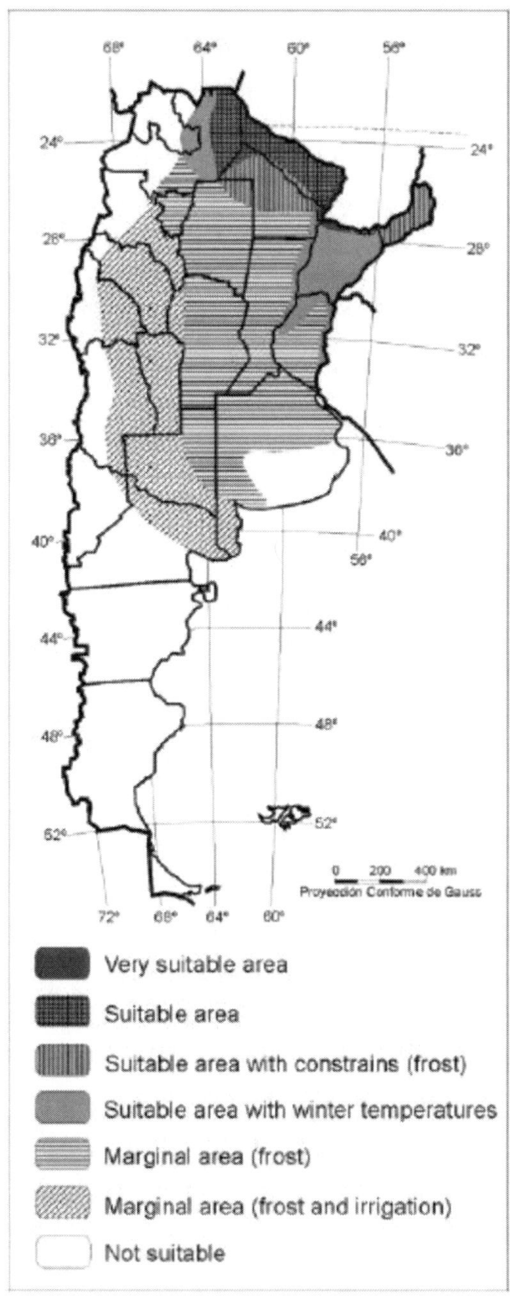

Figure 4. Agroclimatic aptitude for the cultivation of *Moringa oleifera (moringa)*.

4. *Ricinus communis*

Ricinus communis is known by several names including: ricino, castor, tártago, higuerilla, mamoneira, mamona, palma christi, higuereta, castor bean, castor-oil plant.

Castor oil plant's seed is an oleaginous that have been used since remote times with industrial and medicinal reasons, is being used as raw in many countries of the world to produce biodiesel [29].

This plant is original from Africa but it can be found in both wild and cultivated states in all tropical and subtropical countries around the world. In wild conditions this plant is well adapted to arid conditions and it resists long periods of drought, even in the fruit ripening state. However, it produces lighter seeds with lower oil content [5].

Ranging from Cool Temperate Moist to Wet through Tropical Desert to Wet Forest Life Zones, *castor bean* is reported to tolerate annual temperature of 7.0 to 27.8°C (44.6 to 82°F) and annual precipitation of 200 to 4290 mm [9].

Even though it is cultivated as an annual plant, in tropical and subtropical regions it behaves as a perennial plant. In areas with severe frost or very dry seasons, it is managed as an annual culture, in order to avoid the death of the whole plantation.

In our country, it grows spontaneously as weed. Even in the province of Buenos Aires, it could be seen along water streams, also at the shores of the very contaminated rivers and streams of the Matanza-Riachuelo basin.

Its oil constitutes the only commercial source of hydroxylated fat acids, as it has about 85% of ricinoleic acid. *Castor* oil already has its international market as it has more than 700 uses. Therefore, specialists say that the market for this oil is unlimited. Its usage includes medicine and cosmetic use, petroleum substitution in plastic and lubricants, production of fiber optics, bulletproof glass and bone prostheses. It might also be used as antifreeze for aircraft and space rocket combustibles and lubricants.

There are dehiscent and indehiscent genotypes. In case of dehiscent varieties, it turns necessary a higher number of partial crops and the harvest takes place when the bunch has not reached total ripening, with the oil percentage being lower than the one achieved in total ripening [11].

Yields depend on the soil humidity. Thus, with proper watering it yields 900 -1000 kg/ha whilst with a lower watering it yields 300-400 kg/ha [29].

Temperatures higher than 40°C (104°F) produce an abortion of flowers, sexual reversion of feminine flowers to masculine and yield reduction with the consequent loss in oil production. Temperatures lower than 10°C (50°F) produce lower quantity of seeds due to the loss of pollen spreading [10].

The *castor* plant is a tree that survives and can grow in marginal lands. Since it is adapted to live under humid, sub-humid and semi-arid conditions, it appears to be an alternative cultivation for those regions.

In Argentina, castor oil production was historically marginal and by 1989 it was not produced locally any more and Argentina started importing it. The Argentine domestic market for castor oil is 340 tons per year.

There have been castor bean cultivation trials in Misiones, Formosa, Chaco and Salta, which are already producing castor oil [24]. The return of this culture to Argentina would allow in a first stage to substitute imports and even become exporters provided the agroclimatic conditions are promissory of successful crops.

Figure 5 shows the region suitable for cultivation under rainfed conditions. Even though this is an agroclimatic research, it would be necessary to analyze the requirements of the different cultivation in the international market, mainly in the Iberian Peninsula where they have certain germplasms that are resistant to frost, thus enabling to implement cultivations at higher latitudes. Besides, it would be useful to evaluate the oil yields of the genotypes available in our country. Its cultivation will not present adaptation problems provided there are many *castor bean* bushes growing spontaneously up to 40° Southern Latitude [29]

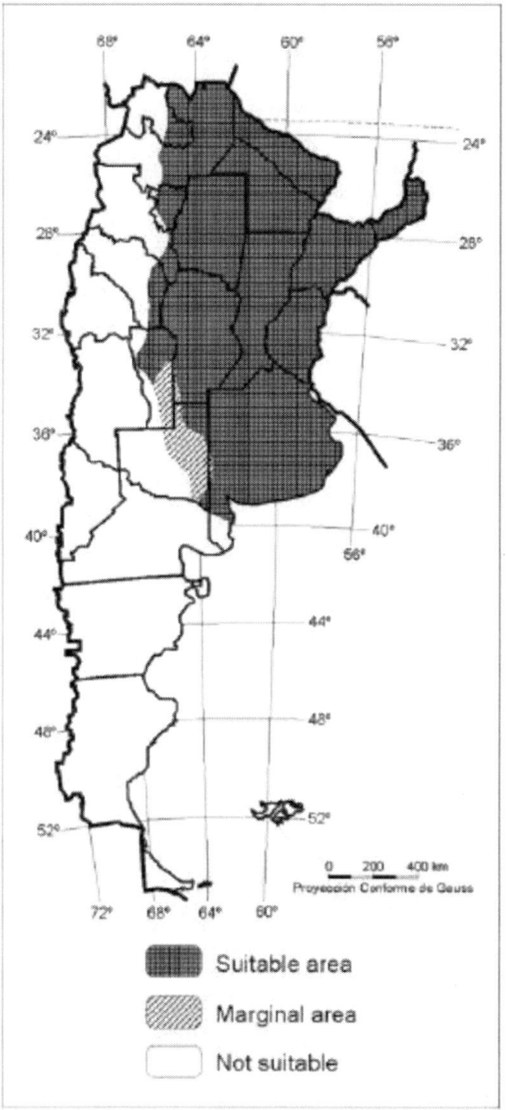

Figure 5. Agroclimatic aptitude for the cultivation of *Ricinus communis (castor bean)*.

5. *Cynara cardunculus*

Cynara cardunculus, belongs to the family of the Compositae. It is known with common names as: *penca, cardon penquero, Castilla thistle, cardoon.*

It was accidentally introduced in Argentine by mid of the XVIII century, possibly with wheat seeds, for its cultivation in the Pampas. It was mistakenly sown with good wheat seeds and then spread by the effect of wind. It grows by the end of summer and beginning of autumn, it flourishes by late spring and early summer and it produces seeds during almost the whole year.

It is considered a weed that grows in the roads, in both cultivated and non cultivated lands. In our country it is considered National Plague and it is mentioned as such by the Vegetal Health Law N° 5.770 [25].

Lignocellulosic materials are currently used for heating and electricity generation [31] and they offer an excellent potential for future bioethanol production.

Biomass production of *cardoon* cultivated lands mostly depends on water availability in spring and during the active growth stage. Trials are being made in Argentina expecting to achieve 17 tons of harvestable biomass by hectare, with an average range of rains of 400-450 mm. With a medium humidity of 15% it would represent 14,5 tons DM.

Its fruits, with high oil content (25%), protein (20%) and fiber may be used for animal feed, oil production or both at the same time. The oil is also fit for human consumption and it may also be used in cosmetics, pharmaceutical products and biodiesel production [12]; [31].

Being a perennial plant, soil erosion is reduced as -year after year- the plant process is integrally developed and it is repeated by an indeterminate number of years, which normally ranges between 6 and 8. Another advantage that presents is the low demand of fertilizers and watering, thus resulting in low implantation cost. However, it should be analyzed the impact on the agricultural economy and if its implantation is feasible from the economic point of view.

Figure 6 shows the potential productive areas in rainfed condition.

6. *Sesamus indicum*

Sesame seeds (*Sesamum indicum*) are one of the world's most important and oldest oilseed crops known to men, with more than 4000 years of cultivation in Asiria and Babylon. Its great adaptation aptitude, it can be cultivated in tropical, subtropical and mild regions.

At worldwide level it draws its importance from the fact that it is a food crop, a raw material for industry, feed for livestock, as well as a leading export crop. *Sesame* yield is highly variable depending upon the growing environment, cultural practices and cultivars.

Sesame, also known as *sesamum, gingelly, sesamo, ajonjoli, benissed, sim-sim* and till is an important annual oilseed crop. It has been cultivated for centuries, particularly in Asia and Africa, for it is high content in edible oil and protein.

Most of *sesame* seeds are used for extraction and production of oil. The oil has a mild smell and pleasant taste and as such, it is a natural salad oil requiring little or no winterization. It is a cooking oil in the form of shortening and margarine, as a soap fat in pharmaceuticals and as a synergist for insecticides.

Dry biotypes grow in regions with medium annual rains ranging between 200 to 400 mm and well drained soils. Excessive humidity damages the culture in any of its development stages. The optimum temperature range for growth, flourishing and ripening is 26 to 30°C (78.8 to 86°F). In regions with warm and strong winds, the plant produces smaller seeds and with lower oil content. For that reason, sesame is cultivated in colder summer regions and mild regions during the coolest months. It is not resistant to frost [13].

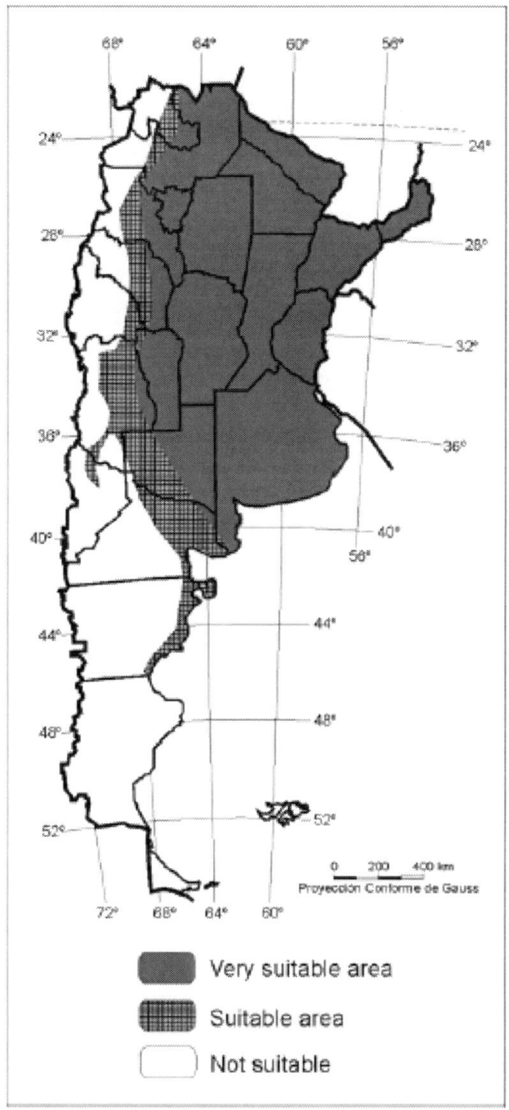

Figure 6. Agroclimatic aptitude for the cultivation of *Cynara cardunculus (cardoon)*.

Seeds have 50-60% of mild taste oil. The oil produced from the first cold pressing is among the most expensive oils for human consumption in the world due to its low cholesterol content and high poly-unsaturated fat acids content: it has almost 47% of oleic acid and 39% of linoleic acid [7]. It has a light yellow color, it is not drying and it can resist high temperatures. The oil has a high stability due to the presence of three natural antioxidants: 'sesamolin', 'sesamin' and 'sesamol' [13]. It is used in salads, elaboration of margarines and pastry.

The oil from the second hot pressing –after being extracted- has a lower quality than cold pressing oil. This oil is used in the production of soaps, paints, cosmetics, inks, pharmaceutical industry, pesticide manufacturing [55] and it may be used in biodiesel production [6].

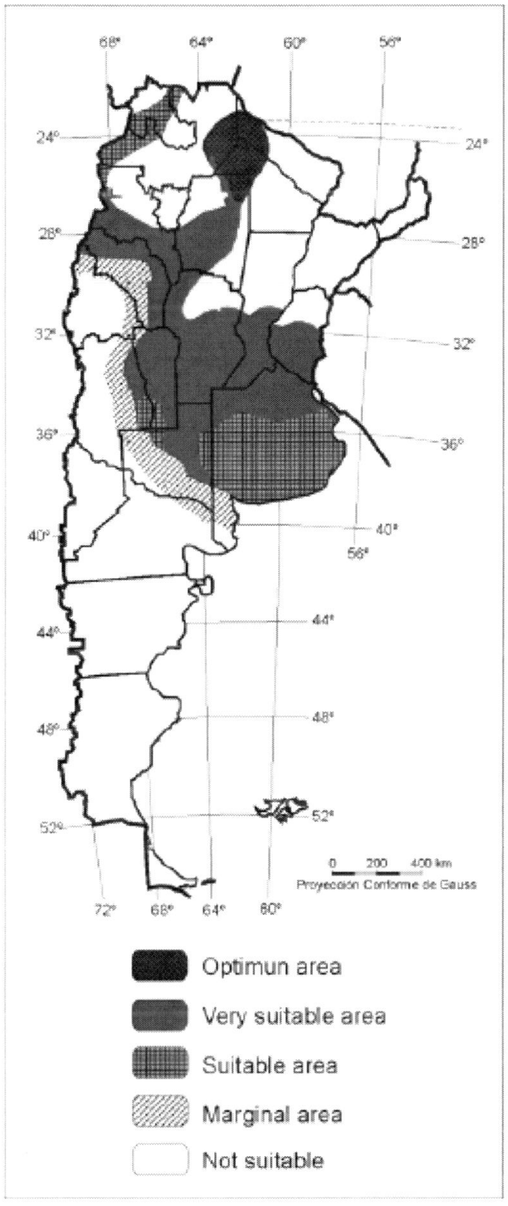

Figure 7. Agroclimatic aptitude for the cultivation of *Sesamum indicum (sesame)*.

Worldwide average productivity ranges 390 kg/ha. Brazil has 20,000 planted hectares and produces 13,000 tons with a yield of 650 kg/ha [4]. However, under optimum conditions yields may reach 1,500 kg/ha. Regretfully, we do not have statistical data in Argentina provided production is made upon contracts, with purchase commitments from food companies, both for domestic consumption or for export to Japan.

It is sowed from October to January according to the latitude. For sowing season definition, it is recommended to take into account the vegetative cycle of the culture and regional rains, planning harvest to coincide with the beginning of the dry season.

The "optimum area" shown in Figure 7 comprises part of 4 provinces: Formosa, Chaco, E of Salta and NE of Santiago del Estero. "Very suitable region" comprises a little area of the province of Salta, part of Catamarca, La Rioja, Santiago del Estero, Córdoba, San Luis, Mendoza, La Pampa, center and south of Santa Fe, Entre Ríos, San Juan and Buenos Aires. "Suitable areas" can be found in part of Mendoza, San Luis, La Pampa, Buenos Aires, Catamarca, Salta and Jujuy. "Suitable regions with humidity constraints" comprise part of San Juan, La Rioja, Mendoza, La Pampa, San Luis and S of Buenos Aires. The latter may improve its condition by implementing complementary watering.

Without complementary watering and depending on hydric conditions in the area defined as "suitable with hydric limitations", producers may sell oil for human consumption or for industrial use, or for both of them.

7. *Guitzotia abyssinica*

Guizotia abyssinica (known as *niger, ramtil, ramtilla, inga seed,* and *blackseed*) represents the greatest source for eatable oil in Ethiopia, various African and Asian countries [35].

Ranging from Warm Temperate Dry to Moist through Tropical Very Dry to Moist Forest Life Zones, *niger* is reported to tolerate annual precipitation of 660 to 1790 mm and annual temperature of 13.6 to 27.5°C (56.5 to 81.5°F)[9].

It is an annual herbaceous specie with spring growth, original from eastern Africa. The oil is fit for soap, paint, lighting, lubricants production and as fragrance holder in parfums [40]. The seed contains 17-20% protein [1], 34-40% carbohydrates, 13.5% fiber and it is an important source of thiamine, riboflavin and niacin [40]. Ethiop origin seeds have about 40% oil, with the following composition of fat acids: 75-80% of linoleic acid, 7-8% palmitic and estearic acid, and 5-8% of oleic acid [34]. Indian biotypes present 25% oleic acid and 55% linoleic acid [47]. The Ethiop seed is better than the Indian for usage in paints, due to its higher content of linoleic acid and the biodiesel obtained through transesterification process has excellent quality. Provided it is a plant with 2 very differentiated biotypes (Ethiop and Indian) as far as their bioclimatic needs are concerned [30], it is necessary to find a medium situation to define its hydric and thermal limits.

The areas defined in Figure 8 as suitable and very suitable from the agroclimatic point of view, show a great coincidence with the lands currently destined to soybean cultivation. Even though both cultures have spring-summer cycles, *niger* may be included in the rotation scheme of soybean. It would be useful to analyze the profitability aspects of this culture.

In case of an abrupt fall in international soybean oil demand, consequence of environment protection measures that the US and the EU are willing to impose, *niger* might replace it in some regions in Argentina [22].

8. *Sinapis alba*

Sinapis alba (*White mustard, kedlock, charlock, mostaza blanca*) is a herbaceous specie of the Brassicaceae family, native of the Mediterranean region and the Crimea.

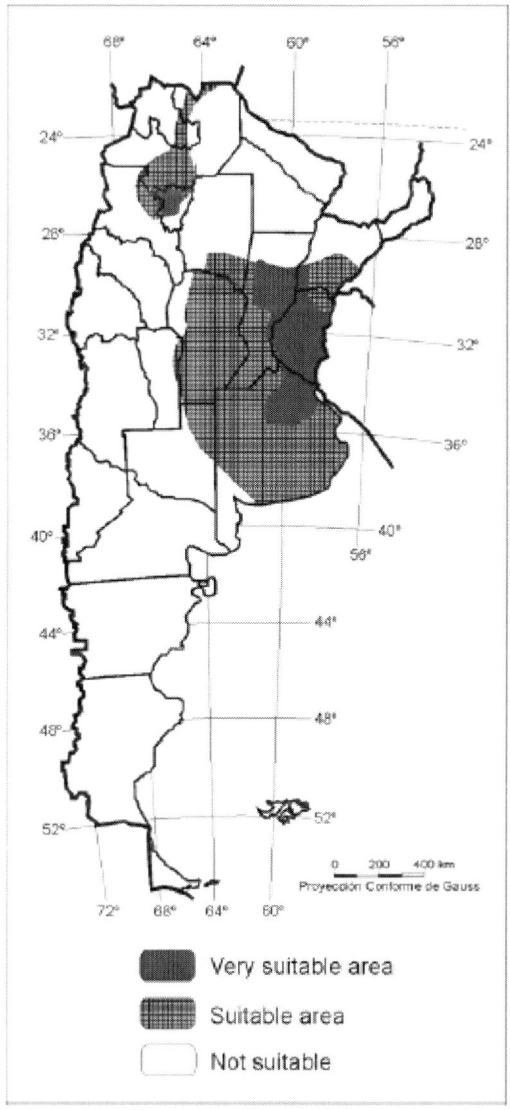

Figure 8. Agroclimatic aptitude for the cultivation of *Guizotia abyssinica* (*niger*).

A quick-growing long-day annual which prefers temperate climates with some humidity. Can withstand high temperatures, but very hot days during flowering and ripening may reduce seed setting and lower quality of seed [9].

It is recognized by its large peak fruits and shaggy lower part, as well as for its mustard smell when squeezed.

The seed has more than 35% of semi-drying oil. From the seeds, a non eatable oil can be obtained if it is not rectified and refined, generally used as lubricant for industrial purposes, as combustible for lightning and also fit for biodiesel.

The *white mustard* is a winter-spring cycle plant that may be cultivated in short cycles, commonly in rotation with other grain cereals, thus enabling the early release of the batch with the possibility of second-crop cultures.

It is traditionally sowed in the south eastern region of the province of Buenos Aires, with an estimated surface of 400 hectares and yields of 500-600 kg/ha. National production is not enough to satisfy the domestic consumption due to the low yields obtained, and this fact does not encourage increasing production surfaces. This process should be accompanied by the incorporation of technology to achieve yields ranging 1000 kg/ha, as in the main producing countries like Canada, Czech Republic and Hungary. Imported grains are mostly from Germany. Mustard is exported to border countries [28].

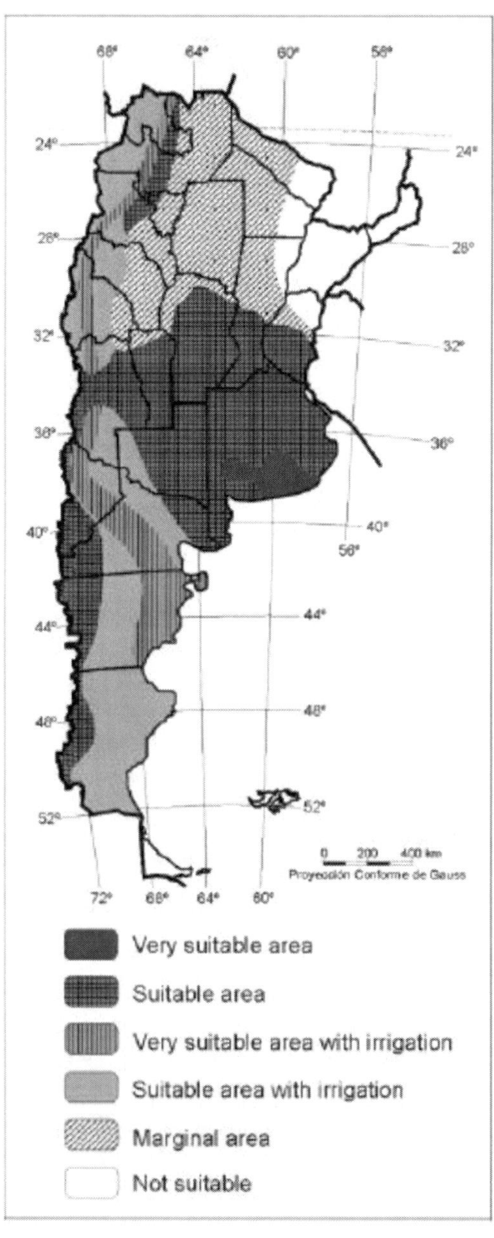

Figure 9. Agroclimatic aptitude for the cultivation of *Sinapis alba (white mustard)*.

Original from Eurasia, in Argentina this plant appears as weed in summer cultivation fields. This means that its introduction as extensive cultivation in areas where it is not cultivated nowadays will not present adaptation problems.

White mustard, the same as Ethiopian mustard (*Brassica carinata*) [27] presents a low oil content (30%), which may need to be substantially increased to >40% in order to be competitive with other biodiesel sources. As *white mustard* is not easily crossed with *B. napus* and there is a little genetic variation, the selection process is quite easy.

Figure 9 shows the "optimum cultivation area" which is the south eastern region of the province of Buenos Aires. The "suitable region" goes beyond the Pampas, including part of the provinces of Río Negro, La Pampa, San Luis and Mendoza. It also comprises the Patagonian valleys of Neuquén, Río Negro, Chubut and Santa Cruz. In the north of the country, it can be found in Salta, Jujuy, Tucumán and Catamarca. In case of incorporation of watering, there are huge semi-arid areas classified as "suitable" or "very suitable with watering" that may be incorporated to production.

9. *Thlaspi arvense*

Thlaspi arvense is known by several names including: *Pennycress, Bastardcress, Fanweed, Stinkweed, Frenchweed, Carraspique (hispanish), Zurrón boliviano.*

Pennycress is an annual species (Crucifer). Worldwide it is possible to find it between 80° LN up to 45° LS [39].

Pennycress is a popular food plant in various parts of the world and is often cultivated in Europe. *T. arvense* is adapted to a wide range of environmental conditions in temperate climates [15]. It may be considered a new culture, although it was known as an old weed since the Bronze Age. It does not represent a big problem for summer crops provided this weed finishes its cycle by late spring and it does not compete with corn or soybean.

In Argentina, it is weed of alfalfa (*Medicago sativa*) and it may cause disorders in cattle. It is a huge problem as weed in grain fields because of its unpleasant smell and taste, damaging not only the crops but also milk and its derivates as it transmits its smell and taste. Flour or food becomes unusable in case of presence of these seeds.

Their multiple uses make an attractive cultivation of it, fundamentally because their seeds contain 20-38% of oil. The high tenor of linoleic acid of the oil indicates that it´s suitable to obtain biodiesel [8].

It is a source of industrial oil because of its range of fatty acids: 16 to 24 carbon chain atoms [15]. When compared to rape (*Brassica napus*), this oil has a higher percentage of erucic acid and an even higher percentage of linoleic acid. Its viscosity is almost the same as rape's, although –at high temperatures- it might be slightly higher. Because of this fact and the similar composition of both oils, this plant might substitute rape oil as lubricant, in paints and varnish and in other products based on drying oils.

As shown in Figure 10, *pennycress* may grow under three humidity conditions: humid, sub-humid and semi-arid to arid regimes. However, owing to economic reasons, its cultivation is not recommended in humid or sub-humid climates.

On the other hand, there is a vast semi-arid and arid region where it could be cultivated without competing for land with traditional oleaginous crops. Consequently, it could be incorporated to marginal lands in the NE of the province of Santa Cruz and part of the provinces of Río Negro, Chubut, Neuquén, Mendoza, San Juan, La Rioja, Catamarca, Salta and Jujuy.

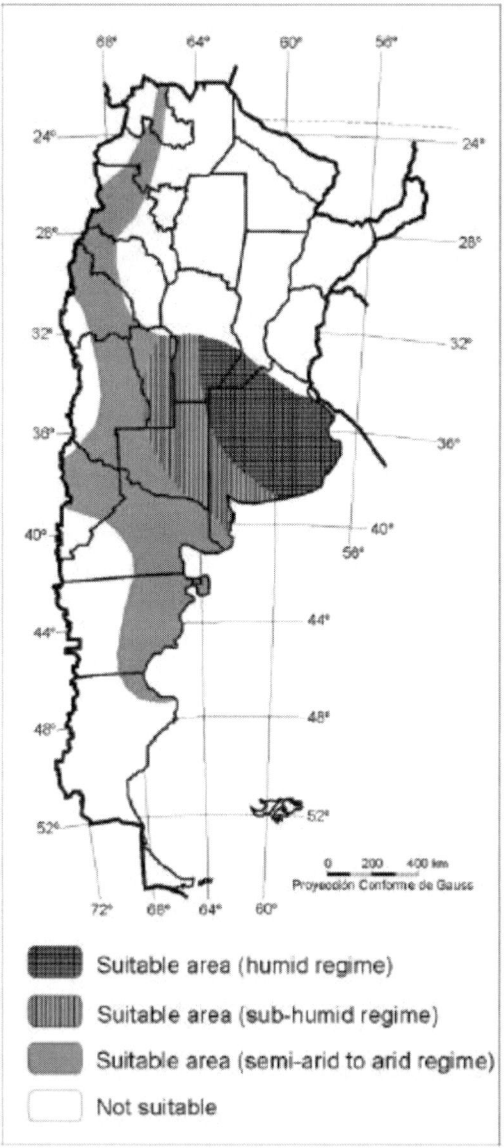

Figure 10. Agroclimatic aptitude for the cultivation of *Thlaspi arvense (pennycress)*.

Obviously, it would be necessary to perform yield trials but the rusticity of this weed, the possibility to employ machinery used for other crops, its good profitability and the chance to obtain an industrial-quality oil are factors that may foster its expansion. However, it would be advisable to appraise the environmetal impact of sowing great extensions of land, as the winds coming from the west may spread the seeds towards areas of traditional agriculture.

This cruciferous presents another advantage: its cultivation would not compete in land with rape as it would be sown in marginal lands at the W and S of the current rape cultivation area.

10. *Crambe abyssinica*

Only a little part of its history as crop is known. Its cultivation probably started in the URSS [42]. There is evidence of experimental research work in Russia, Sweden and Poland after the 2nd world war [51].

Crambe can grow in sites with rainfall in the range of 350 to 1,200 mm and an annual average temperature in the range of 5.7 to 16.2°C (42.3 to 61.2°F) [38].

The *crambe* can be cultivated as a spring or winter specie, depending on the thermal regime of the region where it is going to be planted. It can grow as spring crop (like rape Canola of spring, *Brassica napus*) in Europe or as winter crop (like rape of winter, *B. napus*) in Mediterranean climates [50].

As *crambe* is a new crop, there is only few yield data available. But because of the specific demand of its oil, many European countries are actually making experimental trials to learn about it. Austria reported yield values from 0.97 to 3.33 t/ha with an oil content of 23%–38% [57].

Crambe is a non food oil seed crop. The oil is high in erucic acid and low in polyunsaturated fatty acids. The oil is converted to erucamide which has applications in industry as a slip agent in plastics.

It looks like a promissory crop for Argentina because of its great tolerance to draught and frost, its very short cycle as it blooms at 35 days and it can be harvested at 90 days, and because of its uniform maturity which let a mechanized harvesting.

Lately, Brazilian farmers producing soybean have showed great interest on *crambe* because of its low cultivation costs, mechanized harvesting and because it can be sowed as winter crop in March or April after soybean.

Its oil can be distinguished from other ones because of its high content of erucic acid (50-60%, C22:1), a fatty acid of long chain which has special industrial uses. In U.S.A., the *crambe* has been cultivated to replace the importation of rape of high erucic content from Poland and Canada [23].

Agroclimatic aptitude for the *spring crambe*, is presented in Figure 11, where three sectors can be observed in the "very suitable area": 1) Northwest Argentina (NOA) sub-region includes great part of the Salta, Jujuy, Tucuman and Catamarca provinces, 2) Patagonic sub-region includes south of Neuquen, west of Rio Negro and northwest of Chubut and 3) Pampean plains sub-region, exceeding the limit of the pampean prairie, includes almost all the San Luis, Entre Rios and Cordoba provinces, centre east of Mendoza and south of Santiago del Estero, centre and south of Santa Fe, the half eastern part of La Pampa, Buenos Aires and a little sector of Rio Negro.

The remaining parts of the Chaco plains and Mesopotamia result "suitable with constrains" because of their high summer temperatures. A little area appears as "marginal" because of the irrigation needs or the excessive summer temperatures. To be able to develop with 350 mm of annual precipitation is considered one of the goodness of the crop.

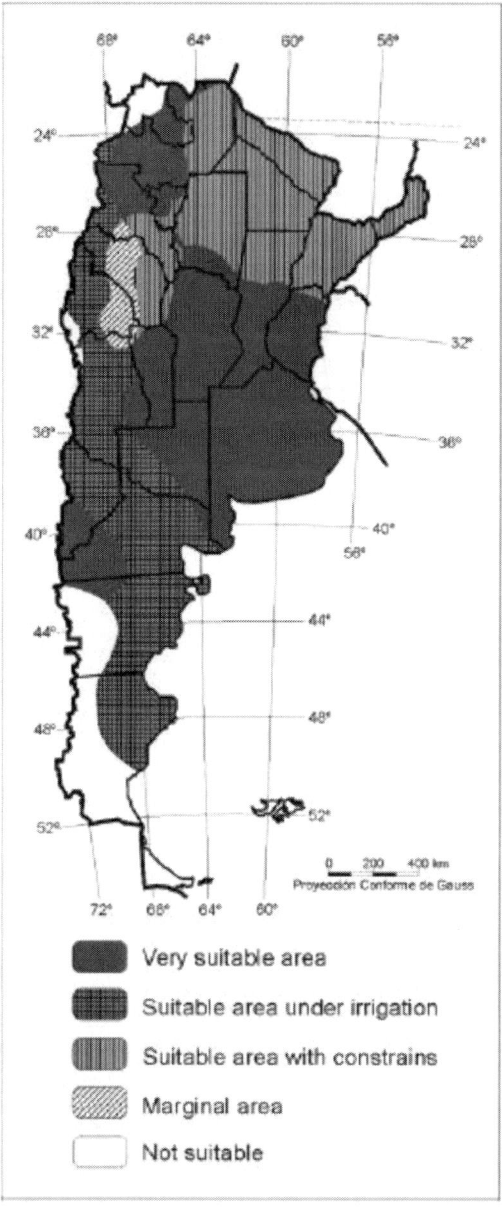

Figure 11. Agroclimatic aptitude for the cultivation of *spring Crambe.*

In this figure, the zones of the country for possible localization of the *spring crambe* in the semiarid Argentinean region in unirrigated conditions (part of La Pampa, San Luis, La Rioja, Catamarca, Mendoza, etc.) are presented.

In Figure 12, which shows the agroclimatic aptitude for the *winter crambe*, it can be observed a different situation, as the mayor part of the national territory has been classified as not suitable. This is mainly due to the intensity of frost in winter.

Probability of damage or crop destruction of once every five years (P: 20%) because of temperatures lower than the level of –6°C (21.2°F) has been included. The "very suitable zone" is located in the south and southeast of Buenos Aires province. The "suitable zone with thermical constraints" because of very high temperatures in spring include almost all the northern and eastern part of Argentina, besides a strip of land located in centre west of the country. The "suitable under irrigation and marginal sectors" are located in part of the Mendoza, San Juan, La Rioja and Catamarca provinces.

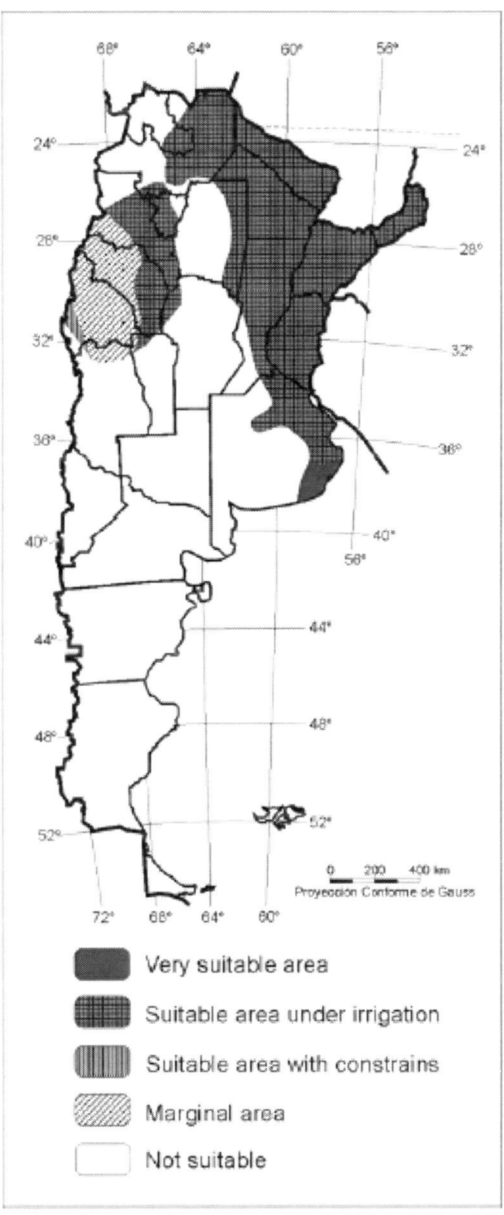

Figure 12. Agroclimatic aptitude for the cultivation of *winter Crambe*.

The attached map shows aptitude for the cultivation of *spring crambe* under Argentine conditions in the major part of the national territory.

Contrarily, the diffusion of *winter crambe* is limited by the incidence of winter frost in sites with intesities higher than –6°C. However the *crambe* could be included into a rotation with soybean in the western and northern regions of Argentina [23].

CONCLUSION

A selection of non traditional oilseed crops was made as potential target crops by using information from previous studies of this working team.

The maps provided may be a useful guidance to those seeking to develop energy crops. They show the best areas for the different crops, as well as non-suitable or suitable with constraints areas. They may be helpful for farmers willing to plant energy crops, users trying to locate suitable areas and governmental offices interested in fostering promotion measures.

It is important to remark and bear in mind that the maps are only indicative and should not be regarded as definitive with respect to individual applications. They provide the basis for future works, at a higher cartographic scale, which will provide detailed regional guidance for the planting of energy crops. There is clear guidance on their use and it is important that this is referred to when looking at the maps.

Of all the annual cultures analyzed, *cardoon, pennycrees* and *spring crambe* are those with higher potential for cultivation in the semi-arid to arid regions in Argentina.

However, as far as *cardoon* is concerned, it should be appraised the environmental impact of its sowing at great extensions provided the western winds may spread the seeds to traditional agricultural regions. It should be remarked that the *cardoon* being a wild weed, capable of adaptation to almost the whole country, it would require a special handling in order to avoid the increment in the use of herbices in bordering areas devoted to traditional agriculture. This is also valid for *pennycress*.

Use of *physic nut, moringa* and *karanja* should be avoided in frost-prone regions. These plants immediately sheds its leaves after mild frost events, leading to very low seed production, while more severe frost is lethal for these species.

Moringa should only be sown in the N of the country due to the high incidence of frost and its culture may not be convenient as biofuel producer provided the suitable and optimum areas are under humid climate conditions, unless it could be used for social or environmental purposes as: erosionated soils recovery, cattle feed, etc.

It would be convenient to try *Karanja* at the Delta and Samborombón Bay regions with plenty of saline, alkaline soils, with drainage problems, etc, (Natracuol typical, Hapludol natric and Hapludol thapto argic) which are not suitable for traditional crops. Both at the Paraná Delta and the Buenos Aires Low Plains, forestation with this species would be a long term sustainable exploitation that may generate sustained income for a very long period of time as the karanja lives as long as 80-100 years.

It would be useful to analyze the requirements of *castor bean* cultivation in the international market, mainly in the Iberian Peninsula where they have frost-resistant germplasms (-8° C) and perform in-situ yield trials, provided they are using Brazilian varieties. Besides, it would be necessary to evaluate the oil yields of the genotypes currently available in our country. The return of the cultivation at great scale in Argentina would enable in a first stage the substitution of oil imports.

Niger could be integrated to the soybean rotation scheme. It would be useful to analyze the profitability aspects of this culture. In case of an abrupt fall in international soybean oil demand, consequence of environment protection measures that the US and the EU are willing to impose, *niger* might replace it in some regions in Argentina.

Sesame cannot currently compete as biodiesel producer due to its low yield in seeds, when compared to other oleaginous varieties.

White mustard presents a low oil volume (30%) which should need to be raised substantially to > 40% in order to be competitive with other biodiesel sources.

Economic and profitability features would obviously play a leading role in motivation to farmers, as in this research we only considered the agro-climatic variables.

Such crops may generate new business opportunities in Argentine rural areas, thus providing additional diversity and innovation and a more sustainable food chain. Non traditional oilseed crops may offer new markets and food marketshare for agriculture. In this respect non traditional oilseed does not differ so much from food crops, although their potential additional environmental benefits they can help to deliver a more sustainable farming sector contributing to society's wider needs.

REFERENCES

[1]   Abebe, M. (1975). Ecophysiology of noog (Guizotia abyssinica Cass). PhD Thesis. University of California, Riverside.

[2]   Achten, W. M. J., Verchot, L., Franken, Y. J., Mathijs, E., Singh, V. P., Aerts, R., and Muys, B. (2008). Jatropha bio-diesel production and use. *Biomass and Bioenergy*, 32, 1063-1084.

[3]   Agarwal, A.K. and Rajamanoharan, K. (2009). Experimental investigations of performance and emissions of Karanja oil and its blends in a single cylinder agricultural diesel engine. *Applied Energy* 86: 106–112.

[4]   Arriel, N.H.C. (1997). Diagnosis and sesame perspective in Brazil. In: Thematic Reunion about oil raw materials in Brazil: diagnosis, perspectives and research priorities. Campina Grande. Proceedings. Campina Grande: EMBRAPA Cotton / MAA / ABIOVE. 119-138.

[5]   Amorim Neto, M. da S.; Araújo, A. E. de; Beltrão, N.E. de M.; Silva, L.C.; and Gomes, D.C. (1999). Zoning and planting season for the castor bean, in the state of Paraiba. Campina Grande: EMBRAPA-CNPA. *Technical Note*, 108. 7p.

[6]   Banapurmath, N.R, Tewari, P.G, and Hosmath, R.S. (2008). Performance and emission characteristics of a DI compression ignition engine operated on Honge, Jatropha and sesame oil methyl esters. *Renewable Energy* 33:1982-1988.

[7]   Beech, D.F. (1985). Sesame: Research possibilities for yield improvement. p. 96–106. In: A. Ashri (ed.). Sesame and safflower status and potentials. *FAO Plant Production and Protection*. Paper 66. Rome, Italy.

[8]   Carr, P.M. Potential of fanweed and other weeds as novel industrial oilseed crops. In: J. Janick and J.E. Simon (eds.), *New crops*. New York. Wiley. 1993. 384-388.

[9]   Duke. J. 1983. *Handbook of Energy Crops*. Published only on the Internet.

[10] EMBRAPA. Castor cultivation. Production Systems 4. (2003). Federative Republic of Brazil. Electronic version. ISSN1678-8710. Available from: http://sistemasdeproducao. cnptia.embrapa.br/FontesHTML/Mamona/CultivodaMamona/importancia.htm

[11] EMBRAPA. (2005). Technical instructions for the cultivation of castor in the state of Matto Grosso do Sul. Production Systems 8. Federative Republic of Brazil. 64 pp.

[12] Encinar, J. M.; Gonzáles, J. F.; Rodríguez, J. J. and Tejedor, A. 2002. Biodiesel fuels from vegetable oils: Transesterification of Cynara cardunculus L. *Oil with ethanol. Energy and Fuels*. 16. 443 – 450.

[13] Falasca, S; Anschau, A. and Galvani, G. (2010). The potential areas for sesame production (Sesamus indicum) in Argentina, raw materials for biodiesel. *Revista AVERMA* 14. 11.36-40.

[14] Falasca, S. and Bernabé, M.A. (2009). Agroclimatic aptitude of Argentina for the cultivation of Jatropha curcas L. Proceedings XVI Brazilian Congress of Agrometeorology. Belo Horizonte, Brazil. Published in CD rom.

[15] Falasca, S. and Bernabé, M. A. (2008) ¿Is it possible to produce Thlaspi arvense for biodiesel in Argentina? *Revista Virtual REDESMA. Biofuels*. V 2 (2). La Paz. Bolivia.

[16] Falasca, S. and Bernabé, M.A. (2009). Agro-climatic zoning of moringa (Moringa oleifera) in Argentina to produce biodiesel and biethanol. *Revista AVERMA* 13. 11.65 – 11.70.

[17] Falasca; S. and Bernabé, M.A. (2009). Potential areas for cultivation of moringa (Moringa oleifera) in Argentina for energy and as a social project. In: XIII World Forestry Congress. Proceedings. FAO. Buenos Aires, Argentina.

[18] Falasca, S. and Bernabé, M.A. (2009). Agro-climatic zoning of moringa (Moringa oleifera) in Argentina to produce biodiesel and bioethanol. In: Proceedings XXXII Congress ASADES. National University of Río Cuarto. Córdoba, Argentina. Published en CD rom.

[19] Falasca, S. and Bernabé, M.A. (2009). The replacement of Jatropha curcas culture by Pongamia pinnata in Argentina. In: Proceedings XVI Congresso Brasileiro de Agrometeorology. Belo Horizonte, Brazil. Published in CD rom.

[20] Falasca; S. and Bernabé, M.A. (2010). The Pongamia (Pongamia pinnata L) feedstock for biodiesel in Argentina and Latin America. In: CIIM 10. International Conference on Mechanical Engineering and Energy. Eastern University. Santiago, Cuba. Published in CD rom.

[21] Falasca, S. and Bernabé, M. A. (2008). Potential uses and delimitation of the area of cultivation of Moringa oleifera in Argentina. *Revista Virtual REDESMA. Biofuels*. Research Area. V 2 (1) La Paz. Bolivia.16pp.

[22] Falasca, S; Flores, N. and Galvani, G. (2010). Delimitation of the growing areas Guitzotia abyssinica (niger) in Argentina as a feedstock for biodiesel. In: XIII Meeting of Agrometeorology and VI Latin American Meeting of Agrometeorology. Southern National University Bahía Blanca, Argentina.

[23] Falasca, S; Flores, N; Lamas, M.A; Carballo, S and Anschau, A. (2010). Crambe abyssinica: an almost unknown crop with a promissory future to produce biodiesel. *International Journal of Hydrogen Energy*. 35. 5808-5812.

[24] Falasca, S. and Ulberich, A. Argentina's potential for the production of castor bean (Ricinus communis var commmunis L. (2007). Available from: http://www.biodiesel. com.ar/download/Ricino_BiodiselArgentinaWeb.pdf.

[25] Falasca, S and Ulberich, A. (2008). The thistle "Cynara cardunculus" agroclimate as an energy crop in semi-arid areas of Argentina. *Revista AVERMA* 12. 35-39.

[26] Falasca; S. and Ulberich, A. (2008). South American bioenergy potential from forest plantations with Jatropha curcas, hieronymi and macrocarpa. *Revista Virtual REDESMA. Biofuels.* 2 (2) 102-114. La Paz. Bolivia.

[27] Falasca, S. and Ulberich, A. (2010). The production of Ethiopian mustard (Brassica carinata) in Argentina as an energy crop. *Revista Geográfica IPGH* N° 148. 7-22.

[28] Falasca, S. and Ulberich, A. (2010).The white mustard (Sinapis alba), a source of biodiesel for Argentina. In: VI Conference on the Environment. UNICEN. Tandil, Argentina.

[29] Falasca, S. and Ulberich, A. (2006). Delimitation of the area of cultivation of castor (Ricinus communis) as a source of biodiesel. In: X National Congress of Cartography. Argentine Center of Cartography. Proceeding. Buenos Aires, Argentina.

[30] FAO. ECOCROP 1. "The crop environmental requirements database". Land and water digital media series 4. FAO. Roma, Italia. (1988).

[31] Fernández, J; Curt, M.D. and Aguado, P.L. (2007). Industrial applications of Cynara cardunculus L. for energy and other uses. *Industrial Crops and Products.* 24 (3) 222-229.

[32] Folkard, G. and Sutherland, J. (1996). Moringa oleifera. *Agroforesty Today* 8 (3): 5-8.

[33] Francis, G; Edinger, R. and Becker, K. (2005). A concept for simultaneous wasteland reclamation, fuel production, and socio-economic development in degraded areas in India: need, potential and perspectives of Jatropha plantations. *Natural Resources Forum*, 29, 12-24.

[34] Geleta, M.; Asfaw, Z.; Bekele, E. and Teshome, A. (2002). Edible oil crops and their integration with the major cereals in North Shewa and South Wello, central highlands of Ethiopia: an ethnobotanical perspective. *Hereditas* 137, 29-40.

[35] Getinet, A. and Sharma, S.M. Niger: Guizotia abyssinica (L. f.) Cass. Promoting the conservation and use of underutilized and neglected crops. (1996). Inst. of Plant Genetics and Crop Plant Res., Gatersleben, and Int. Plant Genetic Resources Inst., Rome. 59 pp. Available from: http://www.ipgri.cgiar.org/publications/pdf/136.pdf.

[36] Guzowski C, and Recalde M. (2008). Renewable energy in Argentina: energy policy analysis and perspectives. *International Journal of Hydrogen Energy* 33(13):3592–5.

[37] Heller, J. (1996). Physic nut, Jatropha curcas. Promoting the Conservation and Use of Underutilized and Neglected Crops. International Plant Genetic Resources Institute (IPGRI), Rome, Italy. 66 pp.

[38] IENICA. Crops database. (2002). Available from: http://www.ienica.net/cropsdatabase.htm; crambe.

[39] Isbell, T. (2007). Status of industrial crops in the U.S. [abstract]. Association for the Advancement of Industrial Crops. p. 2. Pennycress.

[40] Kandel, H. and Porter, P. (Eds.) (2002). Niger (Guizotia abyssinica) (L. f.) Cass. Production in northwest Minnesota. University of Minnesota. Extension Service. St. Paul MN, United States. 28 pp.

[41] Mc Laughlin, SP. (1985) Economic prospects for new crops in the Southwestern United States. *Economic Botany.* 39 (4), 473-48.

[42] Mastebroek, H.D; Wallenburg S.C. and Van Soest, L.J. (1994). Variation for agronomic characteristics in Crambe (Crambe abyssinica Hochst). *Industrial Crops and Products* 2:129–36.

[43] Meera, B; Kumar, S. and Kalidhar, S.B. (2003). A review of the chemistry and biological activity of Pongamia pinnata. *Journal of Medicinal and Aromatic Plant Sciences.* 25(5):441–446.

[44] Mohammed, A.S; Lai, O.M; Muhammad, S.K; Long, K. and Ghazali, H.M. (2003). Moringa oleifera, potentially a new source of oleic acid-type oil for Malaysia. Investing in Innovation. *Bioscience and Biotechnology,* 3: 137-140.

[45] Nair, A.G. (2005). Estimation of carbon sequestered in Pongamia pinnata and Eucalyptus spp. M. Sc. Thesis, Forest Research Institute, Dedhradun, India. P:47.

[46] Nasir; E; Ali, S.I. Flora of West Pakistan: an annotaded catalogue of the vascular plants of West Pakistan and Kashmir Karachy, Pakistan: Fakhri Printing Press.1028 pp. 1972.

[47] Nasirullah, K; Mallika, T; Rajalakshmi, S; Pashupathi, K.S; Ankaiah, K.N; Vibhakar, S; Krishnamurthy, M.N; Nagaraja, K.V. and Kapur, O.P. 1982. Studies on niger (Guizotia abyssinica) seed oil. *J. Food Sci. and Technol.* 19:147-149.

[48] Nautiyal, B.P and Venhataraman, K.G. 1987. Moringa an ideal tree for social forestry. Growing conditions and uses. *Myforest.* 23 (1): 53-58.

[49] Ometto, J.C. Bioclimatology plant. São Paulo. Agronomic Editorial. *Agronomic CERES.* 440 p. 1981.

[50] Oplinger, E.S; Oelke, E.A; Kaminski, A.R; Putnam, D.T; Teynor, T.M; Doll, J.D, et al. Alternative field crops manual. (1991). University of Wiscosin and University of Minesota. Available from: http:// www.hort.purdue.edu/newcrop/afcm/crambe.html.

[51] Papathanasiou, G.A; Lessman, K.J. and Nyquist, W.E. (1966). Evaluation of eleven introductions of Crambe, Crambe abyssinica Hochst. *Agronomy Journal.* 58:587–589.

[52] Ramanchandran, C; Peter, K.V; and Gopalakrishman, P.K. (1980). Drumstick (Moringa oleifera) a multiporpose Indian vegetable. *Economic Botany,* 34 (3): 276-283.

[53] Saydut, A; Duz, M.Z; Kaya, C, Kafadar, A.B, and Hamamci, C. (2008). Transesterified sesame (Sesamum indicum L.) seed oil as a biodiesel fuel. *Bioresource Technology* 99:6656-6660.

[54] Scott, P; Pregelj, L; Chen, N. and Gresshoff, P. (2008). Pongamia pinnata: an untapped resource for the biofuels industry of the future? *Bioenergy Research.* 1 (1) 2-11.

[55] Simon, J.E.; Chadwick and Craker, L.E. (1984). Herbs: An Indexed Bibliography. 1971-1980. The Scientific Literature in Selected Herbs and Aromatic and Medicinal Plants of the Temperate Zone. Archon Books, 770 pp., Hamden, CT. Available from: http://www.purdue.edu/newcrop/med-aro/factsheets/SESAME.html.

[56] Troup, R.S. The silviculture of Indian trees. Oxford, UK: Clarendon Press. 3th. vol. 1195 pp. 1921.

[57] Vollmann, J. and Ruckenbauer, P. (1993). Agronomic performance and oil quality of crambe as affected by genotype and environment. *Bodenkultur.* 44:335–43.

In: Biodiesel: Blends, Properties and Applications   ISBN: 978-1-63117-024-9
Editors: Jorge M. Marchetti and Zhen Fang   © 2014 Nova Science Publishers, Inc.

Chapter 6

STUDY OF BIODIESEL PURIFICATION METHODS

M. Berrios, J. A. Siles, M. A. Martín and A. Martín
Departamento de Química Inorgánica e Ingeniería Química,
Universidad de Córdoba, Campus Universitario de Rabanales,
Edificio Marie Curie, Córdoba, Spain

ABSTRACT

Renewable fuels have now come to play an important role in meeting the world's energy requirements. Due to the economic, environmental and supply problems derived from the use of fossil fuel, liquid biofuels are a growing and interesting complement to petroleum-based fuel. Biodiesel, which consists of long-chain fatty acid methyl esters (FAME) obtained from renewable lipids such as those in vegetable oils or animal fat can be used as both an alternative fuel and an additive for petroleum diesel.

After transesterification reaction, the obtained biodiesel must be purified before being used as diesel fuel in compliance with the EN 14214 Standard. The main disadvantages of using biodiesel are the high cost of the raw material and the need to purify the product, resulting in a very expensive final product in comparison to petrol-diesel. The first drawback to using biodiesel could be overcome by employing a cheap raw material such as used cooking oils or by using microalgae oils. On the other hand, improving product purification is one of the aims of this chapter with a view to fulfilling EN 14214 specifications.

Because FAME cannot be classified as biodiesel until the EN 14214 Standard specifications are met, the purification stage is essential. Depending on the raw material (used cooking oils, crude vegetable oils or refined vegetable oils) and the catalyst used, the purification stage will be more or less difficult. The purity level of the biodiesel has a strong effect on fuel properties and on engine life.

Several studies have examined the purification of biodiesel using different methods: liquid-liquid extraction (with water, acids or organic solvents), adsorption (silica gel, activated carbon, rice hull ash, bentonite, and magnesium silicate), ion exchange (cation or anion resins), molecular distillation and membrane separation.

The aims of this chapter are to revise the different purification methods described in the literature and examine the efficiency of three methods for removing methanol, glycerol and soaps under conditions that have been kept as close to commercial operating practice as possible: (a) adsorption (magnesium silicate and bentonite), (b) liquid-liquid extraction (distilled water, tap water, glycerol), and (c) ion exchange (cation resin). Further objectives are to study water and free fatty acid removal and the effect of these processes on final density, viscosity, glycerides and FAME.

INTRODUCTION

General Background

World energy demands, the price of petroleum and environmental concerns about pollution generated by cars or industrial gases continue to rise. The most feasible way to solve these problems is by using alternative or renewable fuels. Renewable fuels have now come to play an important role in meeting the world's energy requirements. Among renewable fuels, biofuels are defined as liquid or gaseous fuels for the transport sector that are predominantly produced from biomass [1]. Due to the economic, environmental and supply problems derived from the use of fossil fuel, liquid biofuels are a growing and interesting complement to petroleum-based fuel. Biodiesel, which consists of long-chain fatty acid methyl esters (FAME) obtained from renewable lipids such as those in vegetable oils or animal fat can be used as both an alternative fuel and an additive for petroleum diesel [2].

There are several advantages to using biodiesel. It is renewable and biodegradable, generates lower greenhouse emissions, contains little or no sulphur, and mixes in all proportions with petroleum diesel with no engine modifications being required. Furthermore, biodiesel is low in hydrocarbons, CO and particle emissions, acts as a lubricant, and has agricultural and environmental benefits. However, there are also some drawbacks to biodiesel such as its high freezing point (between 0 and $-5°C$), the obstruction of filters due to solvent power, its lower energy capacity than petroleum diesel, and storage problems due to the biodegradability of the product [3].

Transesterification is the most common way to produce biodiesel. In this reaction, triglycerides (the main components of vegetable oils) react with an alcohol to produce fatty acid monoalkyl esters and glycerol. Methanol is the alcohol that is normally used in the process due to its low cost and physical and chemical advantages (polar compound and short chain alcohol). This reaction, in turn, consists of three consecutive reversible reactions with the intermediate formation of diglycerides and monoglycerides. Although the reaction can take place in the absence of a catalyst, this is not economically feasible. The catalysis employed to produce biodiesel can be homogeneous or heterogeneous. Although there are different types of catalysts (acid, alkaline or enzymatic), homogeneous alkaline catalysts are the most widely used in industry to produce biodiesel as they accelerate the process and the reaction conditions are more moderate [4, 5].

The main disadvantages of using biodiesel are the high cost of the raw material and the need to purify the product, resulting in a very expensive final product in comparison to petroleum diesel. The first drawback to using biodiesel could be overcome by employing a cheap raw material such as used cooking oils [6-9]. Improving product purification is one of the aims of this chapter with a view to fulfilling EN 14214 specifications [10].

Methodologies for Biodiesel Production

There are several methodologies for biodiesel production. Some of them use pre-treatments to remove impurities in the raw materials such as free fatty acids (FFA), wax, tocopherols, solids or phosphatides that could interfere or prevent the transesterification reaction. As can be observed in Figure 1, the production methods can be classified according to the system, the catalyst or the agitation used.

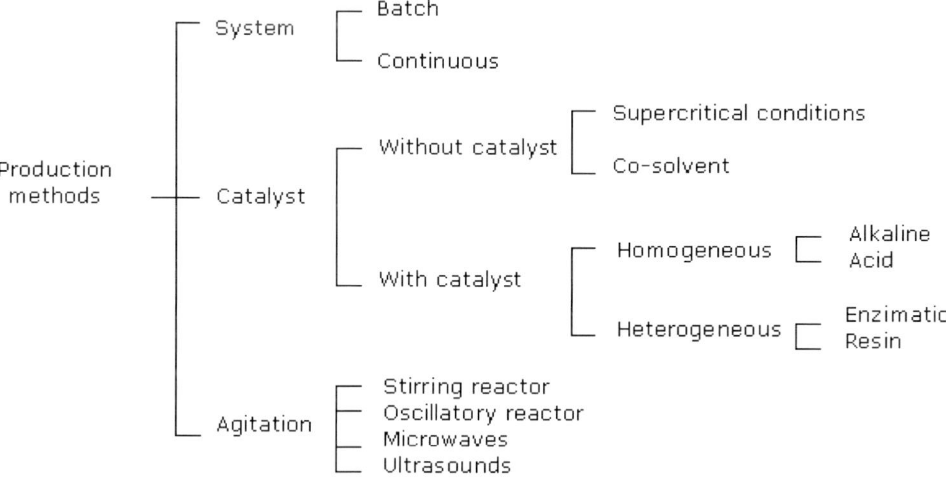

Figure 1. Diagram of production methods for biodiesel.

Separation of Phases

Regardless of the method selected to produce biodiesel after the reaction, the glycerol phase and excess alcohol must be separated and extracted from the reaction mixture. The excess alcohol is removed by means of flash distillation and is returned to the process. The glycerol phase is separated by settling or centrifuging. The two phases must be purified in order to fulfil quality requirements and improve the profitability of the process. Given that neither of these processes is 100% efficient, a final purification stage is needed in order to meet the requirements of the EN 14214 Standard [10].

Need for Purification

Because FAME cannot be classified as biodiesel until the EN 14214 Standard specifications [10] are met, the purification stage is essential. The European Standard EN 14214 on biodiesel quality [10] establishes the quality requirements of FAME for use as automotive fuels. The purity level of the biodiesel has a strong effect on fuel properties and engine life. Non-purified biodiesel contains several impurities such as free glycerol, soap, metals, methanol, FFA, catalyst, water and glycerides (mono-, di- and triglycerides). Table 1 shows the effect of these impurities on biodiesel properties and engine life.

Table 1. Effect of impurities on biodiesel and engines

Impurity	Effect on biodiesel properties	Effect on engine life
Glycerol	Settling problems	Increase aldehydes and acrolein emissions
Methanol	Low values of density and viscosity Low flash point	Corrosion of Al and Zn pieces
Free fatty acids	Low oxidation stability	Corrosion
Water	Hydrolysis (free fatty acids formation	Corrosion Filter blockage (Bacteriological growth)
Glycerides	High viscosity Crystallization	Deposits in the injectors (carbon residue) Filter blockage (sulphated ashes)
Metals (soap, catalyst)	Settling problems (emulsions by soap formation)	Deposits in the injectors (carbon residue) Engine weakening

Several studies have examined the purification of biodiesel using different methods: liquid-liquid extraction (with water, acids or organic solvents), adsorption (silica gel, activated carbon, rice hull ash, bentonite, and magnesium silicate), ion exchange (cation or anion resins), molecular distillation and membrane separation.

The glycerol phase can also be processed in order to maximize the FAME yield and increase the interest of the by-product. The glycerol phase can contain some amounts of FFA, soaps, methanol and traces of FAME. The first stage involves the addition of phosphoric acid or sulphuric acid for breaking down the soaps and producing FFA. Depending on the catalyst used in the transesterification reaction, it is possible to obtain a fertilizer. This is the case, for example, when potassium hydroxide is used as a catalyst and phosphoric acid as a reagent in soap decomposition [11]. Otherwise, the obtained solid is considered a residue. The obtained FFA are insoluble in the glycerol phase and can be easily removed and returned to the FAME production [11]. Refined glycerol has conventional uses in the cosmetic, pharmaceutical and explosive industries, but the required purity levels make the purification process too expensive. Refined glycerol, without high purity, has applications in combustion processes, animal feed, fertilizer production (mixed with animal manure) or methane production by means of anaerobic digestion [12].

Quality Standard Requirements

Quality standards must be met before any type of fuel can be used commercially. Quality standards serve as guidelines for the production process, guarantee the quality of the product and provide the competent authorities and agencies with tools to assess security risks and environmental pollution [13]. Moreover, the automotive and motor industry uses these fuel standards to provide a guarantee that biodiesel is a suitable fuel for vehicles [11].

Biodiesel and petroleum diesel differ greatly in their chemical composition and, as such, in their physical and chemical properties. Although their behaviour in motor applications is similar, the quality requirements are quite different. The quality references for biodiesel are EN 14214 Standard [10] in Europe and ASTM D6751 Standard [14] in the USA. Both standards establish the quality requirements for pure biodiesel and for blends of biodiesel and petroleum diesel in any proportion. The main difference between the two is that the European standard sets more specifications and is more restrictive than the American standard. The quality requirements of the EN 14214 Standard can be observed in Table 2.

Table 2. Generally applicable requirements and test methods

Property	Units	Limits Minimum	Maximum	Test method
FAME content	% (m/m)	96.5	-	EN 14103
Density at 15°C [a]	kg/m^3	860	900	EN ISO 3675 EN ISO 12185
Viscosity at 40°C [b]	mm^2/s	3.50	5.00	EN ISO 3104
Flash point	°C	101	-	EN ISO 2719[c] EN ISO 3679[d]
Sulphur content	mg/kg	-	10.0	EN ISO 20846 EN ISO 20884
Carbon residue (on 10% distillation residue)[e]	% (m/m)	-	0.30	EN ISO 10370
Cetane number		51.0	-	EN ISO 5165
Sulphated ash content	% (m/m)	-	0.02	ISO 3987
Water content	mg/kg	-	500	EN ISO 12937
Total contamination	mg/kg	-	24	EN 12662
Copper strip corrosion (3 h at 50°C)	rating	Class 1		EN ISO 2160
Oxidation stability, 110°C	hours	6.0	-	prEN 15751 EN 14112
Acid value	mg KOH/g	-	0.50	EN 14104
Iodine value	g iodine/100 g	-	120	EN 14111
Linolenic acid methyl ester	% (m/m)	-	12.0	EN 14103
Polyunsaturated (≥4 double bonds) methyl esters	% (m/m)	-	1	Under development
Methanol content	% (m/m)	-	0.20	EN 14110
Monoglyceride content	% (m/m)	-	0.80	EN 14105
Diglyceride content	% (m/m)	-	0.20	EN 14105
Triglyceride content	% (m/m)	-	0.20	En 14105
Free glycerol	% (m/m)	-	0.02	EN 14105 EN 14106
Total glycerol	% (m/m)	-	0.25	EN 14105
Group I metals (Na+K)	mg/kg	-	5.0	EN 14108 EN 14109
Group II metals (Ca+Mg)	mg/kg	-	5.0	EN 14538 EN 14538
Phosphorus content	mg/kg	-	4.0	EN 14107

Notes on the table: % (m/m) is mass/mass fraction. [a] Density may be measured by EN ISO 3675 over a range of temperatures from 20°C to 60°C. Temperature correction shall be made according to the formula given in Annex C of EN 14214. [b] If CFPP is -20 °C or lower, the viscosity shall be measured at -20 °C. The measured value shall not exceed 48 mm^2/s. In this case, EN ISO 3104 is applicable without the precision data owing to non-Newtonian behaviour in a two-phase system. [c] Procedure A to be applied. Only a flash point test apparatus equipped with a suitable detection device (thermal or ionization detection) shall be used. [d] A 2 ml sample and apparatus equipped with a thermal detection device shall be used. [e] ASTM D 1160 shall be used to obtain the 10 % distillation residue.

Some of the requirements are specific for biodiesel, while others are common to both biodiesel and petroleum diesel. In what follows, these requirements are explained [11].

Quality requirements for FAME:

- FAME content

The minimum percentage for monoalkyl esters of fatty acids (in general FAME) in biodiesel is 96.5%. This specification is essential to determine a product as biodiesel and to detect adulterated blends. A FAME content of less than 96.5% can indicate unsuitable operational conditions, undetected peaks in the analytical method of determination (EN 14103 Standard), or high level of compounds different from glycerides (unsaponificable matter, polymers, oxidized esters, etc.).

- Free glycerol content

The maximum admissible percentage of free glycerol is 0.02%. The free glycerol content depends directly on the process, specifically phase separation and ester purification. Moreover, the free glycerol can appear due to the hydrolysis of residual glycerides in biodiesel. The main disadvantage of the presence of free glycerol is that it attracts other polar compounds such as water, monoglycerides and soaps, which damage the injection system of engines.

- Monoglycerids, diglycerids, triglyceride and total glycerol content.

The maximum percentages of these four compounds are 0.80%, 0.20%, 0.20% and 0.25%, respectively. The content of these compounds is closely related to the process and increases the viscosity of the product. As a consequence of increased viscosity, fuel atomization is worse and deposits are formed in injectors and valves.

- Methanol

0.20% is the maximum admissible percentage for methanol content. A high methanol content leads to security risks in the transport and storage of biodiesel. An effective distillation stage is compulsory in the process in order to remove the alcohol present in the mixture.

- Unsaturated fatty acid

These compounds are controlled by means of three specifications under the EN 14214 Standard: iodine value ($<120g$ $I_2/100g$ sample), linolenic acid methyl ester ($<12.0\%$) and polyunsaturated methyl esters ($<1.0\%$). The unsaturated compounds tend to polymerize at high temperatures and show problems of oxidation stability.

- Acid value

This specification is a measure of the mineral acids and the FFA present in a fuel (<0.5 mg KOH/g). The acid value depends on the raw material, the process (acid catalysts, soap decomposition, etc.) and the hydrolysis of methyl esters during storage. High values are related to corrosion and deposits in the engine.

- Phosphorus, alkaline metals and alkaline earth metals contents

The maximum contents of these compounds are 4.0 mg/kg, 5.0 mg/kg and 5.0 mg/kg, respectively. The presence of phosphorous is related to the presence of phospholipids in the raw materials or the use of phosphoric acid in the process. On the other hand, the metallic ions are present in biodiesel as a consequence of the process. The alkaline metals come from the catalyst, while the alkaline earth metals come from the hardness of the wash water. These compounds affect engine performance due to the formation of ashes, the reduced efficiency of oxidation catalytic converters or injection pump blockage.

Quality requirements for biodiesel and petroleum diesel:

- Density at 15°C

Biodiesel density is higher than petroleum diesel density and depends on the fatty acids composition and the purity of the raw material. The limits are 860 and 900 kg/m^3.

- Kinematic viscosity at 40°C

The selected range is from 3.50 to 5.00 mm^2/s. This specification is closely related to fatty acids concentration and their unsaturation level. Other compounds that can influence this specification are residual glycerides and products from oxidative degradation during storage. High kinematic viscosity values cause the inadequate atomatization of fuel and incomplete combustion.

- Flash point

This specification is essential for assessing security risks during the transport and storage of a fuel. For biodiesel, the minimum value is 101°C, twice the value established for petroleum diesel.

- Sulphur content

One of the characteristics that make biodiesel environmental friendly is its low sulphur content, which is limited to 10 mg/kg. The presence of sulphur in a fuel leads to engine wear and the emission of particles and sulphur oxides.

- Carbon residue

This specification refers to the amount of residual carbon matter following the evaporation and pirolysis of a fuel sample under specific conditions. It is related to glycerides, FFA, soaps and catalyst content, and is limited to 0.30%.

- Cetane index

This specification measures the ignition quality of diesel-type fuels. A minimum value of 51 (the same for diesel) is required to ensure correct engine performance.

- Ash

This is a measure of the amount of inorganic pollutants such as abrasive solids, catalyst residues and soluble soaps in the fuel sample. These compounds are oxidized during the combustion and generate ashes, which are related to engine dirt. This specification is limited to 0.02%.

- Water content

Water content is a decisive property in biodiesel quality. Water is brought on during the process and must be reduced to 500 mg/kg or lower by means of drying. The monoalkyl esters are hygroscopic and can absorb large amounts of water. This excess water can promote bacteriological growth and the hydrolysis of esters.

- Total contamination

Total contamination is defined as the insoluble matter retained after filtration of a heated sample through a 0.8µm standardized filter. The maximum value for this specification is 24 mg/kg.

- Copper strip corrosion (3h at 50°C)

This specification shows the tendency of a fuel to cause corrosion in copper, zinc or bronze pieces from engine or storage tanks. Corrosion can be induced by acids or sulphur compounds. The standard establishes a class 1 corrosion pattern (the range is from 0 to 4).

- Oxidation stability

Due to their chemical composition, monoalkyl esters are more sensitive than petroleum diesel to oxidative degradation. This structural alteration can generate sediments, corrosion or blockage in engines. The EN 14214 Standard establishes an induction period of over 6 hours by means of the Rancimat method.

- Cold filter plug point (CFPP)

Although it is not a specification in the EN 14214 Standard, the behaviour of biodiesel at low temperatures is a quality criterion in cold regions. Partial solidification at low temperatures can cause blockage of fuel lines and filters that can lead to problems when starting engines. The pour point can also be used to determine the winter properties of fuel.

PURIFICATION METHODS

As mentioned above, several studies have examined biodiesel purification using different methods: liquid-liquid extraction (with water, acids or organic solvents), adsorption (silica gel, activated carbon, rice hull ash, bentonite, and magnesium silicate), ion exchange (cation or anion resins), molecular distillation and membrane separation. All of these methods provide guidelines for removing the main impurities present in biodiesel in order to fulfil the EN 14214 Standard.

Liquid-Liquid Extraction

Liquid-liquid extraction, also known as solvent extraction and partitioning, is a method to separate compounds based on their relative solubilities in two different immiscible liquids; usually water and an organic solvent. It is an extraction of a substance from one liquid phase into another liquid phase. In the case of biodiesel purification, the substance is an impurity coming from the raw material or due to the production process.

Different types of extraction agents for biodiesel purification can be found in the literature such as distilled water, tap water, acidified water, petroleum ether, sulphuric acid, ortho-phosphoric acid, glycerol, deep eutectic solvent, and so on.

Karaosmanoglu et al. [15] studied three liquid-liquid extraction methods for the purification step of biodiesel originating from rapeseed oil: (a) washing the product with hot distilled water at three temperatures: 50°C, 65°C and 80°C, (b) dissolving the product in petroleum ether and sequential washing with distilled water, and (c) neutralizing the product with sulphuric acid in solid form or dissolved in methanol (1:1). According to their studies, the best purification method was washing with distilled water at 50°C on the basis of biodiesel purity and operational costs. The purity of the biodiesel obtained was about 99%. Since both glycerol and methanol are highly soluble in water, washing with water is very effective in removing both contaminants and, until a few years ago, was the most common method of purification. Due to their water solubility, this method also has the advantage of removing any residual sodium compounds and soap, which are the result of high FFA raw material. But washing with water also has some disadvantages such as generating a highly polluting liquid effluent, significant product loss due to retention in the water phase, or the formation of emulsions when processing used cooking oils or other raw material with a high FFA content [16].

Following one of the methods of Karaosmanoglu et al. [15], Siler-Marinkovic and Tomasevic [17] used a transesterification mixture comprised mainly of FAME and glycerol.

The mixture was filtered through sintered glass and washed with petroleum ether, which was then washed with water until the washings were neutral. The organic phase was dried over sodium sulphate, filtered and evaporated. An additional problem involved in this process is that petroleum ether must be removed from the FAME product, thus increasing the costs of operation and installation.

Berrios and Skelton [3] studied two biodiesels of different origins and different ways of purifying them, namely liquid-liquid extraction with distilled water, tap water and acidified water. They evaluated the efficiency of the removal of methanol and glycerol under a variety of conditions as close to commercial operating practise as possible. The efficiency of soap removal and the effect of the processes on final FFA and oxidation stability were also investigated. The results demonstrated that only the biodiesel purified by means of water washing directly from the glycerol separation fulfilled the requirements of the EN 14214 Standard. Although tap water was found to be the most suitable option in economic terms, it has some disadvantages such as water supply and cost, emulsions, wastewater treatment and drying of the final product.

Predojevic [18] studied the purification of biodiesel from used cooking oils. His aim was to investigate the influence of different purification methods on the properties and yield of FAME. The research evaluated the influence of liquid-liquid extraction with 5 wt% phosphoric acid, and liquid-liquid extraction with hot distilled water (50°C). The properties Predojevic evaluated were kinematic viscosity at 40°C, density at 15°C, acid value, iodine value, water content, saponification value, cetane index, fatty acid composition and methyl ester content (purity and yield). The results showed that the phosphoric acid treatment obtained better yields ($\approx$92%) than the hot water treatment ($\approx$89%). According to Predojevic [18], liquid-liquid extraction with a phosphoric acid treatment is more suitable for purifying the product from used cooking oils than liquid-liquid extraction with hot water.

Glisic and Skala [19] used a combination of the neutralization step and the washing step to purify biodiesel from alkaline transesterification. They selected ortho-phosphoric acid as the most suitable neutralizer of sodium hydroxide. The washing step with hot water was optimized in order to minimize water usage during the combined neutralization-washing stage in biodiesel production. Such a solution would make the biodiesel synthesis process more economical taking into account the lower amount of energy consumed for water evaporation during purification of the final product. In addition, this procedure is more acceptable in terms of its environmental impact as wastewater release is minimized.

Sabudak and Yildiz [20] used the same methods as Berrios and Skelton [3] to purify biodiesel such as hot water washing (50°C-60°C). According to their results, the lowest yield was obtained by water washing. When two-step acid-base transesterification was used, the yield prior to purification was 90.3%. After water washing purification, the results were 95.6%; a value that is lower than the required minimum under the EN 14214 Standard.

Recently, Berrios et al. [21] have studied liquid-liquid extraction using three extraction agents: distilled water, tap water and glycerol. Washing was carried out at room temperature. Three concentrations of extraction agent (5, 10 and 15 wt%) and one or two extraction steps were studied.

The final product was separated by centrifugation. Samples were analysed for methanol, free glycerol, glycerides, water, acid value, soap, density, kinematic viscosity and FAME using standard methods [22-29] to study the effect on these variables. The experimental system can be observed in Figure 2.

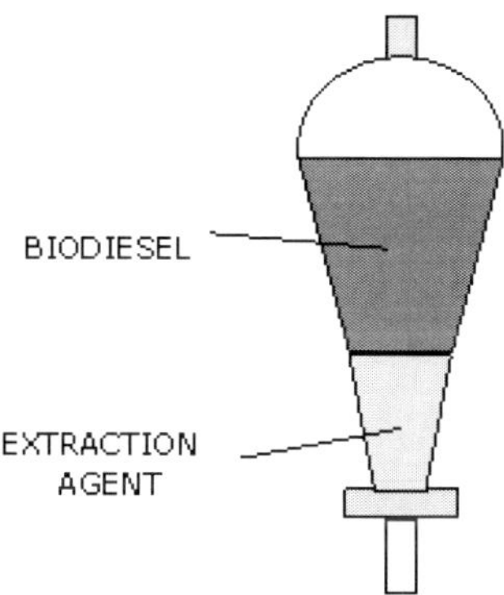

Figure 2. Liquid-liquid extraction system.

When distilled water was employed, a negative effect was observed in the water and the FFA content. The first can be explained by the operational system in which the water used for washing was removed by centrifugation at 4000 rpm for 10 min. A more intensive centrifugation or an additional vacuum drying step could be needed in order to decrease the final water content. As regards FFA content, the partial hydrolysis of FAME to produce FFA could be in order [30]. Although the FFA content increased by about 25%, neither test exceeded the acid value requirements of the EN 14214 Standard [10]. Better results were obtained when the wash water was added in two steps instead of one, with removal percentages for soap, glycerol and methanol of 61%-82%, 55%-84% and 99.4%, respectively, for 5-15 wt% of distilled water. Methanol removal was almost complete under all the conditions tested. This high removal might be due to the polarity of both molecules. Among the conditions tested, the best results with distilled water were reached at two steps and 10 wt% of distilled water. The same results were found when using tap water instead of distilled water.

The best results were obtained at two steps and 10 wt% tap water. Adding water in two steps favoured the distribution of impurities between the phases. When using glycerol as an extraction agent, soap removal was around 69.6-81.8%. Similar to water washing, methanol removal was almost complete (99.2-99.8%). A 27.5% increase over the initial FFA content was found due to a left-side movement in the esterification reaction, which occurred previous to purification.

Although hydrolysis is a problem during storage prior to analysis, the acid value requirements of the EN 14214 Standard were fulfilled [10]. Glycerol was the only purifying agent that had a positive effect on water content. The results were around 26.5%-68.1%, but the best results were achieved in 2-step additions. The fact that both species had the same polarity could have favoured this removal. Increased glycerol content is one disadvantage to this method.

Neither of the conditions tested has fulfilled the requirements of the EN 14214 Standard [10] for glycerol content. Although the 2-step contact impeded glycerol removal, this could be improved by increasing the centrifugation speed or by post-treatment with ion exchange resins.

The best results with glycerol were obtained with a concentration of 15 wt% and the 2-step contact. Figures 3-5 show the removal of methanol, glycerol and soap for liquid-liquid extraction with distilled water, tap water and glycerol, respectively.

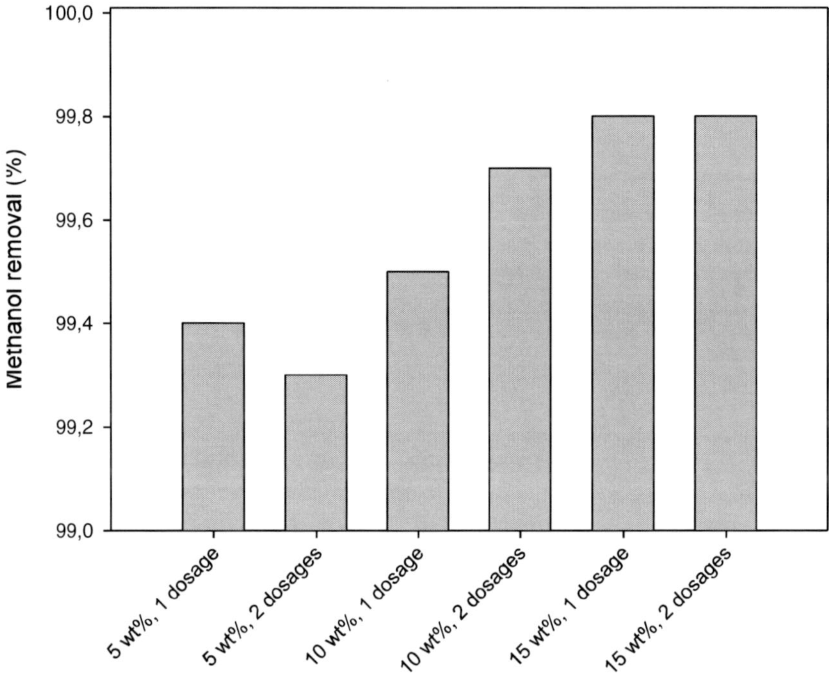

Figure 3. Methanol removal by distilled water washing.

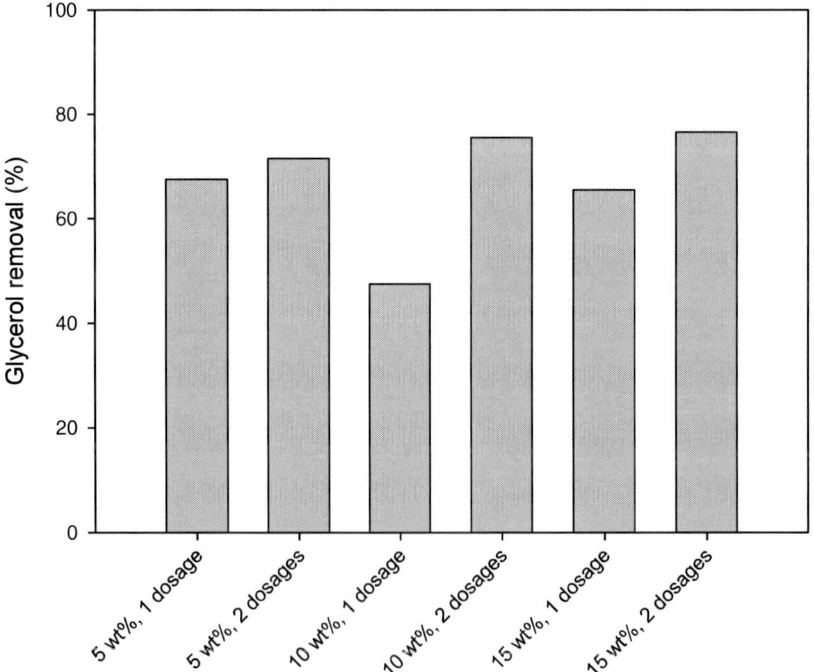

Figure 4. Glycerol removal by tap water washing.

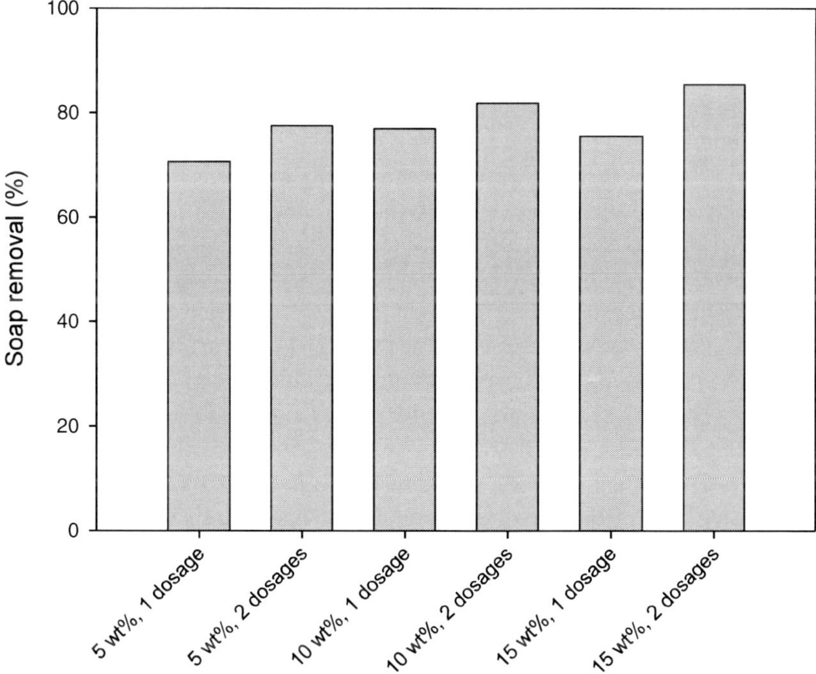

Figure 5. Soap removal by liquid-liquid extraction with glycerol.

Finally, Hayyan et al. [31] and Shahbaz et al. [32] have shown that ionic liquids, specifically deep eutectic solvents, are a novel purification method for removing glycerol from biodiesel. A deep eutectic solvent is a mixture of two or more compounds which has a melting point lower than that of either of its components.

They comprise mixtures of organic halide salts such as choline chloride with an organic compound which is a hydrogen bond donor capable of forming a hydrogen bond with the halide ion such as amides, amines, alcohols, carboxylic acids and many more. In the case of biodiesel purification, the hydrogen bond donor is glycerol. They studied the best ratio of deep eutectic solvent composition and the best biodiesel to deep eutectic solvent ratio in terms of extraction capabilities.

Liquid-liquid extraction methods are an interesting alternative to purify biodiesel in terms of removing impurities. In fact, it is widely used in biodiesel production as a purification process.

However, other aspects must be taken into account such as the subsequent wastewater treatment or the recovery of organic solvents. As mentioned above, water washing has certain disadvantages including the generation of a highly polluting liquid effluent, significant product loss due to retention in the water phase, or the formation of emulsions when processing used cooking oils or other raw material with a high FFA content [16]. Therefore, other environmental-friendly methodologies should be studied in order to improve the biodiesel production process. Using residual glycerol to purify biodiesel could be a promising option.

Adsorption

Adsorption is the adhesion of atoms, ions, biomolecules or molecules of gas, liquid, or dissolved solids to a surface. This process creates a film of the adsorbate (the molecules or atoms being accumulated) on the surface of the adsorbent. It differs from absorption, in which a fluid permeates or is dissolved by a liquid or solid.

Although this research is not about biodiesel purification, it provides some information about the use of adsorbents to purify oily matrixes. In this line, McNeill et al. [33] suggested treating used cooking oils with blends of activated carbon and silica in order to remove FFA and polar compounds (deep fat frying products). They applied adsorption as a purification method for used cooking oils. These same adsorbents (dry washing) could be utilized in biodiesel. They used two samples of used cooking oils with nine different blends of activated carbon and silica. The results demonstrated that the blends of adsorbents were effective in reducing acid value, peroxide value, saturated and unsaturated carbonyls, polar compounds and photometric colour. A combination of the two adsorbents was selected due to the unique adsorptive characteristics of these two compounds. Activated carbon removes impurities by occlusion and by adsorption based on chemical properties, while the adsorptive qualities of silica are based primarily on interaction with exposed silanol groups. They concluded that the blends of activated carbon and silica were most effective in reducing acid value and saturated carbonyls and least effective in removing total polar compounds. Therefore, depending on the composition of the raw material (oil or biodiesel), this combination of adsorbents may or may not be useful.

Özgül-Yücel and Türkay [34] used rice hull ash (RHA) and commercial silica gel to purify FAME by adsorption of FFA under atmospheric conditions. Their studies demonstrated that commercial silica gel was a better adsorbent for purifying FAME than RHA, resulting in less loss of FAME and similar levels of FFA. The adsorption isotherms of FFA were determined with both adsorbents. FFA adsorption displayed Freundlich-type isotherms. The adsorption system was not totally dry; a miscella system with hexane and the adsorbents was created.

Dmytryshyn et al. [35] also used silica gel as a purifying agent. Prior to the silica gel treatment, the methanol was removed from the product with the help of a rota-evaporator at 60°C during 30 minutes. After this initial step, the product was transferred to an Erlenmeyer flask to which silica gel was added. The mixture was then stirred for approximately 20 minutes, and then filtered to remove the silica gel. A layer of sodium sulphate crystals was added to a vacuum filtration funnel, and the ester was filtered through it to remove any traces of water present. These adsorbents, like magnesium silicates, ion exchange resins, silica and RHA comprise the dry washing methods. The main disadvantage of these processes is to find a solution for the solid waste that is generated.

Berrios and Skelton [3] also studied adsorption as a purification method. The selected adsorbent was magnesium silicate. According to their research, at least 0.75wt% is required with a contact time of 10 min in order to decrease the glycerol and soap content. With regards to methanol removal, the results were better than ion exchange resins but not higher than water washing. They also analysed the influence of pre-drying the adsorbent or operating at a high temperature, but no improvement was observed.

Adsorption on silica gel (column chromatography) was also selected by Predojevic [18] to remove the impurities present in biodiesel from used cooking oils. The research compares the efficiency of adsorption versus liquid-liquid extraction. The results showed that silica gel as phosphoric acid treatments obtained the highest yields (≈92%). The average purity level regarding the purification treatments were very similar: 97.41% for silica gel, 97.16% for phosphoric acid, and 97.92% for hot water.

Manuale et al. [36] studied the capacity of transesterification with supercritical methanol to react with different raw materials and a final adsorption-based refining step. The adsorbents used were activated carbon, impregnated activated carbon, diatomaceous earth and silica gel. Refining was performed at room temperature in a gas-tight autoclave with stirring. The adsorbent loading was 0.3–5 wt% and contact time was 720 min. Tests using silica gel as adsorbent were performed with stirring at 90°C in a vacuum (0.21 bar) during 90 min. The best adsorption capacity was displayed by silica. Silica adsorbed as much as 140% FFA and should be the adsorbent of choice for any industrial bleacher.

1 wt% magnesium silicate at high temperature (70°C-80°C) was employed by Sabudak and Yildiz [20] to compare density, viscosity, flash point, water content, acid value, iodine value and methyl ester content with other purification methodologies. Of all the production processes used in the study, the methyl esters yield increased by at least 1% after purification. Specifically, with magnesium silicate and two-step acid-base transesterification, the yield enhanced from 90.3% to 96.9%.

Berrios et al. [47] used two different adsorbents (magnesium silicate and bentonite) to study the purification of biodiesel from used cooking oils. The experiment was performed in a batch reactor with a magnetic regulation speed agitator. Figure 6 shows a diagram of the system. Contact time was 15 min. and the final product was separated by centrifugation. The experiments were run at room temperature with three adsorbent concentrations of 0.50, 0.75 and 1.00 wt% (relative to the mass of the FAME product) and 200, 400 and 600 rpm (agitation speed) as suggested by a previous paper [3]. Several specifications of the EN 14214 Standard [10] were analysed including methanol, free glycerol, glycerides, water, acid value, soap, density, kinematic viscosity and FAME using standard methods [32-39].

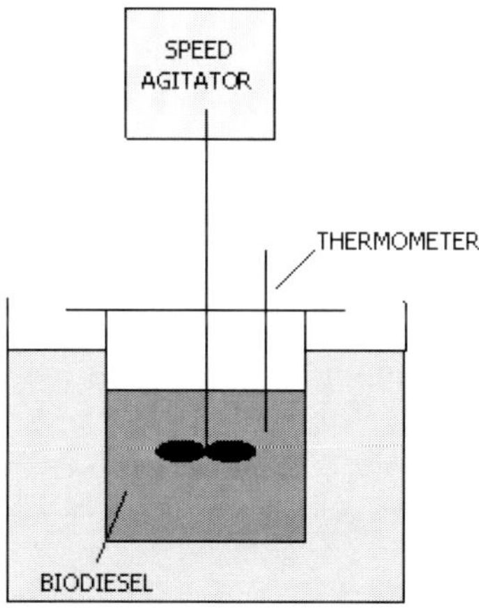

Figure 6. Adsorption system.

Magnesium silicate (Magnesol®) had a positive effect on the removal of soap, methanol and glycerol. Soap and methanol were reduced by around 67%-92% and 98%, respectively, for the 0.50 to 1.00 wt% range of Magnesol®.

 M. Berrios, J. A. Siles, M. A. Martín et al.

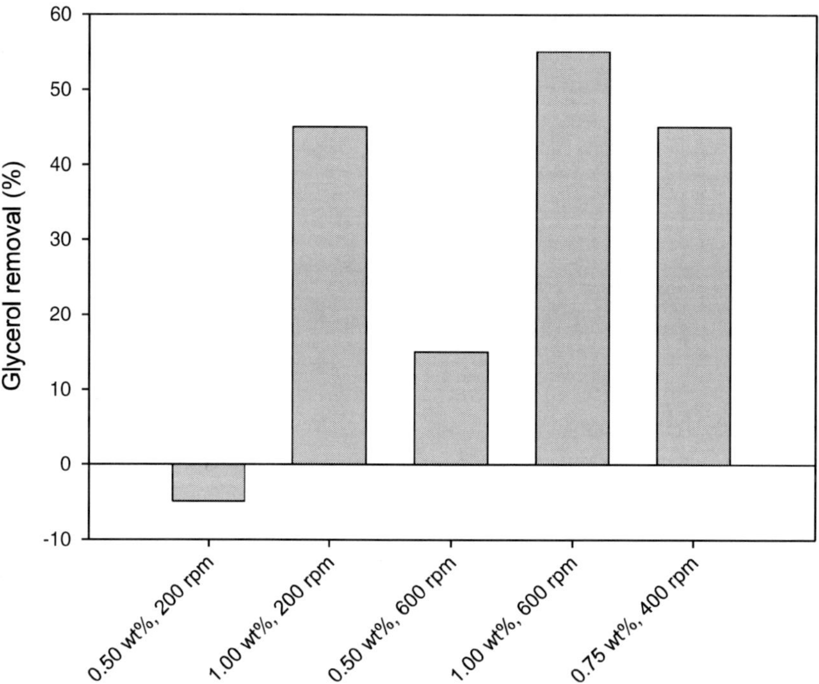

Figure 7. Glycerol removal for magnesium silicate adsorption.

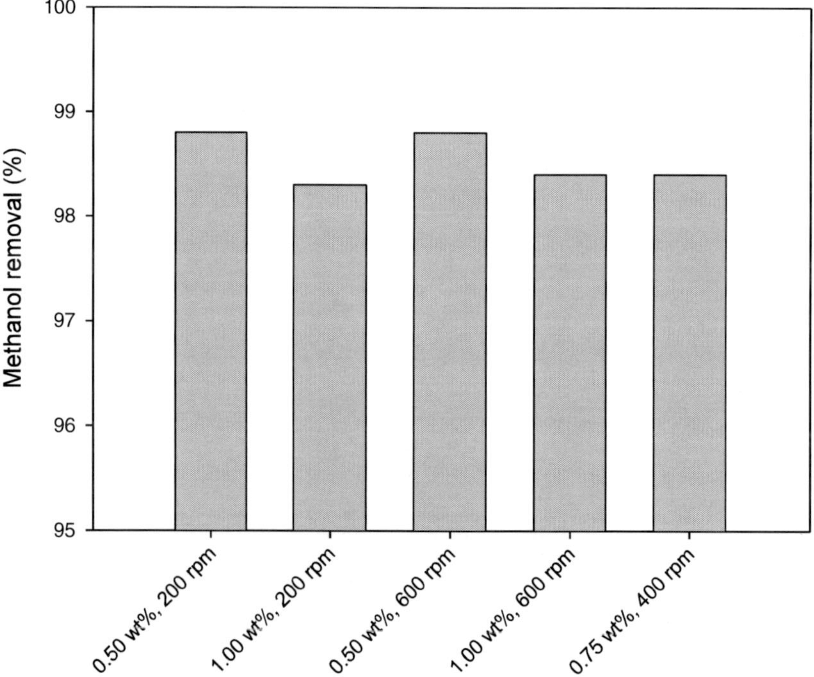

Figure 8. Comparison of adsorption conditions for methanol removal with bentonite.

The reductions in glycerol were lower with percentages ranging from 15% to 55%. The best results with Magnesol® were obtained with a concentration of 1.00 wt% and 600 rpm. Bentonite had a positive effect on acid value and the removal of soap, methanol and glycerol in the 0.50-1.00 wt% range. Methanol was the most highly retained impurity with a 98% removal rate. Soap was reduced by 24%-40% and glycerol by 15%-20%. Although the results obtained with bentonite were slightly lower than with Magnesol®, FFA retention (16-30%) can be useful in samples from high FFA feedstocks. The best results with bentonite were obtained with a concentration of 1.00 wt% and 200 rpm.

The influence of different concentrations of magnesium silicate on glycerol removal can be observed in Figure 7. Figure 8 shows methanol removal by means of bentonite adsorption.

Ion Exchange

Ion exchange is a reversible chemical reaction wherein an ion (an atom or molecule that has lost or gained an electron and thus acquired an electrical charge) from solution is exchanged for a similarly charged ion attached to an immobile solid particle. These solid ion exchange particles are either naturally occurring inorganic zeolites or synthetically produced organic resins. The synthetic organic resins are the predominant type used today because their characteristics can be tailored to specific applications. Ion exchange resins are classified as cation interchangers, which have positively charged mobile ions available for exchange, and anion interchangers, whose changeable ions are negatively charged. Both anion and cation resins are produced from the same basic organic polymers. They differ in the ionizable group attached to the hydrocarbon network. It is this functional group that determines the chemical behaviour of the resin. Resins can be broadly classified as strong or weak acid cation exchangers or strong or weak base anion exchangers.

Ion exchange techniques have been less developed than other purification methods for biodiesel. Berrios and Skelton [3] and Sabudak and Yildiz [24] used cation resins to remove several impurities as methanol, glycerol and soaps. Sabudak and Yildiz [24] also studied the yield obtaining a final value of 98.4%. Berrios and Skelton [3] calculated the resin capacity to compare the relative economics of the processes for a possible industrial application. The ion exchange process brought the free glycerol level down to the specifications in EN 14214 Standard. The capacity of methanol removal for both resins was very low (around 20 L biodiesel/kg resin).

Berrios et al. [21] also compared the ion exchange method with adsorption and liquid-liquid extraction. The ion exchange resin was examined by passing the feed through a resin column contained in a 1.5 cm diameter glass tube fitted with a glass wool plug (Figure 9). The flow was controlled by a metering pump and the outlet restricted to ensure that a head of liquid remained above the resin at all times. The flow was as recommended in the Lewatit trade literature: 1.5-2 BV·h^{-1} (BV: bed volumes) and a liquid hourly space velocity at 2.01 h^{-1}. The temperature of the fixed bed experiment was set at 22°C (room temperature). Samples were initially taken at 1 h intervals. After the first 6 h, samples were taken at 2 h and 4 h intervals until the resin saturation was detected (54 h). The samples were then analysed for methanol, free glycerol and water using standard methods [22-29]. At every 10 h interval of the operation, a bulked sample was analysed in order to determine acid value, soap, density, kinematic viscosity and FAME using standard methods.

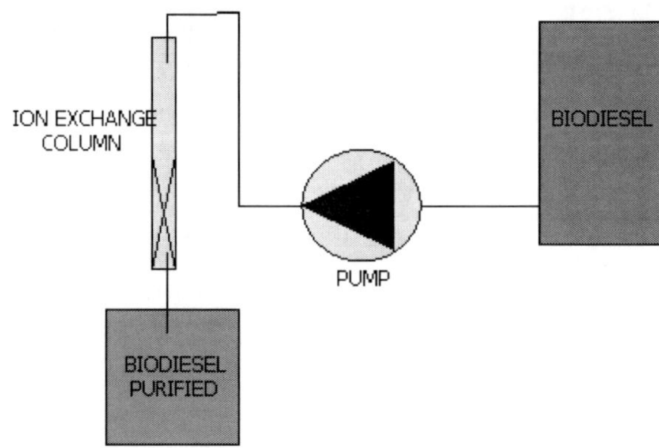

Figure 9. Ion exchange system.

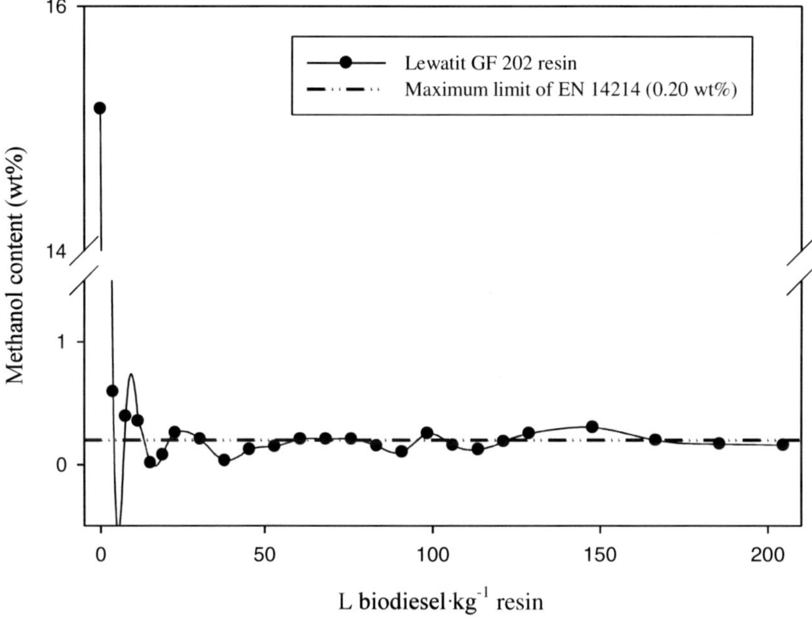

Figure 10. Evolution of methanol content vs. resin capacity.

The tested resin (Lewatit® GF202) showed no influence on density, kinematic viscosity or glycerides, FAME, FFA and water content. Soap, methanol and glycerol removal were 52.2%, 98.8% and 20.2%, respectively. The resin capacity was quite low (205 L treated biodiesel·kg⁻¹ resin) compared to others tested previously [3]. An advantage of this resin is that it can be regenerated and reused, whereas other resins can be used only one time. Although a high volume of methanol is required for regeneration (around 10 BV), it could be interesting to study the use of this methanol in the transesterification process. However, the regeneration of resin is not the purpose of this chapter. The evolution of methanol and glycerol removal versus resin capacity is shown in Figures 10 and 11, respectively. Figure 10 shows that resin saturation in methanol was reached quickly, mainly due to the initial methanol content (15.16 wt%) in the crude biodiesel. Methanol must be removed by flash evaporation before the product is passed through the ion exchange resin column to ensure that the methanol concentration in the feed is similar as suggested by the suppliers. Figure 11 shows that after 100 L treated biodiesel·kg^{-1} resin, some evidence of back washing of glycerol was detected. These results are in line with previous research [3].

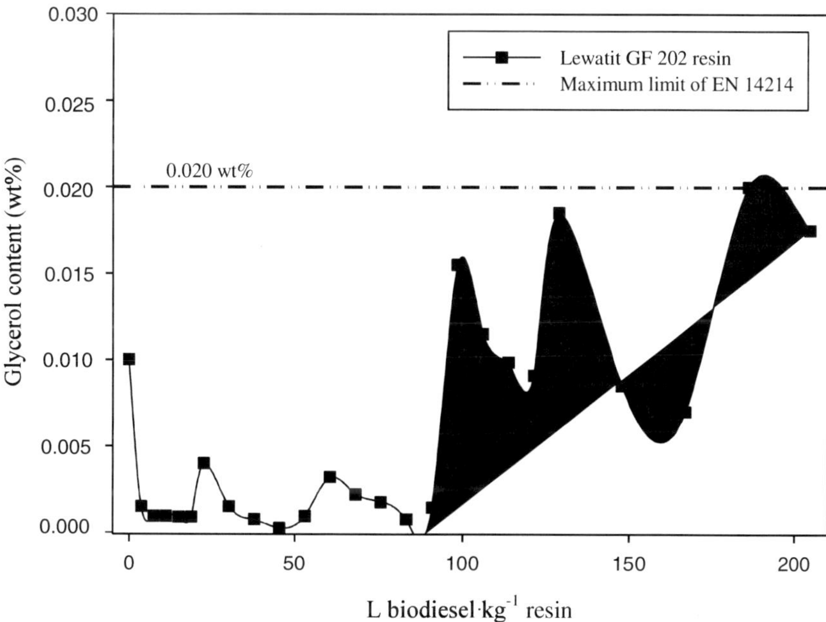

Figure 11. Evolution of glycerol content vs. resin capacity.

Molecular Distillation

Distillation is a purification and separation process that utilizes heat energy, which separates compounds according to a boiling point. Distilled vapours are then cooled and condensed in separate containers. There are a variety of distillation types named for specific features, such as steam distillation, fractional distillation and vacuum distillation. Molecular distillation, also called short-path distillation, is particularly useful in isolating components of heat-sensitive liquids or in separating substances with very high boiling points. Three special features characterize molecular distillation: the mixture is distilled under reduced pressure as it lowers boiling temperatures; the time that substances are exposed to heat is minimized; and the cooling process is sped up. The following features are found on a sophisticated molecular distillation apparatus: a high-vacuum pump improves vaporization; the distance between its heat source and the condenser is kept small; and it employs a motorized wiper to apply a thin film of substance to the heating wall. This means only a small number of molecules undergo distillation at any time. The main uses of distillation are for purposes of stability (for compound heat-sensitivity) and high-boiling applications (the petroleum industry utilizes molecular distillation to isolate high-boiling petroleum fractions such as motor oils).

Molecular distillation was employed to purify FAME from the crude biodiesel by Wang et al. [37] to replace the washing step used to reduce the effluent of the process after transesterification. The yield of the purified biodiesel (FAME) was calculated from the ratio of the mass of the purified biodiesel to that of the crude biodiesel. Their work has demonstrated that the yield of purified biodiesel increased as the evaporator temperature increased, specifically the yield increased from 88.93% to 98.32% from 90°C to 120°C. A modification in the colour was observed when the evaporator temperature increased from colourless to light yellow. Therefore, molecular distillation appears to be an interesting method for biodiesel purification due to the low distillation temperature which ensured no polymerization and the decomposition of FAME to obtain a high yield of final products, and the absence of washing water, making the process more environmentally friendly. In fact, distillation was introduced in biodiesel production via two ways: removal of FFA from the raw materials with a high acid value and purification of FAME at high temperatures (160°C - 220°C).

Membrane Separation Technology

A membrane can be described as a thin barrier between the two bulk phases and it is either a homogeneous phase or a heterogeneous collection of phases. The membrane is a permselective barrier that permits transport of some component but retains others. The flow of material across a membrane is kinetically driven by the application of pressure concentration, vapour pressure, hydrostatic pressure, electrical potential, or temperature. In membrane separations, each membrane has the ability to transport one component more readily than the other due to differences in the physical and chemical properties of the membrane and the permeating components. Furthermore, some components can freely permeate through the membrane, while others will be retained. The stream containing the components that permeate through the membrane is called permeate, while the stream containing the retained components is called retentate. The transport of permeate across the membrane is achieved by applying either mechanical, chemical, electrical or thermal force. In the past few decades, the membrane separation process has become increasingly widespread. It has drawn attention throughout the world, especially in the separation technology field, and has come to compete in many ways with conventional separation techniques such as distillation, adsorption, absorption and extraction. Practically all filtration membranes are based on synthetic organic (polymeric or liquid) or inorganic (ceramic, metal) membranes.

A new purification method is currently being investigated by Wang et al. [38], Saleh et al. [39] and [40] and Atadashi et al. [41]. According to Saleh et al. [40], the use of membrane technology in biodiesel purification to avoid water washing can provide solutions for many environmental problems by recovering valuable products as well as treating effluents and minimizing their harm to the atmosphere. A membrane has a physical barrier that restricts the transport of various chemical species. It can be homogeneous or heterogeneous, symmetric or asymmetric in its structure, and it may be either neutral or may carry positive or negative charges or both. The membrane may have a thickness of between less than 100 nm to more than a centimetre and can be either polymeric or ceramic. Common membrane processes include microfiltration, ultrafiltration, reverse osmosis, electro-dialysis, gas separation, nanofiltration, or per-vaporation. These are largely differentiated by the size of the molecules being separated and/or the pore size of the membrane.

Instead of using the hot water washing step, Wang et al. [38] use a ceramic membrane separation process to purify the product. After removing the methanol, the product is micro-filtered by the ceramic membrane to remove the residual glycerol, catalyst and soap and obtain the final product. The main advantage to this new method is that the steps to purify the biodiesel are simplified and waste is not generated by the washing water. One drawback, however, is that the membrane must be cleaned. The soap and the free glycerol that accumulate on the surface and/or precipitate over the membrane pores are easily removed by the methanol due to its high solubility to these polar substances. After this step, the methanol must be purified in order to reuse it in the cleaning step.

On the other hand, Saleh et al. [39] and [40] used polymeric membranes rather than ceramic membranes. They studied glycerol removal and the influence of the presence of water, soap and methanol on the removal of glycerol. Their results indicate that water has a large effect on removing glycerol from the FAME phase. Unlike conventional separation processes, however, only small quantities of water are needed to form large particles within the dispersed glycerol in FAME. The presence of methanol before introducing the crude FAME to the membrane separation system has a negative effect on glycerol separation by the membrane.

Comparison of Processes

Finally, Berrios et al. [21] compared the different purification methods from the viewpoint of the removal of impurities. All the purification methods tested (liquid-liquid extraction with water or glycerol, adsorption with magnesium silicate or bentonite and ion exchange) can remove soap, methanol and glycerol effectively. None of the methods had an effect on density, kinematic viscosity or FAME and glyceride content. Some of the methods have shown an influence on FFA and water content.

In Figure 12 the removal percentages of soap, methanol, glycerol, FFA and water are compared for all the purification methods. As can be observed, all of the methods were very efficient in removing methanol. The water-based methods achieved the highest glycerol removal (around 80%) due to the similar polarity of water and glycerol. Adsorption with magnesium silicate showed the highest soap removal (90%), although good results (80%) were also obtained with the liquid-liquid extraction methods.

An important factor for comparing purification methods is to determine the appropriate treatment when the purifying agent has completed its function. The adsorbents cannot be reused so they are dispatched to landfill. Because a highly polluting effluent is generated in the liquid-liquid extraction method with water, aerobic sewage treatment or anaerobic digestion is compulsory before disposal. The residual glycerol can be used for several purposes such as distillation for its purification, anaerobic co-digestion (nitrogen and phosphorus contribution are needed) for biogas production and as a raw material for biotechnology processes. Finally, the saturated resin can be regenerated for some time and when the capacity is low, be sent to landfill as adsorbents. The high methanol volume used for the regeneration can be utilized later in the transesterification process. The regeneration residues (traces of glycerol, biodiesel, methanol and salt) can be used in anaerobic co-digestion.

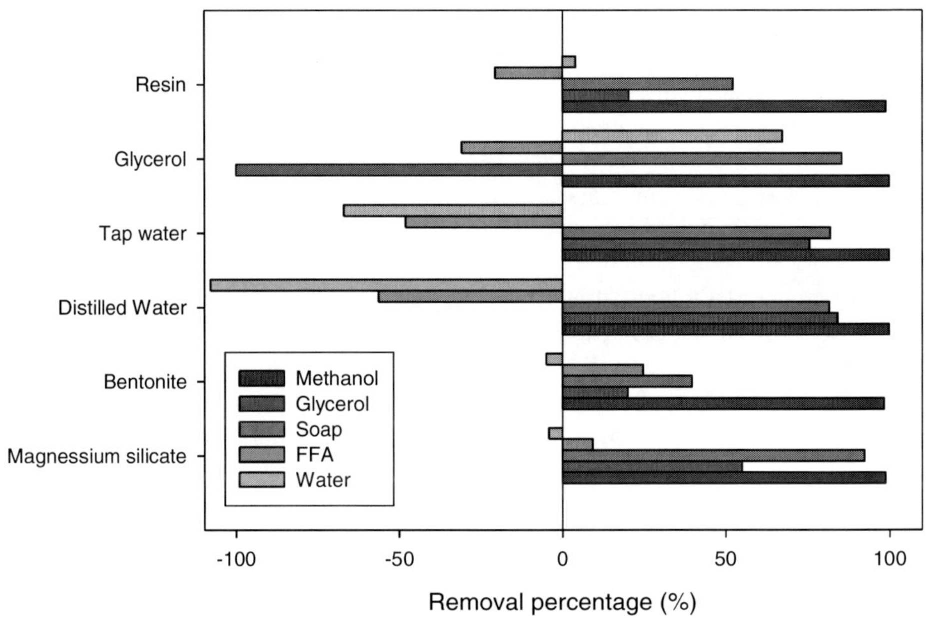

Figure 12. Removal percentages (%) vs. purification methods.

CONCLUSIONS

This chapter has provided a review of the research in biodiesel purification. As mentioned above, the purification stage is essential in biodiesel production in order to obtain a quality product that complies with the EN 14214 Standard. Although each method reviewed is useful for removing some impurities, none of them is 100% efficient for all the impurities. More research is therefore needed in order to establish a purification method that is suitable for all types of biodiesel production processes: one-step alkaline transesterification, two-step alkaline transesterification, two-step acid-alkaline transesterification, supercritical transesterification and enzymatic transesterification, among others.

Among the purification methods discussed in the chapter, liquid-liquid extraction methods are the most conventional. Nowadays, water washing is the most widespread method used. However, it has some disadvantages such as water supply and cost, emulsions, wastewater treatment and drying of the final product. Adsorption and ion exchange techniques are also well-known purification methods, but their results in biodiesel purification have not been fully described in terms of residue disposal or regeneration. Although molecular distillation is used to purify or concentrate several chemical compounds at length, the energy required during the process places limitations on its application in biodiesel production. The new liquid-liquid extraction method with deep eutectic compounds could be an interesting alternative if the price of these compounds is reduced. Deep eutectic compounds based on residual glycerol could be an option to remove glycerol from FAME. Nevertheless, this method does not have an effect on other impurities such as soap, FFA, water or methanol. Finally, the membrane separation technology seems to be the most promising purification method for the future. The main drawback to this method, however, is that the dirt which accumulates on the membrane surface reduces the efficiency of the separation. In conclusion, the most suitable purification method will depend on the biodiesel production process in question.

As mentioned above, to evaluate the purification methods it is essential to employ the appropriate treatment for the purifying agent when it has completed its function. Given that the adsorbents cannot be reused, they are dispatched to landfill. Because a highly polluting effluent is generated in the liquid-liquid extraction method with water, aerobic sewage treatment or anaerobic digestion is compulsory before disposal. Residual glycerol can be used for several purposes such as distillation for its purification, anaerobic co-digestion (nitrogen and phosphorus contribution are needed) for biogas production and as a raw material for biotechnology processes. The saturated resin can be regenerated for some time and when the capacity is low, be sent to landfill as adsorbents. The high methanol volume used for the regeneration can be utilized later in the transesterification process. The regeneration residues (traces of glycerol, biodiesel, methanol and salt) can be used in anaerobic co-digestion.

ACKNOWLEDGMENTS

The authors are grateful to the BIDA, S.A. company for funding this work. The authors would like to thank lab technician Inmaculada Bellido for her collaboration.

REFERENCES

[1] Dermibas, A. Importance of biodiesel as transportation fuel. *Energy Policy*. 2007, *35*, 4661–4670.

[2] Berrios, M.; Siles, J.; Martín, M.A.; Martín, A. A kinetic study of the esterification of FFA in sunflower oil. *Fuel*, 2007, *86*, 2383-2388.

[3] Berrios, M.; Skelton, R.L. Comparison of purification methods for biodiesel. *Chem. Eng. J.*, 2008, *144*, 459-465.

[4] Freedman, B.; Pryde, E.H.; Mounts, T.L. Variables affecting the yields of fatty esters from transesterified vegetable oils. *J. Am. Oil Chem. Soc.*, 1984, *61*, 1638–1643.

[5] Reid, E.E. Studies in esterification IV. The interdependence of limits as exemplified in the transformation of esters. *Am. Chem. J.*, 1911, *45*, 479–516.

[6] Tomasevic, A.V.; Siler-Marinkovic, S.S. Methanolysis of used frying oil. *Fuel Process. Technol.*, 2003, *81*, 1-6.

[7] Wang, Y.; Ou, S.; Liu, P.; Zhang, Z. Preparation of biodiesel from waste cooking oil via two-step catalyzed process. *Energ. Convers. Manage.*, 2007, *48*, 184-188.

[8] Zhang, Y.; Dubé, M.A.; McLean, D.D.; Kates, M. Biodiesel production from waste cooking oil: 1. Process design and technological assessment. *Bioresource Technol.*, 2003, *89*, 1-16.

[9] Zhang, Y.; Dubé, M.A.; McLean, D.D.; Kates, M. Biodiesel production from waste cooking oil: 2. Economic assessment and sensitivity analysis. *Bioresource Technol.*, 2003, *90*, 229-240.

[10] EN 14214:2008. Automative fuels. Fatty acid methyl esters (FAME) for diesel engines. Requirements and test methods.

[11] Mittelbach M.; Remschmidt C. Biodiesel: el manual completo (2007). Editorial Boersedruck Ges.m.b.H, Viena, Austria.

[12] Siles J.A.; Martín M.A.; Chica A.F.; Martín A. Anaerobic digestion of glycerol derived from biodiesel manufacturing. *Bioresource Technol.*, 2009, *10*, 5609-5615.

[13] Prankl H. High biodiesel quality required by European Standards. *Eur. J. Lipid Sci. Technol.*, 2002, *104*, 371-375.

[14] ASTM D6751-09. American specification for biodiesel (B100) (2009).

[15] Karaosmanoglu, F.; Cigizoglu, K.B.; Tüter, M.; Ertekin, S. Investigation of the refining step of biodiesel production. *Energ. Fuel.*, 1996, *10*, 890-895.

[16] Canakci, M.; Van Gerpen, J. Biodiesel production from oils and fats with high free fatty acids. *Trans. ASAE*, 2001, *44*, 1429-1436.

[17] Siler-Marinkovic, S.; Tomasevic, A. Transesterification of sunflower oil in situ, *Fuel*, 1998, *77*, 1389-1391.

[18] Predojevic, Z.J. The production of biodiesel from waste frying oils: A comparison of different purification steps. *Fuel*, 2008, *87*, 3522-3528.

[19] Glisic, S.B.; Skala, D.U. Design and optimization of purification procedure for biodiesel washing. *Chem. Ind. Chem. Eng. Quarterly*, 2009, *15*, 159-168.

[20] Sabudak, T.; Yildiz, M. Biodiesel production from waste frying oils and its quality control. *Waste Manage.*, 2010, *30*, 799-803.

[21] Berrios, M.; Martín, M.A.; Chica, A.F.; Martín, A. Purification of biodiesel from used cooking oils. UNDER REVIEW.

[22] EN 14103:2003. Fat and oil derivates. Fatty Acid Methyl Esters (FAME). Determination of ester and linolenic methyl ester contents.

[23] ISO 12937:2000. Petroleum products. Determination of water. Coulometric Karl Fischer titration method.

[24] EN 14104:2003. Fat and oil derivates. Fatty Acid Methyl Esters (FAME). Determination of acid value.

[25] EN 14105:2003. Fat and oil derivates. Fatty Acid Methyl Esters (FAME). Determination of free and total glycerol and mono-, di-, triglyceride contents.

[26] ISO 3104:1996. Petroleum products. Transparent and opaque liquids. Determination of kinematic viscosity and calculation of dynamic viscosity.

[27] ISO 3675:1996. Crude petroleum and liquid petroleum products. Laboratory determination of density. Hydrometer method.

[28] EN 14110:2003. Fat and oil derivates. Fatty Acid Methyl Esters (FAME). Determination of methanol content.

[29] Van Gerpen, J.; Shanks, B.; Pruszko, R.; Clements, D.; Knothe, G. Biodiesel Analytical Methods. National Renewable Energy Laboratory; NREL/SR-510-36240 (2004).

[30] Ma, F.; Clements, L.D.; Hanna, M.A. The effects of catalyst, free fatty acids, and water on transesterification of beef tallow. *Trans ASAE*, 1998, *41*, 1261-1264.

[31] Hayyan, M.; Mjalli, F.S.; Hashim, M.A.; AlNashef, I.M. A novel technique for separating glycerine from palm oil-based biodiesel using ionic liquids. *Fuel Process. Technol.*, 2010, *91*, 116–120.

[32] Shahbaz, K.; Mjalli, F.S.; Hashim, M.A.; AlNashef, I.M. Using deep eutectic solvents for the removal of glycerol from palm oil-based biodiesel. *J. Applied Sci.*, 2010, *10*, 3349-3354.

[33] McNeill, J.; Kakuda, Y.; Kamel, B. Improving the quality of used frying oils by treatment with activated carbon and silica. *J. Am. Oil Chem. Soc.*, 1986, *63*, 1564-1567.

[34] Özgül-Yücel, S.; Türkay, S. Purification of FAME by rice hull ash adsorption. *J. Am. Oil Chem. Soc.*, 2003, *80*, 373-376.

[35] Dmytryshyn, S.L.; Dalai, A.K.; Chaudhari, S.T.; Mishra, H.K.; Reaney, M.J. Synthesis and characterization of vegetable oil derived esters: evaluation for their diesel additive properties. *Bioresource Technol.*, 2004, *92*, 55-64.

[36] Manuale, D.L.; Mazzieri, V.M.; Torres, G.; Vera, C.R.; Yori, J.C. Non-catalytic biodiesel process with adsorption-based refining. *Fuel*, 2010, doi:10.1016/j.fuel.2010.10.047.

[37] Wang, Y.; Nie, J.; Zhao, M.; Ma, S.; Kuang, L.; Han, X.; Tang, S. Production of Biodiesel from Waste Cooking Oil via a Two-Step Catalyzed Process and Molecular Distillation. *Energ. Fuels*, 2010, *24*, 2104-2108.

[38] Wang, Y.; Wang, X.; Liu, Y.; Ou, S.; Tan, Y.; Tang, S. Refining of biodiesel by ceramic membrane separation. *Fuel Process. Technol.*, 2009, *90*, 422-427.

[39] Saleh, J.; Tremblay, A.Y.; Dubé, M.A. Glycerol removal from biodiesel using membrane separation technology. *Fuel*, 2010, *89*, 2260–2266.

[40] Saleh, J.; Dubé, M.A.; Tremblay, A.Y. Effect of Soap, Methanol, and Water on Glycerol Particle Size in Biodiesel Purification. *Energ. Fuels*, 2010, *24*, 6179–6186.

[41] Atadashi, I.M.; Aroua, M.K.; Aziz, A.A. Biodiesel separation and purification: A review. *Renew. Energ.*, 2011, *36*, 437-443.

In: Biodiesel: Blends, Properties and Applications
Editors: Jorge M. Marchetti and Zhen Fang

ISBN: 978-1-63117-024-9
© 2014 Nova Science Publishers, Inc.

Chapter 7

VALORIZATION OF WASTES AND BY-PRODUCTS DERIVED FROM BIODIESEL MANUFACTURING

J. A. Siles, M. Berrios, M. A. Martín and A. Martín
Inorganic Chemistry and Chemical Engineering Department.
University of Cordoba. Campus Universitario de Rabanales.
Edificio Marie Curie, Cordoba (Spain).

ABSTRACT

Biodiesel fuels have recently drawn much attention due to the various advantages they have over petroleum-based fuels. However, the recent growth of the biodiesel industry has created a surplus of impure glycerol. This has resulted in a dramatic decrease in crude glycerol prices over the last few years and given rise to environmental concerns regarding the disposal of contaminated glycerol. On the other hand, the biodiesel obtained by alkaline transesterification requires purification. Water washing is frequently used; a process that generates large amounts of highly polluted wastewater. Consequently, the biodiesel industry has been forced to implement effective treatment technologies in order to fulfill the severe quality standards related to environmental protection that are currently being developed and to recover valuable compounds contained in wastes and by-products. Anaerobic digestion is an interesting way to treat and revalorize abundant and low-priced glycerol and wastewater streams. This chapter examines the performance and the stability of the anaerobic digestion process of glycerol-containing waste derived from biodiesel manufacturing. The glycerol was previously acidified with phosphoric acid and centrifuged in order to recover the catalyst used in the transesterification reaction (KOH) as agricultural fertilizer (potassium phosphates). The results reveal that pre-treated glycerol has a high level of anaerobic biodegradability (around 100% COD) and that a substantial quantity of methane can be obtained in this way (292 mL CH_4/g COD removed at 1atm, 25°C). In addition, the treatment of wastewater derived from biodiesel manufacturing (428 g COD/L) was studied. Firstly, wastewater was acidified to recover free fatty acids. After centrifuging, the aqueous phase underwent two different physical-chemical pre-treatments to demulsify the remnant organic matter: coagulation-flocculation and electrocoagulation. The effluents of both pre-treatments were subjected to biomethanization. Of the two physical-chemical pre-treatments performed, electrocoagulation was found to be the most suitable.

Furthermore, this chapter describes the anaerobic co-digestion of acidified glycerol and electrocoagulated wastewater after mixing at a proportion that allows the different glycerol and wastewater flows generated in the biodiesel industry to be simultaneously absorbed. Biodegradability was found to be around 100%, while the methane yield coefficient reached a mean value of 310 mL CH_4/g COD removed (1atm, 25°C), thus showing the convenience of treating both wastes simultaneously. This research forms part of what is known as the biorefinery concept; a relatively new term referring to the conversion of feedstock into a host of valuable chemicals and energy with the consequent economic and environmental benefits.

1. INTRODUCTION

Nowadays there is a marked interest in the development of alternative energy resources due to the continuous rise in the price of petroleum, increasing environmental awareness due to the use of non-renewable fuels, dependence on foreign countries and the decline of petroleum reserves (Hansen et al., 2005). As a result, bio-fuels have recently drawn much attention given that they are a good alternative to petroleum-based fuel, are a renewable and non-toxic fuel, lead to a favorable energy balance, and generate a lower amount of harmful emissions (Ito et al., 2005).

Bio-fuels are manufactured using biomass as raw material. In general, bio-fuels may be solids (vegetable waste, urban or industrial organic waste fraction), liquids (bio-alcohol or bio-diesel) or gaseous (biogas, hydrogen, synthesis gas). Of all the bio-fuels available, biodiesel is characterized for its use in internal combustion engines (diesel and Otto engines); a fact that has led to increased world production in the last few years. Following upon international laws such as European Directive 2001/0265, which establishes that by the end of 2010 transport vehicles should contain 5.75% of bio-components derived from renewable resources, the manufacture of biodiesel in Europe has grown. According to the European Biodiesel Board, biodiesel production in 2008 was 7,775,000 metric tons (35.7% higher than in 2007). The main manufacturer of biodiesel was Germany with 2,819,000 tons, followed by France (1,815,000 metric tons) and Italy (595,000 metric tons). Had the aim of the above directive been achieved, the demand for biofuels in Europe would have reached 10 million tons in 2010. On the other hand, the National Biodiesel Board reported a 56% increase in biodiesel production in the United States over the same period, reaching a mean production of 2,649,500 metric tons.

1.1. Biodiesel Manufacturing

Biodiesel is a liquid biofuel produced from vegetable oils and/or animal fats, which is composed of a mixture of mono-alquil esters of long chain fatty acids. The biodiesel density, viscosity and cetane number are very similar to diesel oil, while the ignition point is higher, thus conferring greater security. Consequently, biodiesel may be mixed with diesel-oil for use in engines or even to substitute diesel-oil completely (Moser, 2009).

Among the biodiesel production methods described in the literature, transesterification is one of the most frequent as the characteristics of the final product are quite similar to those of diesel-oil and the manufacturing process is relatively simple (Ma and Hanna, 1999; Balat and

Balat, 2010). Transesterification, also called alcoholysis, is the reversible reaction of one triglyceride (the main component of oils and fats, approximately 98% w/w) with an alcohol (generally methanol or ethanol), which is transformed into methyl ester (biodiesel) and glycerol (Figure 1) (Ma and Hanna, 1999).

$$
\begin{array}{ccccccc}
CH_2\text{-OOC-}R_1 & & & & R_1\text{-COO-}R' & & CH_2\text{-OH} \\
| & & & \xrightarrow{\text{Catalyst}} & & & | \\
CH\text{-OOC-}R_2 & + & 3R'OH & \rightleftharpoons & R_2\text{-COO-}R' & + & CH\text{-OH} \\
| & & & & & & | \\
CH_2\text{-OOC-}R_3 & & & & R_3\text{-COO-}R' & & CH_2\text{-OH} \\
\\
\text{Triglyceride} & & \text{Alcohol} & & \text{Ester} & & \text{Glycerol}
\end{array}
$$

Figure 1. Transesterification of triglyceride with alcohol (Ma and Hanna, 1999).

This reaction is influenced by factors such as the chemical composition of the raw material, the alcohol and alcohol concentration and the catalyst used, as well as by the temperature, pressure and stirring intensity in the reaction medium. Numerous raw materials may be used for biodiesel manufacturing including vegetable oils (maize, sunflower, rapeseed, soybean, coconut or palm), animal fats (pig or cow), residual oils such as used cooking oils or lipids generated by some seaweed species (*Chlorella protothecoides*) (Moser et al., 2009). On the other hand, aliphatic monoalquil alcohols with 1 to 8 carbon atoms might be used as reactant for the transesterification reaction (Demirbas, 2009). However, the addition of the shortest chain alcohols such as methanol, ethanol, propanol or buthanol (Speit, 2008) is more frequent at full scale; especially methanol due to its low cost and physical and chemical advantages (polar alcohol with the shortest chain). In particular, an alcohol:triglyceride molar ratio of 3:1 is required to complete the transesterification reaction, although in practical terms this ratio is usually higher to maximize the process yield (Ma and Hanna, 1999).

Furthermore, acidic or alkaline catalyst or enzymes are utilized in order to increase the reaction rate and yield. Nevertheless, alkaline catalysis is the most frequent in full-scale plants, with the optimal catalyst concentration ranging from 0.005 to 0.350% w/w. Finally, reaction temperature and stirring intensity are essential factors to ensure the proper course of the transesterification reaction.

In general, temperature is maintained at 60°C at atmospheric pressure, while the stirring intensity can vary depending on the manufacturing process implemented (Freedman et al., 1984; Fukuda et al., 2001; Zhang et al., 2003; Moser, 2009).

Consequently, a wide range of manufacturing process designs may be implemented depending on the desired properties of the final product or the availability of resources; factors which strongly influence the generation and characterization of wastes and by-products derived from the main process.

2. BIOREFINERY OF BIODIESEL PRODUCTION

Despite the fact that biodiesel has many advantages over fossil fuels, the manufacturing process requires high energy inputs and the final product is costly. Furthermore, a large volume of impure and low-priced glycerol is produced in the transesterification reaction in addition to highly polluted wastewaters derived from raw materials and biodiesel purification (Balat and Balat, 2010). This constitutes an economic and environmental disadvantage that must be minimized in order to improve the viability of manufacturing this renewable bio-fuel.

Biodiesel produced from used cooking oils has become increasingly widespread, thus allowing the global economic balance to be improved (Zhang et al., 2003). The used cooking oil employed in the production of biodiesel is mainly derived from the catering industry (frying process) and is usually around 60% cheaper than crude vegetable oils or even free, depending on market availability (Predojevic, 2008). Nevertheless, used cooking oils frequently contain high concentrations of free fatty acids (FFA) and other products derived from the frying process that interfere in the biodiesel manufacturing process. The water and some polar compounds contained in waste oils may be removed by water washing and subsequent centrifugation. However, the high acidity of waste oils make a previous esterification step necessary when the biodiesel manufacturing process involves alkaline transesterification (Ruiz-Méndez et al., 2008). Additionally, a filtration step and/or drying step could be required to reduce suspended solids and moisture concentration, respectively (Moser, 2009).

The revalorization and treatment of by-products and wastes derived from biodiesel manufacturing from used cooking oils is described in detail in the sections that follow, as well as the most frequently described methodologies in the literature.

2.1. Exploitation of Glycerol Derived from Biodiesel Manufacturing

At present, pure glycerol has more than 2000 different applications due to its physical-chemical properties: glycerol is water-soluble, pure, odorless and colorless; has a certain sweetness; absorbs and retains water; has the capacity to solve flavors and dyes; has plasticizing properties; protects against fermentation processes; has antioxidant properties; is non-toxic and easily biodegradable; and is a polyol. Pure glycerol is currently applied, for example, as an ingredient of foods, cosmetics and pharmaceuticals. It is used in plastics, resins, cellophane or in the tobacco industry as a hydraulic agent; in heating fluid formulations; and in the paper industry as a humidifier, plasticizer and lubricant (Elvers et al., 1990). However, the rapid growth of the biodiesel manufacturing industry in recent years has led to a large volume of impure glycerol. In fact, the production of crude glycerol exceeds the present commercial demand for purified glycerol and the purification of glycerol (with a view to being sold) generated during the biodiesel manufacturing is not a viable option for the biodiesel industry (Chi et al., 2007). Specifically, the production of 100 kg of biodiesel generates approximately 10 kg of low-priced and impure glycerol (55-90%), thus leading to a marked drop in glycerol prices over the last few years. Consequently, the purification of glycerol is not an economically viable option for the biodiesel industry (Sandun et al., 2007).

Although glycerol may be burnt, with the consequent economic benefit, the implementation of biorefineries that co-produce valuable products, together with the main bio-fuel, has been proposed as a viable option for using glycerol. Additionally, this would prevent any inconveniences due to impurities generated during the glycerol combustion step. In this sense, several procedures have been proposed for the revalorization of glycerol, including the following chemical and biological transformations:

- The conversion of glycerol into propylene glycol and acetone through thermo-chemical processes after recovering the catalyst used in the transesterification reaction (Dasari et al., 2005; Chiu et al., 2006).
- The etherification of glycerol with either alcohols (e.g., methanol or ethanol) or alkenes (e.g., isobutene) and the production of oxygen-containing components, which could have suitable properties for use as fuel or solvents. Karinen and Krause (2006) studied the etherification of glycerol with isobutene in liquid phase with acidic ion exchange resin. The effect of reaction conditions on the system was analyzed and conditions for optimal selectivity toward ether were found to be isobutene/glycerol molar ratio of 3 at 80°C. Five product ethers with wide application at full scale were obtained.
- The microbial conversion (fermentation) of glycerol to 1,3-propanediol, which can be used as a basic ingredient of polyesters (Barbirato et al., 1998; Ito et al., 2005). The biological generation of this compound has been widely described in the literature. In general, this transformation is carried out under discontinuous conditions, although variable inoculum/substrate ratios may be applied depending on the microorganism strain used, either *Klebsiella pneumoniae*, *Citrobacter freundii*, *Clostridium butyricum* or *Enterobacter agglomerans* (Barbirato et al., 1998; Ito et al., 2005)
- Other products such as butanol (Biebl, 2001), propionic acid (Bories et al., 2004), ethanol and formate (Jarvis et al., 1997), succinic acid (Lee et al., 2001), dihydroxyacetone (Bories et al., 1991; Claret et al., 1994), polyhydroxyalkanoates (Koller et al., 2005), docosahexaenoic acid (DHA) (Chi et al., 2007) or oleic, linoleic and oxalic acids (André et al., 2010) have also been obtained using glycerol as a carbon source. Furthermore, Ito et al. (2005) evaluated the biological production of hydrogen and ethanol from glycerol-containing wastes discharged after a biodiesel manufacturing process for using *Enterobacter aerogenes* HU-101. The biodiesel wastes were diluted with a synthetic medium to increase the rate of glycerol utilization and the addition of yeast extract and tryptone to the synthetic medium accelerated the production of hydrogen and ethanol. The yields of hydrogen and ethanol decreased with increasing concentrations of biodiesel wastes and commercially available glycerol (pure glycerol). Moreover, the rates of hydrogen and ethanol production from biodiesel wastes were much lower than those at the same concentration of pure glycerol, in part due to the high salt content in the wastes. In a continuous culture with a packed-bed reactor using self-immobilized cells, the maximum rate of hydrogen production from pure glycerol was 80 mmol/L·h, yielding ethanol at 0.8 mol/mol-glycerol, while the same rate from biodiesel wastes was only 30 mmol/L·h. Nevertheless, using porous ceramics as a support material to

fix cells in the reactor, the maximum hydrogen production rate from biodiesel wastes reached 63 mmol/L·h, obtaining an ethanol yield of 0.85 mol/mol-glycerol. On the other hand, Adhikari et al. (2008) studied the use of glycerol as bio-renewable substrate for hydrogen production using a steam reforming process. Their study focused on nickel-based catalysts with MgO, CeO_2 and TiO_2 supports. Maximum hydrogen yield was obtained at 650°C with MgO supported catalysts, which corresponds to 4 mol of hydrogen out of 7 mol of stoichiometric maximum.

Anaerobic digestion is another alternative for revalorizing abundant and low-priced glycerol streams. This process may be defined as the biological conversion of organic material into a variety of end products including 'biogas' whose main constituents are methane and carbon dioxide (Wheatley, 1990). The advantages of anaerobic digestion include low levels of biological sludge, low nutrient requirements, high efficiency and the production of methane, which can be used as an energy source. The stoichiometry of the anaerobic digestion of glycerol can be summarized as follows (Christensen and McCarty, 1975; McCarty, 1975):

$$C_3H_8O_3 + a\ (NH_3) \rightarrow b\ (CH_4) + c\ (CO_2) + d\ (C_5H_7NO_2) + e\ (NH_4HCO_3)$$

The products of the reaction are methane, carbon dioxide, biomass and ammonic bicarbonate, where a, b, c, d and e are stoichiometric coefficients. In order to obtain the value of these coefficients and determine the methane yield of the anaerobic digestion of glycerol, a mass balance was made taking into consideration a biomass yield of anaerobic bacteria of 0.05 (w/w). The values of a, b, c, d and e were found to be 0.663, 1.648, 0.526, 0.041 and 0.622 mol, respectively. In contrast, when an electron balance was carried out assuming that every electron was used in the methane generation, b reached a value of 1.750. A simple calculation showed that the theoretical methane yield is 94.2%, which is a useful compound due to its caloric power (Lower Caloric Power): 35,793 kJ/m^3, which is equivalent to 9.96 $kW·h/m^3$. Siles et al. (2009) recently evaluated the performance and stability of the mesophilic anaerobic digestion process of glycerol-containing waste derived from a biodiesel manufacturing process consisting of esterification and alkaline transesterification of used cooking oils. The resulting waste was subjected to flash distillation in order to recover the excess methanol added in the transesterification step. Subsequent acidification with phosphoric acid followed by centrifugation allowed the catalyst used in the transesterification reaction (KOH) to be recovered and revalorized as agricultural fertilizer (potassium phosphates). The biomethanization of the remnant liquid phase was developed at laboratory scale using granular and non-granular anaerobic sludge in order to test their activity. A blank test with distilled glycerol was performed, although its full-scale applicability would be unviable due to the high cost of the distillation pre-treatment. Furthermore, due to the high organic concentration in both substrates (acidified glycerol: 1010 g COD/kg; distilled glycerol: 1155 g COD/kg), dilution with clean water was required, even though it would not be necessary at full scale. The addition of nutrients for proper anaerobic digestion was also required, especially for the biomethanization of distilled glycerol. The results demonstrated that the biomethanization of acidified glycerol with granular sludge is the best option for revalorizing glycerol-containing waste under the study conditions. The biodegradability of the acidified glycerol was found to be around 100% (in terms of COD), while the methane yield coefficient reached a mean value of 292 mL CH_4/g COD removed (at 1 atm, 25°C). (Figure 2). Nevertheless, a marked inhibition phenomenon was observed when the organic loading rate was higher than 0.21-0.38 g COD/g VSS·d. Although this rate is not high, this technique could mitigate economic and environmental drawbacks due to the low price and purity of glycerol derived from biodiesel manufacturing.

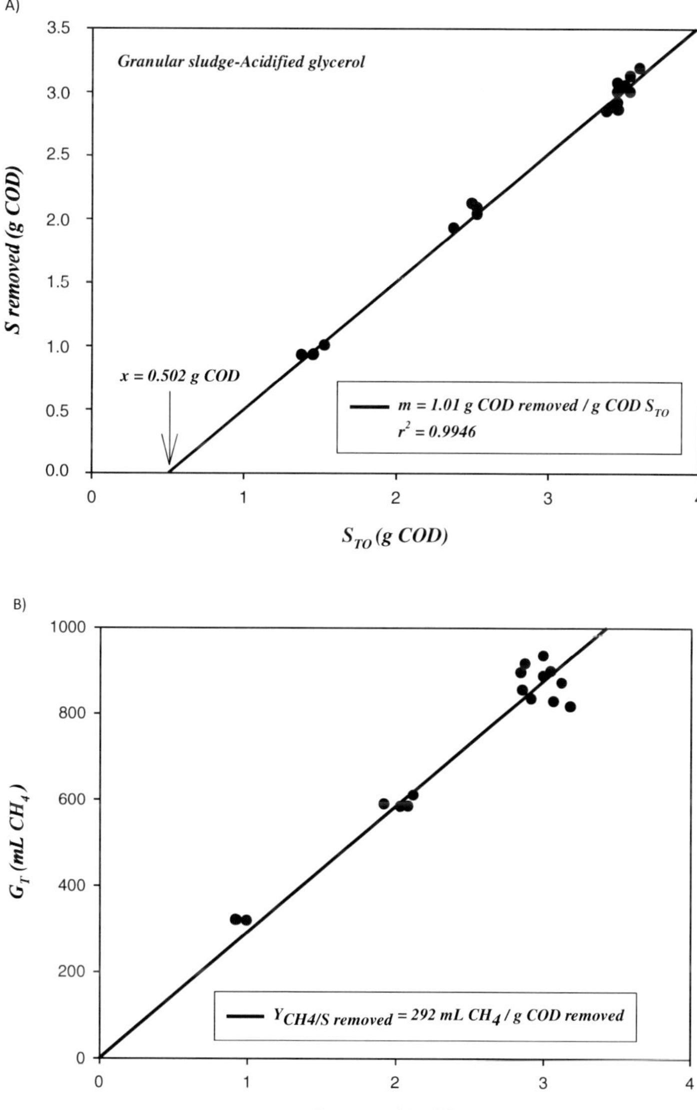

Figure 2. A) Plot of the amount of substrate removed against the substrate added for all the experiments to obtain the biodegradability percentage of the by-product using granular sludge-acidified glycerol. B) Variation of the experimental maximum methane volume produced (G_T) (at 1 atm, 25°C) with the COD consumed to obtain the methane yield coefficient of the process using granular sludge-acidified glycerol (Siles et al., 2009).

2.2. Treatment of Wastewater Derived from Biodiesel Manufacturing

According to the EN 14214 Standard (2008), the biodiesel obtained by transesterification requires purification. Although several purification strategies have been described in the literature (Berrios and Skelton, 2008), the most frequent process consists of washing with clean water (10-15% w/w of water with respect to biodiesel); a process which generates large amounts of highly polluted wastewater. In general, wastewater is generated during biodiesel manufacturing due to biodiesel purification, the washing of used-cooking oil employed as a substrate to manufacture the biodiesel and swilling-down. Table 1 shows the main characteristics of wastewater derived from the biodiesel manufacturing process as described by Jaruwat et al. (2010), who employed used-cooking oils as a raw material in the alkaline transesterification process.

As can be seen, this is a highly polluting wastewater due to its high pH value and high organic matter concentration (measured through COD and BOD). Given that KOH was utilized as a catalyst in the transesterification reaction, the pH of this waste was found to be 9.25-10.76, while the total COD was in the range of 310-590 g/L. Although the nitrogen concentration is not excessively high, the high concentrations of fats and oils determines which wastewater treatment process must be implemented in order to fulfill the spilling limits established under current legislation. Additionally, although the characteristics of wastewaters derived from different manufacturing processes vary widely, the presence of methanol, glycerol and suspended solids is quite frequent (Chavalparit and Ongwandee, 2009) (Figure 3).

The biological treatment of highly polluted wastewaters under aerobic and/or anaerobic conditions is usually carried out for the purification of urban and industrial wastewaters. However, wastewater derived from biodiesel manufacturing is rich in long-chain fatty acids (LCFAs). The LCFAs show acute toxicity towards microbial consortiums by adsorption onto the cell wall/membrane and interfere with the transport or protective function (Rinzema et al., 1994).

In addition, the sorption of a light layer of LCFAs to biomass leads to the flotation of sludge and consequent sludge washout (Rinzema et al., 1989). Hanaki et al. (1981) examined the inhibitory effect of these compounds on the anaerobic digestion process in batch experiments using synthetic substrates. Adding LCFAs to the anaerobic digesters caused a lag period in the production of methane from acetate and in the degradation of the LCFAs and other substrates.

Table 1. Characterization of wastewater derived from biodiesel manufacturing (Jaruwat et al., 2010)

Parameter	Mean value
pH	9.25-10.76
COD (g/L)	310-590
BOD (g/L)	170-300
Oil and fat (g/L)	18-22
TNK (mg/L)	439-464

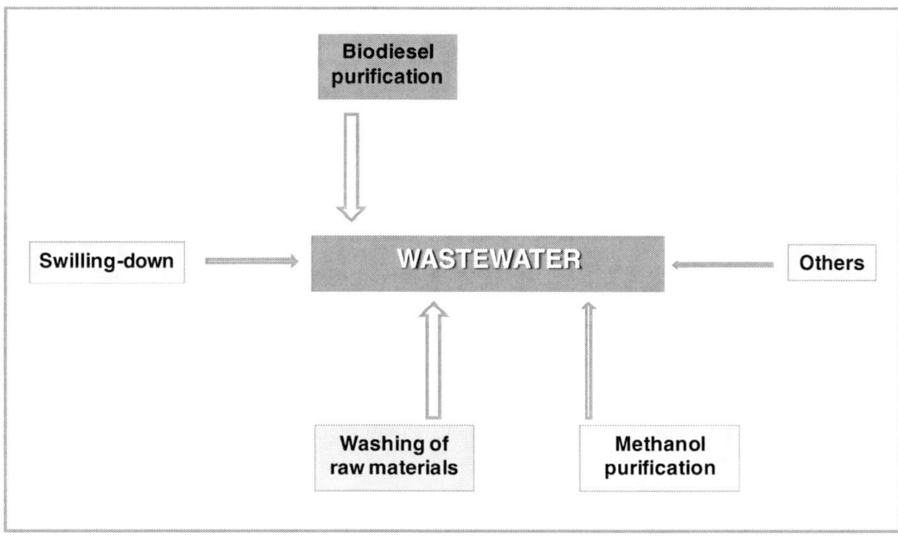

Figure 3. Main sources of wastewater generation in biodiesel manufacturing.

The addition of calcium chloride reduced the inhibitory effect of the LCFAs, but it did not do so after the culture had been exposed to these acids for several hours. Additionally, thermophilic bacteria are more sensitive to LCFAs than mesophilic bacteria, perhaps due to the different composition of its cellular membrane (Hwu and Lettinga, 1997). On the other hand, the inhibitory effect of the compounds contained in wastewater derived from biodiesel manufacturing by alkaline transesterification was also evaluated in aerobic systems. Suehara et al. (2005) studied the aerobic biological treatment of wastewater from biodiesel manufacturing after removing its oily content with hexane. A yeast strain of *Rhodotorula mucilaginosa* was subsequently inoculated in the remnant aqueous phase. The pH was adjusted at 6.8 and some nutrients, including ammonium sulfate, ammonium chloride, yeast extract, KH_2PO_4 and $MgSO_4 \cdot 7H_2O$ were added. However, the biological treatment was markedly inhibited by the presence of some organic compounds contained in the wastewater. No yeast growth was observed for organic matter concentrations higher than 2.14 g volatile solids/L. Given that the wastewater to be treated required dilution to operate under stable conditions, this treatment could be successfully implemented in small biodiesel manufacturing plants, but not at full scale.

More specifically, oleic acid showed similar inhibitory levels to lauric acid, which showed an IC_{50} of 4.3 mM, while the toxicity of caprylic acid was found to be minimum. In contrast, the toxicity of capric and myristic acids was enhanced by the presence of lauric acid (Koster and Cramer, 1987). Nevertheless, LCFA toxicity varies with the type of anaerobic sludge used for the treatment and is more closely related to the physical (specific surface area and size distribution) than to the biological properties of the sludge. Suspended and flocculated sludge, which shows a higher specific surface area, undergoes stronger inhibition than granular sludge (Hwu et al., 1996). Several studies have suggested that the bactericide effect of the LCFAs does not enable the adaptation of methanogenic bacteria (Hanaki et al., 1981; Koster and Cramer, 1987; Angelidaki and Ahring, 1992). However, recent studies based on the degradation of oleic acid in a fixed-bed anaerobic reactor have demonstrated that acclimatization improves the resistance of the sludge to the presence of oleate, thus improving biodegradation capacity in comparison to the sludge in the absence of lipids (Alves et al., 2001a, b). Moreover, the addition of calcium reduces the inhibitory effect of LCFAs. This is likely due to the formation of insoluble salts (Hanaki et al., 1981), but does not prevent the inconvenience of sludge flotation (Alves et al., 2001a, b).

Flotation is a primary treatment that is frequently used to remove oils and fats prior to biological treatments. Nevertheless, this technique is not effective in removing emulsified oils without a previous physical-chemical treatment of coagulation-flocculation (Moursy and Abo-Elela, 1982). For example, Al-Shamrani et al. (2002) described the beneficial effect derived from the addition of aluminum or iron salts on the unstabilization of oily emulsions. A similar study demonstrated that iron chloride is a highly effective coagulant for COD, fat and oils removal Saatci et al. (2001). However, these physical-chemical pre-treatments require the use of dangerous chemicals and relatively complex operational conditions. To mitigate this problem, electrocoagulation has been proposed as a successful pre-treatment for oily wastewater with a subsequent anaerobic treatment. This pre-treatment consists of destabilizing suspended, emulsified or dissolved contaminants in an aqueous medium by introducing an electric current into the medium (Emamjomeh and Sivakumar, 2009). Electrocoagulation has been used successfully for treating a large variety of wastewaters, including solid waste leaching, urban wastewaters, dye wastewaters or those rich in organic compounds such as phenols (Deng and Englehardt, 2007; Bukhari, 2007; Gürses et al., 2002; Feng et al., 2004). Moreover, electrocoagulation has been used to remove oil and to reduce the organic load from restaurant and petrochemical, mechanic and metallurgic industry wastewaters (Xu and Zhu, 2004; Tir and Moulai-Mostefa, 2008). Calvo et al. (2003) conducted an electrocoagulation process to treat different soluble oily wastes with very high COD. Electrocoagulation tests were initially conducted in a continuously operated laboratory-scale unit to reduce COD in the 70%-90% range. For different concentrations of oil, soluble COD removal was found to be in the 49%-78% range when the current rate was a constant 7 A. However, the removal of COD increased to 90% when the current rate was increased to 10 A. Demulsification of oily wastewater by electrocoagulation in a batch cell with horizontally oriented electrodes was also studied by Fouad et al. (2009). This system reduced oil concentration in wastewater from 500 to 6 ppm within 30 min. This pre-treatment could also reduce the polyphenol concentration frequently contained in oily wastewater which inhibits the activity of micro-organisms during biological oxidations (Panizza and Cerisola, 2006). In a similar line, Chavalparit and Ongwandee (2009) studied the treatment of wastewater derived from biodiesel manufacturing by alkaline transesterification of used cooking-oils and palm oils by electrocoagulation with an aluminum anode and graphite cathode. The optimal conditions were found to be pH 6.06, 18.2 V and a reaction time of 23.5 minutes, allowing COD, fat-oil and suspension solids removals of 55, 98 and 97%, respectively. The COD removal was not too high due to the reduced removal of methanol and glycerol contained in wastewater.

The implementation of this treatment technique allows a higher pollutant removal rate to be reached. Furthermore, the operational system is simple, does not require the addition of dangerous chemicals, and generates less sludge, thus achieving economic and environmental benefits (Tezcan Ün et al., 2006; Chavalparit and Ongwandee, 2009). Siles et al. (2010) evaluated the treatment of wastewater derived from a biodiesel manufacturing process (alkaline transesterification of used cooking-oils) with total COD of 428 g/L and pH 10.35. Specifically, the wastewater is derived from the purification of biodiesel, the washing of used cooking oil, the distillation column where the methanol utilized in the production process is purified and from swilling-down. The wastewater was acidified and centrifuged, thus permitting the wastewater to be demulsified and the FFA to be recovered and re-circulated into the biodiesel manufacturing process. This phase accounted for 3.80% (w/w) of the waste. On the other hand, the aqueous phase, in which total and soluble COD was found to be 250 ± 10 and 210 ± 10 g/L, respectively, was neutralized and two different physical-chemical pre-treatments were carried out at laboratory scale: electrocoagulation and coagulation-flocculation. pH was neutralized as it is one of the most restrictive parameters in the coagulation step (Company, 2000). A 45% reduction in global COD was observed with the acidification-electrocoagulation pre-treatment (eight aluminum electrodes with a surface area of 80.5 cm^2 to which a direct current of 12 V, 1.5 A was applied for 30 min), while this same figure was 63% for the acidification-coagulation-flocculation pre-treatment (0.2 mL/L coagulant; 2 mL/L flocculant). In a similar treatment of wastewater derived from biodiesel manufacturing, Jaruwat et al. (2010) reported that the combined treatment of acidification and electro-oxidation (4.28 mA/cm^2) reduced BOD levels by more than 95%.

Subsequently, the resulting aqueous phases derived from the electrocoagulation and coagulation-flocculation pre-treatments were subjected to mesophilic anaerobic digestion using granular sludge. Integrated acidification-electrocoagulation and biomethanization treatments of wastewater derived from biodiesel manufacturing were found to be more efficient than combined acidification-coagulation-flocculation and biomethanization treatments. The results showed that the first treatment allowed a global COD removal of 99%, while the combination of acidification-coagulation-flocculation with biomethanization showed a global percentage of 94%. Moreover, although the methane yield coefficient was similar in both cases, the allowed OLRs and kinetics of methane production were found to be considerably lower for the second treatment. This may be explained by the higher sodium or aluminum concentrations added during the pre-treatment.

2.3. Anaerobic Co-Digestion of Glycerol Derived from Biodiesel Manufacturing with Other Organic Wastes

To minimize the addition of nutrients described by Siles et al. (2009) for the anaerobic digestion of glycerol derived from biodiesel manufacturing, the anaerobic co-digestion of electrocoagulated wastewater and acidified glycerol was also studied at laboratory scale (Siles et al., 2010). The electrocoagulated wastewater was characterized and mixed with glycerol at a proportion of 85 to 15 (COD). In the biodiesel industry, lower amounts of glycerol are generated than wastewater. As a result, the percentage selected for the mixture allows the different flows to be absorbed. The mixture showed a high level of anaerobic biodegradability (around 100%), permitting a substantial quantity of methane to be obtained (310 mL CH_4/g COD removed, at 1atm and 25°C) (Figure 4). The methane production kinetic remained constant at 0.63 mL CH_4/(g VSS·g COD·h) in the OLR range studied. These results indicate that the anaerobic co-digestion of glycerol and wastewater derived from biodiesel manufacturing is a good alternative for treating and revalorizing both wastes.

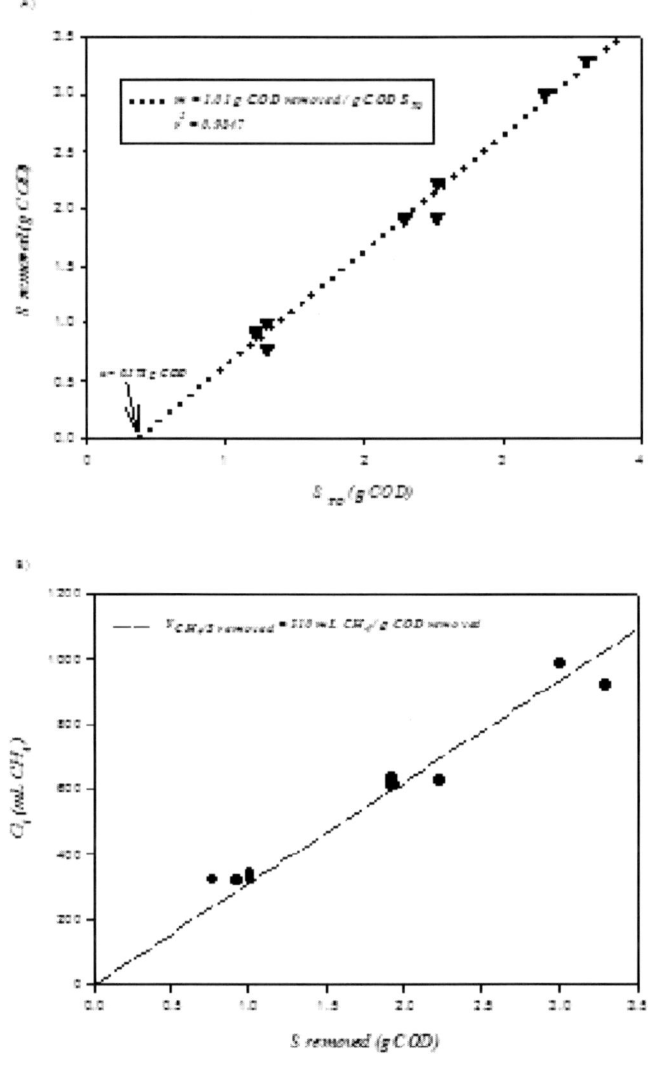

Figure 4. A) Plot of the amount of substrate removed against the substrate added for all the experiments to obtain the biodegradability percentage of the mixture. B) Variation of the experimental maximum methane volume produced (G_T) (at 1 atm, 25 °C) with the COD consumed to obtain the methane yield coefficient (Siles et al., 2010).

On the other hand, an optimization protocol for maximizing methane production by anaerobic co-digestion of several wastes was carried out by Álvarez et al. (2010). A linear programming method was utilized to set up different blends aimed at maximizing the total substrate biodegradation potential (L CH$_4$/kg substrate) or the biokinetic potential (L CH$_4$/kg substrate·d). In order to validate the process, three agro-industrial wastes were employed: pig manure, tuna fish waste and biodiesel waste (mainly glycerol produced in the transesterification of triglycerides with methanol and sodium hydroxide as catalyst). The results were then validated by experimental studies in discontinuous assays. The highest biodegradation potential (321 L CH$_4$/kg COD) was reached with a mixture composed of 84% pig manure, 5% fish waste and 11% biodiesel waste, while the highest methane production rate (16.4 L CH_4/kg COD d) was obtained with a mixture containing 88% pig manure, 4% fish waste and 8% biodiesel waste. Linear programming proved to be a powerful, useful and easy-to-use tool for estimating methane production in co-digestion units where different substrates can be fed. The study demonstrates the suitability of anaerobic co-digestion of glycerol derived from biodiesel manufacturing with other organic wastes. However, as glycerol-containing waste was not pre-treated, phosphate salts could not be recovered and revalorized as fertilizer.

CONCLUSION

Biodiesel is a bio-fuel that has many advantages over fossil fuels. Nevertheless, the rapid growth of the biodiesel industry in recent years has led to the generation of a large volume of impure and low-priced glycerol and highly polluting wastewater, making re-valorization and conventional treatments economically unviable. The implementation of biorefineries to produce valuable products from undesirable by-products and/or wastes could be advantageous for the biodiesel industry.

Figure 5 shows a flowchart of a possible biodiesel production process in which a biorefinery approach could be implemented.

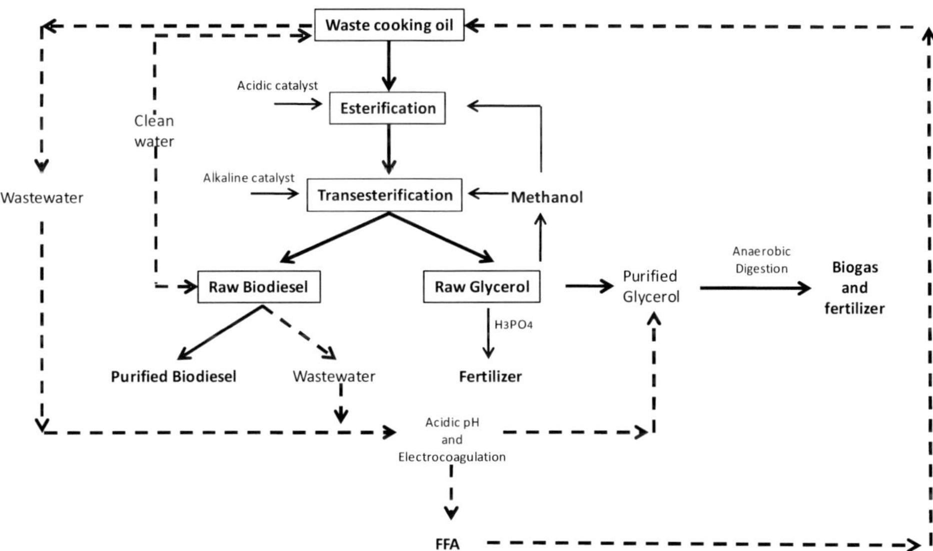

Figure 5. Simplified scheme of a biodiesel manufacturing process implemented from a biorefinery approach.

As can be seen, by using waste cooking-oils as raw material and after carrying out the esterification and sequencing transesterification reactions with acid and alkaline catalyst, respectively, raw biodiesel and glycerol are obtained. The implementation of by-products, waste treatments and revalorization processes could provide numerous benefits. These include the generation of a considerable volume of methane, which can be used for electricity and/or heat purposes without complex pre-treatments due to the absence of sulfur in the substrates to be treated; and a highly stable fertilizer after anaerobic digestion, which is rich in nutrients, remnant non-biodegradable organic compounds, and excess sludge. Additionally, potassium phosphate salts could be obtained, free fatty acids could be recirculated in the form of raw material for biodiesel manufacturing, and the excess methanol added for the esterification and transesterification reactions could be reutilized. The proposed process could improve the viability of biodiesel manufacturing with the consequent economic and environmental benefits.

REFERENCES

Adhikari, S.; Fernando, S.D.; Haryanto, A. Hydrogen production from glycerin by steam reforming over nickel catalysts. *Renew. Energy*, 2008, *33 (5)*, 1097-1100.

Al-Shamrani, A.A.; James, A.; Xiao, H. Destabilisation of oil-water emulsion and separation by dissolved air flotation. *Water Res.*, 2002, *36*, 1503-1512.

Álvarez, J.A.; Otero, L.; Lema, J.M. A methodology for optimising feed composition for anaerobic co-digestion of agro-industrial wastes. *Bioresource Technol.*, 2010, *101(4)*, 1153-1158.

Alves, M.M.; Mota, J.A.; Pereira, R.M.; Pereira, M.A., Mota, M. Effects of lipids and oleic acid on biomass development in anaerobic fixed-bed reactors. Part I: biofilm growth and activity. *Water Res.*, 2001a, *35 (1)*, 255–263.

Alves, M.M.; Mota, J.A.; Pereira, R.M.; Pereira, M.A.; Mota, M. Effects of lipids and oleic acid on biomass development in anaerobic fixed-bed reactors. Part II: Oleic acid toxicity and biodegradability. *Water Res.*, 2001b, *35 (1)*, 264–270.

André, A.; Diamantopoulou, P.; Philippoussis, A.; Sarris, D.; Komaitis, M.; Papanikolaou, S. Biotechnological conversions of bio-diesel derived waste glycerol into added-value compounds by higher fungi: production of biomass, single cell oil and oxalic acid. *Ind. Crops Products*, 2010, *31 (2)*, 407-416.

Angelidaki, I.; Ahring, B.K. Effects of free long-chain fatty acids on thermophilic anaerobic digestion. *Appl. Microbiol. Biotechnol.*, 1992, *37*, 808–812.

Balat, M.; Balat, H. Progress in biodiesel processing. *Appl. Energy*, 2010, *87(6)*, 1815-1835.

Barbirato, F.; Himmi, E.H.; Conte, T.; Bories, A. 1,3-propanediol production by fermentation: An interesting way to valorize glycerine from the ester and ethanol industries. *Ind. Crops Products*, 1998, *7*, 281-289.

Berrios, M.; Skelton, R.L. Comparison of purification methods for biodiesel. *Chem. Eng. J.*, 2008, *144(3)*, 459-465.

Biebl, H. Fermentation of glycerol by Clostridium pasteurianum- batch and continuous culture studies. *J. Ind. Microbiol. Biotechnol.*, 2001, *27*, 18-26.

Bories, A.; Claret, C.; Soucaille, P. Kinetic study and optimisation of the production of dihydroxyacetone from glycerol using *Gluconobacter oxydans*. *Proc. Biochem.*, 1991, *26*, 243-248.

Bories, A.; Himmi, E.; Jauregui, J.J.A.; Pelayo-Ortiz, C.; Gonzales, V.A. Glycerol fermentation with Propionibacteria and optimisation of the production of propionic acid. *Sci. Aliments*, 2004, *24*, 121-135.

Bukhari, A. Investigation of the electrocoagulation treatment process for the removal of total suspended solids and turbidity from municipal wastewater. *Bioresource Technol.*, 2007, *99*, 914–921.

Calvo, L.S.; Leclerc, J.P.; Tanguy, G.; Cames, M.C.; Patemotte, G.; Valentin, G.; Rostan, A.; Lapicque, F. An electrocoagulation unit for the purification of soluble oil wastes of high COD. *Environ. Prog.*, 2003, *22 (1)*, 57-65.

Chavalparit, O.; Ongwandee, M. Optimizing electrocoagulation process for the treatment of biodiesel wastewater using response surface methodology. *Journal of Environmental Sciences*, 2009, *21*, 1491-1496.

Chi, Z.; Pyle, D.; Wen, Z.; Frear, C.; Chen, S. A laboratory study of producing dosahexaenoic acid from biodiesel-waste glycerol by microalgal fermentation. *Proc. Biochem.*, 2007, *42*, 1537-1545.

Chiu, C.W.; Dasari, M.A.; Sutterlin, W.R.; Suppes, G.J. Removal of residual catalyst from simulated biodiesel's crude glycerol for glycerol hydrogenolysis to propylene glycol. *Ind. Eng. Chem. Res.*, 2006, *45*, 791-795.

Christensen, D.R.; McCarty, P.L. Multi-process biological treatment model. *J. Water Pollution Control Fedn.*, 1975, *47*, 2652-2664.

Claret, C.; Salmon, J.M.; Romieu, C.; Bories, A. Physicology of Gluconobacter oxydans during dihydroxyacetone production from glycerol. *Appl Microbiol. Biotechnol.*, 1994, *41*, 359-365.

Company, J. Coagulantes y floculantes aplicados en el tratamiento de aguas (Coagulants and flocculants for wastewater treatment), first ed. Gestió i promoció editorial, S.L., Barcelona, 2000.

Dasari, M.A.; Kiatsimkul, P.P.; Sutterlin, W.R.; Suppes, G.J. Low-pressure hydrogenolysis of glycerol to propylene glycol. *Appl. Catal. A-Gen.*, 2005, *281*, 225-231.

Demirbas, A. Biodiesel from waste cooking oil via base-catalytic and supercritical methanol transesterification. *Energy Convers. Manage.*, 2009, *50*, 923-927.

Deng, Y.; Englehardt, J.D. Electrochemical oxidation for landfill leachate treatment. *Waste Manage.*, 2007, *27*, 380–388.

Elvers, B.; Hawkins, S.; Weinheim, G. S. *Ullmann's encyclopaedia of industrial chemistry*, 5th ed.; NY:VCH., Basel (Switzerland), Cambridge, New York, 1990.

Emamjomeh, M.M.; Sivakumar, M. Review of pollutants removed by electrocoagulation and electrocoagulation/flotation processes. *J. Environ. Manage.*, 2009, *90*, 1663-1679.

EN 14214:2008. Automotive fuels. Fatty acid methyl esters (FAME) for diesel engines. Requirements and test methods.

European Biodiesel Board (http://www.ebb-eu.org/biodiesel.php) (accessed November 2010).

Feng, C.; Sugiura, N.; Maekawa, T. Performance of two new electrochemical treatment systems for wastewaters. *Journal of Environmental Science and Health*, 2004, *39(9)*, 2533–2543.

Fouad, Y.O.A.; Konsowa, A.H., Farag, H.A.; Sedahmed, G.H. Performance of an electrocoagulation cell with horizontally oriented electrodes in oil separation compared to a cell with vertical electrodes. *Chem. Eng. J.*, 2009, *145(3)*, 436-440.

Freedman, B.; Pryde, E.H.; Mounts, T.L. Variables affecting the yields of fatty esters from transesterified vegetable oils. *JAOCS*, 1984, *61(10)*, 1638-1643.

Fukuda, H.; Kondo, A.; Noda, H. Biodiesel fuel production by transesterification of oils. *J. Biosci. Bioeng.*, 2001, *92(5)*, 405-416.

Gürses, A.; Yalçin, M.; Dogar, C. Electrocoagulation of some reactive dyes: A statistical investigation of some electrochemical variables. *Waste Manage.*, 2002, *22*, 491–499.

Hanaki, K.; Mastsuo, T.; Nagase, M. Mechanism of inhibition caused by long chain fatty acids in anaerobic digestion process. *Biotechnol. Bioeng.*, 1981, *23*, 1591–1610.

Hansen, A.C.; Zhang, Q.; Lyne, P.W.L. Ethanol-diesel fuel blends-a review. *Bioresource Technol.*, 2005, *96*, 277-285.

Hazimah, A.H.; Ooi, T.L.; Salmiah, A. Recovery of glycerol and diglycerol from glycerol pitch. *J. Oil Palm. Res.*, 2003, *15*, 1-5.

Hwu, C.S.; Donlon, B.; Lettinga, G. Comparative toxicity of long chain fatty acid to anaerobic kludges from various origins. *Water Sci. Technol.*, 1996, *34 (5–6)*, 351–358.

Hwu, C.S.; Lettinga, G. Acute toxicity of oleate to acetate-utilizing methanogens in mesophilic and thermophilic anaerobic sludges. *Enzyme Microb. Technol.*, 1997, *21*, 297–301.

Ito, T.; Nakashimada, Y.; Senba, K.; Matsui, T.; Nishio, N. Hydrogen and Ethanol production from Glycerol-containing wastes discharged after biodiesel manufacturing process. *J Biosci Bioeng.*, 2005, *100 (39)*, 260-265.

Jaruwat, P.; Kongjao, S.; Hunsom, M. Management of biodiesel wastewater by the combined processes of chemical recovery and electrochemical treatment. *Energ. Convers. Manage.*, 2010, *51*, 531-537.

Jarvis, G.N.; Moore, E.R.B.; Thiele, J.H. Formate and ethanol are the major products of glycerol fermentation produced by a *Klebsiella planticola* strain isolated from red deer. *J. Appl. Microbiol.*, 1997, *83*, 166-174.

Karinen, R.S.; Krause A.O.I. New biocomponents from glycerol. *Appl Catal A Gen.*, 2006, *306*, 128-133.

Koller, M.; Bona, R.; Braunegg, G.; Hermann, C.; Horvat, P.; Kroutil, M.; Martinez, J.; Neto, J.; Pereira, L.; Varila, P. Production of polyhydroxyalkanoates from agricultural waste and surplus materials. *Biomacromolecules*, 2005, *6(2)*, 561-565.

Koster, I.W.; Cramer, A. Inhibition of methanogenesis from acetate in granular sludge by long-chain fatty acids. *Appl. Environ. Microbiol.*, 1987, *53 (2)*, 403–409.

Lee, P.C.; Lee, W.G.; Lee, S.Y.; Chang, H.N. Succinic acid production with reduced by-product formation in the fermentation of *Anaerobiospirillum succiniciproducens* using glycerol as a carbon source. *Biotechnol. Bioeng.*, 2001, *72*, 41-48.

Ma, F.; Hanna, M.A. Biodiesel production: a review. *Bioresource Technol.*, 1999, *70*, 1-15.

McCarty, P.L. Stoichiometry of biological reactions. *Prog. Water Technol.*, 1975, *7*, 157-172.

Moser, B.R. Biodiesel production, properties, and feedstocks. *In Vitro Cell. Dev. Biol.–Plant.*, 2009, *45*, 229-266.

Moursy, A.S.; Abo-Elela, S.I. Treatment of oily refinery wastes using a dissolved air flotation. *Environmental International*, 1982, *7*, 267-270.

National Biodiesel Board (http://www.biodiesel.org/pdf_files/fuelfactsheets/ Production_Graph_Slide.pdf) (accessed November 2010).

Predojevic, Z.J. The production of biodiesel from waste frying oils: a comparison of different purification steps. *Fuel*, 2008, *87*, 3522-352.

Rinzema, A.; Alphenaar, A.; Lettinga, G. The effect of lauric acid shock loads on the biological and physical performance of granular sludge in UASB reactors digesting acetate. *J. Chem. Tech. Biotechnol.*, 1989, *46*, 257–266.

Rinzema, A.; Boone, M.; van Knippenberg, K.; Lettinga, G. Bactericidal effect of long chain fatty acids in anaerobic digestion. *Water Environ. Res.*, 1994, *66*, 40–49.

Ruiz-Méndez, M.V.; Marmesat, S.; Liotta, A.; Dobarganes, M.C. Analysis of used frying fats for the production of biodiesel. *Grasas y Aceites.* 2008, *59(1)*, 45-50.

Saatci, Y.; Hasar, H.; Cici, M. Treatability of vegetable oil industry effluents through physical-chemical methods. *Resenius Environmental Bulletin*, 2001, *10*, 854-858.

Sandun, F.; Sushil, A.; Kiran, K.; Ranjitha, B. Glycerol based automotive fuels from future biorefineries. *Fuel*, 2007, *86(17-19)*, 2806-2809.

Siles, J. A.; Martín, M. A.; Chica, A.F.; Martín, A. Anaerobic digestion of glycerol derived from biodiesel manufacturing. *Bioresource Technol.*, 2009, *100 (23)*, 5609-5615.

Siles, J. A.; Martín, M. A.; Chica, A.F.; Martín, A. Anaerobic co-digestion of glycerol and wastewater derived from biodiesel manufacturing. *Bioresource Technol.*, 2010, *101(16)*, 6315-6321.

Speit, J.G. *Sythetic fuels handbook: properties, process, and performance.* New York: McGraw-Hill, 2008.

Suehara, K.; Kawamoto, Y.; Fujii, E.; Kohda, J.; Nakano, Y.; Yano, T. Biological treatment of wastewater discharged from biodiesel fuel production plant with alkali-catalyzed transesterification. *J. Biosci. Bioeng.*, 2005, *100(4)*, 437-442.

Tezcan, Ü.Ü.; Ugur, S.; Koparal, A.S.; Bakır, Ö.Ü. Electrocoagulation of olive mill wastewaters. *Sep. Purif. Technol.*, 2006, *52*, 136–141.

Tir, M.; Moulai-Mostefa, N. Optimization of oil removal from oily wastewater by electrocoagulation using response surface method. *J. Hazard. Mater.*, 2008, *158*, 107–115.

Wheatley, A. Anaerobic Digestion: A Waste Treatment Technology. Elsevier, London, 1990.

Xu, X.; Zhu, X. Treatment of refectory oily wastewater by electrocoagulation process. *Chemosphere*, 2004, *56*, 889-894.

Zhang, Y.; Dubé, M.A.; McLean, D.D.; Kates, M. Biodiesel production from waste cooking oil: 1. Process design and technological assessment. *Bioresource Technol.*, 2003, *89*, 1-16.

In: Biodiesel: Blends, Properties and Applications   ISBN: 978-1-63117-024-9
Editors: Jorge M. Marchetti and Zhen Fang   © 2014 Nova Science Publishers, Inc.

Chapter 8

APPLICATION OF SUPERCRITICAL FLUIDS EXTRACTION AND METHYLATION ON PRODUCTION OF BIODIESEL FROM PLANT SEEDS

Chao-Rui Chen[1,2], Chieh-Ming J. Chang[1,], Wei-Heng Chen[1] and Ching-Hung Chen[1]*

[1]Department of Chemical Engineering,
National Chung Hsing University, Taiwan, ROC
[2]Chemical Engineering Division,
Institute of Nuclear Energy Research, Taiwan, ROC

ABSTRACT

This study examined the effects of pressure, temperature and solvent to solid ratio (SSR) on extraction efficiency of triglycerides from powdered *Jatropha curcas* seeds and *Aquilaria crassna* seeds by using supercritical carbon dioxide (SC-CO$_2$) extraction. According to the results of response surface methodology, pressure presented more effective in increasing the recovery of triglycerides, but both temperature and the SSR value are important in obtaining high concentration of triglycerides from two plant seeds that are useful for biodiesel materials. The experimental equilibrium solubilities of these two oils dissolved in SC-CO$_2$ at temperatures of 318–338 K and pressures of 200–350 bar were also investigated. The results indicated that the solubility of the oils linearly increased with the density of SC-CO$_2$ and the experimental data were successfully correlated by the Chrastil equation. Finally, this study demonstrated that subcritical hydrolysis followed by supercritical methylation of the SC-CO$_2$ extracted oil is a feasible two-step process in producing a 98.5% purity level of biodiesel from *Jatropha curcas* kernels.

Keywords: *Jatropha curcas*, *Aquilaria crassna*, SC-CO$_2$ extraction, biodiesel, subcritical hydrolysis, supercritical methylation.

---

* E-mail: cmchang@dragon.nchu.edu.tw

INTRODUCTION

Jatropha curcas (JC), a plant that occurs naturally in the wild, can grow without irrigation under a broad spectrum of rainfall. The plant itself is believed to prevent and control soil erosion and can be used as a living fence or to reclaim carbon credit for the country [1]. In addition, as a pressed cake, it can be used as fertilizer, and digested to produce biogas (CH_4) from its organic waste products. JC is also reported to host few pests and diseases [2], but this may change when it is grown on commercial plantations with regular irrigation and fertilization. From the point of view of applicability for biofuel production, many investors, policy makers and green project developers are interested in using JC to tackle the challenges of energy supply shortage and greenhouse gas emission reduction [3]. The kernel of JC forms a large proportion of the seed and accounts for 61.3±3.1%. There was a large variation in the contents of crude protein (19–31%), lipid (43–59%), neutral detergent fiber (3.5–6.1%), and ash (3.4–5.0%). The gross energy of the kernels was relatively similar (28.5–31.2 MJ/kg) [4, 5]. JC oil contains more than 75% unsaturated fatty acid, this was reflected in the pour and cloud point of the oil.

Aquilaria crassna in Thailand remains only in protected areas where it is heavily poached. It has been traditionally harvested sustainably by shaving wood from living trees; the wood is either chopped from near the base to search for the darkened heartwood (the agarwood), or it is removed with special knives which also stimulates the development of more agarwood in the fresh wound [6]. Species of Aquilaria (Thymelaeaceae) have been harvested for millenia in South and Southeast Asia for their resinous agarwood (aloewood, gaharu wood, or mai hom in Thai) from which cosmetic and pharmaceutical products are made [7]. A few studies have examined the distribution, density, and growth rates of some Indonesian species of Aquilaria [8].

Biodiesel produced from natural resources is one of alternative methods to fulfill the energy demands of the world, and it is better than diesel fuel in terms of sulfur content, flash point, aromatic content and biodegradability [9]. Biodiesel consists of fatty acid methyl esters (FAMEs) which are produced through trans-esterification. One of the important values for selecting FAMEs as biodiesel is the cetane number (CN); higher values indicate better ignition quality. The CN for Jatropha biodiesel is on the high end being 52.3 [10]. For trans-esterification in supercritical methanol or ethanol, no catalyst is required and nearly complete conversions can be achieved in very short periods due to the single phase reaction [11-13]. Cao et al. have carried out trans-esterification of soybean oil in supercritical methanol in the absence of catalyst demonstrated that besides alkali-catalyzed and acid-catalyzed processes, supercritical methylation is an alternative method [14]. Kusdiana and Saka have further discussed the effects of water on biodiesel fuel production by supercritical methanol treatment [15]. After the advent of single-step supercritical methylation process, a few researchers developed two-step one. Hydrolysis and subsequent supercritical methanol treatment process appears to be a promising alternative to the single-step supercritical methanol treatment by Minami and Saka. They concluded that hydrolysis of triglycerides results in the liberation of free fatty acids that can then be methylated more easily [16]. This article elucidates supercritical carbon dioxide extraction of triglycerides from Jatropha seeds followed by subcritical hydrolysis and supercritical methylation to obtain a high quality biodiesel.

SUPERCRITICAL CARBON DIOXIDE EXTRACTION OF TRIGLYCERIDES FROM PLANT SEEDS

Some studies have shown SC-CO_2 extractions of triglycerides from powdered rice bran [17, 18], wine lees [19], turtle fish oil [20], soybean [21], sunflower oil [22], alga [23], corn [24], wheat germ oil [25], rosehip seed, loquat seed and physic nut [26].

Quantification of Free Fatty Acids and Triglycerides

HPLC quantification of five free fatty acids (FFAs) was performed using a reverse-phase analytical column (YMC RP-18, 5 μm, 250 mm × 4.6 mm, Japan). The column was linked to a UV/Vis detector (785A, PerkinElmer, USA) by an HPLC pump (Series 410, PerkinElmer, USA). The column temperature was maintained at 313 K, and the injection volume of the sample was 20 μL. The samples were detected at a wavelength of 240 nm for the five FFAs (palmitic acid, palmitoleic acid, stearic acid, oleic acid and linoleic acid). The mobile phase of the mixed solvent of 85% acetonitrile, 5% methanol and 10% de-ionized water with 1% acetic acid was used. The regression coefficients (R^2) of five calibration curves of free fatty acids were all greater than 0.999.

The stoichiometric coefficient of the triglyceride to its FAME is 1:3 and the molecular weight of the triglyceride is three fold of the FAME. In this study, the conversion of methyl esterification of oil compounds including the TG, DG, MG and FFAs in the extracted oil attains 99% and the amount of the FFAs were almost zero in the extracted oil measured by a HPLC method. Thus, the weight of total triglycerides is equal to that of total FAME. GC quantification of seven triglycerides was performed in a 30 m × 0.5 μm non-polar capillary column (DB-5, JandW, USA) in a gas chromatograph (GC-14B, Shimadzu, Japan). The column temperature was set to 443 K initially and programmed to increase by 5 K/min to 488 K. Then it was set to increase by 2 K/min to 496 K. Finally, the column temperature was increased by 1 K/min until it reached 503 K. The injection volume was 1 μL with 3.4:1 of split ratio. The temperature of flame ionization detector was set at 553 K. The R^2 of seven calibration curves of methylated triglycerides were all at least 0.99. Figures 1 and 2 reveal the GC spectra of Jatropha and Crassna seed oil, respectively.

SC-CO_2 Extractions of *J. Curcas* Oil

After 16 h of Soxhlet *n*-hexane extraction (i.e., SSR = 275) from 30 g of *J. curcas* seed powdered by a high speed grinder to obtain particulates with particle size less than 0.84 mm through an international 20 mesh screen sieve, the maximal concentration of triglycerides (C_{TG}) in the extracted oil was 595.2 mg/g_{ext} and total yield was 49.21% [27]. The data were considered to represent 100% recovery of triglycerides from *J. curcas* seed powder.

Figure 3 displays a schematic flow diagram of the SC-CO_2 extraction. Thirty grams of *J. curcas* seed powder was packed into a 250 mL stainless steel tubular extractor with an inside diameter of 2.5 cm and a height of 54 cm.

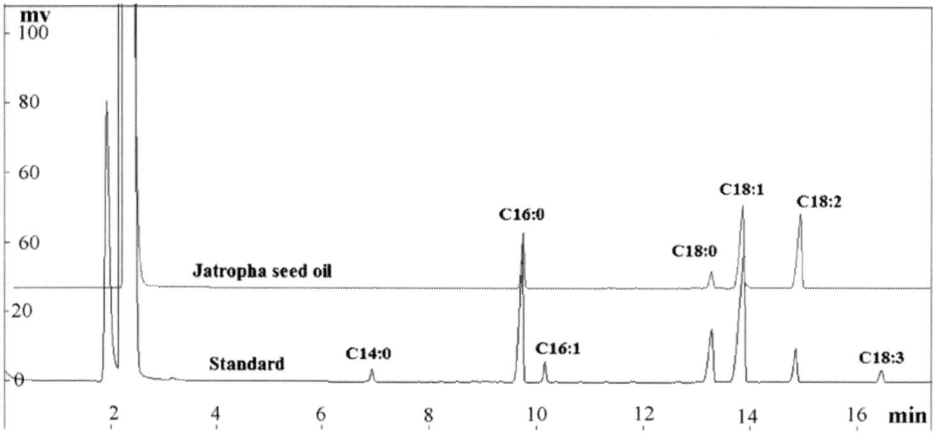

Figure 1. GC spectra of seven fatted acid methyl esters standards (A) and *J. curcas* seed oil (B) produced after methyl esterification [27].

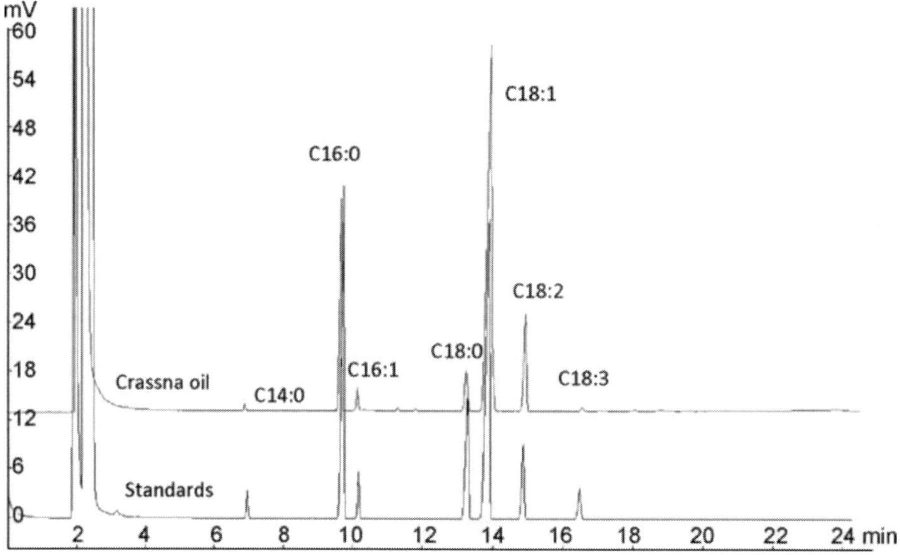

Figure 2. GC spectra of seven fatted acid methyl esters standards (A) and *A. crassna* seed oil (B) produced after methyl esterification [28].

Liquid CO_2 was flowed from a cylinder (1) in which was inserted a siphon-tube, and passed through a cooling bath (3) set at 277 K. The CO_2 was then compressed to the desired pressure using a syringe pump (100DX, ISCO, USA) (5); it was then heated to supercritical conditions using a double-pipe heat exchanger (8) and a re-boiler (10-2). CO_2 flowed upward into the extractor (10-1) at a flow rate of 25 mL/min to extract the oil. The constant pressure was manually adjusted by the first back-pressure regulators (12-1) located at the outlet. Following the extraction, the oil-laden CO_2 was driven into a 130 mL separator (14) and expanded through a spiral-type nozzle by a drop in pressure that was regulated by the second regulator (12-2) to maintain at 50 bar and 303 K. Low-pressure CO_2 was measured using a wet gas meter (W-NKDa-1B, Shinagawa, Japan) (15) and was subsequently returned to ambient conditions. Precipitates (around 15% of the extracted oil) that had washed out with *n*-hexane from the back-pressure regulators and tube lines were then mixed with the collected extract (85% oil) in preparation for analysis.

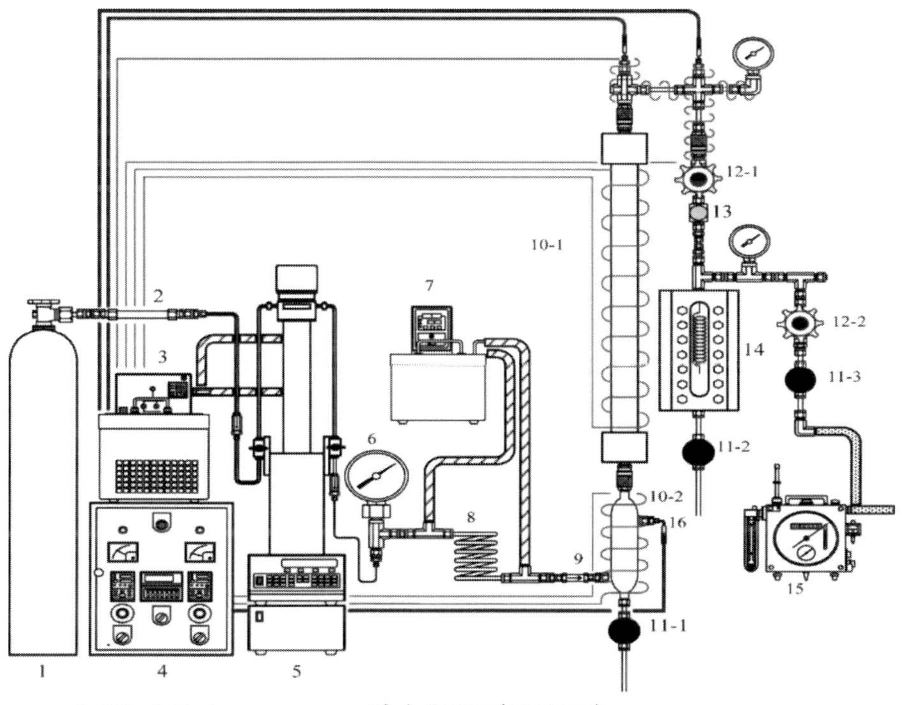

1. CO_2 Cylinder
2. CO_2 Cleanup column
3. Cold liquid circulator
4. Temperature controller
5. High pressure pump
6. Pressure gauge
7. Hot liquid circulator
8. Preheater
9. Check valve

10-1. Extraction vessel
10-2. Reboiler
11-1 ~ 11-3. Metering valve
12-1 ~ 12-2. Back pressure regulator
13. Needle valve
14. Separator
15. Wet gas meter
16. Thermocouple

Figure 3. Schematic flow diagram of SC-CO_2 extraction of *Jatropha curcas* and *Aquilaria crassna* seed oil [27].

Figure 4 plots the effect of pressure and temperature on the recovery of total triglycerides in the extract following the SC-CO_2 extraction of 30 g of *J. curcas* seed powder. Five hours of extraction with 3750 g of CO_2 at 350 bar and 333 K yielded 97.62% recovery of Jatropha triglycerides, but the extraction efficiency declined as the temperature decreased. A retrograde phenomenon of the extraction efficiency of total triglycerides occurred at 270 bar when the temperature decreased from 333 to 313 K. A common hypothesis concerning this phenomenon suggests that below 270 bar, the density of CO_2 substantially influences the efficiency. Above 270 bar, the vapor pressure of the Jatropha triglycerides was the dominating variable affecting efficiency. Stahl et al. [21] observed this crossover phenomenon with SC-CO_2 extraction of soybean oil and Wang et al. [18] also noted the crossover with SC-CO_2 extraction of rice bran oil. Furthermore, data present the C_{TG} value of the Soxhlet *n*-hexane oil is less than that of the SC-CO_2 oil indicated that the SC-CO_2

extraction process is superior to that of organic solvent extraction process in terms of energy and solvent consumption.

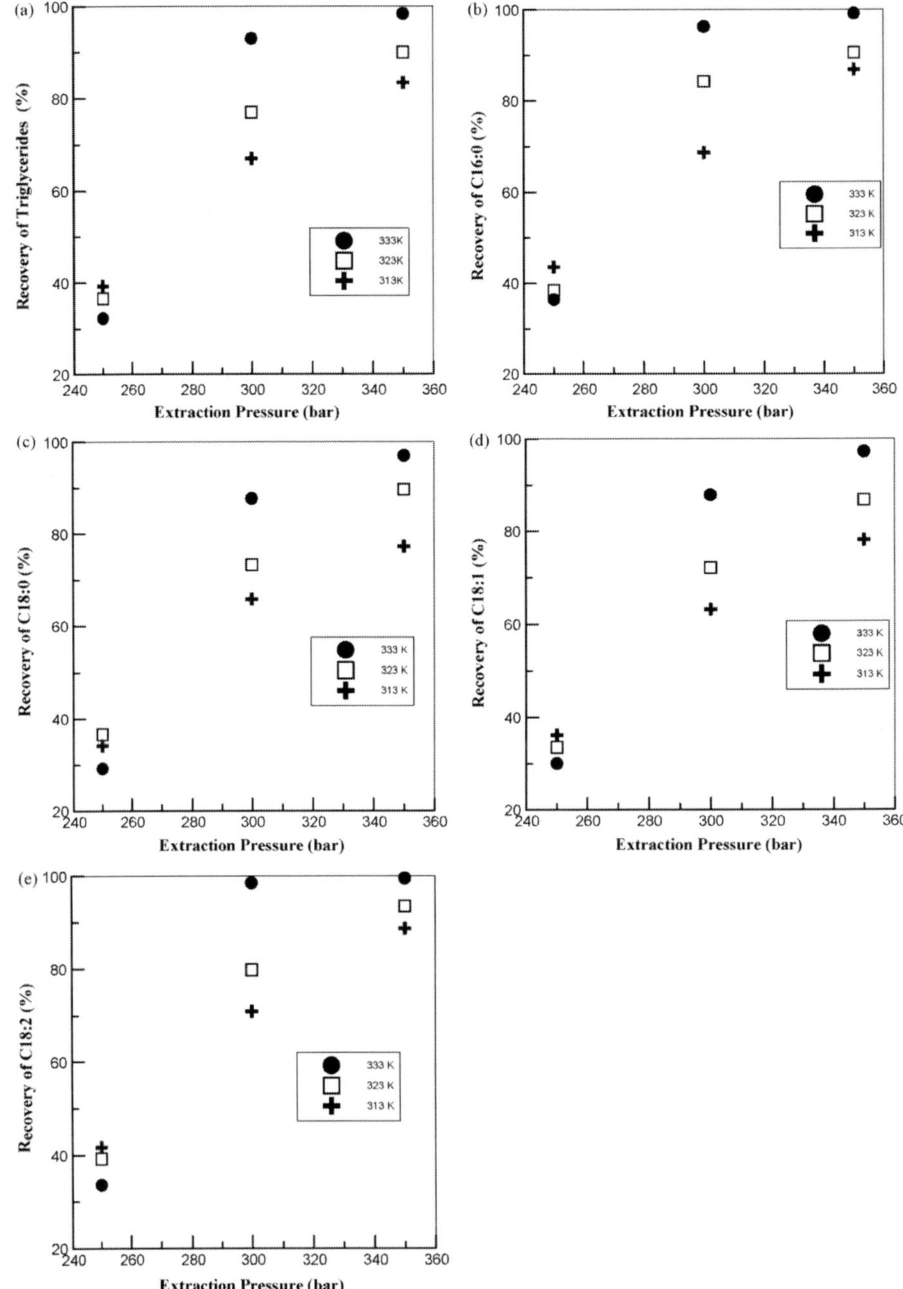

Figure 4. Effect of pressure and temperature on recovery of (a) triglycerides; (b) palmitic acid (C16:0); (c) strearic acid (C18:0); (d) oleic acid (C18:1) and (e) linoleic acid (C18:2) [27].

Response surface methodology (RSM) based on the central composite scheme for three design variables (temperature, pressure and SSR value) with seven axial points, seven factor points and one center point was employed for the SC-CO$_2$ extraction process. Table 1 presents experimental data on the RSM-designed SC-CO$_2$ extractions of triglycerides from 30 g of *J. curcas* seed powder at temperatures ranging from 313 to 333 K, pressures ranging from 250 to 350 bar and SSR (W$_{CO2}$/W$_{feed}$) values ranging from 65 to 125.

Three major responses are concentrations of triglycerides (C$_{TG}$), total yield (TY) and recovery of triglycerides (R$_{TG}$). The maximal C$_{TG}$ reached 705.6 mg/g$_{ext}$ (run 5 in Table 1); the highest TY and R$_{TG}$ reached 43.51% and 97.62%, respectively, under SC-CO$_2$ conditions of 333 K, 350 bar and an SSR value of 125 (run 11 in Table 1). Figure 5 shows the resulting concentrations of triglycerides which represent the quality of the extracted oil that changed with pressure, temperature and SSR value at a center point (i.e., P = 300 bar, T = 323 K and SSR = 95).

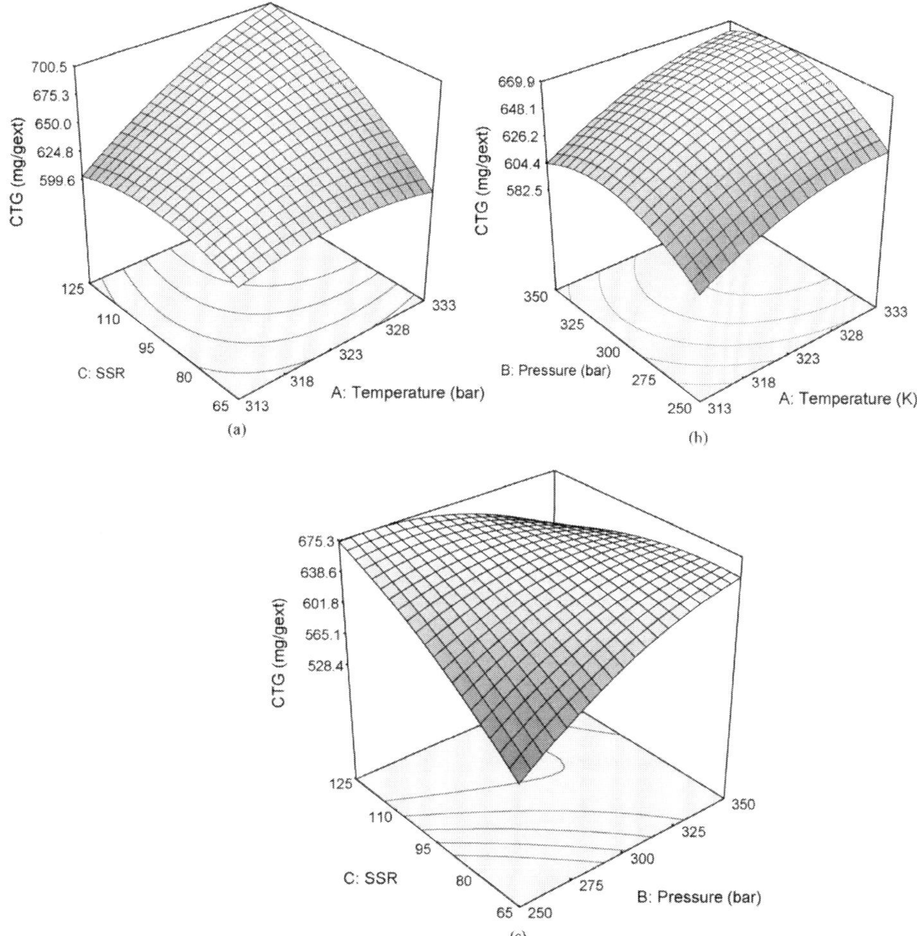

Figure 5. Three-dimensional representation of the concentration of triglycerides (C$_{TG}$) using SC-CO$_2$ extraction of *Jatropha curcas* seed oil (*F*-testing: R^2 = 0.9481, S.D. = 21.46). (a) Pressure: 300 bar; (b) SSR value: 95 and (c) temperature: 323 K [27].

Table 1. RSM-designed SC-CO$_2$ extractions of 30 g of powdered _J. curcas_ seed

Run #	P (bar)	T (K)	SSR	TY (%)	C_{TG} (mg/g)	$C_{C16:0}$ (mg/g)	$C_{C18:0}$ (mg/g)	$C_{C18:1}$ (mg/g)	$C_{C18:2}$ (mg/g)	R_{TG} (%)	β_{TG}
1(F)	250	313	65	15.28	508.0	80.3	31.5	205.2	191.0	26.49	1.73
2(F)	250	333	65	13.05	511.7	81.9	30.7	204.7	194.4	22.81	1.75
3(F)	250	313	125	19.88	630.4	99.8	38.9	253.2	238.5	42.80	2.15
4(A)	250	323	95	16.42	620.4	98.0	39.0	251.2	232.2	34.79	2.12
5(F)	250	333	125	19.74	705.6	112.9	42.3	282.2	268.1	47.57	2.41
6(A)	300	323	125	40.68	641.1	102.6	38.5	256.4	243.6	89.05	2.19
7(C)	300	323	95	35.24	639.4	106.5	40.2	256.3	236.4	76.94	2.18
8(A)	300	313	95	32.22	609.4	95.1	39.7	245.6	229.0	67.04	2.08
9(A)	300	323	65	24.42	640.8	98.8	34.3	244.9	226.8	53.43	2.19
10(A)	300	333	95	40.08	679.6	107.0	42.2	274.4	256.0	93.00	2.32
11(F)	350	333	125	43.51	657.1	101.1	43.4	281.8	230.7	97.62	2.24
12(A)	350	323	95	40.76	646.6	99.0	42.5	269.4	235.7	89.98	2.21
13(F)	350	333	65	38.54	631.3	101.0	37.9	252.5	239.9	83.08	2.16
14(F)	350	313	65	31.52	637.7	102.0	38.3	255.1	242.3	68.63	2.18
15(F)	350	313	125	42.84	547.5	87.6	32.9	219.0	208.1	80.09	1.87

T: temperature; P: pressure; SSR: solvent and solid ratio = (W_{CO2}/W_{feed}); TY: total yield; C_{TG}: concentration of triglycerides in extract; R_{TG}: recovery of triglycerides = [(TY×W_{feed}×C_{TG}/1000) / (TY×W_{feed}×C_{TG}/1000)$_{Soxhlet}$]×100%; β_{TG}: concentration factor of triglycerides = R_{TG} / TY.

From a statistical analysis, the main factor that affects the concentration of triglycerides is temperature and SSR value. Pressure effect is the major factor that influences both the total yield and the recovery of triglycerides. Besides, the experimental data are almost identified with the RSM predicted data for obtaining the maximum recovery at 340 bar, 330 K and the SSR value of 110.

SC-CO$_2$ Extractions of _A. Crassna_ Oil

The schematic flow diagram and the procedure of SC-CO$_2$ extractions of _A. crassna_ oil are similar to that of _J. curcas_ oil. After 16 h of Soxhlet _n_-hexane extraction (i.e., SSR value of 275) from 30 g of powdered _A. crassna_ seed, the maximal C_{TG} in the extracted oil was 522.4 mg/g$_{ext}$ with total yield of 38.21% [28]. The data were also considered to represent 100% recovery of triglycerides from _A. crassna_ seed powder. The Soxhlet _n_-hexane extracted oil contains more than 78% unsaturated fatty acids and the oleic acid is the major component.

Table 2. RSM-designed SC-CO$_2$ extractions of 30 g of powdered *A. crassna* seed

Run #	P (bar)	T (K)	SSR	TY (%)	C_{TG} (mg/g)	$C_{C14:0}$ (mg/g)	$C_{C16:0}$ (mg/g)	$C_{C16:1}$ (mg/g)	$C_{C18:0}$ (mg/g)	$C_{C18:1}$ (mg/g)	$C_{C18:2}$ (mg/g)	$C_{C18:3}$ (mg/g)	R_{TG} (%)	β_{TG}
1(A)	250	323	90	22.31	548.7	2.7	80.1	8.2	35.1	373.1	47.7	1.6	61.33	2.75
2(F)	250	313	60	17.90	460.9	2.3	67.3	6.9	29.5	313.4	40.1	1.4	41.33	2.31
3(F)	250	333	60	16.45	560.7	2.8	81.8	8.4	35.8	381.2	48.7	1.6	46.21	2.81
4(F)	250	333	120	32.94	517.9	2.5	75.6	7.7	33.1	352.1	45.0	1.5	85.47	2.59
5(F)	250	313	120	29.78	482.2	2.4	70.4	7.2	30.8	327.8	41.9	1.4	71.94	2.42
6(A)	300	313	90	30.77	514.9	2.5	75.1	7.7	32.9	350.1	44.7	1.5	79.37	2.58
7(A)	300	333	90	31.03	576.3	2.8	84.1	8.6	36.8	391.8	50.1	1.7	89.59	2.89
8(A)	300	323	60	25.64	518	2.5	75.6	7.7	33.1	352.2	45.0	1.5	66.54	2.60
9(C)	300	323	90	33.25	534.8	2.6	78.0	8.1	34.2	363.6	46.5	1.6	89.08	2.68
10(A)	300	323	120	35.48	520.1	2.6	75.9	7.8	33.2	353.6	45.2	1.5	92.45	2.61
11(F)	350	313	60	29.12	466.4	2.3	68.0	6.9	29.8	317.1	40.5	1.4	68.04	2.34
12(A)	350	323	90	33.48	538.8	2.6	78.6	8.0	34.4	366.3	46.8	1.6	90.37	2.70
13(F)	350	333	120	35.44	542.9	2.7	79.2	8.1	34.7	369.1	47.2	1.6	96.39	2.72
14(F)	350	313	120	34.81	547.1	2.7	79.8	8.2	35.0	372.0	47.5	1.6	95.41	2.74
15(F)	350	333	60	30.59	499.7	2.4	72.9	7.4	31.9	339.7	43.4	1.4	76.58	2.50

T: temperature; P: pressure; SSR: solvent and solid ratio = (W_{CO_2}/W_{feed}); TY: total yield; C_{TG}: concentration of triglycerides in extract; R_{TG}: recovery of triglycerides = $[(TY \times W_{feed} \times C_{TG}/1000) / (TY \times W_{feed} \times C_{TG}/1000)_{Soxhlet}] \times 100\%$; β_{TG}: concentration factor of triglycerides = R_{TG} / TY.

Experimental data of preliminary SC-CO$_2$ extractions based on a fixed temperature of 333 K and pressure of 350 bar show that the recovery of triglyceride increased with CO$_2$ flow rate and SSR value, it reached 96.51% at the CO$_2$ flow rate of 25 mL/min and the SSR value of 140. Response surface methodology based on the central composite scheme for three design variables (temperature, pressure, SSR value) was also employed for this system. Table 2 presents experimental data on the RSM-designed SC-CO$_2$ extractions of triglycerides from 30 g of *A. crassna* seed powder at temperatures ranging from 313 to 333 K, pressures ranging from 250 to 350 bar and SSR values ranging from 60 to 120. The maximal C$_{TG}$ reached 576.3 mg/g$_{ext}$ which temperature, pressure and SSR value set at 333 K, 300 bar and 90, respectively. The C$_{TG}$ values mainly increased with temperature and the SSR value. The highest TY and R$_{TG}$ reached 35.44% and 96.39%, respectively, under SC-CO$_2$ conditions of 333 K, 350 bar and an SSR value of 120. Table 2 also lists the maximal concentration enhancement of triglycerides (β_{TG}) reached 2.89. The generated response surfaces developed using fitted quadratic polynomial equations obtained from regression analysis were shown in Figure 6. Effects of the SSR value and pressure on the R$_{TG}$ values seem to be more significant than the temperature effect; these findings are similar to those from TY.

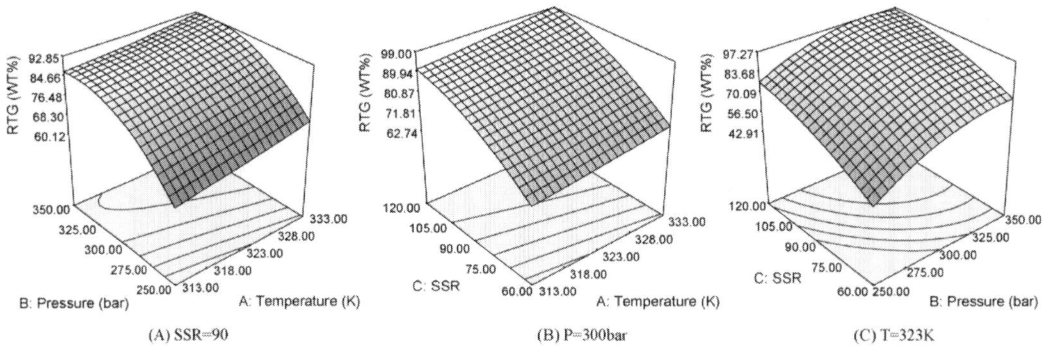

Figure 6. Three-dimensional representation of recovery of triglycerides (R$_{TG}$) using SC-CO$_2$ extraction of *Aquilaria crassna* seed oil (*F*-testing: R^2 = 0.9814, S.D. = 3.95). (a) SSR value: 90; (b) pressure: 300 bar; (c) temperature: 323 K [28].

SOLUBILITY OF PLANT OILS IN SUPERCRITICAL CARBON DIOXIDE

Figure 7 displays a schematic flow diagram of the solubility measurement system [29] which was modified from SC-CO$_2$ extraction apparatus shown in Figure 3. The oil was evenly dispersed into a 150 mL equilibrium cell by the mechanical force of the SC-CO$_2$ fluid and a flow distributor made by θ type stainless steel packing materials, which consisted of a few hollow cylinders made by 60 mesh stainless steel sieves of the 24 mm × 5 mm size. The experiment was carried out at 350 bar and 338 K for the mixture of 15 mL oil and SC-CO$_2$. The effect of flow rate and total volume of CO$_2$ were examined in the preliminary experiments. The suitable flow rate of CO$_2$ is important to prevent the extracted oil entrained to the separator and a long-vertical tube above the equilibrium cell is a buffer zone in avoiding this entrainment. Since enough CO$_2$ consumption and adequate equilibrium time were suitable equilibrium conditions, a flow rate of 5 mL/min at supercritical condition and a total amount CO_2 of 50 L at one atmospheric condition were chosen as two experimental conditions in subsequent high-pressure solubility examinations. The extracted oil compounds in the SC-CO_2 phase are considered as most of triglycerides. Table 3 presents solubility data for the binary system of oil + CO_2 calculated by a gravimetric method under each operation condition. The uncertainty of the solubility is less than 5% based on error contributions of the residue oil collection, electronic balance and GC quantification. From the experimental results, the solubility data and C_{TG} values did increase with increasing temperature and pressure. The highest solubility (29.8 mg/L) and C_{TG} (902.8 mg/g) of *J. curcas* oil was obtained at 338 K and 350 bar. The C_{TG} value concentrated from 880.6 mg/g to the higher value due to the partial low-polar property of triglycerides. Consequently, low-polar solutes surrounded with a few CO_2 molecules to form clusters in various sizes, and this phenomenon has been proposed by Morita and Kajimoto [30]. A similar phenomenon was observed with the *A. crassna* oil.

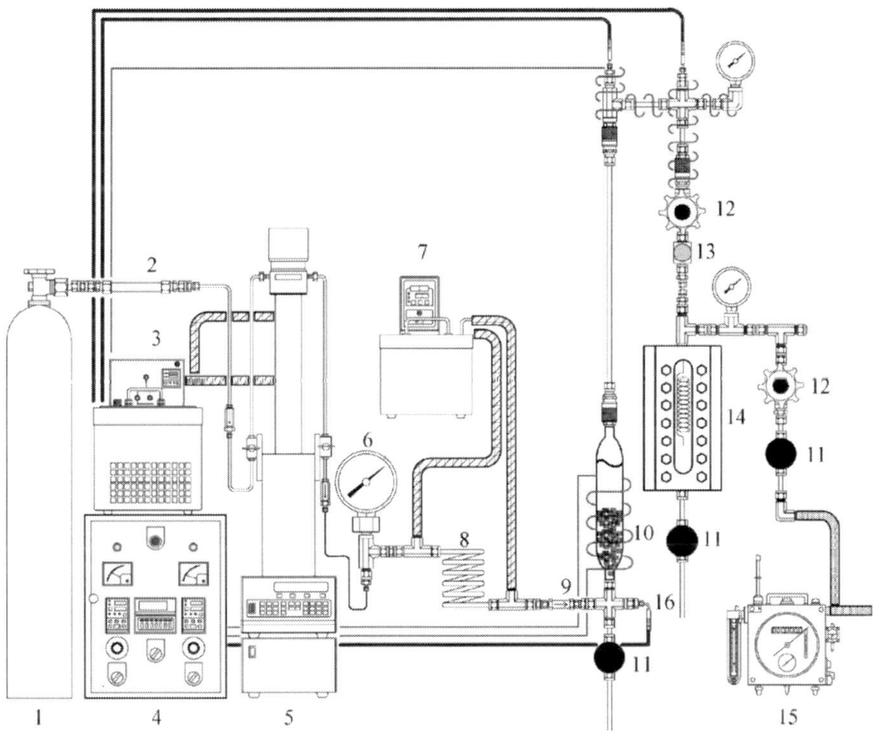

1.	CO_2 cylinder	9.	Check valve
2.	CO_2 ceanup column	10.	Equilibrium cell filled with θ rings
3.	Cold liquid circulator	11.	Metering valve
4.	Temperature controller	12.	Back pressure regulator
5.	High pressure pump	13.	Needle valve
6.	Pressure gauge	14.	Separator
7.	Hot liquid circulator	15.	Wet gas meter
8.	Preheater	16.	Thermocouple

Figure 7. Schematic flow diagram of the solubility measurement system [29].

The solubility data were correlated by utilizing Chrastil equation shown in Eq. (1) to obtain the corresponding parameters.

$$\ln S = k \ln \rho + a / T + b \tag{1}$$

where S (kg/m^3) is the solubility of the solute in the supercritical phase, ρ (kg/m^3) is the density of the pure supercritical fluid, T (K) is the operating temperature, k is an association constant, a is a function of the sum of the enthalpies of solvation and vaporization of the solute, and b is a function of the molecular weights of the solute and solvent. This model is expressed elsewhere by Chrasil [31] and Adachi and Lu [32], describing a solute associates with k molecules of supercritical solvent to form a complex in the equilibrium system. Physical properties of CO_2 were adopted from the National Institute of Standards and Technology (NIST) Chemistry WebBook on-line database.

Table 3. Solubility data of 15 mL of plant seeds oil and 50 L of SC-CO$_2$ as well as correlation data by Chrastil equation

T (K)	P (bar)	ρ_{CO2} (g/L)	W_{feed} (g)	W_R (g)	S_{oil} (mg/L)	C_{TG} (mg/g)	S^*_{oil} (mg/L)
Jatropha curcas							
338	200	692.68	11.7583	10.8788	17.6	883.6	17.2
338	250	761.94	11.6936	10.5871	22.1	889.7	22.2
338	300	808.95	11.7848	10.4603	26.5	897.8	26.0
338	350	844.57	11.8012	10.3122	29.8	902.8	29.2
328	200	755.52	11.6941	10.7406	19.1	887.3	19.1
328	250	811.37	11.7795	10.6635	22.3	893.9	23.0
328	300	850.83	11.6890	10.3865	26.1	897.3	26.1
328	350	881.71	11.7682	10.3347	28.7	901.5	28.7
318	200	813.52	11.6857	10.6622	20.5	890.2	20.2
318	250	857.14	11.7945	10.6120	23.7	896.4	23.2
318	300	890.33	11.7386	10.4436	25.9	897.0	25.7
318	350	917.12	11.7564	10.3394	28.3	899.6	27.8
Aquilaria crassna							
338	200	692.68	11.9663	11.1368	16.6	769.5	16.0
338	250	761.94	11.8496	10.7836	21.3	772.6	20.8
338	300	808.95	11.7388	10.4608	25.6	779.7	24.5
338	350	844.57	11.8972	10.4797	28.4	786.3	27.6
318	200	813.52	11.7931	10.8121	19.6	771.1	19.0
318	250	857.14	11.8865	10.7530	22.7	775.8	21.9

| 318 | 300 | 890.33 | 11.6396 | 10.3976 | 24.8 | 778.2 | 24.3 |
| 318 | 350 | 917.12 | 11.9582 | 10.5882 | 27.4 | 784.3 | 26.3 |

W_{feed}: weight of feed oil; W_R: weight of residue oil; S_{oil}: solubility of oil in $CO_2 = (W_{feed} - W_R)/50$; C_{TG}: concentration of triglycerides in extract oil; S^*_{oil}: calculated solubility by Chrastil equation; Concentration of triglycerides in *Jatropha curcas* oil $C_{TG} = 880.6$ mg/g and in *Aquilaria crassna* oil $C_{TG} = 768.9$ mg/g; CO_2 flow rate = 5 mL/min.

The solubility data obtained from the experiments were highly correlated with the Chrastil equation, as shown in Figure 8. From the Chrastil correlation, high values of the association constant k for oils agreed well with high solubilities in the SC-CO_2 phase. The association constant can be obtained from the slope of the logarithmic plot of solubility versus density of the CO_2. Table 4 presented the parameter values of Chrastil equation for two plant seed oils. The k values of *J. curcas* oil and *A. crassna* oil are 2.6953 and 2.7369, respectively. This is a highly consistent result that on average, one molecule of oil can be surrounded by 2.7 molecules of carbon dioxide.

Table 4. Parameters of Chrastil equation for *Jatropha curcas* and *Aquilaria crassna* oils

	Parameters		
	k	a	b
Jatropha curcas	2.6953	-1431.25	-28.68
Aquilaria crassna	2.7369	-1464.31	-29.70

k is an association constant; a is a function of the sum of the enthalpies of solvation and vaporization of the solute; b is a function of the molecular weights of the solute and solvent.

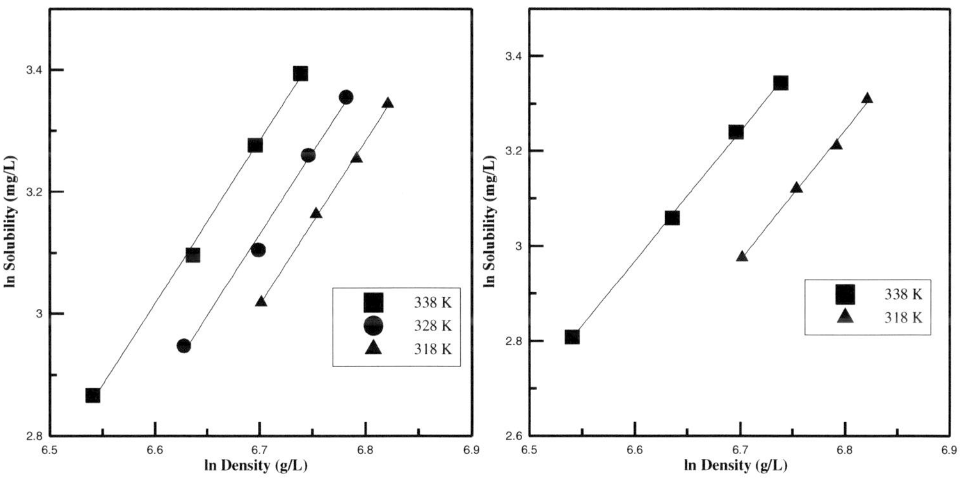

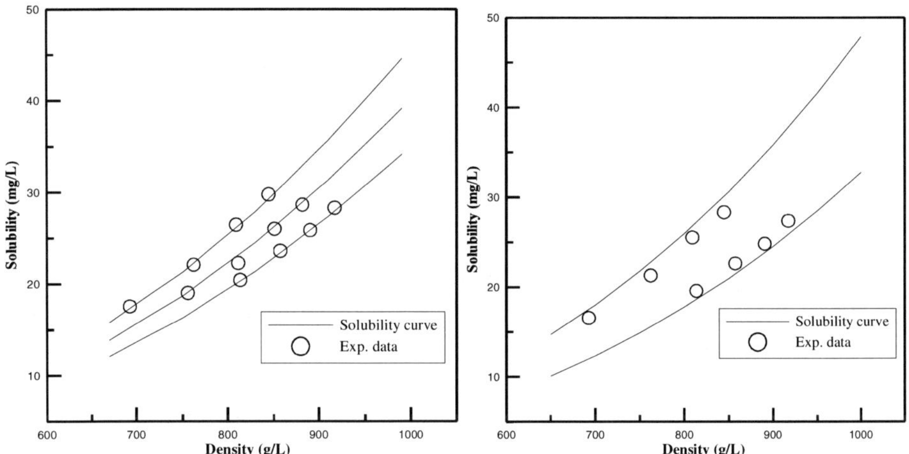

Figure 8. Solubility data correlated by densities of carbon dioxide. The data were collected in the environment of 50 L of supercritical carbon dioxide and 15 mL of (a) *Jatropha curcas* oil and (b) *Aquilaria crassna* oil [29].

BIODIESEL PRODUCTION USING SUBCRITICAL HYDROLYSIS AND SUPERCRITICAL METHYLATION

An international standard that describes the minimum requirements for biodiesel indicate that a minimum 96.5% of FAME is necessary to run vehicles as well as only minor trace compounds are mono-, di-, triglyceride, total glycerine, etc., should be found in the fuel [33]. It is concluded that high concentrations of triglycerides in the extract and high conversions of methylation lead to a high quality biodiesel.

Hydrolysis and Methylation of *J. Curcas* Oil

Figure 9 shows a schematic flow diagram for both the subcritical hydrolysis and the supercritical methylation processes [34]. For the subcritical hydrolysis process, 100 mL of de-ionized water and 0.25 mL of 99% acetic acid were put into a 1 L stainless steel tubular reactor (4520, Parr Instruments Co., USA). The autoclave was heated by an external heating mantle to attain the desired temperature (523 or 543 or 563 K) and manipulated by a temperature controller (4842, Parr Instruments Co., USA) at ± 2 K. Then, 10 mL of degummed and dewaxed Jatropha oil was pumped into the reactor by a HPLC pump (PU-1580, Jasco, Japan). The operating pressure was maintained by a high pressure N_2 cylinder. For the supercritical methylation process, a certain amount of methanol three times more of the oil was put into the reactor. When the desired temperature was reached, 10 mL of the hydrolyzed oil was pumped into the autoclave by the HPLC pump. At the end of each experiment, the lower layer was a mixture of methanol and water; the upper layer of the reacted solution was collected and the residual methanol was removed using a vacuum rotary evaporator. The content of free fatty acids and triglycerides in the upper layer of oil was quantified by HPLC and GC methods, respectively.

The hydrolysis reaction occurred at 110 bar and 543 K for 1 h, the conversion of triglycerides (X_{TG}) in the hydrolyzed oil was 92.1% calculated from the maximum content of FFAs in the hydrolyzed oil, said 813.4 mg/g dividing by the concentration of total fatty acids in the hydrolyzed oil, said 883.2 mg/g. A two-factor (time and temperature) response surface methodology was employed to find the effect of factors on the hydrolysis conversion. The results indicated the reaction temperature seemed to have an equally significant effect on subcritical hydrolysis as time did. The methylation reaction occurred at 110 bar and 563 K for 15 min, the conversion of free fatty acids (X_{FFA}) was 99.0%. The maximum content of fatty acid methyl esters in the methylated oil was determined to be 985.0 mg/g by using the GC method that is higher than those of Minami and Saka [35]. A three-factor RSM experimental design with a central composite scheme for the supercritical methylation of the hydrolyzed oil was adopted to find the dependent variables (i.e., X_{FFA}). The resulting X_{FFA} value that represented the quality of the methylated oil varied with time, SSR value (MeOH versus FFA) and temperature. In this supercritical methylation process, the X_{FFA} value increased with time and temperature, but decreased slightly as the SSR value increased. The increase suggested a positive effect of time and temperature on rate. The decrease may have been due to the free fatty acids acting as reactants during the methylation reaction; the large SSR value therefore resulted in a low FFAs concentration to reduce the rate of the methylation.

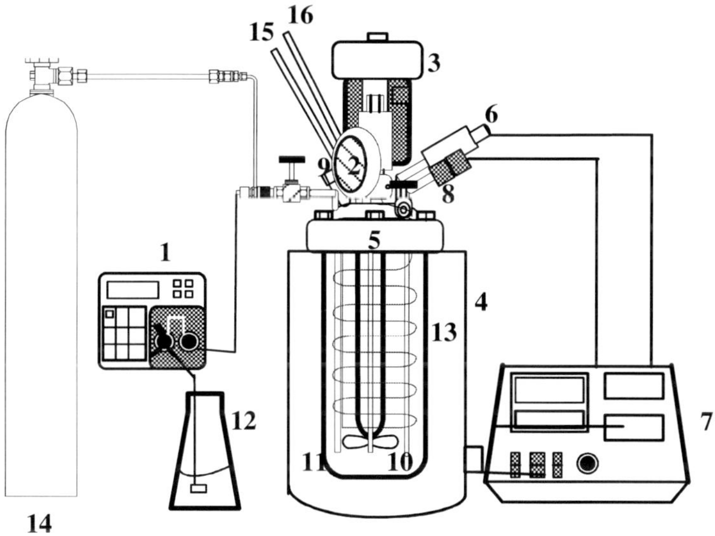

1.	LC pump	9.	Safety rapture disc
2.	Pressure gage	10.	Internal string system
3.	Motor	11.	Inlet tube
4.	Heater	12.	Sample bottle
5.	Reactor	13.	Cooling Coil
6.	Pressure detector	14	N_2 cylinder
7.	Temperature controller	15	Cooling water in
8.	Thermocouple	16	Cooling water out

Figure 9. Schematic flow diagram of subcritical hydrolysis and supercritical methylation reactions [34].

Determination of Rate Constant and Activation Energy of Reactions

Figure 10(A) shows the kinematic relationship between the conversion of triglycerides via hydrolysis and time at 523, 543 and 563 K. Only high temperatures resulted in high rates of hydrolysis, the conversion reached a maximum of 92.1% with a deep black opaque color at where the concentration of FFAs in water might lower than that of TG to drive the hydrolysis towards the forward side. Within 1 h of hydrolysis at 543 K and 110 bar, the X_{TG} reached 91.6% with see-through red colored oil. The experimental data $(1-X_{TG})$ were plotted with reaction time as shown in Figure 10(B) to determine the rate constant (k) of the hydrolysis. The rate constant at 563 K was 0.056 min^{-1} and was obtained from the slope of the first order reaction. The activation energy of the hydrolysis was 68.5 kJ/mole obtained by an Arrhenius plot of ln k versus 1/T as shown in Figure 10(C). Figure 11 displays the kinematic relationship between the conversion of free fatty acids via methylation and time at 523, 543 and 563 K.

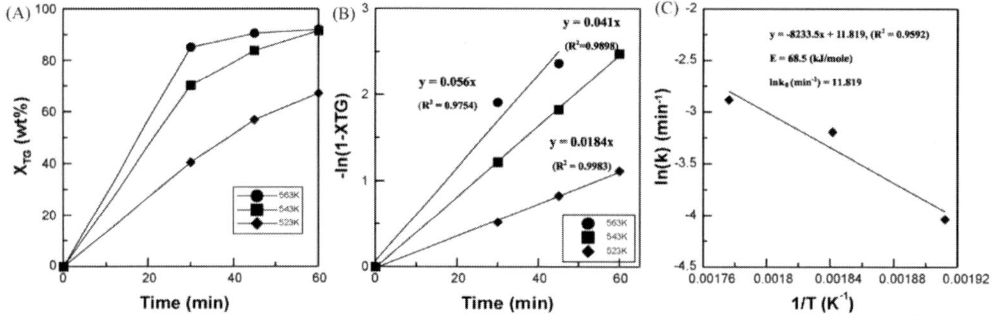

Figure 10. Kinematic relationship between conversion of triglycerides (X_{TG}) via hydrolysis and time at 523, 543 and 563 K. (A) Kinetics curves of hydrolysis; (B) determination of the rate constant of the hydrolysis reaction; (C) Arrhenius plot [34].

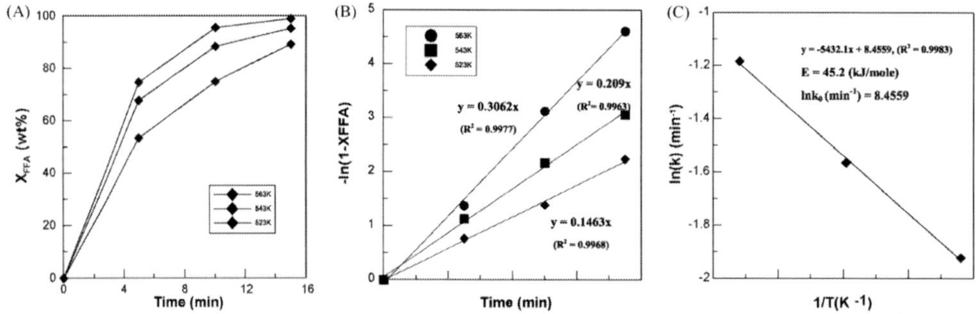

Figure 11. Kinematic relationship between the conversion of free fatty acids (X_{FFA}) via methylation and time at 523, 543 and 563 K. (A) Kinetics curves of methylation; (B) determination of the rate constant of the methylation reaction; (C) Arrhenius plot [34].

The rate constant of the methylation process at 563 K was 0.3062 min^{-1}, and the activation energy of the methylation was 45.2 kJ/mole. The density and viscosity of the methylated oil are 0.8663 g/cm^3 and 4.1239 mPa s (cP), measured by Anton-Paar densitometer and viscometer, respectively. The viscosity of this methylated oil meets the international standard for biodiesel criteria, Max. 4.5 mPa s (cP). In addition, the effect of carbon dioxide addition on the supercritical methylation procedure was also studied. The results indicated that methylation temperature could be reduced by adding a certain amount of supercritical carbon dioxide.

CONCLUSION

This study investigated designed supercritical carbon dioxide extraction of *J. curcas* and *A. crassna* oil from powdered plant seeds using a response surface methodology. Changing pressures, SSR values and temperatures in SC-CO_2 extractions indicated that pressure and the SSR value are more effective than temperature in enhancing the extraction efficiency of triglycerides. However, two main factors that affect the concentration of triglycerides are temperature and the SSR value. This study also provided useful information of solubility data as well as a semi-empirical correlation for the application of SC-CO_2 extraction of triglycerides from the plant seeds of *J. curcas* and *A. crassna*. The solubility of oil was enhanced by increasing both temperature and pressure and the solubility data were successfully correlated with the Chrastil equation. For the production of biodiesel using supercritical fluids technology, the experimental data indicated that the two-step process of hydrolysis and subsequent methylation was a more suitable pathway for producing biodiesel from Jatropha kernel. The activation energy necessary for hydrolysis was higher than that for methylation which indicated hydrolysis was the rate determining step for biodiesel production.

ACKNOWLEDGMENTS

The authors would like to thank the National Science Council of the Republic of China, Taiwan for financially supporting this research (NSC 98-2221-E-005-053-MY3) and (NSC99-2622-B-005-005-CC2). This work is also supported in part by the Ministry of Education, Taiwan, ROC, under the ATU plan.

REFERENCES

[1]   G. Francis, R. Edinger, K. Becker, A concept for simultaneous wasteland reclamation, fuel production, and socioeconomic development in degraded areas in India: need, potential and perspectives of Jatropha plantations, *Natural Resources Forum* 29 (2005) 12–24.

[2]   G.M. Gubitz, M. Mittelbach, M. Trabi, Exploitation of the tropical oil seed plant *Jatropha curcas* L., *Bioresource Technology* 67 (1999) 73–82.

[3]   K. Openshaw, A review of *Jatropha curcas*: an oil plant of unfulfilled promise, Biomass and Bioenergy 19 (2000) 1–15.

[4] H.P.S. Makkar, K. Becker, F. Sporer, M. Wink, Studies on nutritive potential and toxic constituents of different provenances of *Jatropha curcas*, *Journal of Agricultural and Food Chemistry* 45 (1997) 3152–3157.

[5] E.T. Akintayo, Characteristics and composition of *Parkia biglobbossa* and *Jatropha curcas* oils and cakes, *Bioresource Technology* 92 (2004) 307–310.

[6] L. Zhang, W.Y. Brockelman, M.A. Allen, Matrix analysis to evaluate sustainability: the tropical tree *Aquilaria crassna*, a heavily poached source of agarwood, *Biological Conservation* 141 (2008) 1676–1686.

[7] I.H. Burkill, Dictionary of the economic products of the malay peninsula, Ministry of Agriculture and Co-operatives, Kuala Lumpur, Malaysia, 1966, pp. 1583–1587.

[8] J.V. LaFrankie, Population dynamics of some tropical trees that yield non-timber forest products, *Economic Botany* 48 (1994) 301–309.

[9] Srivastava, R. Prasad, Triglycerides-based diesel Fuels, *Renewable and Sustainable Energy Reviews* 4 (2000) 111–133.

[10] M. Mohibbe Azam, A. Waris, N.M. Nahar, Prospects and potential of fatty acid methyl esters of some non-traditional seed oils for use as biodiesel in India, *Biomass Bioenergy* 29 (2005) 293–302.

[11] G. Madras, C. Kolluru, R. Kumar, Synthesis of biodiesel in supercritical fluids, *Fuel* 83 (2004) 2029–2033.

[12] D. Kusdiana, S. Saka, Kinetics of transesterication in rapeseed oil to biodiesel fuel as treated in supercritical methanol, *Fuel* 80 (2001) 693–698.

[13] V. Rathore, G. Madras, Synthesis of biodiesel from edible and non-edible oils in supercritical alcohols and enzymatic synthesis in supercritical carbon dioxide, *Fuel* 86 (2007) 2650–2659.

[14] W. Cao, H. Han, J. Zhang, Preparation of biodiesel from soybean oil using supercritical methanol and co-solvent, *Fuel* 84 (2005) 347–351.

[15] D. Kusdiana, S. Saka, Effects of water on biodiesel fuel production by supercritical methanol treatment, *Bioresource Technology* 91 (2004) 289–295.

[16] E. Minami, S. Saka, Kinetics of hydrolysis and methyl esterification for biodiesel production in two-step supercritical methanol process, *Fuel* 85 (2006) 2479–2483.

[17] C.R. Chen, C.H. Wang, L.Y. Wang, Z.H. Hong, S.H. Chen, W.J. Ho, C.M.J. Chang, Supercritical carbon dioxide extraction and deacidification of rice bran oil, *Journal of Supercritical Fluids* 45 (2008) 322–331.

[18] C.H. Wang, C.R. Chen, J.J. Wu, L.Y. Wang, C.M.J. Chang, W.J. Ho, Designing supercritical carbon dioxide extraction of rice bran oil that contain oryzanols using response surface methodology, *Journal of Separation Science* 31 (2008) 1399–1407.

[19] J.J. Wu, J.C. Lin, C.H. Wang, T.T. Jong, H.L. Yang, S.L. Hsu, C.M.J. Chang, Extraction of antioxidative compounds from wine lees using supercritical fluids and associated anti-tyrosinase activity, *Journal of Supercritical Fluids* 50 (2009) 33–41.

[20] L.H. Chang, C.T. Shen, S.J. Hsieh, S.L. Hsu, H.C. Chang, C.M.J. Chang, Recovery and enhancement of unsaturated fatty acids in soft-shelled turtle fish oil using supercritical carbon dioxide and associated catalase release activity, *Separation and Purification Technology* 64 (2008) 213–220.

[21] E. Stahl, S.E. Chutz, H.K. Mangold, Extraction of seed oils with liquid and supercritical carbon dioxide, *Journal of Agricultural and Food Chemistry* 28 (1980) 1153–1157.

[22] M.J. Cocero, L. Calvo, Supercritical fluid extraction of sunflower seed oil with CO_2–ethanol mixtures, *Journal of the American Oil Chemists Society* 73 (1996) 1573–1578.

[23] R.L. Mendes, B.P. Nobre, M.T. Cardoso, A.P. Pereira, A.F. Palavra, Supercritical carbon dioxide extraction of compounds with pharmaceutical importance from microalgae, *Inorganica Chimica Acta* 356 (2003) 328–334.

[24] E. Rónyai, B. Simándi, S. Tömösközi, A. Deák, L. Vigh, Z. Weinbrenner, Supercritical fluid extraction of corn germ with CO_2–ethyl alcohol mixture, *Journal of Supercritical Fluids* 14 (1998) 75–81.

[25] P. Shao, P. Sun, Y. Ying, Response surface optimization of wheat germ oil yield by supercritical carbon dioxide extraction, *Food and Bioproducts Processing* 86 (2008) 227–231.

[26] S. Machmudah, M. Kondo, M. Sasaki, M. Goto, J. Munemasa, M. Yamagata, Pressure effect in supercritical CO_2 extraction of plant seeds, *Journal of Supercritical Fluids* 44 (2008) 301–307.

[27] W.H. Chen, C.H. Chen, C.J. Chang, Y.H. Chiu, D. Hsiang, Supercritical carbon dioxide extraction of triglycerides from *Jatropha curcas* L. seeds, *Journal of Supercritical Fluids* 51 (2009) 174–180.

[28] W.H. Chen, C.H. Chen, C.J. Chang, B.C. Liau, D. Hsiang, Supercritical carbon dioxide extraction of triglycerides from *Aquilaria crassna* seeds, *Separation and Purification Technology* 73 (2010) 135–141.

[29] P.Y. Chen, W.H. Chen, S.M. Lai, C.J. Chang, Solubility of Jatropha and Aquilaria oils in supercritical carbon dioxide at elevated pressures, *Journal of Supercritical Fluids* 55 (2011) 893–897.

[30] Morita, O. Kajimoto, Solute–solvent interaction in nonpolar supercritical fluid: a clustering model and size distribution, *Journal of Physical Chemistry* 94 (1990) 6420–6425.

[31] J. Chrastil, Solubility of solids and liquids in supercritical gases, *Journal of Physical Chemistry* 86 (1982) 3016–3021.

[32] Y. Adachi, B.C.Y. Lu, Supercritical fluid extraction with carbon dioxide and ethylene, *Fluid Phase Equilibria* 14 (1983) 147–156.

[33] R.E. Babcock, E.C. Clausen, M. Popp, W.B. Schulte, Yield characteristics of biodiesel produced from chicken fat-tall oil blended feedstocks, Mackblackwell Rural Transportation Center, MBTC-2092, annual report 2007–2008, Fayetteville, Arkansas, 2008.

[34] C.H. Chen, W.H. Chen, C.J. Chang, S.M. Lai, C.H. Tu, Biodiesel production from supercritical carbon dioxide extracted Jatropha oil using subcritical hydrolysis and supercritical methylation, *Journal of Supercritical Fluids* 52 (2010) 228–234.

[35] E. Minami, S. Saka, Kinetics of hydrolysis and methyl esterification for biodiesel production in two-step supercritical methanol process, *Fuel* 85 (2006) 2479–2483.

In: Biodiesel: Blends, Properties and Applications
Editors: Jorge M. Marchetti and Zhen Fang

ISBN: 978-1-63117-024-9
© 2014 Nova Science Publishers, Inc.

Chapter 9

CHARACTERIZATION OF EMISSIONS AND INDOOR AIR QUALITY OF PUBLIC TRANSPORT BUSES USING ALTERNATIVE FUELS

Ashok Kumar, Akhil Kadiyala, Dinesh Somuri,
Kaushik K. Shandilya, Srikar Velagapudi
and Vinay Kumar V. Nerella*
Department of Civil Engineering, The University of Toledo,
Toledo, Ohio, US

ABSTRACT

This Chapter presents a summary of the results obtained from a field program using the Toledo Area Regional Transit Authority (TARTA) public transport buses, running on biodiesel blend and ultra-low sulfur diesel (ULSD) fuels. This TARTA research program is divided into two studies: exhaust emissions and indoor air quality.

A comprehensive analysis of pollutant levels released from the tailpipe exhaust emissions resulting from the use of different biodiesel blends under different operating modes is included. Indoor air quality (IAQ) in buses operating on B20 grade biodiesel and ULSD has been studied for more than two years. Study of vehicular pollutant exposure resulted in concluding that it is safe for the passengers as well as drivers, with the indoor pollutant levels being well below the available health guidelines. Particulate matter filters collected between 2007 and 2009 were analyzed using computer controlled environmental scanning electron microscopy and energy dispersive X-ray spectrometry (SEM/EDS) to characterize shape and aspect ratio of the fine particulate matter.

Keywords: Indoor Air Quality, Mobile Exhaust Emissions, Public Transport Buses, Biodiesel, and Ultra Low Sulfur Diesel.

---

* Phone: (419) 530-8120/8136. Email: akumar@utoledo.edu

INTRODUCTION

Study of vehicular exhaust emissions is important as they are one of the major sources of ground level pollution and have a major impact on public health as observed by Soehodho and Taufick (2005). Some studies that have observed the factors influencing exhaust emissions using biodiesel fueled vehicles are Dorado et al. (2003), Wang et al. (1997), Mazzoleni et al. (2007), McCormick et al. (2000), Agarwal (2006), Carrarreto et al. (2004), The United States Environmental Protection Agency Report (2002), Younglove et al. (2005), Chen et al. (2007), Brodrick et al. (2002), and Storey et al. (2003). From these studies as reported in the literature it has been observed that the effects of biodiesel blends on exhaust emission pollutant levels of urban transit buses during their regular service have not been studied (Shandilya and Kumar, 2009).

The importance of studying indoor air quality in a transit microenvironment is equally valuable and is vindicated by the fact that people spend about 87% of their time indoors and 5-6% of their daily time commuting, mostly between their workplace and their residence as observed by Klepeis et al. (2001) in the National Human Activity Pattern Survey (NHAPS). Many vehicular studies have identified the various factors influencing indoor air quality levels using regression analysis, observed pollutant trends, and studied the exposure levels to commuters [Zhang et al. (2006), Chan et al. (2002), Chan et al. (2003), Diapouli et al. (2007), Chan and Liu (2001), Gulliver and Briggs (2004), Gomez-Perales et al. (2004), Bremauntz and Ashmore (1995), Clifford et al. (1997), Chan et al. (2002), Chan et al. (1999), Kuo et al. (1999), Chan (2003), Alm et al. (1999), Adams et al. (2002), Fitz et al. (2003), Rodes et al. (1998), Fruin (2003), Praml and Schierl (2000), Sabin et al. (2005), Shandilya and Kumar (2009)]. However, all these studies have used conventional diesel fuels, and none of them have used regression tree analysis, an advanced statistical method to determine the important factors affecting vehicular indoor air quality levels except for a few studies by Kadiyala (2008) and Kadiyala and Kumar (2008a, 2008b). Also, none of these studies have characterized shape and aspect ratio of the fine particulate matter in a biodiesel fueled bus (Shandilya et al., 2009; Shandilya, 2002).

The purpose of this research is to study the effects of using alternative diesel fuels on vehicular emissions and indoor air quality of public transport buses in Toledo, Ohio. This study also aims to determine the various factors affecting vehicular exhaust emission levels and indoor air quality inside the bus compartment and to characterize the fine particulate matter collected indoors.

None of the studies reported in the literature have conducted such field programs for more than two years since it is very time-consuming and expensive. This field program tries to fill the knowledge gap on the environmental impacts of biodiesel fueled vehicles. This chapter provides an overview of the salient finding of the field program along with a discussion on the environmental impact of biodiesel on indoor air quality and exhaust emissions in public transport buses.

METHODOLOGY

This section provides an overview of the criteria used for selecting the test run buses and experimental setup used to monitor indoor pollutant levels and exhaust emissions.

Exhaust Emission Study

Selection of Test Routes and Buses for the Study

TARTA has been the "Ride of Toledo" since 1970 and has over 40 routes in the metropolitan area. The test route selected for this study was Route # 20T, which runs from TARTA central garage (A) to Franklin Park Mall (B) as can be seen in Figure 1. The route selected is entirely urban with length of the route being 4.7 miles and the testing time was approximately 15-17 minutes. The testing on all the selected buses was carried out in such a way that one bus was tested for each day, and the driver remained the same for all the tested buses. The fleets selected for the test are the Bluebird built 300 series buses having a 2005 Cummins (ISB) engine and Thomas built 500 series buses having 2003 Mercedes Benz (MBE) engine. Both the 300 series and 500 series engines are turbocharged, having an equal number of combustion chambers with 500 series buses having engine capacity of 7.2 liters, while 300 series buses have 5.9 liters. Both the fleets have exhaust gas recirculation (EGR) and employed with common rail direct fuel injectors. Idle-engine emission testing was conducted in an open space outside the garage with the engine in idle mode (i.e., acceleration and the speed will be zero). The duration of each test cycle was 15 minutes.

Soy based biodiesel with different mixes: 0% (B0 or ULSD), 5% (B05), 10% (B10), and 20% (B20), with the base fuel as ULSD, are used as test fuels. The fuel tank for each selected bus was filled with B0, B05, B10, or B20 and allowed to run for four hours on the road, before the actual testing was started, so that the entire fuel system was rinsed with the required fuel and was ready for the testing. The emission test was repeated twice, on consecutive days, on each bus to assure good to test repeatability.

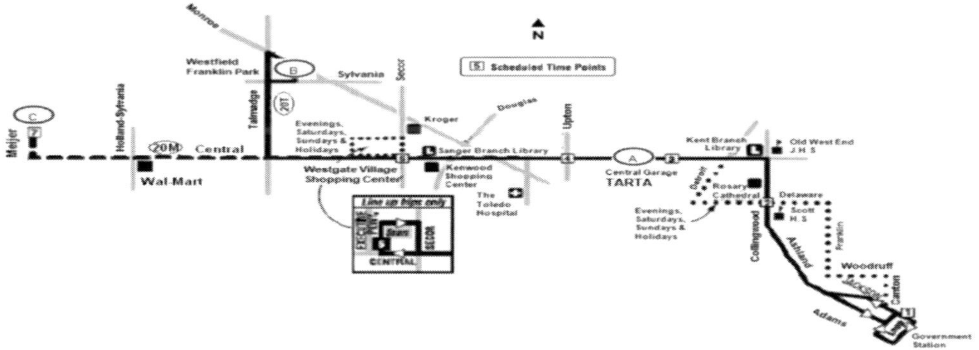

Figure 1. TARTA Test Route Selected for Exhaust Emission and IAQ Study.

Instrumentation and Experimental Setup

TESTO 350XL is the portable emission measurement system (PEMS) used for collecting exhaust emission data. The instrument measures continuously up to six gases: O_2, CO_2, CO, SO_2, NO, and NO_2 with calculated NO_X, and a temperature sensor with an integrated thermoelectric cooler for continuous temperature compensation for accurate measurement. The units of O_2 and CO_2 are given in the percentage of gas present in the volume of air analyzed by the sensor. The values of O_2 and CO_2 are automatically calculated by the analyzer in terms of volume percent (%). The instrument was set up for one-second concentration measurement and connected to the laptop to download the data simultaneously. The system was installed on the rear seat of the buses with power sourced from the bus power control panel at the rear of the bus. The probe of the instrument was clamped to the tail pipe and the instrument connected to a laptop to download the data simultaneously during the run. The collected data was then exported into an Excel spreadsheet. The power was supplied to the instruments from the vehicle's power control panel. Figure 2 shows the experimental setup used for carrying out the exhaust emission study.

The desired engine parameters are collected by connecting the on-board diagnostic (OBD) unit to the engine computer module (ECM) of the bus and to the laptop. The analyzer set-up was connected to the exhaust pipe and both the engine parameter readings and exhaust analyzer readings are measured simultaneously.

Figure 2. Experimental Setup for Exhaust Emission Study.

Indoor Air Quality Study

Selection of Test Routes and Buses for the Study

Two buses (ID 506 and 536) were selected from the 500 series of the TARTA fleet with 506 running on biodiesel (B20) and 536 on ultra low sulfur diesel. The buses were run on a pre-assigned route daily and the location of the buses during the run was identified by the GPS unit located inside the bus. The route selected was #20 which runs from the TARTA garage (A) to the Meijer store (C) on Central Avenue and returns (refer to Figure 1).

Instrumentation and Experimental Setup

Continuous monitoring of particulate matter in the buses was done using Grimm 1.108 and TSI Dust Trak^{TM} 8520 instruments. Temperature and relative humidity were measured simultaneously along with the gases using YES Plus instruments. The gases monitored using Yes Plus instrument were carbon dioxide (CO_2), carbon monoxide (CO), nitric oxide (NO), nitrogen dioxide (NO_2) and sulfur dioxide (SO_2). Power is continuously supplied to the instruments from the adapters connected to the bus. A wired mesh box is provided to safeguard the instruments which are held in place by Velcro attachments. The experimental setup required to monitor IAQ levels can be seen from Figure 3. Refer to Kadiyala et al. (2010) and Kadiyala (2008) for more details on experimental setup and test protocol adopted by the researchers.

Figure 3. Experimental Setup for IAQ Study.

Data collection for IAQ included downloading the data from instruments, obtaining meteorological data, and monitoring the real time variables. The data downloaded from all the instruments were set for one-minute intervals that are then averaged to one hour for analysis. The filters were changed periodically and the instruments were calibrated on a regular basis. The independent variables considered for studying behavior of in-vehicle gases were ambient temperature, ambient relative humidity, wind speed, wind direction, precipitation, visibility, indoor temperature, indoor relative humidity, passenger count, bus status (bus position/door position - Idle/Open, Idle/Close, Run/Close), number of cars and buses/trucks ahead, while ambient $PM_{2.5}$ concentrations were also used additionally for modeling particulate matter. Meteorological data were downloaded from the National Climatic Data Center (NCDC, 2008) and ambient $PM_{2.5}$ concentrations were obtained from the EPA. The passenger data, vehicles in front of the bus, and the bus status were monitored using the cameras present in the bus which made a video recording during the run. One can observe from Figure 3 the position of cameras showing vehicle indoors, traffic ahead, and door status. Kadiyala et al. (2010), Kadiyala and Kumar (2011), and Kadiyala (2008) provided a more detailed explanation on the instrument maintenance procedures adopted and monitoring of the real-time variables.

Particulate Matter Filter Analyses

SEM Analyses: The particle analysis was carried out by use of the ESEM coupled to an automatic computer imaging system having a resolution of 10 nm. The size, morphology, and shape of the particles were analyzed with the SEM. Seven samples from the biodiesel engine, with respect to particle size, distribution, and appearance were analyzed. The filters were prepared for an electron microscopic investigation. The ESEM (Quanta 3D) with an image analyzer located in the Electron Micro-Beam Analysis Laboratory at the University of Michigan in Ann Arbor, Michigan was used for the analysis of samples collected from the TARTA buses.

Full filters were mounted on bulk sample holders with silver adhesive. The fields for counting and size distribution analysis were randomly selected. The interaction of the electron beam with the sample produces various effects that can be monitored with suitable detectors. The resulting signals include secondary electrons and backscattered electrons that are used for imaging. The secondary, backscattered, and x-ray signals were collected in synchronization with the position of the electron beam to provide highly detailed spatial and compositional information of microscopic features. However, PM_{10} characterization will require more detailed quantification. Signal strength is proportional to particle volume. More experimental details are given in Shandilya and Kumar (2010a and 2010b).

RESULTS AND DISCUSSION

Exhaust Emission Study

On-Road Exhaust Emissions

Effect of Biodiesel Concentrations (B0, B05, B10,and B20)

The real-world on-road exhaust emission concentrations from transit buses equipped with MBE and Cummins engines, fuelled with B0, B05, B10, and B20 were measured. The emission variation of the monitored pollutants with biodiesel blends can be seen in Figure 4. CO emission levels were observed to decrease with an increase in percentage of biodiesel in the base fuel. The reduction was found to be 38% with MBE engines, from B0 to B20; and 22% for Cummins engines from B05 to B10. The emission levels of NO_X were observed to increase proportionally with the biodiesel concentration in the base fuel. The percentage of increase was about 15% from B0 to B20 for MBE engines; and 10% for Cummins engines from B0 to B10. CO_2 exhaust emissions reduced by 11% from B0 to B20 for MBE engines and increased up to 23% from B05 to B10 for Cummins engines. SO_2 emissions are a result of combustion of fuels containing sulfur and the concentrations were observed to decrease by 5% from B0 to B20 for buses equipped with MBE engines.

On-road instantaneous emission models were developed, using regression analyses of MINITAB 15® software, for the monitored pollutants. But the development of these regression models was restricted to B20, as the engine data obtained for B05 and B10 was not continuous due to the difficulty in maintaing the OBD connections on the day of test. These models help in understanding and predicting the on-road emission behavior of the public transport buses and the influence of each operating variable on pollutant concentrations. The values of R^2 and adjusted R^2 explain the predictability of the model and the higher these values are, better is the predictability of the model. All the models showed good predictability with R^2 and adjusted R^2 being in the range of 81-98 and 62-95, except for CO_2 that has R^2 and adjusted R^2 values of 71.3 and 59.8 respectively. Refer to Nerella and Kumar (2009) for a detailed discussion on the developed regression models.

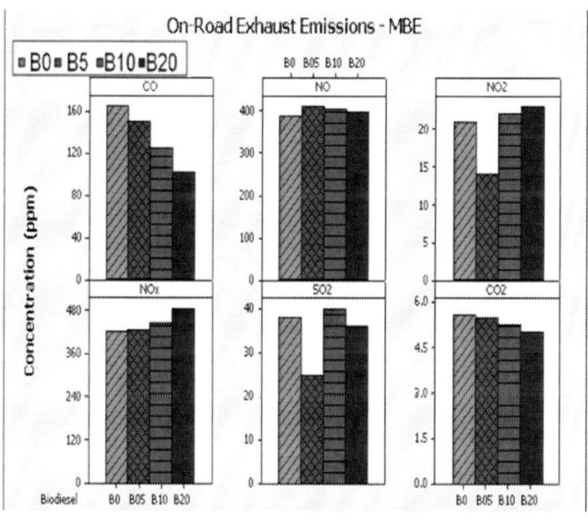

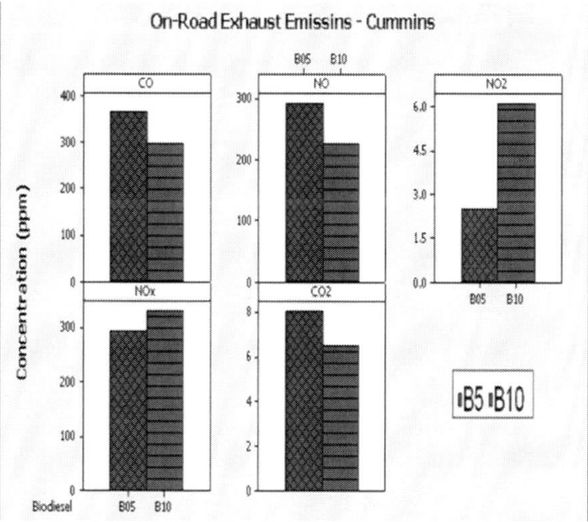

Figure 4. Biodiesel Impacts on Exhaust Emissions from Different Engines During On-Road Tests.

Idle-engine Exhaust Emissions

Effect of Biodiesel Concentrations (B0, B05, B10,and B20)

The idle exhaust emissions of an MBE engine, fuelled with B0, B05, B10, and B20, were measured and analyzed. The variation of exhaust emission levels with different blends of biodiesel can be seen in Figure 5.

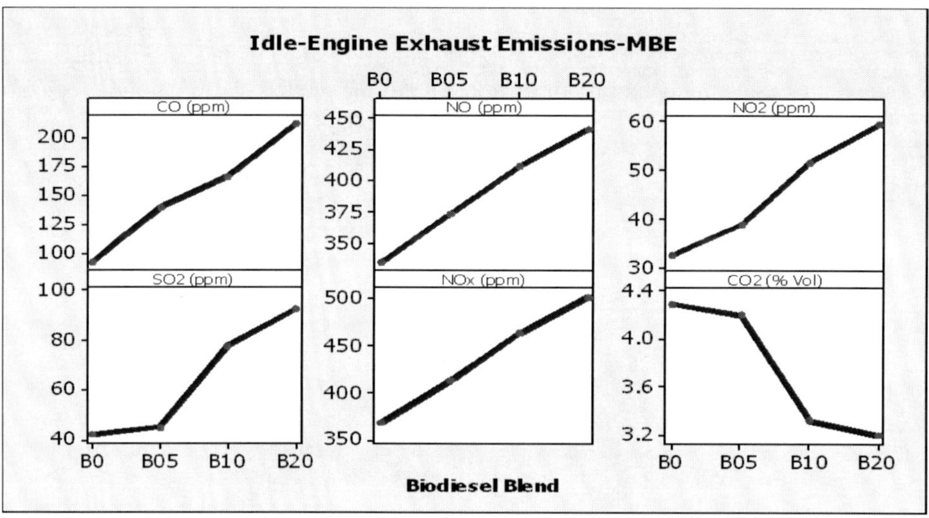

Figure 5. Effect of Biodiesel on Idle Exhaust Emissions.

All the monitored pollutants, expect CO_2, were observed to increase in proportion to the biodiesel concentration in the base fuel. Furthurmore, the CO_2 emission reduction was proportional to the percent of biodiesel in the base fuel. There was only 2% reduction of CO_2 from B0 to B05. However, the reduction was significant from B05 to B10 (21%) and a reduction of 4% was observed from B10 to B20. The emission concentrations of CO (128%) and SO_2 (121%) have increased significantly from B0 to B20. From these results, one can observe that idle-engine emissions increase with the increase of percent of biodiesel in the base fuel. Regression models for idle exhaust emissions, B0, B05, B10, and B20, were developed using MINITAB 15[®] by Nerella and Kumar (2009).

Indoor Air Quality

Trends

Particle Numbers: Figure 6 provides the seasonal variation of particulate numbers in B20 and ULSD buses. Particulate number concentration levels tend to be same for both biodiesel and ULSD buses most of the time. There are corresponding peaks and lows in the particulate number concentrations depending on several conditions. Early morning peaks are a result of more buses kept in idling position in the garage before the test bus starts on the route. Also, higher particle numbers are found at stop signals and during periods of higher ventilation, i.e., doors are open. Particulate number concentrations are mainly higher in the winter season as a result of insufficient ventilation and slow moving traffic. For both B20 and ULSD buses, it was observed that the particle numbers are higher in the winter season mainly due to low air exchange rate. In-vehicle particulate number concentrations for diesel buses with aerodynamic diameter > 0.30μm was observed in the range of 15,000 – 350,000 particles per liter. Vijayan et al. (2007) have discussed the characterization of sub-1μm particulate matter.

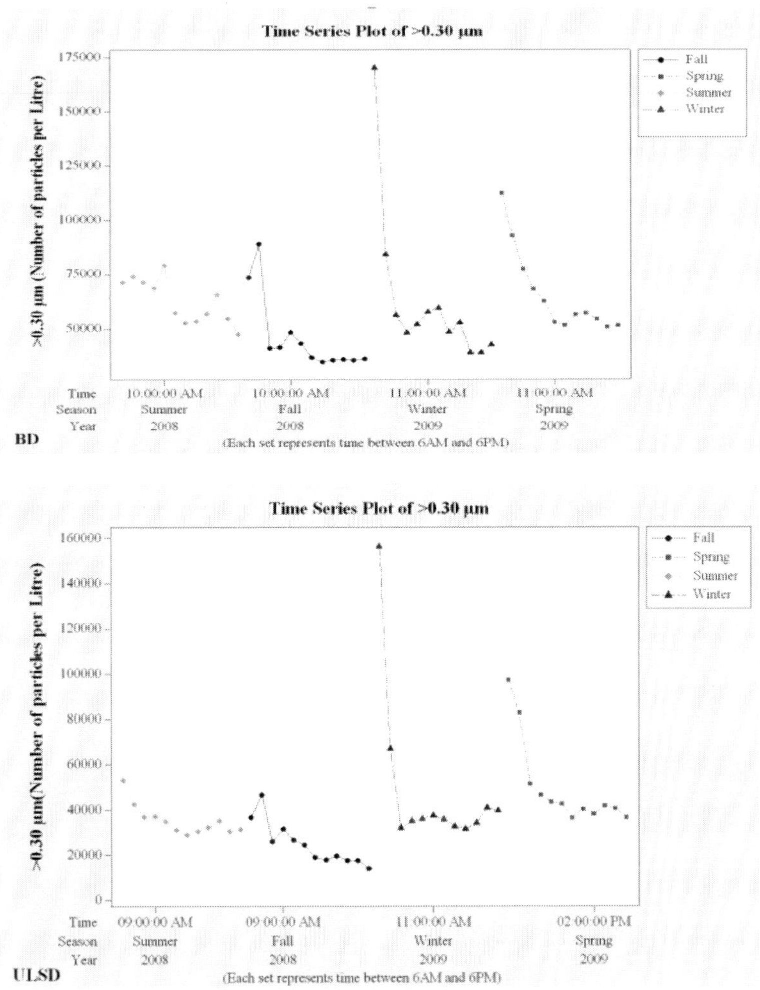

Figure 6. Seasonal Variation of Particle Numbers in B20 and ULSD Buses.

CO₂ Concentrations: Vehicular CO_2 levels were observed to be mainly influenced by passengers inside the bus. Higher concentrations of CO_2 were observed during periods of peak hours with higher vehicular traffic and passenger ridership. CO_2 concentrations were higher for the B20 bus when compared to the ULSD bus due to higher lead vehicular traffic as observed from the video analysis. Vijayan and Kumar (2008) and Kadiyala and Kumar (2008b, 2009) have used regression and regression tree analysis respectively to determine the factors influencing vehicular CO_2 levels. The methodology involved in development of regression and regression tree modeling was discussed in detail by Kadiyala and Kumar (2008a). Figure 7 presents the seasonal variation of CO_2 levels in both the test buses. CO_2

levels are observed to be higher during the fall for BD and ULSD buses due to more passenger ridership during fall compared to other seasons. The concentrations are higher during the spring than winter due to the more passenger ridership during spring as observed in video analysis. Passenger ridership predominated ventilation settings as can be seen in Figure 7. Refer to Kadiyala and Kumar (2011) and Kadiyala (2008) for detailed discussion on trends and factors influencing vehicular CO_2 levels.

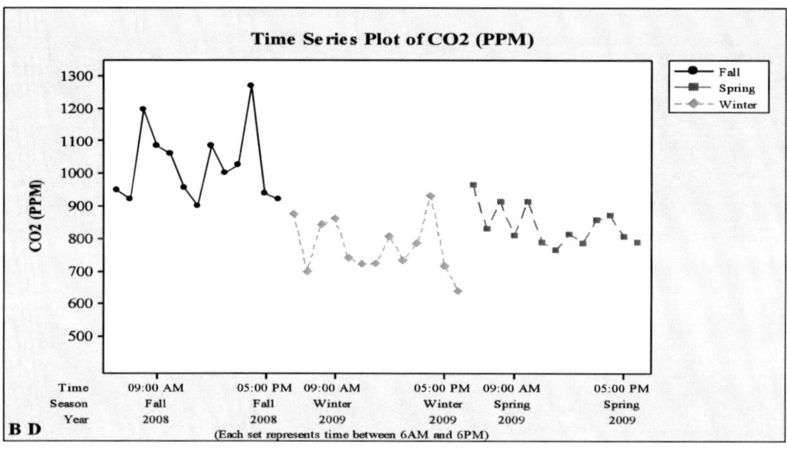

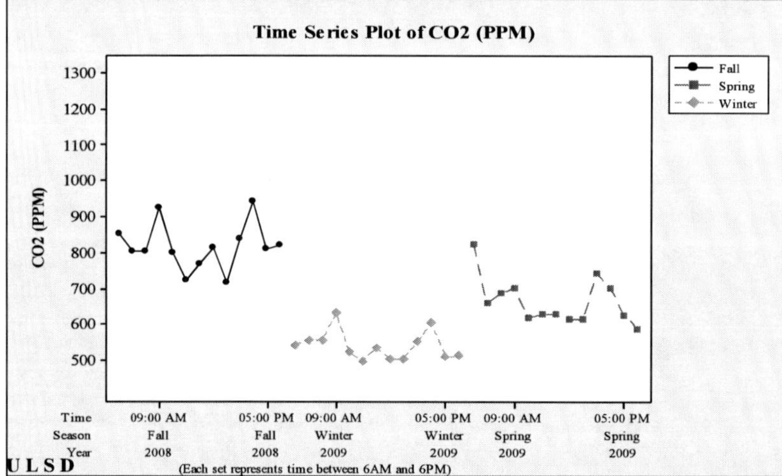

Figure 7. Seasonal Variation of CO_2 in B20 and ULSD Buses.

CO Concentrations: Figure 8 shows the seasonal variation of CO in both the test buses. CO levels are observed to be higher in the winter for the B20 bus and in the spring for the ULSD bus. It was observed from the video analysis that lead vehicular traffic and ventilation settings mainly affected CO levels. Refer to Vijayan and Kumar (2008) and Kadiyala and Kumar (2008b, 2009) for determining the factors influencing CO levels with regression and regression tree analysis respectively. Refer to Kadiyala and Kumar (2011) and Kadiyala (2008) for a detailed discussion on trends and factors influencing vehicular CO levels.

NO and NO₂ Concentrations: The seasonal variations of oxides of nitrogen (NO and NO_2) are shown in the Figure 9. The concentrations were observed to be negligible for the majority of the time. The NO_2 concentrations were observed to be higher during the winter compared to other seasons due to low air exchange. Peaks are observed during periods of higher vehicular traffic and concentration levels varied with ventilation settings.

The NO concentrations are observed to be higher during fall season compared to spring and winter seasons in the ULSD bus due to the penetration of pollutants as windows are kept open for longer durations.

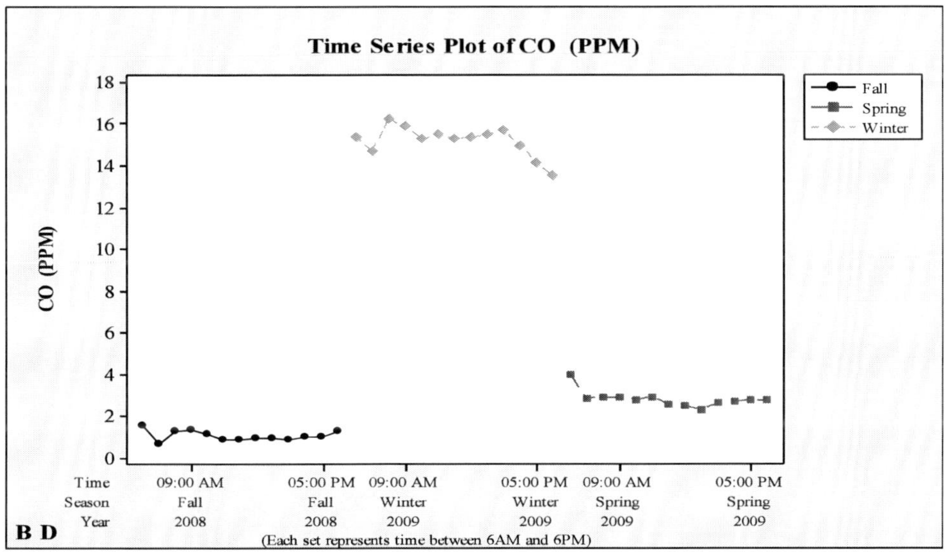

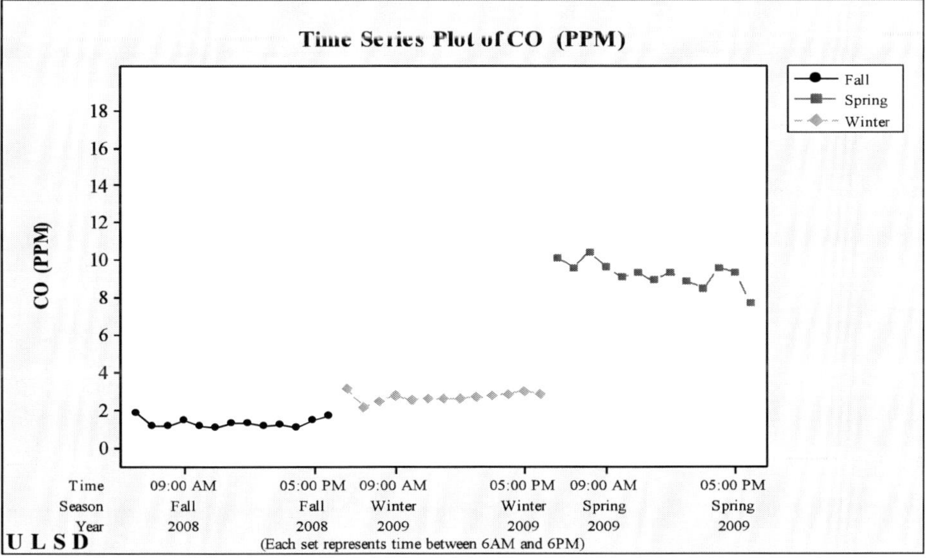

Figure 8. Seasonal Variation of CO in B20 and ULSD Buses.

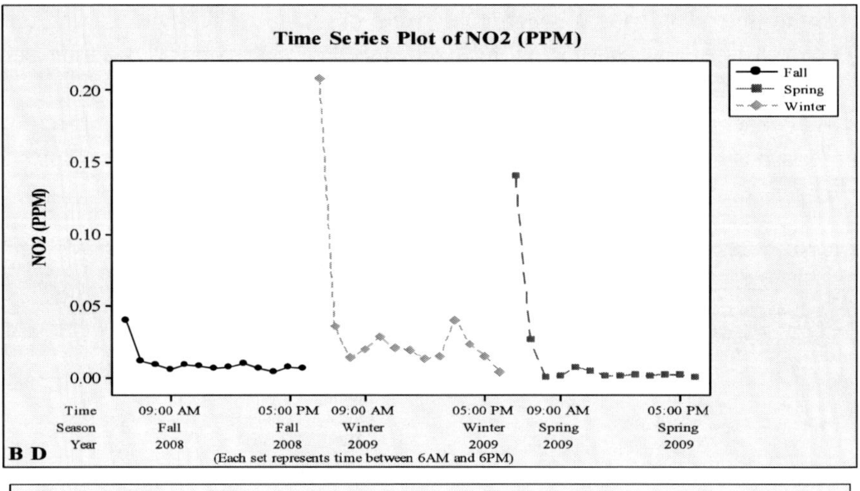

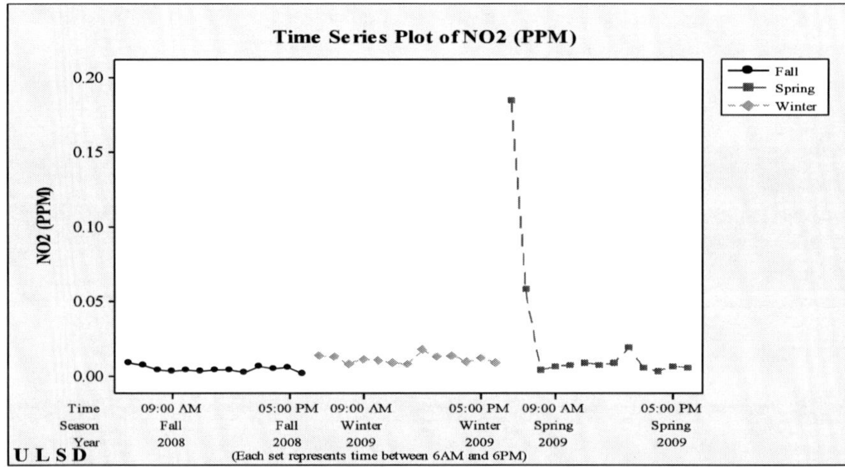

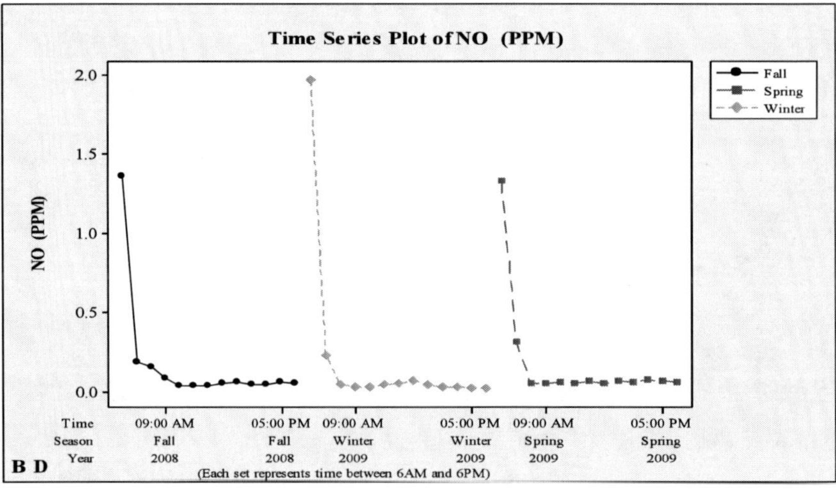

Figure 9. (Continued).

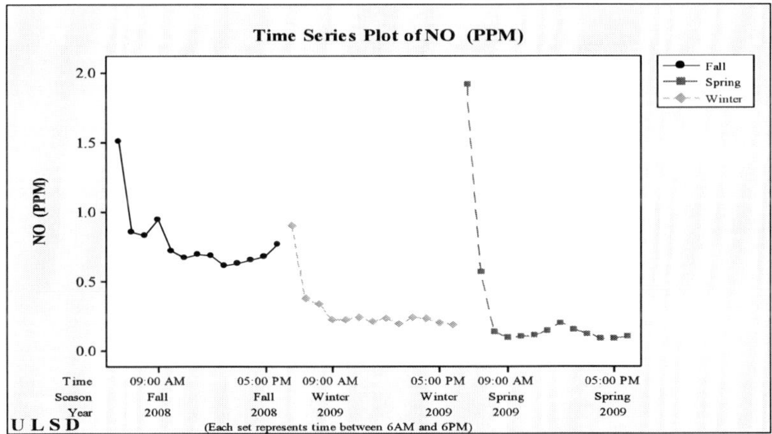

Figure 9. Seasonal Variation of NO_2 and NO in B20 and ULSD Buses.

Vijayan and Kumar (2008) and Kadiyala and Kumar (2008b, 2009) have developed regression and regression tree models and identified influential factors affecting vehicular NO and NO_2 levels. Refer to Kadiyala and Kumar (2011) and Kadiyala (2008) for more information on trends and factors influencing oxides of nitrogen in both the buses.

SO_2 Concentrations: Figure 10 presents the seasonal variation of SO_2 in B20 and ULSD buses. The seasonal variations of SO_2 are shown in the Figure 10. The SO_2 concentrations are observed to be lower during winter for BD bus when compared to other seasons as a result of lower air exchange.

The concentrations are observed to be lower during spring for the USLD bus because of the lower vehicular traffic. Vijayan and Kumar (2008) and Kadiyala and Kumar (2008b, 2009) have used regression and regression tree analysis respectively and determined the influential factors affecting vehicular SO_2 levels. Kadiyala and Kumar (2011) and Kadiyala (2008) provided more information on trends and factors influencing sulfur dioxide in both the buses.

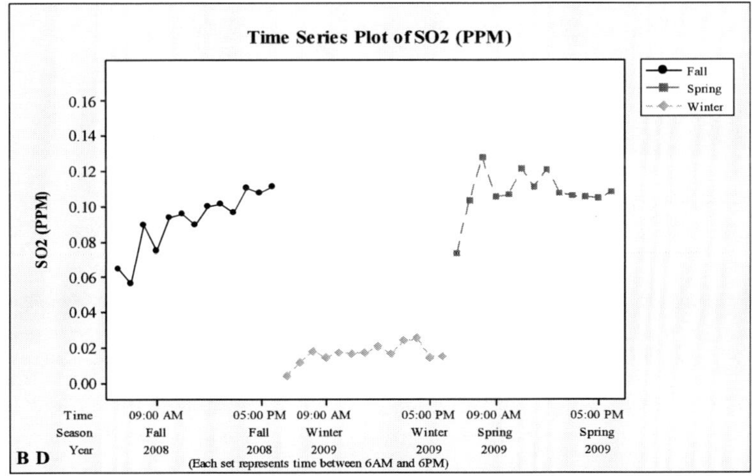

Figure 10. (Continued).

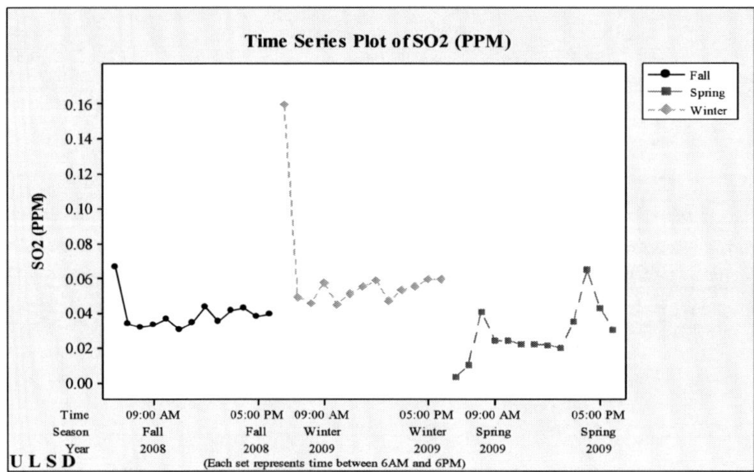

Figure 10. Seasonal Variation of SO_2 in B20 and ULSD Buses.

Kadiyala et al. (2010) have performed statistical tests to compare the driver's/commuter's exposure in both the test buses and observed that it is safe to travel in public transport buses operating on biodiesel and ultra low sulfur diesel.

Particulate Matter Filter Analysis

Size Distribution: The observations made from size distribution analysis of 17 airborne PM samples distributed among five different buses, fueled over 22 months by B20 and ULSD, are discussed here. The peaks in uni-modal or bimodal distribution were found at 0.1, 0.2, 05, and 0.6μm. In summer, the particle size distribution was multimodal for both buses. In the spring, buses fueled with B20 had uni-modal particle size distribution and the ULSD-fueled bus had multimodal PSD. In the winter, for both buses results are not enough to draw any conclusion. In the fall, buses fueled with B20 blend had uni-modal and bimodal particle size distribution, and the ULSD-fueled bus had uni-modal and multimodal PSD. The peaks in multimodal distribution have shown varied patterns. The peak for filter paper collected inside the B20 bus was found at small particle sizes, so one can conclude that smaller particles dominate inside the B20 bus. The ultrafine mode is near 0.1μm and the coarse mode is beyond 0.3μm. The uni-modal and bi-modal peak lies in the ultrafine mode region. Multi-mode peaks were found mostly in the coarse mode region.

Particle Shape Analysis: Different shapes of particles were observed from particulate matter filter analysis done in the lab. The various shapes observed are square, pentagon, hexagon, agglomerate, sphere, column or stick, oblong, and triangle with some unknown formations.

Aspect Ratio: The aspect ratio found for different filters collected inside the B20 bus were in the average value range of 2.4-3.6, while standard deviation value ranged between 0.9 and 7.4. The aspect ratio found for different filters collected inside the ULSD bus were in the average value range of 2.3-2.9, while standard deviation value ranged between 1 and 7.3. An elongated particle is characterized by length longer than the width of the particle, i.e., high aspect ratio (>1).

Refer to Shandilya and Kumar (2010a and 2010b) for more information on the particulate matter filter analysis and a detailed discussion on characterization of particulate matter size, shape, and aspect ratio.

CONCLUSIONS

An experimental protocol that can be used to continuously monitor vehicular exhaust emission levels and in-vehicle pollutant levels for a longer period of time has been developed. This setup could be used in any public transport system without having to worry about safety of the instruments. This research helped to determine the trends for various monitored pollutants and identified the factors influencing pollutant levels using different statistical analyses. A study of "Exposure" revealed that it was safe to travel in public transport buses operating on alternative fuels as none of the pollutant levels exceeded the available guidelines with the exception of CO. Precautionary measures like keeping the windows closed during high traffic would certainly reduce the commuters exposure to CO below the guidelines. Particulate matter filter analysis provided more insight into the in-vehicle PM characterizations.

ACKNOWLEDGMENTS

The authors would like to thank the United States Department of Transportation (US DOT), Toledo Area Regional Transit Authority (TARTA), and MIOH-UTC for grants awarded to the Intermodal Transportation Institute (ITI) of The University of Toledo. The authors would also like to express their sincere gratitude to the TARTA management and the employees for their continued interest and involvement in this work. The views expressed in this Chapter are those of the authors and do not represent the views of the funding organizations.

BIOGRAPHICAL SKETCH

Ashok Kumar (*akumar@utnet.utoledo.edu*) is a professor and chairman of Civil Engineering at the University of Toledo, Toledo, Ohio. Before coming to Toledo he worked for Syncrude Canada Ltd. as an atmospheric physicist where he developed, planned, and conducted studies related to the dispersion of emissions of a tar sands plant. Dr. Kumar received his Bachelors of Science in Engineering (Honors) from Aligarh University in India (1970), his Masters of Applied Science from University of Ottawa, Canada (1972) and his Ph.D. from the University of Waterloo, Canada (1978). He is registered as a Professional Engineer in the Province of Alberta, Canada and a Diplomat of the American Academy of Environmental Engineers. He is a fellow member of Air and Waste Management Association (A and WMA). Air and Waste Management Association presented him with L. A. Ripperton Award for distinguished achievement as an educator in the field of air pollution control in June 2003. His research work has focused on finding innovative solutions to fundamental and applied problems in air quality, risk analysis, and environmental data analysis.

Akhil Kadiyala (akhilkadiyala@gmail.com) is a Ph.D. student in the Department of Civil Engineering at The University of Toledo. His research work involves the prediction of contaminant concentrations inside the bus using experimental data and theoretical models.

Kaushik Shandilya (*kshandi@rockets.utoledo.edu*) is a Ph.D. candidate at the Department of Civil Engineering at The University of Toledo, Toledo, Ohio. His research interests include atmospheric chemistry, morphology, green buildings, environmental analytical chemistry, monitoring and modeling, urban public transportation, indoor air quality, emission measurement, and ambient air quality.

Vinay Kumar V. Nerella (*nvinayk@rockets.utoledo.edu*) completed his MS degree in the Department of Civil Engineering at The University of Toledo in May 2010. His MS thesis is entitled "An Analysis on Vehicular Exhaust Emissions from Transit Buses Running on Alternative Fuels". He received his undergraduate degree, Bachelor of Engineering (BE), from Andhra University in India (2006). His research interests include air pollution, exhaust emission analysis, sustainability, pollution prevention, risk modeling, green engineering, indoor air quality, industrial ventilation, energy efficiency, and GIS applications for environmental engineering.

Dinesh Chandra Somuri (dsomuri@rockets.utoledo.edu) is in the process of completing his MS degree in the Department of Civil Engineering at The University of Toledo. His MS thesis is entitled "Study of In-vehicle Particulate Number Concentrations for Buses Running with ULSD and Bio-diesel Fuels". He received his undergraduate degree, Bachelor of Technology (B.Tech), from Acharya Nagarjuna University in India (2006). His research interests include indoor air quality in vehicles, emission analysis, particulate matter, dispersion and risk modeling, sustainability, industrial ventilation, energy efficiency, and ground water modeling.

Srikar Velagapudi (svelaga@rockets.utoledo.edu) is in the process of completing his MS degree in the Department of Civil Engineering at The University of Toledo. His MS thesis is entitled "Characterization of Indoor Air Pollutants Monitored over Three Years in Public Transport Buses Running on Bio-diesel and ULSD Fuels". He received his undergraduate degree, Bachelor of Technology (B.Tech), from Acharya Nagarjuna University in India (2007). His research interests include air pollution, vehicular emission analysis, quality control engineering, dispersion modeling, risk modeling, sustainability, industrial ventilation, energy efficiency, and ground water modeling.

REFERENCES

Adams, H.S., Nieuwenhuijsen, M.J., Colvile, R.N., Older, M.J., and Kendall, M. 2002. Assessment of road users' elemental carbon personal exposure levels, London, UK. *Atmospheric Environment*. 36: 5335–5342.

Agarwal, A.K. 2006. Biofuels (alcohols and biodiesel) applications as fuels for internal combustion engines. *Progress in Energy and Combustion Science*. 33: 233-271.

Alm, S., Jantunen, M.J., and Vartiainen, M. 1999. Urban commuter exposure to particulate matter and carbon monoxide inside an automobile. *Journal of Exposure Analysis and Environmental Epidemiology.* 9: 237-244.

Bremauntz, F.A.A., and Ashmore, M. 1995. Exposure of Commuters to carbon monoxide in Mexico City–1. Measurement of in-vehicle concentrations. *Atmospheric Environment,* 29: 525-532.

Brodrick, J., Dwyer, H.A., Farshchi, M., Harris, D.B, and Jr. King, F.G. 2002. Effects of engine speed and accessory load on idling emissions from heavy-duty diesel truck engines. *Journal of Air and Waste Management Association.* 52: 1026-1031.

Carraretto, C., Macor, A., Mirandola, A., Stoppato, A., and Tonon, S. 2004. Biodiesel as alternative fuel: experimental analysis and energetic evaluations. *The Science of the Total Environment.* 29: 2195-2211.

Chan, A. 2003. Commuter exposure and indoor–outdoor relationships of carbon oxides in buses in Hong Kong. *Atmospheric Environment.* 37: 3809–3815.

Chan, L.Y., Chan, C.Y., and Qin, Y. 1999. The effect of commuting microenvironment on commuter exposures to vehicular emission in Hong Kong. *Atmospheric Environment.* 33: 1777-1787.

Chan, L.Y., Lau, W.L., Lee, S.C., and Chan, C.Y. 2002. Commuter exposure to particulate matter in public transportation modes in Hong Kong. *Atmospheric Environment.* 36: 3363–3373.

Chan, L.Y., Lau, W.L., Wang, X.M., and Tang, J.H. 2003. Preliminary measurements of aromatic VOCs in public transportation modes in Guangzhou, China. *Environment International.* 29: 429–435.

Chan, L.Y., Lau, W.L., Zou, S.C., Cao, Z.X., and Lai, S.C. 2002. Exposure level of carbon monoxide and respirable suspended particulate in public transportation modes while commuting in urban area of Guangzhou, China. *Atmospheric Environment.* 36: 5831–5840.

Chan, L.Y. and Liu, Y.M. 2001. Carbon monoxide levels in popular passenger commuting modes traversing major commuting routes in Hong Kong. *Atmospheric Environment.* 35: 2637-2646.

Chen, C., Huang, C., Jing, Q., Wang, H., Pan, H., Li, L., Zhao, J., Dai, Y., Huang, H., Schipper, L., and Streets, D.G. 2007. On-road emission characteristics of heavy-duty diesel vehicles in shanghai. *Atmospheric Environment.* 41(26): 5334-5344.

Clifford, M.J., Clarke, R., and Riffat, S.B. 1997. Driver's Exposure to carbon monoxide in Nottingham, UK. *Atmospheric Environment.* 31: 1003–1009.

Diapouli, E., Grivas, G., Chaloulakou, A., and Spyrellis, N., 2007. PM_{10} and ultrafine particles counts in-vehicle and on-road in the Athens Area. Water, *Air, and Soil Pollution: Focus.* 8: 89–97.

Dorado, M.P., Ballesteros, E., Amal, J.M., Gomez, J., and Lopez, F.J. 2003. Exhaust emissions from a diesel engine fuelled with Transesterified Waste Olive Oil. *Fuel.* 82 (11): 1311-1315.

EPA Draft Technical Report: A comprehensive analysis of biodiesel impacts on exhaust emissions. 2002. EPA-420-P-02-001. U.S. Environmental Protection Agency.

Fitz, D.R., Winer, A.M., Colome, S., Behrentz, E., Sabin, L.D., Lee, S.J., Wong, K., Kozawa, K., Pankratz, D., Bumiller, K., Gemmill, D., Smith, M. 2003. *Characterizing the Range of Children's Pollutant Exposure During School Bus Commutes.* Final Report. Contract No. 00-322. California Air Resources Board: Sacramento, CA.

Fruin, S. 2003. *Black Carbon Concentrations Inside vehicles: Implications for Refined Diesel Particulate Matter Exposures.* Ph.D. Thesis. University of California, Los Angeles, CA.

Gomez-Perales, J.E., Colvile, R.N., Nieuwenhuijsen, M.J., Fernandez-Bremauntz, A., Gutierrez-Avedoy, V.J., Paramo-Figueroa, V.H., Blanco-Jimenez, S., Bueno-Lopez, E., Mandujano, F., Bernabe-Cabanillas, R., and Ortiz-Segovia, E. 2004. Commuters' exposure to $PM_{2.5}$, CO, and benzene in public transport in the metropolitan area of Mexico City. *Atmospheric Environment.* 38:1219–1229.

Gulliver, J., and Briggs, D.J. 2004. Personal exposure to particulate air pollution in transport microenvironments. *Atmospheric Environment.* 38:1-8.

Kadiyala, A. 2008. *Identification of Factors Affecting Contaminant Levels and Determination of Infiltration of Ambient Contaminants in Public Transport Buses Operating on Biodiesel and ULSD Fuels.* M.S. Thesis, University of Toledo, Toledo, OH.

Kadiyala, A. and Kumar, A. 2008a. Application of CART and Minitab Software to Identify Variables Affecting Indoor Concentration Levels. *Environmental Progress.* 27(2): 160-168.

Kadiyala, A., and Kumar, A. 2008b. *Study of Factors Affecting In-Vehicle Pollutant Levels in the City of Toledo Public Transport Buses Running on Biodiesel.* Paper #508, Proceedings of the 101st A and WMA Annual Conference, Portland, Oregon.

Kadiyala, A., and Kumar, A. 2009. *Study of In-Vehicle Pollutant Behavior in Public Transport Buses Running on Alternative Fuels.* Paper #571. Proceedings of 102nd A and WMA Annual Conference, Detroit, MI.

Kadiyala, A., and Kumar, A. 2011. Study of In-Vehicle Pollutant Variation in Public Transport Buses Operating on Alternatie Fuels in the City of Toledo, Ohio. *The Open Environmental and Biological Monitoring Journal.* 4:1-20

Kadiyala, A., Kumar, A., and Vijayan, A. 2010. Study of Occupant Exposure of Drivers and Commuters with Temporal Variation of In-Vehicle Pollutant Concentrations in Public Transport Buses Operating on Alternative Diesel Fuels. *The Open Environmental Engineering Journal.* 3: 55-70.

Klepeis, N.E., Nelson, W.C., Ott, W.R., Robinson, J.P., Tsang, A.M., Switzer, P., Behar, J.V., Hern, S.C., and Engelmann, W.H. 2001. The National Human Activity Pattern Survey (NHAPS): a resource for assessing exposure to environmental pollutants. *Journal of Exposure Analysis and Environmental Epidemiology.* 11: 231–252.

Kumar, A., and Nerella, V.V.K. 2009. Experimental Analysis of Emissions from Transit Buses Fuelled with Bio-diesel. *The Open Environmental Engineering Journal.* 2: 81-96.

Kuo, H.W., Wei, H.C., Liu, C.S., Lo, Y.Y., Wang, W.C., Lai, J.S., and Chan, C.C. 1999. Exposure to volatile organic compounds while commuting in Taichung, Taiwan. *Atmospheric Environment,* 34: 3331-3336.

Mazzoleni, C., Kuhns, D.H., Moosmuller, H., Witt, J., Nussbaum, J.N., Oliver Chang, M.C., Parthasarathy, G., Nathagoundenpalayam, K.S., Nikolich, G., and Watson, G.J. 2007. A case study of real-world tailpipe emissions for school buses using a 20% biodiesel blend. *Science of the Total Environment.* 385: 146-159.

McCormick, R.L., Graboski, M.S., Alleman, T.L., and Yanowitz, J. 2000. Idling emissions from heavy-duty diesel and natural gas vehicles at high altitude. *Journal of Air and Waste Management Association.* 50(11): 1992-1998.

National Climatic Data Center. See http://cdo.ncdc.noaa.gov/ulcd/ULCD (accessed July 2008).

Praml, G., Schierl, R. 2000. Dust Exposure in Munich Public Transportation: A Comprehensive 4-year Survey in Buses and Trams. *International Archives of Occupational Environmental Health.* 73: 209-214.

Rodes, C., Sheldon, L., Whitaker, D., Clayton, A., Fitzgerald, K., Flanagan, J., DiGenova, F., Hering, S., Frazier, C. 1998. *Measuring Concentrations of Selected Air Pollutants Inside California Vehicles.* Final Report. Contract No. 95-339. California Air Resources Board: Sacramento, CA.

Sabin, L., Kozawa, K., Behrentz, E., Winer, A., Fitz, D., Pankratz, D., Colome, S., Fruin, S. 2005. Analysis of Real-Time Variables Affecting Children's Exposure to Diesel-Related Pollutants during School Bus Commutes in Los Angeles. *Atmospheric Environment.* 39: 5243–5254.

Shandilya, K.K. 2002. *Characterization and Speciation of Fine Particulates in Ambient Air in Delhi.* M.E. Thesis. Malaviya National Institute of Technology, Rajasthan University, Jaipur, India.

Shandilya, K.K., Gupta, A.B., and Khare, M. 2009. Defining Aerosols by physical and chemical characteristics. *Indian Journal of Air Pollution Control,* March, 9(3): 107-126.

Shandilya, K.K., and Kumar, A. 2009. Analysis of research studies on exhaust emission from the heavy duty diesel engine fueled by biodiesel. Chapter 2, Handbook of Environmental Research, Nova Publishers, NY.

Shandilya, K.K., and Kumar, A. 2010a. Qualitative Evaluation of Particulate Matter inside Public Transit Buses Operated by Biodiesel. *The Open Environmental Engineering Journal.* 3, 13-20.

Shandilya, K.K., and Kumar, A. 2010b. Morphology of Single Inhalable Particle inside Public Transit Biodiesel Fueled Bus. *Journal of Environmental Sciences.* 22(2): 263-270.

Soehodho, S., and Taufick, S. 2005. Study On Correlation Between Motor Vehicle Emission and Public Health. *Proceedings of the Eastern Asia Society for Transportation Studies.* 5: 1841-1856.

Storey, J.M.E., Thomas, J.F., Sr. Lewis, S.A., Dam, T.Q., Edwards, K.D., DeVault, G.L., and Retrossa, D.J. 2003. Particulate matter and aldehyde emissions from idling heavy-duty diesel trucks. *Society of Automotive Engineers.* Paper No. 2003-01-0289.

Vijayan, A., and Kumar, A. 2008. *Characterization of Indoor Air Quality Inside Public Transport Buses Using Alternative Diesel Fuels.* Proceedings of TRB Conference, Washington, D.C.

Vijayan, A., Kumar, A. and Abraham, M.A. 2007. Characterization of Sub-1μm Particulate Matter Distribution inside Public Transport Buses. Paper #528. Proceedings of the 100th A and WMA Annual Conference, Pittsburgh, Pennsylvania.

Wang, W., Clark, N.N., Lyons, D.W., Yang, R.M., Gautam, M., Bata, R.M., and Loth, J.L. 1997. Emissions comparisons from alternative fuel uses and diesel buses with a chassis dynamometer testing facility. *Environmental Science and Technology.* 31: 3132-3137.

Younglove, T., Scora, G., and Barth, M. 2005. Designing on-road vehicle test programs for the development of effective vehicle emission models, *Journal of the Transportation Research Board.* 1941: 51-59.

Zhang, G.S., Li, T.T., Luo, M., Liu, J.F., Liu, Z.R., and Bai, Y.H. 2006. Air pollution in the microenvironment of parked new cars. *Building and Environment.* 43: 315–319.

In: Biodiesel: Blends, Properties and Applications   ISBN: 978-1-63117-024-9
Editors: Jorge M. Marchetti and Zhen Fang   © 2014 Nova Science Publishers, Inc.

Chapter 10

BIODIESEL PRODUCTION USING CATION-EXCHANGE RESIN AS HETEROGENEOUS ACID CATALYST

Yaohui Feng[1], Jianxin Li[*1,2] and Benqiao He[1]
[1]The State Key Lab of Hollow Fiber Membrane Materials and Processes,
School of Material and Chemical Engineering,
Tianjin Polytechnic University, Tianjin, P. R. China
[2]Shanghai Advanced Research Institute,
Chinese Academy of Sciences, Shanghai, P. R. China

ABSTRACT

Biodiesel production through free fatty acids (FFA) esterification with methanol using cation-exchange resin as heterogeneous acid catalyst is becoming a hotspot. In this work, three types of cation-exchange resins (NKC-9, 001×7 and D61) were employed to prepare biodiesel via the batch esterification of FFA from acidified oils generated from waste frying oils. According to this study, the catalytic activity of NKC-9 was higher than that of 001×7 and D61. The FFA conversion by NKC-9 increased with increasing in the amount of catalyst, reaction temperature and time and methanol/oil molar ratio. The maximal conversion of reaction was approximately 90.0%. Further, NKC-9 resin exhibited good reusability in batch esterification. Gas chromatography-mass spectrometry analysis revealed that the production was simplex and mainly composed of C16:0 (palmitic), C18:2 (linoleic), and C18:1 (oleic) acids of methyl ester, respectively. Furthermore, continuous esterification of acidified oil with methanol was carried out with NKC-9 resin in a fixed bed reactor with an internal diameter of 25 mm and a height of 450 mm to produce biodiesel. The results showed that the FFA conversion increased with increases in methanol/oil mass ratio, reaction temperature and catalyst bed height, whereas decreased with increases in initial water content in feedstock and feed flow rate. The FFA conversion kept over 98.0% during 500 h of continuous esterification processes under 2.8:1 methanol to oleic acid mass ratio, 44.0 cm catalyst bed height (the amount of packed resin 87.5 g), 0.62 ml/min feed flow rate and 65 °C reaction temperature, showing a much high conversion and operational stability as compared with the batch esterification. Furthermore, the loss of sulfonic acid groups from NKC-9 resin into the

---

[*] Corresponding Author. Tel. :+86 22 24528072 ; fax : +86 22 24528055, Email: jxli0288@yahoo.com.cn (J.X. Li)

production was not found during continuous esterification. In sum, NKC-9 resin as heterogeneous acid catalyst shows the potential commercial applications to esterification of FFA.

Keywords: Cation-exchange resin; Biodiesel; Esterification; Fixed bed reactor; Heterogeneous acid catalyst.

1. INTRODUCTION

Biodiesel is a nonpetroleum-based fuel that consists of fatty acid alkyl esters (usually methyl esters, FAME) derived from either the transesterification of triglycerides (TG) or the esterification of free fatty acids (FFA) with low molecular weight alcohols [1]. This fuel is one of the promising substitutes for the petroleum derivates as it is technically feasible, economically competitive, environmentally acceptable and readily available [2]. On the other hand, the biodiesel similar to petroleum-based diesel can be used as fuel in compression ignition engines [3]. Normally, biodiesel can be produced from refined oil and methanol by transesterification in the presence of a homogeneous basic catalyst, such as potassium or sodium hydroxide or metal compounds [4]. However, it is difficult to transesterify the oils with high FFA content using the commercially available alkaline catalyst because of soap formation. Canakci and Van Gerpan [5] found the transesterification would not occur if FFA content in the oil was beyond 3%. They developed a 2-step pretreatment reaction which can reduce the acid level of the high FFA feedstocks to <1%.

Esterification by strong acid catalyzing is a typical method of producing biodiesel from high FFA oil [6]. When strong acidic catalysts, e.g., sulfuric or hydrochloric acid, are used, the esterification is very fast and has a high conversion. However, it has a number of drawbacks such as the existence of side reactions, equipment corrosion, additional neutralization of the reaction mass and difficult separation of catalyst from the production stream. Recently, methods of enzyme catalyst and supercritical methanol of catalyst-free have been proposed and investigated in order to overcome the drawbacks of homogenous acid catalysis. However, the methods have also been limited due to the high cost and unstable activity of enzyme and reaction conditions of high temperature and high pressure needed in the supercritical method [7-8].

Biodiesel production from high FFA feedstocks through heterogeneously catalytic process using ion-exchange resins has attracted considerable attention because resins possess greater advantages over homogeneous catalyst, enzyme and supercritical methanol in eliminating equipment corrosion, diminishing wastewater effluent and reducing process cost [9-12]. Commonly ion-exchange resins are composed of copolymers of divinylbenzene, styrene and sulfonic acid groups grafted on the benzene [being the active site] [13]. Their catalytic activity strongly depends on their swelling properties because the swelling capacity controls reactants accessibility to the acid sites and, therefore, affects their overall reactivity [14]. Furthermore, cation-exchange resins can offer better selectivity towards the desired product(s) and better reusability compared to homogeneous acid catalysts [15]. Guerreiro et al. [16] studied the catalytic properties of Nafion, a perfluorinated ion-exchange resin known as a very strong Brönsted acid. A very long induction period of 100h occurred because swelling for the resin in the soybean is negligible. Özbay et al. [13] investigated various ion-exchange resins for the esterficiation of waste cooking oils and methanol, but the FFA conversion was less than 50% for either resin.

On the other hand, resins are suitable for long-term operation and industrial scale production [17]. Therefore, a continuous process for biodiesel production in a fixed bed reactor packed with resins is becoming a hotspot. Liu et al. [18] reported a cation-exchange resin possessed high catalytic activity in the continuous esterification of feedstock containing a high content of FFA with methanol in a fixed bed reactor, whereas the operational stability of fixed bed reactor gradually decreased. The biodiesel yield was less than 90% when operational time was over 28 h. Shibasaki-Kitakawa et al. [19] carried out a continuous transesterification of triolein with ethanol in fixed bed reactor packed with anion-exchange resin (PA 306s resin) as heterogeneous base catalyst. Though a very high conversion was achieved, the operational stability of fixed bed reactor was not further investigated.

The aim of this present study is to investigate the activities of three cation-exchange resins, NKC-9, 001×7 and D61, for preparation of biodiesel. The resin with best catalytic properties is selected for batch esterification of acidified oil. The effects of the amount of catalyst, reaction temperature, reaction time and methanol/oil molar ratio on the FFA conversion are examined. Moreover, the characterizations of feedstock and production are also analyzed. Based on the batch mode, the continuous esterification of acidified oil with methanol by the resin in a fixed bed reactor was studied so as to provide the data for its commercial applications. The operational stability of continuous esterification by the resin in the fixed bed reactor was also conducted.

2. EXPERIMENTAL

2.1. Materials

The two kinds of oil feedstocks from waste fried oils were supplied by Hubei Haolin Bioenergy Company, China. The acid values of oils were 13.7 and 36.0 mg KOH/g, respectively. Filter paper with an average pore size of 5 μm was used to remove impurities from the oil feedstocks before use. The other chemicals (analytical grade) were purchased from Tianjin Kermel chemical reagent Co., Ltd., China.

Cation-exchange resins (NKC-9, 001×7 and D61) were purchased from the Tianjin NanKai Hecheng SandT Co., Ltd., China. NKC-9 is in H^+ form and 001×7 and D61 are in Na^+ forms. 001×7 and D61 were firstly pretreated by washing with deionized water to eliminate some impurities and then transformed with 1M HCl solution to be replaced with sodium ions. Secondly, 001×7 and D61 resins continued to be washed with deionized water to pH~7, and then washed with methanol to swell the resins. At the same time, methanol can displace water remained in resins. Finally, 001×7 and D61 resins were stored under airtight conditions before use. NKC-9 was also swollen in absolute ethanol for 30 min. This is because about 100% swelling ratio of NKC-9 easily results in reactor plugging or a very high pressure drop of liquid flow [12]. Then the swollen resin was dried for 24 h at 90 °C in an oven so as to remove the residual ethanol. Physical properties of ion-exchange resins as received are summarized in Table 1. The total exchange capacity and bead size of three resins are comparable. The average pore diameter of NKC-9 (~56 nm) is larger than that of D61 (~11.3 nm). The water content of NKC-9 is the lowest among the three resins.

Table 1. Characteristics of cation-exchange resins

Property	NKC-9	001×7	D61
Structure*	MR	G	MR
Matrix	Styrene, divinyl-benzene	Styrene, divinyl-benzene	Styrene, divinyl-benzene
Form	H^+	Na^+	Na^+
Total exchange capacity			
(a) mmol/g (dry)	$\geq$4.7	$\geq$4.5	$\geq$4.2
(b) mmol/g (wet)	$\geq$1.5	$\geq$1.8	$\geq$1.4
Surface area (m^2/g)	77	—	83.9
Average pore diameter (nm)	56	—	11.3
Bead size (%)	0.4~1.25mm $\geq$95	0.315~1.25mm $\geq$95	0.315~1.25mm $\geq$95
Moisture content (%)	$\leq$10	46~52	50~60
Density (wet)			
True (g/ml)	1.20~1.30	1.24~1.28	1.15~1.25
Apparent (g/ml)	0.70~0.80	0.77~0.87	0.75~0.85

*MR: macroeticular structure. G : gel structure.

2.2. Batch Esterification

The batch esterification reactions were carried out in a three-necked batch reactor (250 ml) equipped with a reflux condenser to prevent the escape of methanol and a mechanical agitation. The three-necked reactor was placed in a temperature-controlled jacket. 10 g acidified oil (13.7 mg KOH/g) was poured into the three-necked batch reactor and preheated. After reaching desired temperature, 0.6~2.4 g cation-exchange resin and 1.2~8.7 g methanol were added into the reactor to carry out esterification for desired time. On completion of this reaction, the production was poured into a separating funnel for separating the excess methanol. The excess methanol with impurities moved to the top layer and was removed. The lower layer was separated by reduced pressure distillation to purify the product.

2.3. Continuous Esterification in a Fixed Bed Reactor

The continuous esterification was performed in a fixed bed reactor. The scheme of the fixed bed reactor was shown in Figure 1. The reactor was composed of a water-jacketed stainless steel column with an internal diameter of 25 mm and a height of 450 mm. The column was packed with pretreated cation-exchange resin. Methanol and oil (36.0 mg KOH/g) were mixed and preheated in a feedstock tank. The reactant was fed to the inlet of reactor using a peristaltic pump. The reaction temperature was controlled by a thermostat water bath to keep constant temperature with an error of ± 0.5 °C. The temperature difference between the inlet and the outlet was below 0.5 °C during all the runs. 10 ml effluent was collected at each hour of reaction time when the effluent solution from the outlet of reactor was introduced into an accumulation tank. The purification of sample was consistent with batch mode.

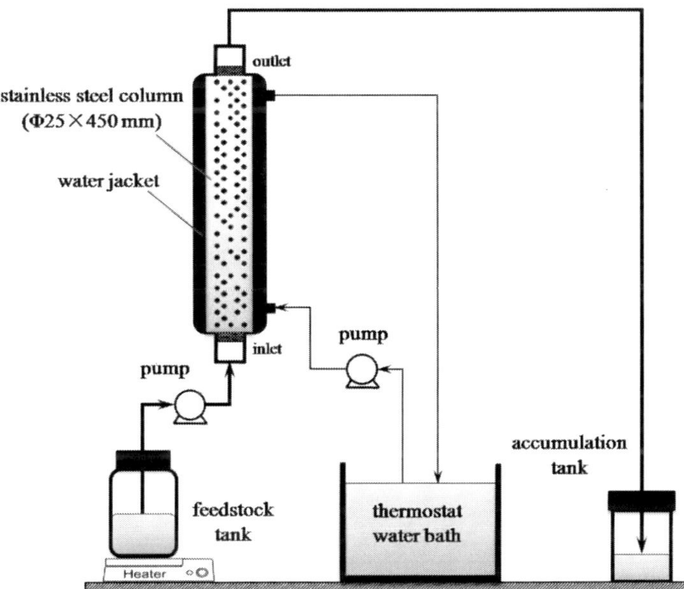

Figure 1. Scheme of the experimental apparatus used for the continuous esterification.

2.4. Analysis of the Samples

The samples were analyzed by titration to measure the acid value of the product. The measurement of acid value was carried out as the following procedure. 49 ml hot ethanol was firstly neutralized by 0.1M KOH solution, 0.5 ml phenolphthalein as indicator. 0.25 g sample was added into the neutralized ethanol to fully dissolve by heating. The sample was then titrated by 0.1 M KOH solution. The volume of KOH consumed by the FFA was recorded and acid value was calculated using the following equation [20]:

$$S = \frac{56.1 \times C \times V}{M} \tag{1}$$

In Eq. (1), S: acid value (mg KOH/g); C: concentration of the KOH was used for titration, M; V: volume of KOH employed for titration, ml; M: weight of the sample taken to analyze, g.

The FFA conversion is defined as the ratio of change of acid value of the acidified oil before and after reaction to the initial acid value for batch mode [21]. In the continuous mode, the FFA conversion is defined as the ratio of change of acid value of the reactor inlet and outlet streams to the inlet stream acid value. Water content was tested by the Karl Fischer water tester with a sensitivity of 500 ppm (KF-1A, Shanghai Precision and Scientific Instrument Co., Ltd., Shanghai, China). Sulfur content was detected by Micro-coulometric Analyzer (WK-2D, Jiangfen Electroanalytical Instrument CO., Ltd., Jiangsu, China). The main chemical compositions in the feedstock and production and the conversion efficiency for each individual fatty acid were investigated by gas chromatography–mass spectrometry (GC–MS) (6890 N GC/5973 MS, Agilent Technologies).

3. RESULTS AND DISCUSSION

3.1. Batch Mode

3.1.1. Comparison of Different Cation-Exchange Resins

The FFA conversions for different catalysts in the batch mode were investigated at the conditions of (i) amount of catalyst 20 wt.%, (ii) mole ratio of methanol to oil 6:1, (iii) reaction temperature of 64 °C, and (iv) reaction time of 2 h. FFA conversions were 79.7, 32.2 and 10.3% by using NKC-9, 001×7 and D61 as catalysts, respectively. The catalytic activities of resin were arranged as NKC-9>001×7> D61. This order of catalytic activities could be explained according to the nature of the resins. NKC-9 has a larger average pore diameter than that of 001×7 and D61 resin, which is beneficial to the reactants to access the active sites of the resin [22]. The small pores in 001×7 and D61 resins might increase mass transfer resistance and decrease the catalytic activity. The presence of water has a detrimental effect on the activity of cation-exchange resins [23]. Since NKC-9 with lowest water content can adsorb the water produced in esterification, it could decrease the concentration of water in the reaction system, thus promote the esterification toward ester formation. Therefore, the esterification in the next experiments was conducted by using NKC-9 as the catalyst.

3.1.2. Effect of Batch Reaction Parameter

The effect of catalyst amount was studied at the conditions of mole ratio of methanol to oil 6:1, reaction temperature of 64 °C and reaction time of 2 h. The catalyst amount was arranged in 6~24 wt.%. The FFA conversion increased considerably with an increase in the amount of catalyst. When the amount of catalyst was 6 wt.% and 20 wt.%, the FFA conversion was 58.9% and 89.3%, respectively. The conversion hardly increased with an increase in catalyst amount from 20 wt.% to 24 wt.%, which is consistent with the report by Saha and Sharma [24].

The temperature was varied from 60 °C to 68 °C to study its effect on FFA conversion. The other conditions were methanol to acidified oil at 7:1 molar ratio, 20 wt.% NKC-9 catalyst and reaction time of 2 h. A significant increase in FFA conversion was observed with an increase in the temperature from 60 °C to 66 °C. It can be explained as the increase in temperature would result in the increase in reaction rate and equilibrium constant for an endothermic reaction. While FFA conversion decreased with an increase in the temperature from 66 °C to 68 °C, because there is a loss of methanol at high reaction temperature [3]. The maximum FFA conversion obtained at 66 °C was approximate 90.0%.

The molar ratio of reactants significantly affects the FFA conversion as well as production cost of biodiesel. The effect of the range of 1:1 ~ 7:1 molar ratio of methanol to acidified oil on the conversion was investigated at the conditions of 20 wt.% NKC-9 catalyst at 64 °C and reaction time of 2 h. When the molar ratio of methanol to oil was 1:1, the FFA conversion was only 66.3%. And the FFA conversions rose from 74.3% to 87.5% with an increase of the molar ratio of methanol to oil from 2:1 to 5:1. The maximum FFA conversion was achieved at the molar ratio of 6:1. However, with further increase in molar ratio, the FFA conversion almost kept stable. The same trend was described by Yadav and Thathagar [23].

The relationship between conversion and reaction time was also investigated. The reaction conditions were the amount of NKC-9 20 wt.%, reaction temperature 64 °C and methanol/oil molar ratio 3:1. The results show that the reaction time had a marked effect on the esterification. When reaction time was 4 h, the FFA conversion was 95.4%. Then the rate of the reaction was declined after 4 h [19]. Therefore, about 4 h of reaction time is suitable for the esterification.

3.1.3. Orthogonal Experiments

Table 2 lists the schedules of the orthogonal test in which the key parameters including the amount of catalyst (A), reaction temperature (B), reaction time (C) and methanol/oil molar ratio (D) were selected as four factors. Each factor had three levels to be optimized. FFA conversion was taken as the index point to evaluate the extent of reaction under different factors and levels. Detailed scheme of orthogonal test is listed in Table 3.

Table 2. Factors and levels selected for orthogonal experiments

Level	A: amount of catalyst (wt.%)	B: reaction temperature(°C)	C: reaction time (h)	D: methanol/oil molar ratio
1	16	64	1	2:1
2	18	65	2	3:1
3	20	66	3	4:1

Table 3. Detailed schemes of orthogonal test and FFA conversion

Series No.	Level of A	Level of B	Level of C	Level of D	Conversion (%)
1	1	1	1	1	44.1
2	1	2	2	2	66.3
3	1	3	3	3	87.3
4	2	1	2	3	73.9
5	2	2	3	1	83.6
6	2	3	1	2	50.8
7	3	1	3	2	84.0
8	3	2	1	3	52.2
9	3	3	2	1	66.6
K1	65.9	67.3	49.0	64.8	$\overline{Y}$ =67.6
K2	69.4	67.4	68.9	67.0	
K3	67.6	68.2	85.0	71.1	Q_T=2027.1
Q_j	18.4	3.9	1951.2	61.4	

From the mathematical processing of the data, some conclusions were made. Firstly, the factor of reaction time shows the most notable influence on index point. Methanol to acidified oil molar ratio was then the relatively remarkable factor comparing with the amount of catalyst and reaction temperature. Secondly, the highest FFA conversion (approximately 90.0%) was obtained when the conditions of 18 wt.%, 66 °C, 4:1 and 3 h were selected.

In Table 3 the "Q_j" refers to sum of squares of deviations, "$\overline{Y}$" refers to total mean and "Q_T" refers to total sum of squares of deviations.

3.1.4. Analysis of Feedstock and Reaction Product

The compositions of feedstock were converted into their corresponding FAME in reaction product by esterification with NKC-9 as heterogeneous acid catalyst. The feedstock and FAME compositions were determined by GC-MS. The main compositions in feedstock and No.3 product in orthogonal experiments by MS analysis were listed in Tables 4 and 5, respectively.

Table 4. MS analysis of feedstock

Peak	Retention Time (min)	Constituents	Fatty Acid
A	16.818	palmitic acid methyl ester	Palmitic (C16:0)*
B	17.121	palmitic acid	palmitic (C16:0)
C	17.370	palmitic acid ethyl ester	palmitic (C16:0)
D	18.342	oleic acid methyl ester	oleic (C18:1)
E	18.407	stearic acid methyl ester	stearic (C18:0)
F	18.810	oleic acid	oleic (C18:1)
G	19.089	oleic acid ethyl ester	oleic (C18:1)

*: the first number stands for the number of the carbon in the compound; the second number stands for the number of C=C double bond in the compound.

Table 5. MS analysis of No.3 sample in orthogonal experiments

Peak	Retention time (min)	Constituents	Fatty Acid
α	19.961	palmitic acid methyl ester	palmitic (C16:0)
β	22.415	linoleic acid methyl ester	linoleic (C18:2)
γ	22.510	oleic acid methyl ester	oleic (C18:1)
δ	22.890	stearic acid methyl ester	stearic (C18:0)

It was found that the compositions of feedstock consisted of (i) palmitic acid methyl ester, (ii) palmitic acid, (iii) palmitic acid ethyl ester, (iv) oleic acid methyl ester, (v) stearic acid methyl ester, (vi) oleic acid, and (vii) oleic acid ethyl ester in Table 4. However, the FAME compositions in reaction product merely contained four constituents, (i) palmitic acid methyl ester, (ii) linoleic acid methyl ester, (iii) oleic acid methyl ester and (iv) stearic acid methyl ester in Table 5. It could be seen by the compassion of Tables 4 and 5 that palmitic acid and oleic acid in the feedstock were transformed into corresponding methyl ester by esterification. At the same time, transesterification also happened [16]. Palmitic acid ethyl ester and oleic acid ethyl ester were transformed into palmitic acid methyl ester and oleic acid methyl ester, respectively. Furthermore, there was no change for methyl esters in the feedstock, such as palmitic acid methyl ester, oleic acid methyl ester and stearic acid methyl ester.

3.1.5. Reusability

The reusability of NKC-9 was investigated. It was found that the conversion of FFA kept unchanged, even slightly increased with runs at the first 10 runs. This may be because the more fragments of the resin were formed due to breakdown of resin particles under mechanical agitation, which increases the surface area of the resin catalyst to enhance the catalytic property. However, FFA conversion obviously decreased after the 10th run mainly due to the loss of resin in the process of taking sample or separating product. In the experiment, we found that the particles of the resin after reusing 10 times were far finer than the initial particles. During the reusing of the resin, the fragmentation and loss should occur at the same time. The fragmentation of the resin helps to improve the catalytic property; while the loss of the resin decreases the catalytic property (please see Section 3.1.2). At the first 10 runs, the effect of fragmentation might surpass the loss of the resin particles, showing high FFA conversion. And after the 10th run, the loss of the resin might be predominant due to too fine particles of the resin, exhibiting low FFA conversion. Nevertheless, about 60.0% FFA conversion can be obtained even when esterification reaction of the 15th run was finished. Therefore, the advantage of esterification using NKC-9 as solid acid catalyst was obvious. Nevertheless, it is necessary to decrease the loss of the resin or to add new resin during esterification in order to keep high conversion [10].

3.2. Continuous Mode

3.2.1. Effect of Continuous Reaction Parameter

The effect of initial water content in the feedstock on FFA conversion by NKC-9 resin in the fixed bed reactor was studied at the conditions of 1.12 ml/min feed flow rate, 1.25:1 methanol/oil mass ratio, 65 °C reaction temperature and 33.0 cm catalyst bed height (the amount of packed resin 70.3 g). FFA conversion almost linearly decreased from 94.3% to 55.3% with an increase in water content from 0.3 wt.% to 5.0 wt.%. It indicates that the presence of water hindered the ester formation with the esterification. This is because the water molecules were absorbed on the active sites ($-SO_3H$) of the resin surface to form a water layer to block the accessibility of reactants to active sites [25]. As a consequence, it is necessary to relieve the initial water content in the feedstock in order to obtain a high FFA conversion.

Flow rate is one of the key parameters during the continuous esterification carried out in the fixed bed reactor. The conversion decreased from 94.4% to 81.1% with an increase of flow rate from 0.82 to 2.32 ml/min. A significant decrease in FFA conversion was observed when the feed flow rate was over 1.12 ml/min. It implies that the flow rate is a function of the reaction time. The results obtained are consistent with the report by Santacesaria et al. [12].

The change of FFA conversion with methanol/oil mass ratio was also investigated. The FFA conversion increased from 89.5% to 94.3% with an increase in the mass ratio of methanol to oil from 0.35:1 to 1.25:1. According to LeChatelier's Principle, the excessive methanol was used to shift the equilibrium of reversible reaction toward the direction of ester formation [26]. However, with further increase in methanol/oil mass ratio, the FFA conversion almost kept stable. The similar results were described by Feng et al. [10].

The reaction temperature as one of the most important parameters effects on the FFA conversion during the esterification. The FFA conversion markedly increased from 17.1% to 94.0% with an increase in temperature from 25 °C to 65 °C. Obviously, a high temperature results in the increase in reaction rate and equilibrium constant for an endothermic reaction (Feng et al., 2010). However, when the reaction temperature was over 65 °C, there was no oil in the effluent solution from the outlet of reactor, and only methanol and a small amount of water flew out. This is because a large amount of methanol steam would be generated to block the outlet of reactor when the reaction temperature is over methanol boiling point (64.5 °C). Finally, it brings to the oil hard to flow out from the outlet of reactor [27].

The catalyst bed height in the fixed bed reactor is associated with the residence time during continuous esterification. The FFA conversion increased rapidly from 75.1% to 96.2% with the increase of the catalyst bed height from 11.0 (the amount of packed resin 24.2 g) to 44.0 cm (87.5 g). Especially, FFA conversion was over 90% when catalyst bed height was over 33 cm (70.3 g). It demonstrates that a high catalyst bed height provides a longer reaction time and more active sites to promote the reaction between oil and methanol at a given flow rate [7].

3.2.2. Conversion Efficiency for Each Individual Fatty Acid

The main compositions in the feedstock by GC–MS were: (i) palmitic acid, (ii) palmitic acid methyl ester, (iii) oleic acid, (iv) oleic acid methyl ester, (v) stearic acid methyl ester, (vi) eicosenoic acid methyl ester as shown in Table 6. The conversion efficiency for each individual fatty acid was also investigated. The each individual fatty acid content decreased with an increase in FFA conversion. The content of palmitic acid decreased from 4.46 to 1.28 wt.% at 52.9% of FFA conversion. That is to say, the conversion efficiency of palmitic acid was 71.3%, which is higher than that of the oleic acid (46.9%). It was noted that there was a little amount of linoleic acid (0.86 wt.%) produced when FFA conversion rose to 85.6%. It dropped to 0.71 wt.% in the sample when FFA conversion further increased to 96.2%. Ultimately, the obtained biodiesel compositions were (i) palmitic acid methyl ester, (ii) linoleic acid methyl ester, (iii) oleic acid methyl ester, (iv) stearic acid methyl ester, (v) eicosenoic acid methyl ester. These results are consistent with the batch mode work (please see Section 3.1.4) [10].

Table 6. MS analysis of feedstock

Peak	Retention Time (min)	Constituents
A	16.166	palmitic acid methyl ester
B	16.617	palmitic acid
C	17.892	oleic acid methyl ester
D	18.091	stearic acid methyl ester
E	18.330	oleic acid
F	20.305	eicosenoic acid methyl ester

3.2.3. Operational Stability of Esterification in Fixed Bed Reactor

Based on the above investigations, the operational stability of esterification in the fixed bed reactor was studied. The operation conditions were methanol to oleic acid (with an acid value of 200.0 mg KOH/g) mass ratio of 2.8:1, catalyst bed height of 44.0 cm (the amount of packed resin 87.5 g), feed flow rate of 0.62 ml/min and reaction temperature of 65 °C. The FFA conversions were over 98.0% during 500 h of continuous running, showing a much high conversion and excellent operational stability. Firstly, this is because the fixed bed reactor packed with resin could successfully avoid the breakdown of resin particles under mechanical agitation in the batch mode. Consequently, the catalytic property kept unchanged with long time operation as a result of no loss of the resin catalyst [10]. Secondly, the conversion could be easily controlled by adjusting the methanol to oil ratio, feed flow rate, reaction temperature and catalyst bed height. It is also noted that the conversion obtained in a continuous mode was higher than the batch reaction reported in our previous paper [10]. This is because water produced in the esterification was in time removed with the mixture flowing in the case of higher methanol/oil ratio. The excessive methanol promoted not only continuous esterification but also the extraction of water on the resin surface [28]. Consequently, the water content was very low in the fixed bed reactor. Therefore, continuous esterification in fixed bed reactor packed with NKC-9 resin has the advantage over the batch reaction.

Further, it may also be noted that the sulfur content in biodiesel is a great important parameter during the application of biodiesel. The lost of $-SO_3H$ groups on the solid catalyst may result in an increase in sulfur content in the biodiesel production during the esterification with the solid catalyst. Here, the sulfur content in the biodiesel obtained by the esterification of acidified oil carried out with NKC-9 resin was 154 ppm, which was less than both that in the feedstock (192 ppm) and the ASTM standard D6751 (<500 ppm). The reason is because the NKC-9 resin in the fixed bed reactor could absorb sulfur compounds in the feedstock. It indicates that there was no or very little loss of sulfonic acid groups ($-SO_3H$) from NKC-9 resin into the production of biodiesel during continuous esterification in the fixed bed reactor [29].

CONCLUSION

Three cation-exchange resins (NKC-9, 001×7 and D61) were used to examine their reactivity for the preparation of biodiesel via batch esterification of acidified oil and methanol. Catalytic activity of NKC-9 was higher than that of 001×7 and D61. Highest FFA conversion (~90.0%) was obtained by NKC-9. Gas chromatography-mass spectrometry analysis showed that the production was simplex and mainly composed of C16:0 (palmitic), C18:2 (linoleic), and C18:1 (oleic) acids of methyl ester. Furthermore, NKC-9 resin shows excellent reusability in batch esterification. Based on the batch mode, continuous esterification of acidified oil with methanol was carried out with NKC-9 resin as a heterogeneous solid acid catalyst in a fixed bed reactor to produce biodiesel. FFA conversion increased with increases in methanol/oil mass ratio, reaction temperature and catalyst bed height, whereas decreased with increases in initial water content in oil feedstock and feed flow rate. At the same time, NKC-9 resin exhibits a much high FFA conversion and operational stability. The loss of sulfonic acid groups from NKC-9 resin into the production during continuous esterification was not found. Therefore, NKC-9 resin shows the potential commercial applications to esterification of FFA.

ACKNOWLEDGMENTS

The authors acknowledge the financial support by National High Technology Research and Development Program of China (Grant No.2009AA03Z223), Tianjin Natural Science Foundation (Grant No.08JCYBJC26400 and 08JCZDJC24000).

REFERENCES

[1]   McNeff C. V., McNeff L. C., Yan B., Nowlan D. T., Rasmussen M., Gyberg A. E., Krohn B. J., Fedie R. L., Hoye T. R., 2008. A continuous catalytic system for biodiesel production. *Appl. Catal. A: Gen.*, 343, 39-48.

[2]   Srivastava A., Prasad R., 2000. Triglycerides-based diesel fuels. Renew. *Sust. Energy Rev.*, 4, 111-133.

[3]   Ramadhas A. S., Jayaraj S., Muraleedharan C., 2005. Biodiesel production from high FFA rubber seed oil. *Fuel*, 84, 335-340.

[4]   Marchetti J. M., Miguel V. U., Errazu A. F., 2007. Possible methods for biodiesel production. *Renew. Sust. Energy Rev.*, 11, 1300-1311.

[5]   Canakci M., Van Gerpen J., 2001. Biodiesel production from oils and fats with high free fatty acids. Trans. ASAE, 44, 1429-1436.

[6]   Canakci M., Van Gerpen J., 1999. Biodiesel production via acid catalysis. *Trans. ASAE*, 42, 1203-1210.

[7]   Halim S. F. A., Kamaruddin A. H., Fernando W. J. N., 2009. Continuous biosynthesis of biodiesel from waste cooking palm oil in a packed bed reactor: Optimization using response surface methodology (RSM) and mass transfer studies. *Bioresour. Technol.*, 100, 710-716.

[8]   Lee J. S., Saka S., 2010. Biodiesel production by heterogeneous catalysts and supercritical technologies. *Bioresour. Technol.*, 101, 7191-7200.

[9]   Chen X., Xu Z., Okuhara T., 1999. Liquid phase esterification of acrylic acid with 1-butanol catalyzed by solid acid catalysts. *Appl. Catal. A: Gen.*, 180, 261-269.

[10]   Feng Y. H., He B. Q., Cao Y. H., Li J. X., Liu M., Yan F., Liang X. P., 2010. Biodiesel production using cation-exchange resin as heterogeneous catalyst. *Bioresour. Technol.*, 101, 1518-1521.

[11]   Grob S., Hasse H., 2006. Reaction kinetics of the homogeneously catalyzed esterification of 1-butanol with acetic acid in a wide range of initial compositions. *Ind. Eng. Chem. Res.*, 45, 1869-1874.

[12]   Santacesaria E., Tesser R., Di Serio M., Guida M., Gaetano D., Agreda A. G., 2007. Kinetics and mass transfer of free fatty acids esterification with methanol in a tubular packed bed reactor: A key pretreatment in biodiesel production. *Ind. Eng. Chem. Res.*, 46, 5113-5121.

[13] Özbay N., Oktar N., Tapan N. A., 2008. Esterification of free fatty acids in waste cooking oils (WCO): Role of ion-exchange resins. *Fuel,* 87, 1789-1798.

[14] Lotero E., Liu Y. J., Lopez D. E., Suwannakarn K., Bruce D. A., Goodwin J. G., 2005. Synthesis of biodiesel via acid catalysis. *Ind. Eng. Chem. Res.,* 44, 5353-5363.

[15] Liu Y., Lotero E., Goodwin J. J. G., 2006. A comparison of the esterification of acetic acid with methanol using heterogeneous versus homogeneous acid catalysis. *J. Catal.,* 242, 278-286.

[16] Guerreiro L., Castanheiro J. E., Fonseca I. M., Martin-Aranda R. M., Ramos A. M., Vital J., 2006. Transesterification of soybean oil over sulfonic acid functionalised polymeric membranes. *Catal. Today,* 118, 166-171.

[17] Lu P., Yuan Z., Li L., Wang Z., Luo W., 2010. Biodiesel from different oil using fixed-bed and plug-flow reactors. *Renew. Energy,* 35, 283-287.

[18] Liu Y., Wang L., Yan Y., 2009. Biodiesel synthesis combining pre-esterification with alkali catalyzed process from rapeseed oil deodorizer distillate. *Fuel Process. Technol.,* 90, 857-862.

[19] Shibasaki-Kitakawa N., Honda H., Kuribayashi H., Toda T., Fukumura T., Yonemoto T., 2007. Biodiesel production using anionic ion-exchange resin as heterogeneous catalyst. *Bioresour. Technol.,* 98, 416-421.

[20] Shi W. Y., He B. Q., Ding J. C., Li J. X., Yan F., Liang X. P., 2010. Preparation and characterization of the organic-inorganic hybrid membrane for biodiesel production. *Bioresour. Technol.,* 101, 1501-1505.

[21] Marchetti J. M., Errazu A. F., 2008. Comparison of different heterogeneous catalysts and different alcohols for the esterification reaction of oleic acid. *Fuel,* 87, 3477-3480.

[22] Teo H. T. R., Saha B., 2004. Heterogeneous catalysed esterification of acetic acid with isoamyl alcohol: kinetic studies. *J. Catal.,* 228, 174-182.

[23] Yadav G. D., Thathagar M. B., 2002. Esterification of maleic acid with ethanol over cation-exchange resin catalysts. *React. Funct. Polym.,* 52, 99-110.

[24] Saha B., Sharma M. M., 1997. Reaction of dicyclopentadiene with formic acid and chloroacetic acid with and without cation-exchange resins as catalysts. *React. Funct. Polym.,* 34, 161-173.

[25] Russbueldt B. M. E., Hoelderich W. F., 2009. New sulfonic acid ion-exchange resins for the preesterification of different oils and fats with high content of free fatty acids. *Appl. Catal. A: Gen.,* 362, 47-57.

[26] Su C. H., Fu C. C., Gomes J., Chu I. M., Wu W. T., 2008. A heterogeneous acid-catalyzed process for biodiesel production from enzyme hydrolyzed fatty acids. *AIChE J.,* 54, 327-336.

[27] Feng Y., Zhang A., Li J., He B., 2011. A continuous process for biodiesel production in a fixed bed reactor packed with cation-exchange resin as heterogeneous catalyst. *Bioresour. Technol.,* 102, 3607-3609.

[28] Chen F., Kawasaki J., Naka Y., 2004. Computational Studies of the Countercurrent Multistage Extraction-Coupled Esterification Process of Oleic Acid with Methanol Using Excess Methanol as Extractant. *Chem. Eng. Res. Des.,* 82, 599-604.

[29] Zhu M. L., He B. Q., Shi W. Y., Feng Y. H., Ding J. C., Li J. X., Zeng F. D., 2010. Preparation and characterization of PSSA/PVA catalytic membrane for biodiesel production. *Fuel,* 89, 2299-2304.

In: Biodiesel: Blends, Properties and Applications   ISBN: 978-1-63117-024-9
Editors: Jorge M. Marchetti and Zhen Fang   © 2014 Nova Science Publishers, Inc.

Chapter 11

DESIGN AND OPTIMIZATION OF REACTIVE DISTILLATION SEQUENCES WITH THERMAL COUPLING WITH MINIMUM NUMBER OF REBOILERS

María Vázquez-Ojeda[1], Juan Gabriel Segovia-Hernández[1,], Salvador Hernández[1] and Rafael Maya-Yescas[2]*

[1]Universidad de Guanajuato, Campus Guanajuato,
Departamento de Ingeniería Química, Guanajuato, Gto., Mexico
[2]Universidad Michoacana de San Nicolás Hidalgo,
Facultad de Ingeniería Química, Morelia, Mich., México

ABSTRACT

The esterification of lauric acid and methanol is explored using thermally coupled distillation sequences with side columns and with minimum number of reboilers. The product of the esterification can be used as biodiesel. This is a major step forward since thermally coupled reactive distillation sequences with side columns and with minimum number of reboilers offer significant benefits, such as: reductions on both capital investment and operating costs due to the absence of the reboilers, higher conversion and selectivity as no products are recycled in the form of reflux or boil-up vapors, as well as no occurrence of thermal degradation of the products due to a lower temperature profile in the column. In the traditional process, lauric acid is fed as saturated liquid, while the methanol is fed as saturated vapor stream to the main column, then we can take heat from vapor stream to deliver it to the system without a reboiler. Rigorous simulations were performed in AspenTech AspenONE engineering suite to design this novel reactive distillation process and evaluate the technical and economical feasibility. The results indicate that the energy consumption of the complex distillation sequences with a side column can be reduced significantly by varying operational conditions, in comparison with conventional reactive distillation sequences.

NOMENCLATURE

$D \equiv$ Diameter

$P \equiv$ Pressure top

$Q_C \equiv$ Condenser duty

$Q_B \equiv$ Reboiler duty

$T_C \equiv$ Condenser temperature

$T_B \equiv$ Reboiler temperature

$T_D \equiv$ Dew point temperature

$P_B \equiv$ Presure bottom

$RR \equiv$ Reflux ratio

$R_D \equiv$ Distillate rate

$R_B \equiv$ Bottoms rate

$F_{FA} \equiv$ Flow of fatty acid

$F_M \equiv$ Flow of methanol

$S \equiv$ Number of stages

$S_{FA} \equiv$ Feed stage fatty acid

$S_M \equiv$ Feed stage metanol

$S_{II} \equiv$ Feed stage column II / Rectifier / Stripper

$S_R \equiv$ Number of reactive stage

$S_{RI} \equiv$ Starting reactive stage

$S_{FV} \equiv$ Interconnection stage vapor phase

$S_{LV} \equiv$ Interconnection stage liquid phase

$F_{FV} \equiv$ Interconnection flow vapor phase

$F_{LV} \equiv$ Interconnection flow liquid phase.

INTRODUCTION

The world is presently confronted with the twin crises of fossil fuel depletion and environmental degradation. Indiscriminate extraction and lavish consumption of fossil fuels have led to reduction in underground-based carbon resources. The search for alternative fuels, which promise a harmonious correlation with sustainable development, energy conservation, efficiency and environmental preservation, has become highly pronounced in the present context. The fuels of bio-origin can provide a feasible solution to this worldwide petroleum crisis. Gasoline and diesel-driven automobiles are the major sources of greenhouse gases emission (Johansson and McCarthy, 1999; Kesse, 2000; Cao, 2003). Scientists all around the world have explored several alternative energy resources, which have the potential to quench the ever-increasing energy thirst of today's population. Various biofuel energy resources explored include biomass, biogas (Murphy and McCarthy, 2005), primary alcohols, vegetable oils, biodiesel, etc. These alternative energy resources are largely environment-friendly but they need to be evaluated on case-to-case basis for their advantages, disadvantages and specific applications. Some of these fuels can be used directly while others need to be formulated to bring the relevant properties closer to conventional fuels (Agarwal, 2007).

Biodiesel, the common name for fatty acid methyl esters (FAMEs), is a liquid fuel obtained from the transesterification of vegetable (or animal) oils. It involves simply the reaction, under very mild conditions, between vegetable oil and typically a large excess of alcohol of methanol, in the presence of an acid or basic catalyst, which produces FAMEs as the main product and glycerol as a byproduct (Di Felice et al., 2008). Biodiesel has drawn significant attention due to increasing environmental concern and diminishing petroleum reserves. Biodiesel is a renewable and environmentally friendly energy. Application of this energy not only can significantly reduce the pollution generated from petroleum based diesel oil but also can lessen the dependence on petroleum. Presently, biodiesel is produced commercially in Europe and USA to reduce air pollution and the net emission of greenhouse gas. Surplus edible oils, such as rapeseed oil, soybean oil or castor oil, used cooking oil, rice bran oil, and algae, among others, are used as raw materials for biodiesel (Wahlen et al., 2008; Zhang et al., 2008; Lu et al., 2009; Peña et al., 2009). Outstanding benefits of biodiesel are the high cetane numbers that it naturally achieves and the lack of polluting heteroatoms like sulfur or nitrogen. Cetane number increase and sulfur/nitrogen content reduction are the most capital intensive hydrotreatment processes that a conventional refinery has to handle, a fact that lays the ground for the biodiesel niche in the near-future refining business (Stiefel and Dassori, 2009). Also, biodiesel is safe, renewable, nontoxic, and biodegradable; it contains no sulfur and is a better lubricant.

The current manufacturing biodiesel processes, however, have several disadvantages: shifting the equilibrium to fatty esters by using an excess of alcohol that must be separated and recycled, making use of homogeneous catalysts that require neutralization (causing salt waste streams), expensive separation of products from the reaction mixture, and high costs due to relatively complex processes involving one to two reactors and several separation units. Therefore, to solve these problems, some authors (for example, Kiss et al., 2008) have developed a sustainable biodiesel production process based on reactive distillation using acid catalysts. Reactive distillation integrates reaction and separation in one unit. This intensifies mass transfer and allows in situ energy integration while simplifying the process flowsheet and operation. However, combining the two operations is possible only if the reactions show reasonable conversion and selectivity data at pressures and temperatures that are compatible with the distillation conditions. The reduction in the number of processing units and the direct heat integration between reaction and separation can reduce capital investment as well as utility costs. Increased overall conversion, as well as improved selectivity in competing reactions, can be achieved in reactive distillation by the continuous removal of products from the reaction zone of equilibrium limited reactions (Ciric and Gu, 1994). Other process improvements can be realized such as reducing byproduct formation, improving raw material usage, and overcoming chemical equilibrium limitations.

Design issues for reactive distillation systems are significantly more complex than those involved in ordinary distillation (Doherty and Buzad, 1992). Catalyst selection, liquid holdup on each tray, and position of feeds become important design considerations. Reaction often occurs in the liquid holdup so that the reaction volume is a major design parameter, and constant molar overflow cannot be assumed. Also, a single feed may not be appropriate and a distributed feed must be considered. Thermally coupled distillation systems (TCDS) are obtained through the implementation of interconnecting streams (one in the vapor phase and the other one in the liquid phase) between two columns; each interconnection replaces one condenser or one reboiler from one of the columns, thus providing potential savings in capital investment. Furthermore, through a proper selection of the flow values for the interconnecting streams of TCDS, one can obtain significant energy savings (consequently, reductions in CO_2 emissions) with respect to the energy consumption of conventional distillation sequences. There is a considerable amount of literature on the analysis of the relative advantages of TCDS for ternary separations with equilibrium and nonequilibrium stage models (Triantafyllou and Smith, 1992; Hernández and Jiménez, 1996; Hernández and Jiménez, 1999; Dünnebier and Pantelides, 1999; Yeomans and Grossmann, 2000; Emtir et al., 2003; Olujic et al., 2003; Hernández et al., 2006; Abad-Zárate et al., 2006, among others). These studies have shown that those thermally coupled distillation schemes are capable of achieving typically 30% of energy savings compared to the conventional schemes. Recently some studies have reported the possibility of implementing the process of reactive distillation in thermally coupled distillation sequences. In these conditions it is possible to combine the kindness of the reactive distillation and the savings of energy widely reported for the systems with thermal coupling (Barroso-Muñoz et al. 2007; Wang et al. 2008; Hernández et al. 2009). Thus, in this paper, the production of biodiesel by esterification of methanol and lauric acid is studied using conventional, thermally coupled distillation sequences with side columns and thermally coupled distillation sequences with a side column with minimum number of reboilers (Figure 1 -3).

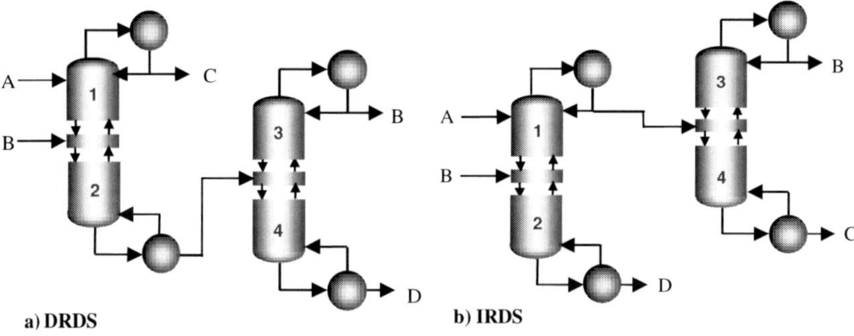

Figure 1. Conventional reactive distillation sequences: (a) direct (DRDS), and (b) indirect (IRDS).

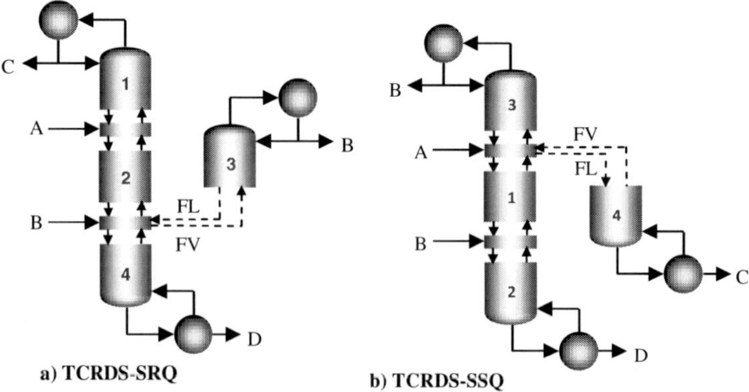

Figure 2. Thermally coupled reactive distillation sequences: (a) with side rectifier (TCRDS-SRQ) and (b) with side stripper (TCRDS-SSQ).

The results indicate that the energy consumption of the complex distillation sequences with a side column and minimum number of reboilers can be reduced significantly by varying operational conditions, in comparison with conventional and thermally coupled systems when reactive systems are included.

PROBLEM STATEMENT

There are three basic methods to produce fatty esters from oils/fats: 1) base catalyzed trans-esterification, 2) acid catalyzed esterification, and 3) enzymatic catalysis. The first method is the most frequently used. However, due to the escalating costs of fatty raw materials, the current trend is to use less expensive alternatives such as animal fat, waste cooking oil from catering premises, or waste vegetable oil. The problem with waste oils is the very high content of free fatty acids (FFA) that lead to soap formation in a conventional base catalyzed process. Therefore, in order to avoid production loss and soap associated problems, the FFA's must be completely converted first to fatty esters by esterification. Moreover, the conventional biodiesel processes employ liquid catalysts, such as H_2SO_4, NaOH or methoxides. The problem is that homogeneous catalysts require neutralization, washing, separation, recovery, and waste disposal operations with severe economical and environmental penalties. To solve this problem we propose an alternative fatty esterification process based on thermally coupled reactive distillation with minimum number of reboilers using acid as catalysts and therefore eliminating the additional separation steps and the salt waste streams, thus simplifying the downstream processing.

The alternative technology proposed in this work offers advantages compared to conventional reactive distillation, as for example diminution in the use of energy in the system and low temperature profile in the reactive separation column, due to the elimination of reboilers to avoid the thermal degradation of the fatty esters products. As a result, the process presents reductions in both capital investment and operating costs due to the absence of a reboiler (no product vapors return to the column). This process approach based on reactive distillation is particularly suitable for treating waste oil or animal fat, including tri-glycerides with up to 100% free fatty acids (FFA).

In this work we analyzed three different types of reactive distillation systems: the conventional case, the thermally coupled arrangement and the thermally coupled arrangement with minimum number of reboilers. Each of these cases is analyzed in the direct and indirect structures.

We took as base designs of the distillation sequences for conventional and thermally coupled reactive systems designs, those obtained by Miranda-Galindo et al., 2011, see Figures 1 and 2. Table 1 shows the design parameters of the configurations of conventional and thermally coupled distillation schemes. Based on the thermally coupled systems we proceeded to remove the maximum amount of reboilers to generate alternative distillation systems. In the case of thermally coupled distillation sequence with side rectifier was possible to remove all the reboilers of the system. In the case of thermally coupled distillation sequence with side stripper is only feasible to eliminate the reboiler of the main column and must remain the reboiler in the side stripper (Figure 3).

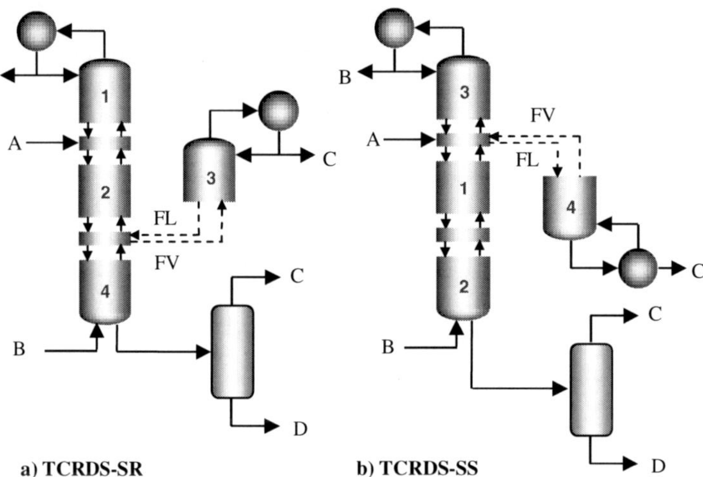

Figure 3. Thermally coupled arrangement with minimum use of reboilers: (a) with side rectifier (TCRDS-SR) and (b) with side stripper (TCRDS-SS).

Table 1. Parameters of the configurations of conventional and Thermally Coupled Distillation Schemes

Sequences	DRDS		IRDS		TCRDS-SRQ		TCRDS-SSQ	
Operating Specifications	Column I	Column II	Column I	Column II	Column main	Rectifier	Column main	Stripper
RR	2.6665	0.06072	2.3100	1.6953	2.0568	0.3468	19.5116	6.1894
R_D [lbmol/h]	100.0240	19.3662	119.3477	19.8149	101.0767	18.2670	19.0155	19.2013
R_B [lbmol/h]	119.9760	100.6098	100.6523	99.5328	100.6563	1.5974	100.6398	100.3446
F_{FA} [lbmol/h]	100.0000	-	100.0000	-	100.0000	-	100.0000	-
F_M [lbmol/h]	120.0000	-	120.0000	-	120.0000	-	120.0000	-
S	21	7	10	15	16	5	22	2
S_{FA}	4	-	3	-	3	-	10	-
S_M	20	-	10	-	15	-	22	-
S_{II}	-	4	-	9	-	-	-	-
S_{FV}	-	-	-	-	12	5	7	1
S_{LV}	-	-	-	-	11	5	7	1
F_{FV} [lbmol/h]	-	-	-	-	19.8644		19.2013	
F_{LV} [lbmol/h]	-	-	-	-	1.5974		119.5460	
S_R	20	-	7	-	15	-	21	-
S_{RI}	2	-	3	-	3	-	3	-

Strictly, the design of the conventional and reactive thermally coupled distillation sequences could be modeled through superstructures suitable for optimization procedures with mathematical programming techniques. However, the task is complicated and is likely to fail to achieve convergence. In this case, to overcome the complexity of the simultaneous solution of the tray arrangement and energy consumption within a formal optimization algorithm, we decoupled the design problem in two stages: (1) tray configuration; (2) energy efficient design (optimal energy consumption). The first stage of our approach begins with the development of preliminary designs for the alternative systems with minimum number of reboilers based on the designs proposed by Miranda-Galindo et al., 2011.

After the tray arrangement for the alternative coupled designs have been obtained, an optimization procedure is used to minimize the total heat duty supplied to the schemes, taking into account the constrains imposed by the required purity of the product streams. The degrees of freedom that remain after the design specifications and tray arrangement are used to obtain the operating conditions that provide minimum total energy consumption. Three degrees of freedom that remain for each integrated sequence are the interconnecting flow (vapor or liquid, depending of the scheme) and both pressures of the feed flowrates. The search procedure provides the optimal values of the interconnecting flows and pressures to minimize the energy consumption for the reaction-separation system. The design is successful if it meets the product specifications.

The optimization strategy can be summarized as follows: (a) A base design for the complex schemes is obtained with a value of pressure for each feed flowrate (alcohol and fatty acid). (b) A value for each interconnecting flow (vapor or liquid) is assumed. (c) A rigorous model for the simulation of alternative coupled schemes with the proposed tray arrangement is solved. In this work Aspen Plus was used for that purpose. If the product compositions are obtained, then the design is kept; otherwise, proper adjustments must be made. (d) One value of interconnecting flow is changed, going back to step (c) until a local minimum in energy consumption for the assumed values of the pressure for each feed flowrate is detected. (e) The value of a pressure of alcohol is modified (the pressure of the fatty acid is constant), going back to step (c) until the energy consumption is minimum. (f) The value of the pressure of the fatty acid is modified, going back to step (c) until the minimum energy consumption supplied to the system is obtained. This result implies that an optimum value has been detected for the design of the complex schemes.

CASE OF STUDY

The esterification process, studied in this work, can be represented conceptually by equation 1.

Alcohol + Fatty Acid $\leftrightarrow$ Ester + Water   (1)

This equilibrium reaction can be favored if the products are removed as the reaction proceeds. An additional problem may present itself, depending on the acid and the alcohol used, binary or ternary homogeneous azeotropes can be formed in the reactive system. For highly non-ideal systems, heterogeneous azeotropes can be formed. These key factors must be considered in order to select the appropriate thermodynamic model when the system is studied with process simulators. For this class of reactive systems, thermodynamic models such as NRTL can be used to calculate vapor-liquid or vapor-liquid-liquid equilibrium. For this study (esterification of methanol and lauric acid), we selected the NRTL model.

Table 2. Kinetic parameters for the pseudohomogeneous kinetic model of the esterification reaction

Reaction	k^0 [mol/g·s]	E_A [kJ/mol]
Esterification	9.1164×10^5	68.71
Hydrolysis	1.4998×10^5	64.66

The systems include two feed streams; the first is lauric acid with a flow of 45.0 kmol/h as saturated liquid, and the second is methanol with a flow of 54.48 kmol/h as saturated vapor. Through an optimization study where the objective was to reduce the total energy load of the system, we obtained the optimum operating pressure of the feed flowrates. The design objective is a process for high-purity fatty ester, over 99.9% mass fraction, suitable for applications in cosmetics, detergents, surfactants or biodiesel application. It is important to highlight that this equilibrium reaction is usually catalyzed using sulfuric acid or p-toluensulfonic acid. The kinetic model (see Table 2) reported in Steinigeweg and Gmehling (2003) was used.

RESULTS

The first stage of our approach begins with the development of preliminary designs for the alternative systems with minimum number of reboilers based on the designs proposed by Miranda-Galindo et al., 2011 (Table 1).

The strategy for the design of thermally coupled sequences with minimum number of reboilers, followed by a search procedure on the interconnecting streams to optimize the design for energy consumption was conducted. A typical optimization search for the TCRDS-SR is shown in Figure 4, where the combination of interconnecting flowrate and pressure that provide the minimum energy consumption can be detected. The tray arrangements and some parameters for that sequence after the optimization task are given in Table 3. The results obtained for energy consumption and total annual cost for each case study are summarized in Tables 4-6. In general, the TCRDS-SR presented energy savings in the range between 15-30% in contrast to conventional and thermally coupled distillation sequences (Table 4). Importantly, the TCRDS-SR does not have any energy as vapor, as the scheme has no reboiler.

The objective function was set to minimize the consumption of cooling water in condensers. For the TCRDS-SR, optimal interconnecting flowrate is 16.0 lbmol/h, while operation pressure is 1 atm (Figures 4-5). Design objective was considered a mass fraction purity of 99.9 % for biodiesel and 99.6% for methanol and water.

Table 3. Parameters of Thermally Coupled Distillation Schemes with minimum number of reboilers

Secuencias	TCRDS-SR		TCRDS-SS	
Operating Specifications	Column I	Rectifier	Column I	Stripper
RR	1.4720	0.7432	36.0000	2.2482
R_D [lbmol/h]	100.0835	13.0986	11.0302	78.8609
R_B [lbmol/h]	106.8180	2.9014	107.8308	101.1391
F_{FA} [lbmol/h]	100.0000	-	100.0000	-
F_M [lbmol/h]	120.0000	-	120.0000	-
S	16	5	23	3
S_{FA}	3	-	10	-
S_M	16	-	23	-
S_{FV}	11	5	7	1
S_{LV}	4	5	7	1
F_{FV}[lbmol/h]	16.0000		78.8610	
F_{LV} [lbmol/h]	2.9014		180.0000	
S_R	16	-	18	-
S_{RI}	2	-	6	-

Table 4. Conventional reactive distillation sequences: information about total annual cost

Sequences	DRDS		IRDS	
Data	Column I	Column II	Column I	Column II
D [m]	0.5480	0.3984	1.0990	0.3360
S	21	7	10	15
P [atm]	1.0000	1.0000	1.0000	1.0000
Q_C [kW]	-1886.8014	-94.0816	-2052.4111	-238.6789
Q_B [kW]	1.6437	919.9382	962.8183	258.5297
T_C [°C]	99.9766	64.5559	83.7987	65.0037
T_B [°C]	125.1838	245.3968	243.6071	103.2329
T_D [°C]	124.9446	145.1890	242.9673	102.5260
P_B [atm]	1.2160	1.1374	1.1291	1.1300
	Costs [$/year]			
Equipment	119,659.57	60,550.05	126,742.78	71,793.34
Utility	13,526.45	206,907.34	230,321.80	37,779.40
TOTAL	400,643.41		466,637.32	

Table 5. Thermally Coupled Reactive Distillation Sequences: information about total annual cost

Sequences	TCRDS-SRQ		TCRDS-SSQ	
Data	Column I	Rectifier	Column I	Stripper
D [m]	0.8294	0.2047	0.6047	0.1922
S	16	5	22	2
P [atm]	1.0000	1.0000	1.0000	1.0000
Q_C [kW]	-1596.6677	-109.5481	-1738.1438	0.0000
Q_B [kW]	628.1472	0.0000	626.4624	96.4604
T_C [°C]	98.0707	64.5418	64.8576	97.3713
T_B [°C]	243.1699	144.6501	250.0000	98.8525
T_D [°C]	167.9259	71.6979	177.1609	97.3713
P_B [atm]	1.1180	1.1975	1.3010	1.0077
	Costo [$/year]			
Equipment	128,919.28	18,710.47	150,345.81	15,205.18
Utility	152,078.88	772.02	152,698.20	13,468.33
TOTAL	300,481		331,718	

Table 6. Thermally Coupled Reactive Distillation Sequences with minimum number of reboilers: information about total annual cost

Sequences	TCRDS-SR			TCRDS-SS		
Data	Column I	Rectifier	Flash	Column I	Stripper	Flash
D [m]	0.7063	0.2034	1.0631	0.9262	0.3167	1.0670
S	16	5	-	23	3	-
P [atm]	1.0000	1.0000	1.0000	1.0000	1.0000	1.0000
Q_C [kW]	-1281.3391	-101.85112	-	-1812.5305	0.0000	-
Q_B [kW]	0.0000	0.0000	611.01	0.0000	407.8718	578.92
T_C [°C]	99.9070	64.6788	-	64.6771	97.4002	-
T_B [°C]	177.1936	131.6704	240.0000	171.6022	114.4295	240.0000
T_D [°C]	193.3727	78.0306	142.6200	186.2989	106.8437	171.6000
P_B [atm]	1.6807	1.6807	-	1.6807	1.6807	-
	Costo [$/year]					
Equipment	102,040.66	20,271.42	12,772.21	178,862.12	27,749.92	12,826.07
Utility	9,030.04	717.78	136,984.81	12,773.53	56,949.32	129,789.67
TOTAL	281,817.91			426,963.00		

Figure 4 shows the dependence of energy consumption with respect to the operating pressure of the column and the pressure of the feed streams. This figure indicates that the minimum energy consumption is achieved when the pressures of the feed streams are the same as the operational pressure of the column. Now, when the operational pressures of the feed streams are different than the operating pressure of the column , the energy consumption does not change significantly. The same procedure was repeated for operational pressures of 0.5, 1.0, 1.5 and 2.0 atm, similar results to those mentioned previously are obtained. Figure 4 shows that the pressure that minimizes the energy consumption is 0.5 atm.

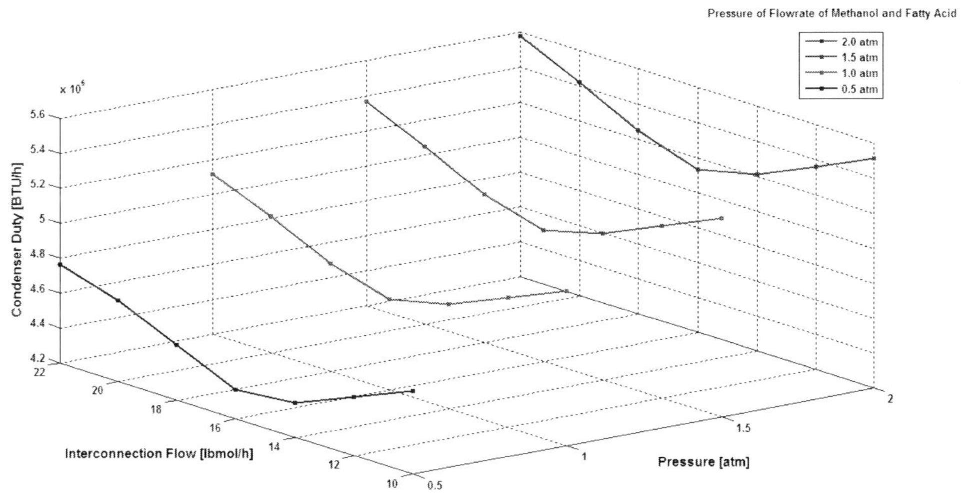

Figure 4. Condenser duty as a function of the interconnecting flow, operating pressure and pressure of feed flowrate of methanol, case TCRDS-SR.

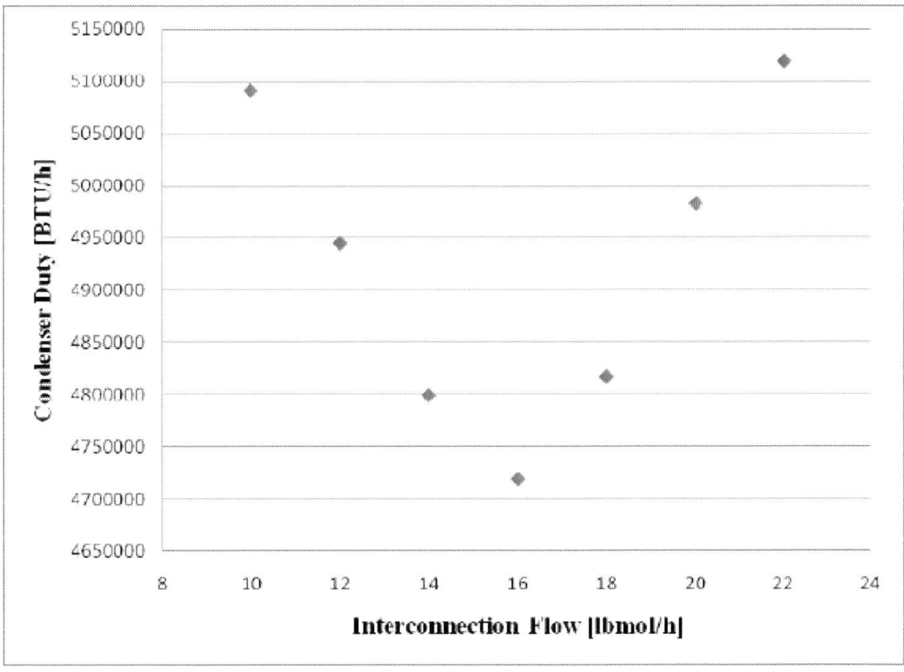

Figure 5. Condenser duty as a function of the interconnecting Flow, case TCRDS-SR.

However, it is well known that the operation in vacuum conditions substantially increases the total annual cost of the distillation column. As a result, the optimum operational pressure of 1 atm was chosen.

Figure 5 displays the optimization search of the TCRDS-SR option. The curve shows an interesting effect of the search variables. The design is sensitive, in terms of its energy consumption, to changes in the interconnecting flowrates. An implication of this observation has to do with operational considerations. In some regions of interconnecting flowrates, minor changes in the operating conditions of the thermally coupled schemes can lead to a significant deterioration of its energy consumption. The control design of this system appears to be an important task to develop.

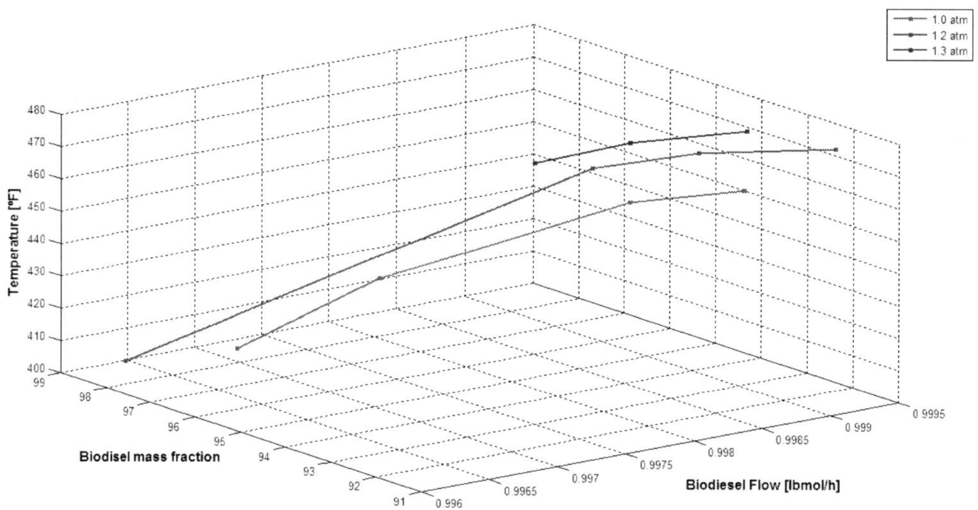

Figure 6. Temperature flash operation as a function of the flow and mass fraction of biodiesel for different operational pressures in flash, case TCRDS-SR.

Given the nature of the system, it is necessary to place an additional flash to achieve the purity of 99.9% biodiesel in mass fraction. Therefore it is necessary to optimize the operational conditions of the flash to minimize energy requirements. The flash must be operated at a temperature of 464 ° F and 1 atm of pressure to obtain a mass fraction purity of 99.9 % for biodiesel as shown in Figure 6. When the flash is operated at a pressure of 1 atm, the operating temperature is lower (464 ° F) obtaining a mass fraction purity of 99.9%. As it can be seen, the optimal pressure of the flash is identical to the operating pressure of the column to minimize energy consumption.

Table 3 shows the design parameters of TCRDS-SS. When the optimization curves (Figure 7) are analyzed, it is observed the dependence of energy consumption with respect to the operational pressure of the column and the pressure of the feed streams. Similar to the case TCRDS-SR, the value of the operational pressure leading to minimal energy consumption is when the pressures of the feed streams are the same as the operating pressure of the column; when the operational pressures of the feed streams are different to those of the column, the energy consumption does not change significantly. The optimal operational conditions found for this configuration are: operation pressure of the column, 1 atm; interconnecting flowrate, 180 lbmol/h. Figures 8 and 9 show the dependence of the energy consumption in the reboiler and condenser with the interconnecting streams. Clearly, the shape of the optimization curves show sensitivity of the energy consumption to changes in operating conditions. It is important to emphasize that the value of the interconnecting flowrate that minimizes energy consumption in the condenser and reboiler is the same.

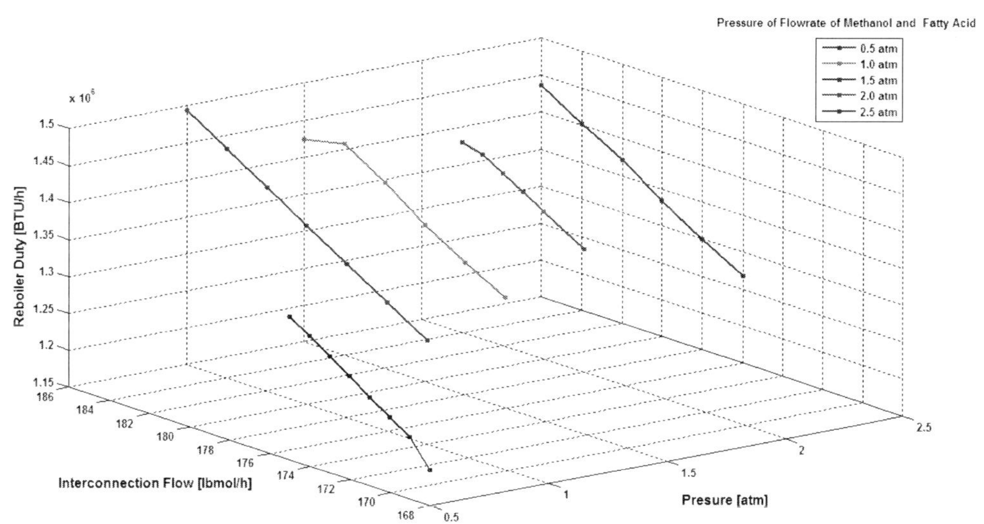

Figure 7. Reboiler duty as function of the operating pressure and interconnection flow, case TCRDS-SS.

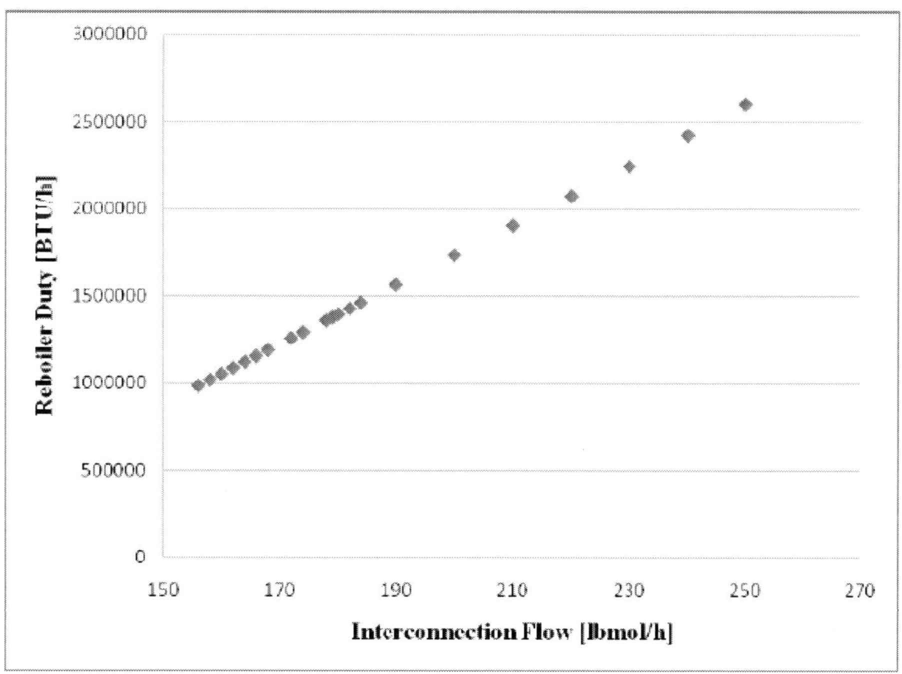

Figure 8. Reboiler duty as function of the interconnecting flow, case TCRDS-SS.

In this particular case, TCRDS-SS only presents savings in energy consumption in comparison with conventional reactive distillation sequences (~10—25%), but the configuration does not present energy savings when it is compared to thermally coupled columns.

Table 7. Total annual cost of sequences

Sequences	Equipment [UDS/year]	Energy consumption [UDS/year]	Total annual cost [UDS/year]
DRDS	180,210	220,434	400,643
IRDS	198,536	268,101	466,637
TCRDS-SRQ	147,630	152,851	300,481
TCRDS-SSQ	165,551	166,167	331,718
TCRDS-SR	135,084	146,733	281,817
TCRDS-SS	219,438	199,513	418,951

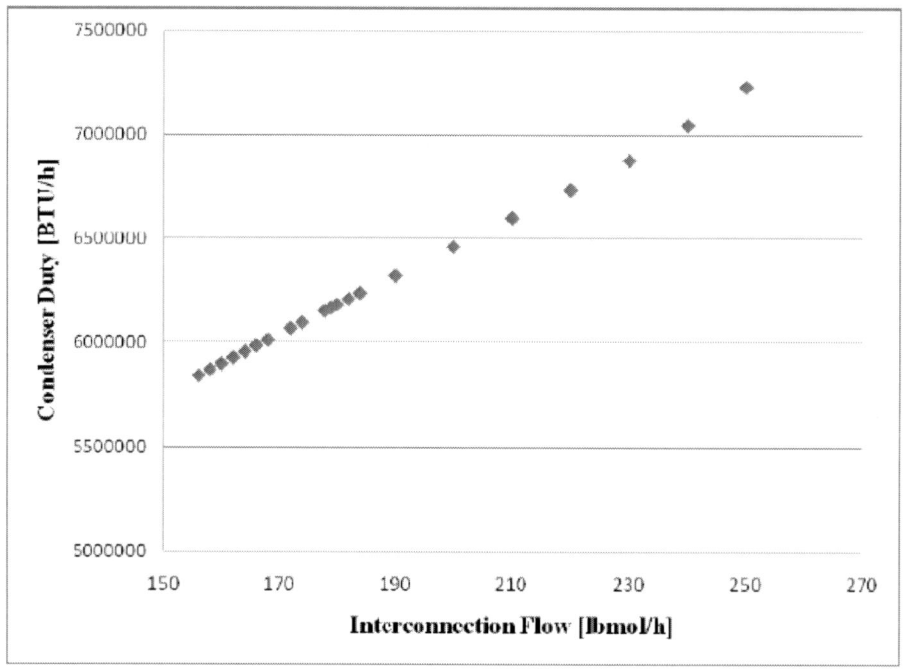

Figure 9. Condenser duty as function of the interconnection flow in the main column, case TCRDS-SS.

Finally, Table 7 summarizes the total annual costs of all systems tested. These costs were calculated using the method of Guthrie (Turton et al., 2004). As it can be seen, only the TCRDS-SR system has advantages in reductions of both energy consumption and equipment costs when compared to conventional and thermally coupled (without removal of reboilers) reactive distillation systems. Table 7 shows that a second option would be TCRDS-SR design which also presents significant savings in energy consumption. This is an interesting situation because the elimination of reboiler in a thermally coupled distillation column would seem that would lead to a design for minimum energy consumption and total annual cost. However according to the results of this work, the presence of a rectifier or side stripper influences in the energy consumption and equipment costs even when carrying out the decrease of the reboilers in the design.

CONCLUSIONS

This study uses computer aided engineering tools such as Aspen Plus for the development of an alternative biodiesel process based on reactive distillation. Two thermally coupled distillation sequences with side columns (stripper or rectifier) and minimum number of reboilers were utilized to carry out the production of the fatty ester compound. The parametric studies showed that the alternative coupled reactive arrangements column requires a lower amount of energy in contrast to the conventional and traditional coupled systems. However, only the TCRDS-SR presents advantages in the total annual cost in comparison with all sequences studied. This process improves the biodiesel production and dramatically reduces the total energy consumption of the process. The major benefit of this alternative process is low operating costs due to the integrated design with no reboilers in comparison with conventional and thermally coupled reactive distillation sequences. According to these results, it can be concluded that thermally coupled distillation sequences with minimum number of reboilers and a side rectifier option presents significant advantages over the classical process used to produce biodiesel.

REFERENCES

Abad-Zarate, E. F., Segovia-Hernández, J.G., Hernández, S., Uribe-Ramírez, A. R., 2006, A Short Note on Steady State Behavior of a Petlyuk Distillation Column by Using a Nonequilibrium Stage Model, *The Can. J. Chem. Eng.*, 84, 381.

Agarwal, A.K., 2007, Biofuels (Alcohols and Biodiesel) Applications as Fuels for Internal Combustion Engines, *Progress in Energy and Combustion Science*, 33, 233.

Barroso – Muñoz, F.O., Hernández, S., Segovia - Hernández, J.G., Hernández – Escoto, H., Aguilera – Alvarado, A. F., 2007, Thermally Coupled Distillation Systems: Study of an Energy – Efficient Reactive Case, *Chem. Biochem. Eng. Q. J.*, 21, 115.

Cao, X., 2003, Climate Change and Energy Development: Implications for Developing Countries, *Resour Policy*, 29, 61.

Ciric, A. R., Gu, D., 1994, Synthesis of Nonequilibrium Reactive Distillation Processes by MINLP Optimization, *AIChE J*, 40, 1479.

Di Felice, R., De Faveri, D., De Andreis, P., Ottonello, P., 2008, Component Distribution between Light and Heavy Phases in Biodiesel Processes, *Ind. Eng. Chem. Res.*, 47, 7862.

Doherty, M.F., Buzad, G., 1992, Reactive Distillation by Design, *Chemical Engineering Research Design*, 70, 448.

Dünnebier, G., Pantelides, C., 1999, Optimal Design of Thermally Coupled Distillation Columns, *Ind. Eng. Chem. Res.*, 38, 162.

Emtir, M., Mizsey, P., Rev, E., Fonyo, Z., 2003, Economic and Controllability Investigation and Comparison of Energy-Integrated Distillation Schemes, *Chem. Biochem. Eng. Q.*, 17, 31.

Hernández, S., Jiménez, A., 1996, Design of Optimal Thermally-coupled Distillation Systems Using a Dynamic Model, *Trans IChemE Part A*, 74, 357.

Hernández, S., Jiménez, A., 1999, Design of Energy-Efficient Petlyuk Systems, *Comput. Chem. Eng.*, 23, 1005.

Hernández, S., Segovia-Hernández, J.G., Rico-Ramírez, V., 2006, Thermodynamically Equivalent Distillation Schemes to the Petlyuk Column for Ternary Mixtures, *Energy*, 31, 1840.

Hernández, S., Sandoval – Vergara, R., Barroso – Muñoz, O.F., Murrieta – Dueñas, R., Hernández – Escoto, H., Segovia - Hernández, J.G., Rico – Ramírez, V., 2009, Reactive Divided Wall Distillation Columns: Dynamic Simulation and Implementation in a Pilot Plant, *Chemical Engineering and Processing: Process Intensification*, 48, 250.

Johansson, T., McCarthy, S., 1999, Global Warming Post-Kyoto: Continuing Impasse or Prospects for Progress?, *Energy Dev. Rep. Energy*, 69.

Kiss, A.A., Dimian, A., Rothenberg, G., 2008, Biodiesel by Catalytic Reactive Distillation Powered by Metal Oxides, *Energy and Fuels*, 22, 598,

Kesse, D, G., 2000, Global Warming—Facts, Assessment, Countermeasures, *J. Pet Sci. Eng.*, 26, 157.

Lu, H., Liu, Y., Zhou, H., Yang, Y., Chen, M., Liang, B., 2009, Production of Biodiesel from Jatropha Curcas L. Oil, *Computers and Chemical Engineering*, 33, 1091.

Miranda, G. E.Y., Segovia-Hernández, J.G., Hernández, S., Gutierrez-Antonio, C., Briones-Ramirez, A., 2011, Reactive Thermally Coupled Distillation Sequences: Pareto Front, *Ind. Eng. Chem. Res.*, 50, 926.

Murphy, J.D., McCarthy, K., 2005, The Optimal Production of Biogas for Use as a Transport Fuel in Ireland, *Renew Energy*, 30, 2111.

Olujic, Z., Kaibel, B., Jansen, H., Rietfort, T., Zich, E., Frey, G., 2003, Distillation Column Internals/Configurations for Process Intensification, *Chem. Biochem. Eng. Q.*, 17, 301.

Peña, R., Romero, R., Martínez, S.L., Ramos, M.J., Martínez, A., Natividad, R., 2009, Transesterification of Castor Oil: Effect of Catalyst and Co-Solvent, *Ind. Eng. Chem. Res.*, 48, 1186.

Steinigeweg, S., Gmehling, J., 2003, Esterification of a Fatty Acid by Reactive Distillation, *Ind. Eng. Chem. Res.*, 42, 3612.

Stiefel, S., Dassori, G., 2009, Simulation of Biodiesel Production through Transesterification of Vegetable Oils, *Ind. Eng. Chem. Res.*, 48, 1068.

Triantafyllou, C., Smith, R., 1992, The Design and Optimization of Fully Thermally Coupled Distillation Columns, *Trans IChemE Part A.*, 70, 118.

Turton, R., Bailie, R. C., Whiting, W. B., y Shaeiwitz, J. A. (2004). Analysis, Synthesis and Design of Chemical Processes, 2ª Ed., Prentice Hall, PTR, New Jersey, EEUU., pp. 197,198.

Wahlen, B.D., Barney, B.M., Seefeldt, L.C., 2008, Synthesis of Biodiesel from Mixed Feedstocks and Longer Chain Alcohols Using an Acid-Catalyzed Method, *Energy and Fuels*, 22, 4223.

Wang, S.J., Wong, D.S.H., Yu, S.W., 2008, Design and Control of Transesterification Reactive Distillation with Thermal Coupling, *Computers and Chemical Engineering*, 32, 3030.

Yeomans, H., Grossmann, I. E., 2000, Disjunctive Programming Models for the Optimal Design of Distillation Columns and Separation Sequences, *Ind. Eng. Chem. Res.*, 39, 1637.

Zhang, L., Guo, W., Liu, D., Yao, J., Ji, L., Xu, N., Min, E., 2008, Low Boiling Point Organic Amine-Catalyzed Transesterification for Biodiesel Production, *Energy and Fuels*, 22, 1353.

In: Biodiesel: Blends, Properties and Applications
Editors: Jorge M. Marchetti and Zhen Fang

ISBN: 978-1-63117-024-9
© 2014 Nova Science Publishers, Inc.

Chapter 12

SIMULTANEOUS REDUCTION OF NOX AND SMOKE USING WATER EMULSION OF KARANJI OIL METHYL ESTER IN A DI DIESEL ENGINE WITH EXHAUST GAS RECIRCULATION

Sukumar Puhan, A. Gopinath and K. Sairam*
Department of Mechanical Engineering,
Vel Tech Multi Tech Engineering College, Avadi, Chennai, India

ABSTRACT

Among the different possible sources, the methyl esters of vegetable oils, known as biodiesel is receiving increasing interest because of their low environmental impact. In addition biodiesel fuel may not require any significant modification of existing diesel engines. The present investigation focuses on the study on simultaneous reduction of NOx and smoke by water emulsion of Karanji oil methyl ester in a single cylinder, direct injection diesel engine with exhaust gas recirculation. The quantity of EGR was varied from 5% (by vol.) to 15% (by vol.) in steps of 5% (by vol.). The quantity of water was varied from 5% (by vol.) to 15% (by vol.) in steps of 5% (by vol.). Water-emulsified diesel fuel with EGR technology has been proven to reduce nitrogen oxides (NOx) and smoke simultaneously at relatively low cost compared to other pollution-reducing strategies. From the test results, it is observed that water emulsion and EGR technology results in a substantial reduction of NO_x by 66% and smoke emissions by 50% with a compromise of 2.5% on thermal efficiency.

Keywords: Karanji Oil Methyl Ester, Water Emulsion, EGR, NO_x.

---

* E-mail ID: sp_anna2006@yahoo.co.in, Phone: +91-44-222427019, Mobile: +91-9444489013

1. INTRODUCTION

Diesel engines are very attractive for the transportation vehicle since they produce high torque output, high thermal efficiency, and consume relatively cheap fuel. Vegetable oils can be used as a fuel for running diesel engines. The number of vegetable oils like karanji oil, rapeseed oil, cottonseed oil, etc has been tested as fuels in diesel engines [2, 3, 4, 7, 11, and 13]. The major problem with the direct use of karanji oil as a fuel in diesel engines is their high viscosity, lower volatility and lower calorific value [4, 13]. Nevertheless, diesel engines produces high pollution mainly oxides of nitrogen (NO_x) and particulate matter (PM). In order to curb this problem, the addition of water is proven to be one of the effective approaches to reduce pollutant formation while at the same time an economic way of reducing NO_x and PM [7, 12, and 15].

Exhaust gas recirculation is one of the well known methods to reduce NO_x formation. Gases can be recirculated either in hot condition called as hot EGR or in cold condition called as cold EGR. The basic principle of the EGR is diversion of some of the exhaust gas back into the intake stream of the engine. This will reduce the intake oxygen concentration and increase the specific heat of the intake charge due to replacement of intake oxygen and nitrogen with CO_2 and H_2O. Due to this, peak combustion temperature will be reduced and this ultimately reduces the NO_x formation [5, 10, and 14]. In the present investigation, cold EGR was adopted.

1.1. Characteristics of Karanji Oil

Karanji oil is extracted from a seed of the Karanji tree, the botanical name for which is "Pinnata Pongama" or "Pongamia", which belongs to the family of Leguminous. Karanji oil or Pongamia oil is a deep yellow or reddish brown in colour. It grows all over the country, from the coastline to the hill slopes. It needs very little care and cattle do not browse it. Ten trees can yield 400 litres of oil, 1200 kg of biomass as green manure per year. It can be used for diesel engines in place of imported diesel [4]. The karanji plant with leaves, buds and seeds is shown in Figure1.

Figure 1. Karanji plant with leaves, buds and seeds.

1.2. Transesterification Process

The molecular weight of karanji oil is 803 gm/mole. Every kilogram of karanji oil requires 240 grams of methanol (6:1 molar ratio to oil). A block diagram illustrating the process of producing biodiesel is given in Figure 2. For transesterification process in laboratory, 1 kg of karanji oil was taken in a round bottom flask; 8 gm of NaOH was dissolved in 240 gm of methanol in a separate vessel, which was poured into round bottom flask while stirring the mixture continuously using constant stirring mechanism.

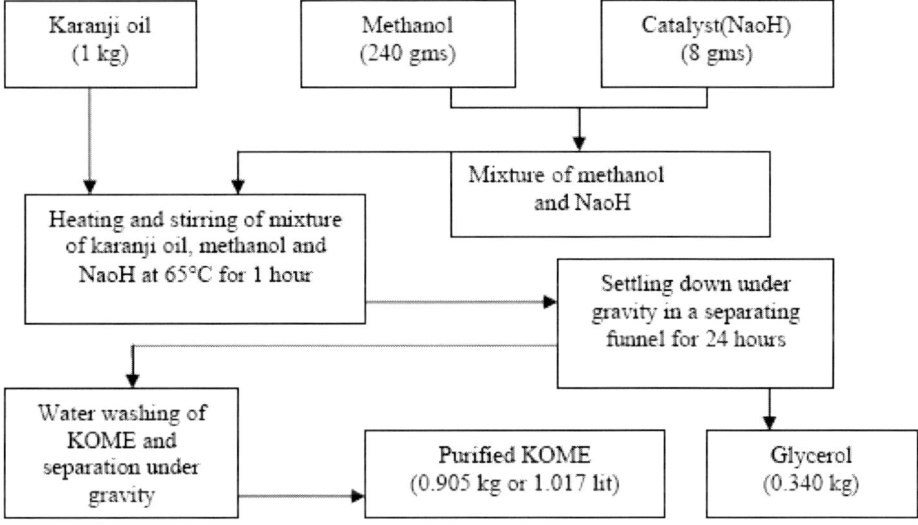

Figure 2. Block diagram illustrating the process of producing karanji oil methyl ester.

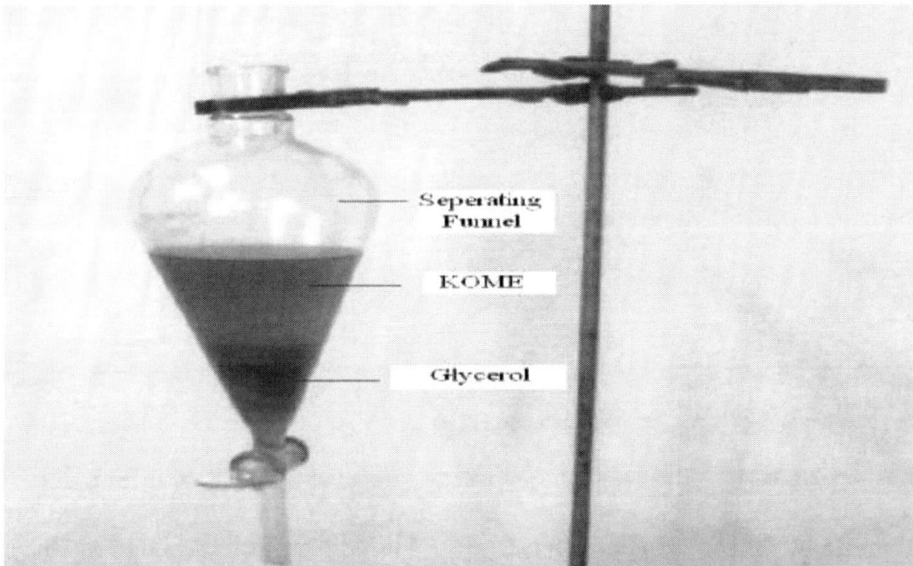

Figure 3. Photographic view of karanji oil methyl ester and glycerol seperation under gravity.

The mixture was stirred and maintained at 65°C for 1 hour and then allowed to settle down under gravity in a separating funnel. In the separating funnel, ester formed on the upper layer and glycerol formed on the lower layer. The photographic view of the KOME and glycerol separation under gravity in the separating funnel is shown in Figure 3. About 0.340 kg glycerol was separated from the mixture. The separated ester was mixed twice with 0.25 kg of hot water and allowed to settle under gravity for 24 hours. The catalyst got dissolved in water, which formed in the lower layer and was separated. About 0.905 kg of purified ester was obtained at the end. The time required for biodiesel production was about 48 hours [14].

1.3. Types of Emulsions

There are two basic types of emulsions in which water and fuel are combined. Water blend fuel (WBF) is a fuel continuous emulsion (water in fuel type). A second type is an aqueous or water continuous emulsion (fuel in water type). When using WBF the engine fuel system recognizes the liquid as KOME fuel because the water droplet (less than 1 micron in size) is encapsulated within a KOME fuel. Water blend KOME fuel also offers fuel lubricity and corrosion characteristics [6]. There are two types of emulsion. One is fuel-in-water type in which the fuel is the dispersed phase and water is the dispersion medium. The other one is water-in-fuel type is shown in Figure 4 in which water is the dispersed phase and fuel is the dispersion medium.

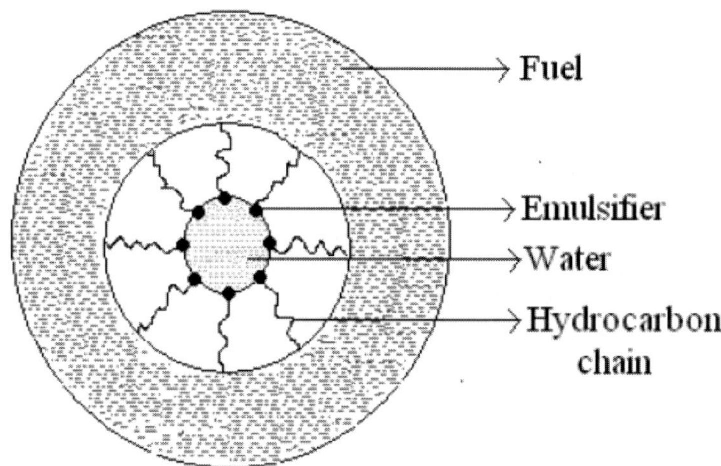

Figure 4. Water-in-fuel type.

1.4. Micro-Explosion

The fundamental reason for the reduced particulate formation is attributed to a phenomenon known as micro-explosion. The micro explosion in water blend KOME fuel is shown in Figure 5. Micro-explosion in water blend KOME fuel is the result of instantaneous vaporization of the water droplets within the fuel droplet as the fuel is exposed to increasing in cylinder temperature during injection. Once the mean temperature of the fuel droplet increases above the boiling point of water, the water quickly and violently evaporates, breaking the droplet into smaller droplets, which results in a more complete vaporization and turbulent mixing of the fuel. Surfactants and other components (e.g., Ethanol or Methanol) enhance the micro-explosion process. Any interaction that reduces the transportation of lower boiling compounds to the surface reduces the rate of diffusion. This process increases the probability of rapid boiling and bubble formation, which leads to enhanced micro-explosion of the fuel droplet. As the water blend fuel enhances fuel atomization, it leads to the selective reduction of the insoluble carbon fraction of the particulate [6].

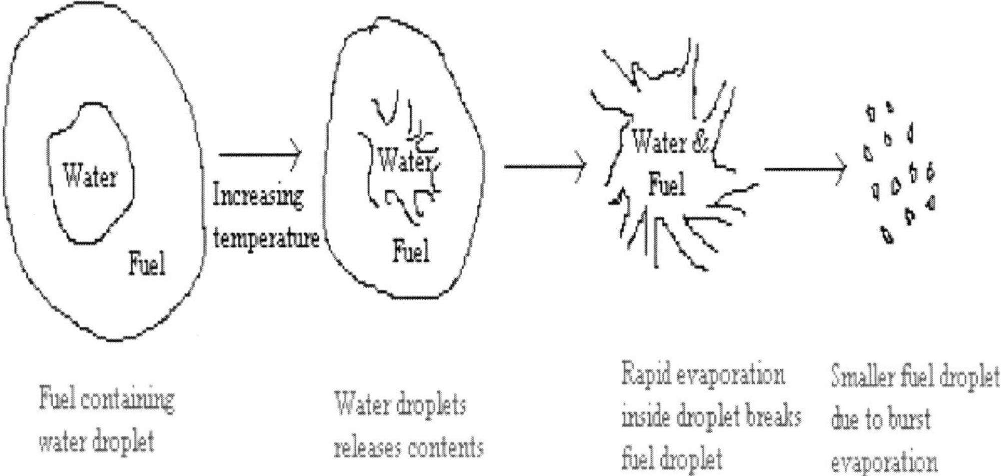

Figure 5. Micro-Explosion.

1.5. Types of Emulsifiers

The following are the different types of emulsifiers that are available. They are 1. Anionic 2. Cationic 3. Non- ionic. Only non ionic emulsifying agents are suggested for preparing emulsified fuel for engine applications owing to its non reactive and non corrosive nature without any source for secondary pollutants formation in engines.

1.6. Emulsifier Used in the Present Work

Using a surfactant SPAN 20 the water KOME emulsion was prepared. The amount of surfactant added was 1% by weight and the stability time was about 18-20 hours. The photographic view of the water-KOME emulsion preparation in magnetic stirrer with hot plate apparatus is shown in Figure 6. It consists of a magnetic stirrer, hot plate, holder for holding the mixing jar and a knob for adjusting the current and temperature. By varying the current, the speed of magnetic stirrer can be increased or decreased. The measured quantities of KOME, water and surfactant were poured into the mixing jar and the mixture was stirred for half an hour at 65°C. A good emulsion was prepared by this method and emulsion was injected into the engine using the conventional diesel fuel injector.

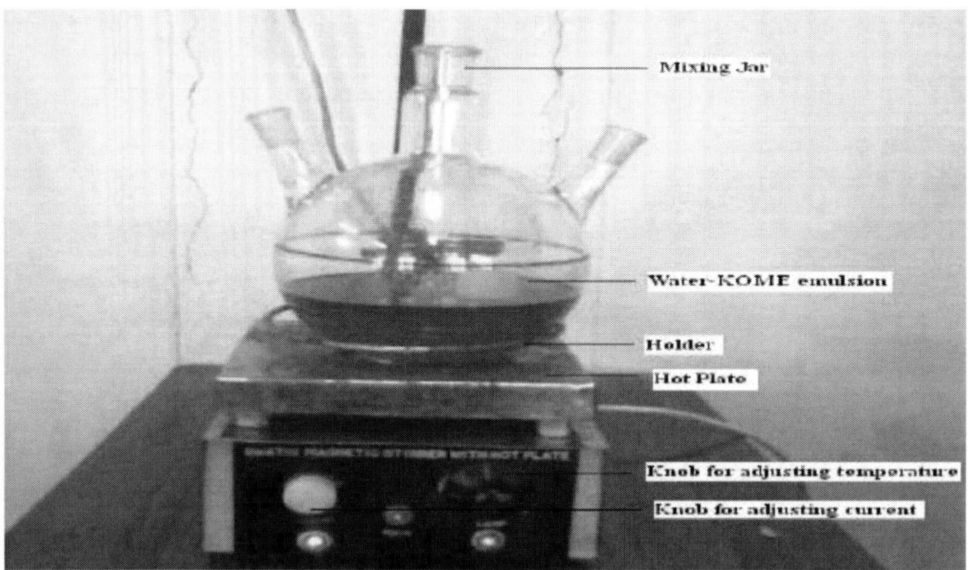

Figure 6. Photographic View Of The Water- Karanji Oil Methyl Ester Emulsion Preparation.

2. PRESENT WORK

The aim of this work was to reduce the NO_x and smoke emissions when water-KOME emulsion is used as the fuel in a DI diesel engine with EGR. Initially the experiments were conducted with the water-KOME emulsion as a fuel in a DI diesel engine without EGR. The quantity of water was varied from 5 to 15% in steps of 5% (by vol.). This quantity of water addition was selected based on the results reported in literatures. From the test results, the optimised water emulsion of karanji oil methyl ester was chosen as a fuel and further experiments were conducted in a DI diesel engine with EGR. The quantity of EGR was varied from 5 to 15% in steps of 5% (by vol.). This quantity of EGR variation was selected based on the results reported in the literatures. The optimum quantity of water addition in KOME and optimum quantity of EGR to be used in a DI diesel engine for the reduction of NO_x and smoke emissions has been studied. Through out the experiment, the static injection timing was set at 23° bTDC (before Top Dead Centre), as that of conventional diesel engine.

2.1. Experimental Setup and Methodology

The schematic of the experimental setup is shown in Figure 7. The specifications of the engine are given in Table 3. A single cylinder 4-stroke air cooled diesel engine having a compression ratio of 17.5:1 and developing 4.4 kW at 1500 rpm was used. The engine was coupled to a swinging type electrical dynamometer for loading the engine.

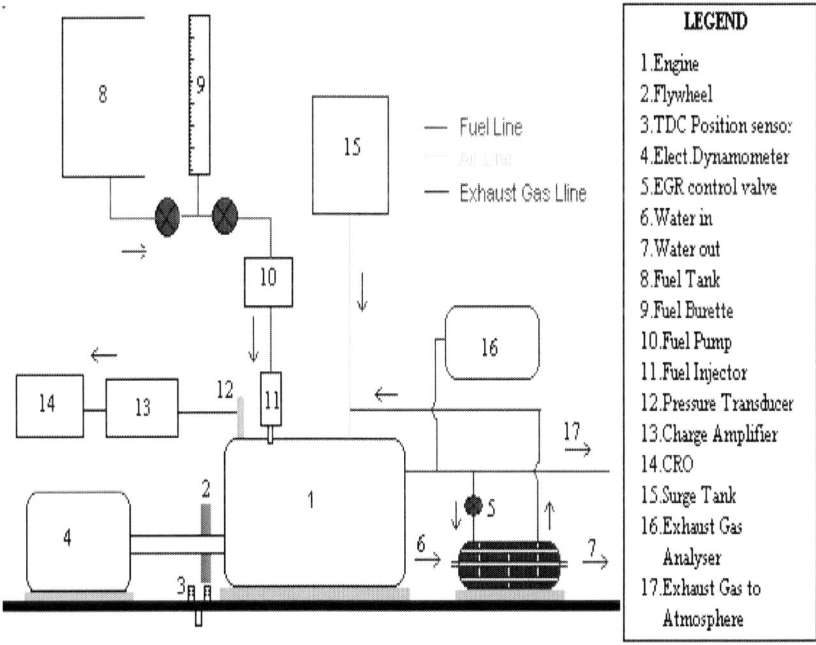

Figure 7. Schematic Of The Experimental Setup.

Table 1. Fatty Acid Composition of Karanji Oil

Sl. No	Fatty acid	Structure	Formula	Wt %
1	Palmitic	16:0	$C_{16}H_{32}O_2$	3.7–7.9
2	Stearic	18:0	$C_{18}H_{36}O_2$	2.4–8.9
3	Lignoceric	24:0	$C_{24}H_{48}O_2$	1.1–3.5
4	Oleic	18:1	$C_{18}H_{34}O_2$	44.5–71.3
5	Linoleic	18:2	$C_{18}H_{32}O_2$	10.8–18.3

Table 2. Properties of tested fuels

Properties	Diesel	KOME + 5% water emulsion	KOME + 10% water emulsion	KOME + 15% water emulsion
Calorific value (MJ/Kg)	44.32	44.02	43.91	43.76
Viscosity (cS)	22.5	24.9	29.8	33.6
Specific gravity	0.8456	0.8512	0.8536	0.856
Flash point (°c)	45	52	61	70
Fire point (°c)	64	73	80	88

Table 3. Engine specifications

Make	Kirloskar
No. of Cylinder	Single Cylinder
Stroke	4 Stroke
Type of Cooling	Air Cooled
Ignition	Compression Ignition
Fueling	Diesel
Bore	87.5 mm
Stroke	110 mm
Compression Ratio	17.5:1
Speed	1500 rpm
Brake Power	4.4 kW
Injection pressure	200 bar
Injection timing	23 ° bTDC

Table 4. Specifications of exhaust gas analyzer

Pollutant	Instrument	Type	Operating principle	Range
CO	QROTECH	QRO- 401	NDIR	0.00- 9.99 %
HC	QROTECH	QRO- 401	NDIR	0 –9999 ppm
CO2	QROTECH	QRO- 401	NDIR	0 – 20.0 %
O_2	QROTECH	QRO- 401	Electrochemical Method	0.00- 25.0 %
NO_X	QROTECH	QRO- 401	Electrochemical Method	0-9999ppm

Exhaust gas analyzer was used to measure the HC, CO, CO2 and NOx levels in the exhaust which operates on the principle of non dispersive infra red method and electrochemical method. The specifications of exhaust gas analyzer are given in Table 4. Smoke levels were obtained using a standard Bosch smoke measuring apparatus. The specifications of exhaust gas analyzer are given in Table 4.

A cold EGR set up was fitted to the exhaust pipe at one end and to the inlet manifold at the other, to deliver the recirculated exhaust gas back to the engine via a control valve is shown in Figure 8. The EGR rate is defined as the ratio between the amounts of exhaust gas recirculated to the total charge intake

$$\text{EGR rate} = \frac{[CO2]_{\text{Intake Gas}} - [CO2]_{\text{Ambient}}}{[CO2]_{\text{Exhaust gas}} - [CO2]_{\text{Ambient}}}$$

The exhaust gas CO_2 level is measured first and the required CO_2 level in the intake charge is calculated based on the required EGR rate. Then EGR valve is opened until the required CO_2 level reached in the intake charge.

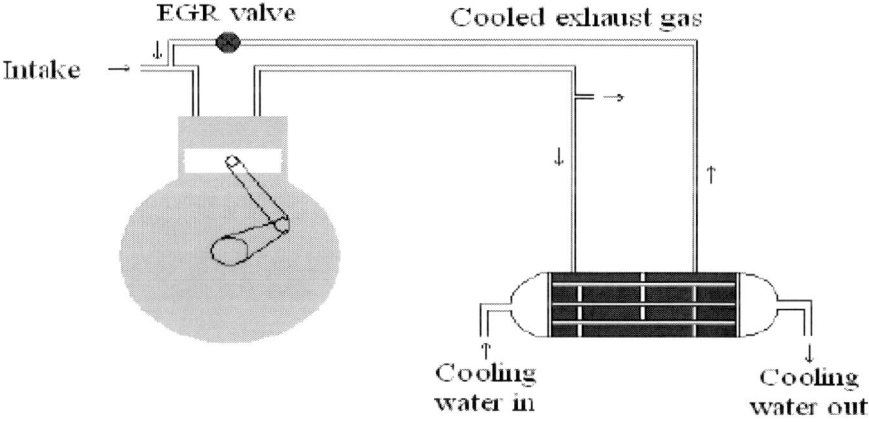

Figure 8. Cold egr setup.

3. RESULTS AND DISCUSSION

Figure 9 shows the variation of brake specific energy consumption with load. It is observed from the figure that 15% water emulsion of KOME has higher specific energy consumption as compared to diesel fuel. At full load, the BSEC of diesel fuel is 13.5 MJ/kWh where as for 15% water emulsion of KOME is 16.6 MJ/kWh. As calorific value becomes less, more fuel needed to produce targeted power (equivalent energy), thus the A/F ratio was shifted to fuel rich mixture.

There is also a direct correlation between fuel calorific value and fuel energy consumption in which as the calorific value becomes lower and it leads to higher specific energy consumption as shown in Figure 9.

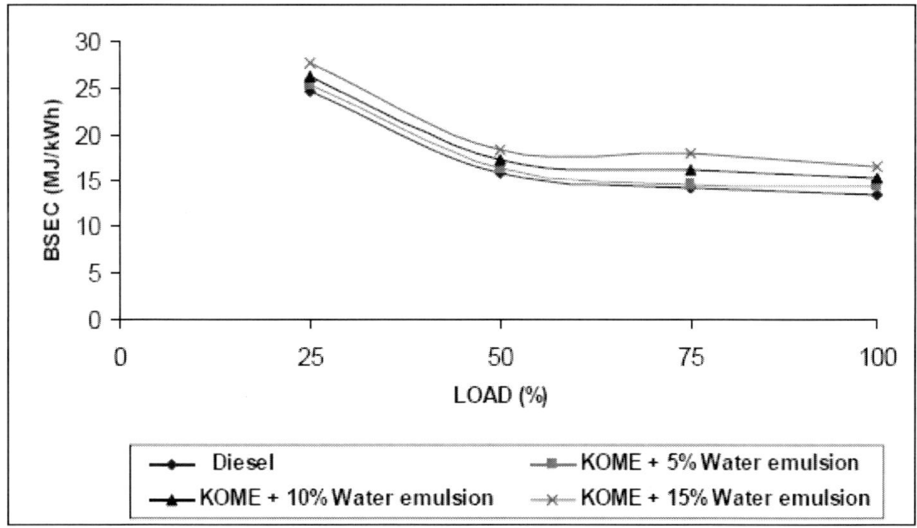

Figure 9. Variation of brake specific energy consumption with load.

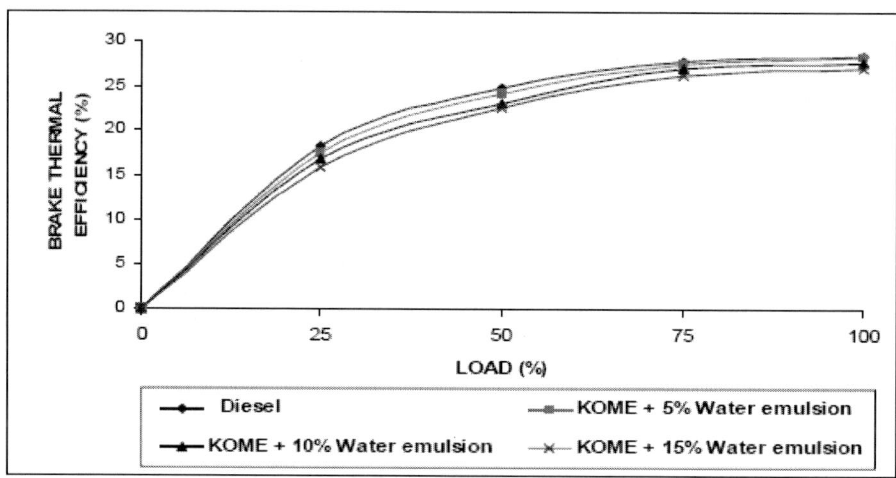

Figure 10. Variation of brake thermal efficiency with load.

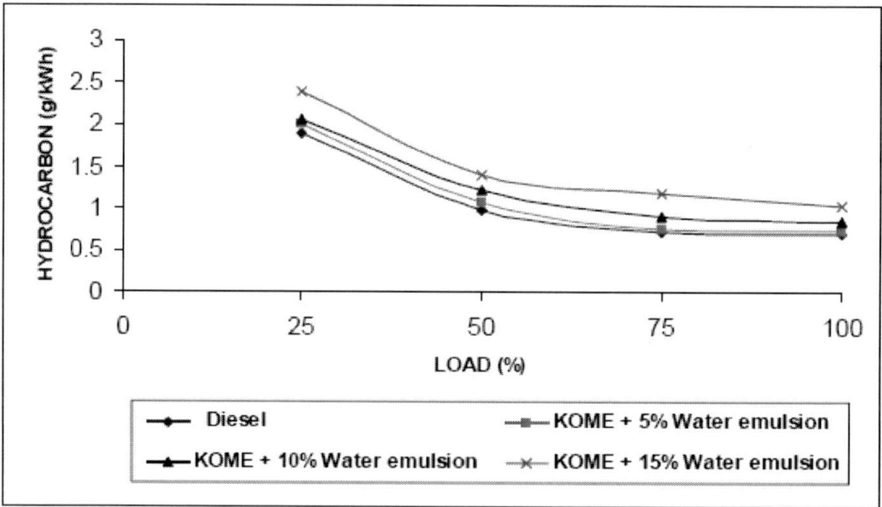

Figure 11. Variation of hydrocarbon with load.

Figure 10 depicts the variation of brake thermal efficiency with load for diesel and various amount of water addition in karanji oil methyl ester. It is observed from the figure that the brake thermal efficiency of diesel fuel is higher as compared to water emulsion of KOME. At full load, the brake thermal efficiency of the diesel fuel is 28.5% where as for 15% water emulsion of KOME is 27%. The brake thermal efficiency of the engine reduces by 1.5% and this may due to the lower calorific value and presence of more amount of water content in 15% water emulsion of KOME.

Generally the effect of fuel viscosity on fuel spray quality would be expected to increase in HC and CO emissions [7]. The variation of unburnt hydrocarbon emission with load is shown in Figure 11. It is observed from the figure that 15 % water emulsion of KOME shows higher hydrocarbon emission as compared to diesel fuel. At full load, the hydrocarbon emission in the case of diesel fuel is 0.701g/kWh where as for 15% water emulsion of KOME it is 1.031 g/kWh. This may probably be due to higher viscosity of 15% water emulsion of KOME.

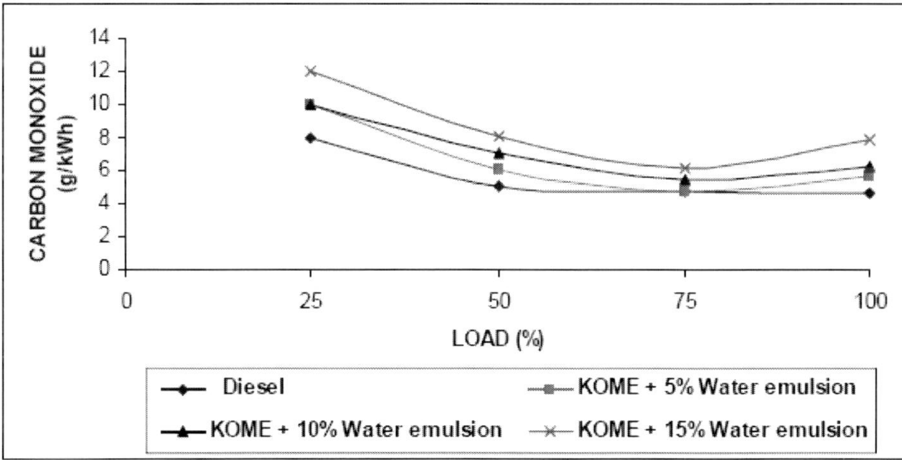

Figure 12. Variation of carbon monoxide with load.

Figure 12 shows the variation of carbon monoxide emission with load. It is observed from the figure that 15% water emulsion KOME gives higher carbon monoxide emission as compared to diesel fuel. At full load the carbon monoxide emission in the case of diesel fuel is 4.63 g/kWh where as for 15% water emulsion of KOME it is 7.82 g/kWh. This trend can be explained by the reason of phenomenon called incomplete combustion, which means that increased level of water addition to fuel decreases the burning efficiency and this leads to increase in CO emission.

The variation of oxides of nitrogen with load for diesel and various amount of water addition in KOME is depicted in Figure 13. It is observed from the figure that the level of NO_x emission decreases as the amount of water present in KOME increases. At full load, the NO_x emission of diesel fuel is 10.11 g/kWh where as for 15% water emulsion of KOME it is 6.73 g/kWh. The NO_x emission of the engine reduces by 33% under full load when using 15% water emulsion of KOME. This is because of the presence of more water in KOME which reduces the combustion temperature inside the cylinder and this ultimately reduces the oxides of nitrogen emission.

Figure 14 portrays the variation of smoke intensity with load. It is observed from the figure that the smoke intensity decreases as the amount of water present in the KOME increases. At full load, the smoke intensity in the case of diesel fuel is 5BSN where as for 15% water emulsion of KOME it is 2.3 i.e., 54% decrease in smoke emission. This is due to the phenomenon of microexplosion and increase in excess air ratio i.e., the presence of oxygen molecule in the fuel and water leads to decrease in smoke emissions.

Figure 15 depicts the variation of exhaust gas temperature with load. It is observed from the figure that the exhaust gas temperature decreases as the amount of water present in KOME increases. This is due to lower calorific value of fuel. At full load, the exhaust gas temperature in the case of diesel fuel is 372° where as for 15% water emulsion of KOME it is 360°C.

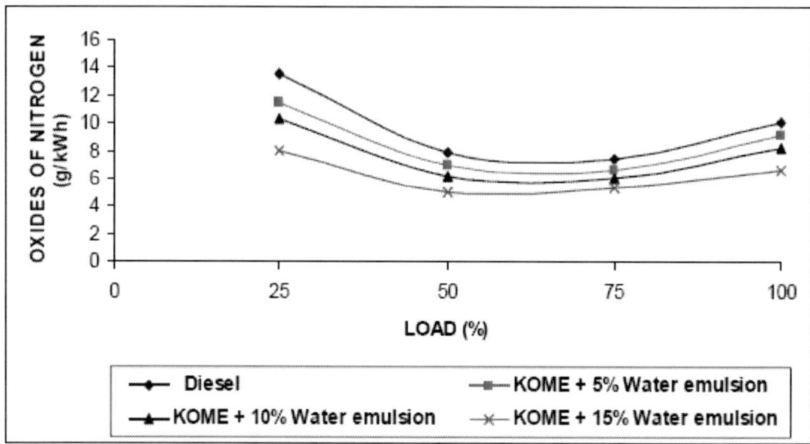

Figure 13. Variation of oxides of nitrogen with load.

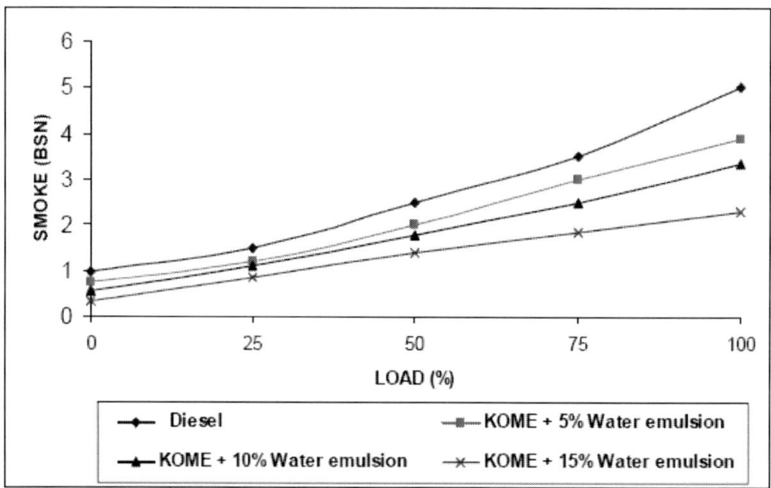

Figure 14. Variation of smoke with load.

From the experiments conducted, it is noticed that the use of 15% water emulsion of KOME reduces the NOx from 10.11 g/kWh to 6.73 g/kWh and smoke from 5 BSN to 2.3 BSN with a compromise of 1.5% of thermal efficiency as compared to that of diesel fuel. This investigation was further continued by using 15% water emulsion of KOME as a fuel in a DI Diesel engine with EGR for further reduction of NOx.

Figure 16 shows the variation of brake specific energy consumption with load at various EGR rate. It is observed from the figure that the increase in EGR rate shows that there is a slight increase in BSEC. At full load, the BSEC in the case of diesel fuel is 13.5 MJ/kWh where as for 15% water emulsion of KOME at 15% EGR rate it is 17.1 MJ/kWh i.e., about 7% increase in energy consumption. This may probably be due to lower energy content of the fuel which has resulted in more fuel consumption.

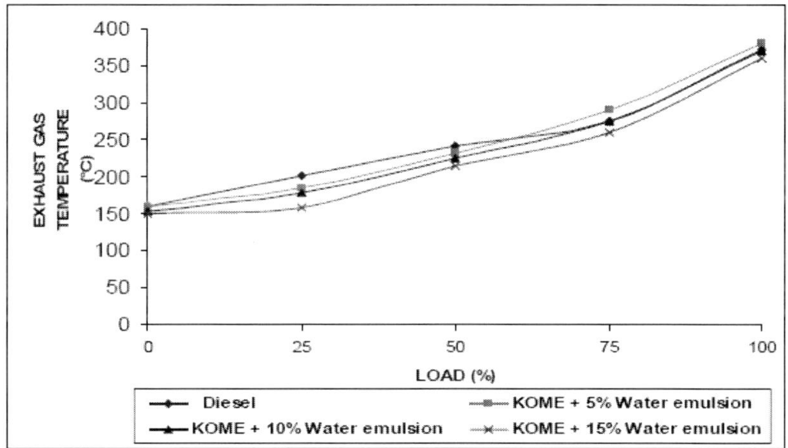

Figure 15. Variation of exhaust gas temperature with load.

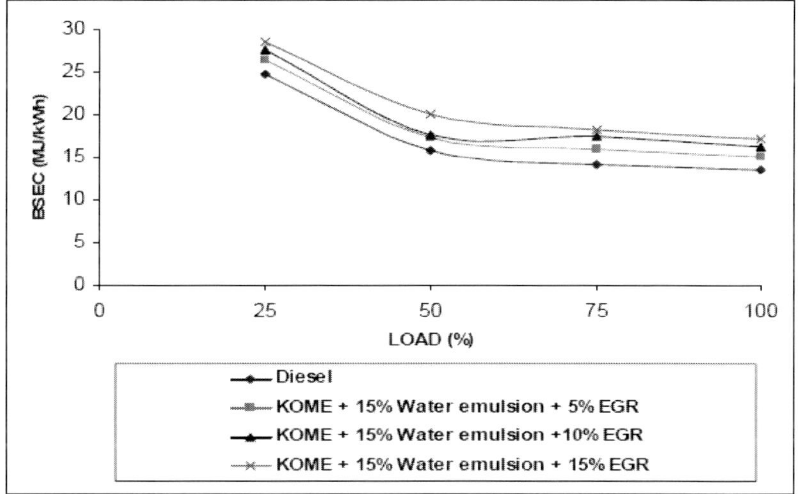

Figure 16. Variation of brake specific energy consumption with load.

The variation of brake thermal efficiency with load at various EGR rate is shown in Figure 17. From the test result, it is observed from the figure that brake thermal efficiency decreases with an increasing EGR rate. At full load, the maximum thermal efficiency of the diesel fuel is obtained as 28.5% where as for 15% water emulsion of KOME it is 26% at 15% EGR rate. Thus the brake thermal efficiency of the engine reduces by 2.5% under full load at 15% EGR rate. This is due to the replacement of oxygen by cooled exhaust gas in the intake air which causes drastic reduction in combustion temperature which leads to reduction in thermal efficiency.

Figure 18 depicts the variation of hydrocarbon emission with load at various EGR rate. It can be seen from the figure that the increase in EGR rate increases the hydrocarbon emission drastically. At full load condition, the hydrocarbon emission at 15 % EGR rate is increased to 61.25 % as compared to that of diesel fuel.

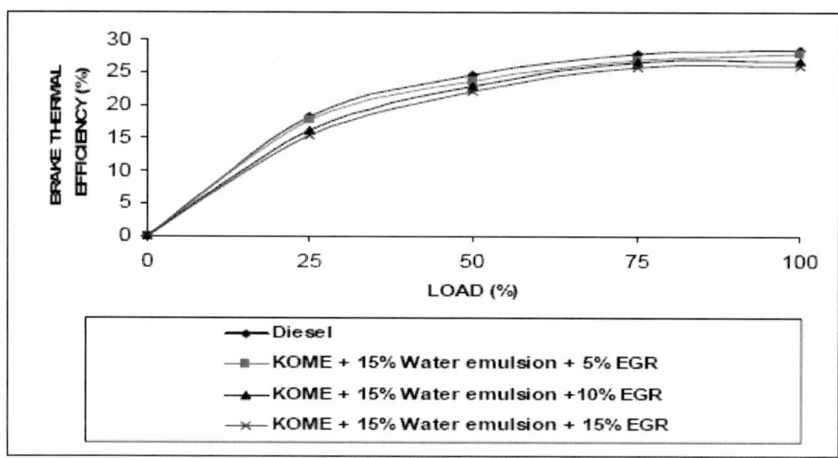

Figure 17. Variation of brake thermal efficiency with load.

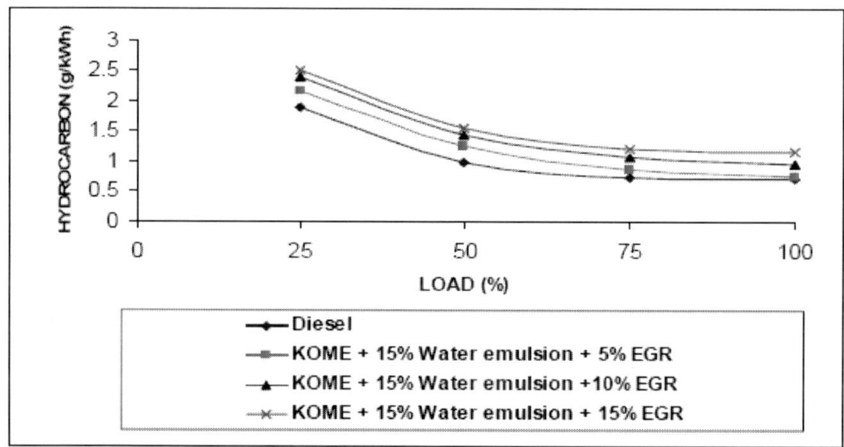

Figure 18. Variation of hydrocarbon with load.

The variation of carbon monoxide emission with load is depicted in Figure 19. It can be seen from the figure that the CO emission of the diesel fuel is 4.63 g/kWh at full load and 21.94 g/kWh for the 15 % water emulsion of KOME at 15% EGR rate. The carbon monoxide emission is high for 15% water emulsion of KOME at 15% EGR rate as compared to that of diesel fuel.

Figure 20 shows the variation of oxides of nitrogen with load at various EGR rate. It is observed from the figure that the NOx emission decreases rapidly with increasing EGR rate. At full load, the NOx emission in the case of diesel fuel is 10.11 g/kWh where as for 15% water emulsion of KOME at 15% EGR rate it is 3.47 g/kWh. The NOx emission of the engine reduces by 66% under full load at 10% EGR rate with a compromise of 2.5% of thermal efficiency. This is due to reduction in intake oxygen by the replacement of intake oxygen and nitrogen by carbondioxide and water. Due to this peak combustion temperature was reduced and this ultimately reduces the NOx formation.

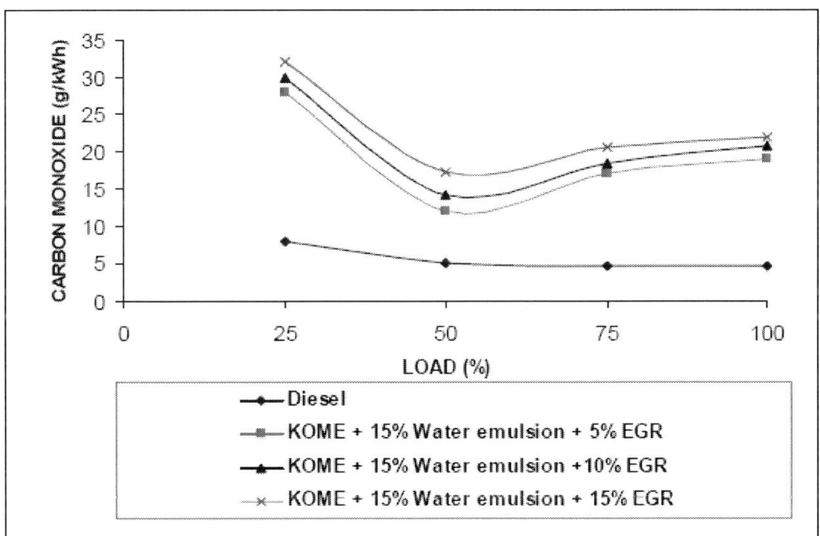

Figure 19. Variation of carbon monoxide with load.

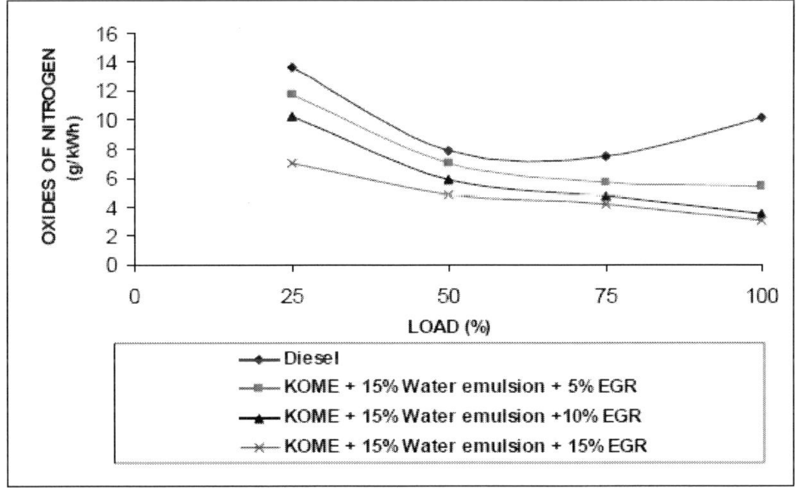

Figure 20. Variation of oxides of nitrogen with load.

The variation of smoke intensity with load at various EGR rate is depicted in Figure 21. It is observed from the figure that the smoke intensity increases with the increasing EGR rate at full load, the smoke intensity in the case of diesel fuel is 5BSN where as for 15% water emulsion of KOME at 10% EGR rate it is 2.5BSN. The smoke intensity of the engine reduces by 50% under full load using 15% water emulsion of KOME as a fuel at 10% EGR rate with a penalty of 2.5% of thermal efficiency.

The increase in HC, CO and smoke is due to the reduced oxygen content of the cylinder charge [14]. Here in these trends are due to the replacement of oxygen by CO_2 and H_2O which in turns to reduce the oxygen molecule in the intake air leads to increase in HC, CO and smoke. The effects on smoke are more difficult to explain. Smoke production in diesel engines is a very complex phenomenon involving a number of conflicting factors. Generally speaking a decrease in premixed combustion would tend to increase smoke and vice versa.

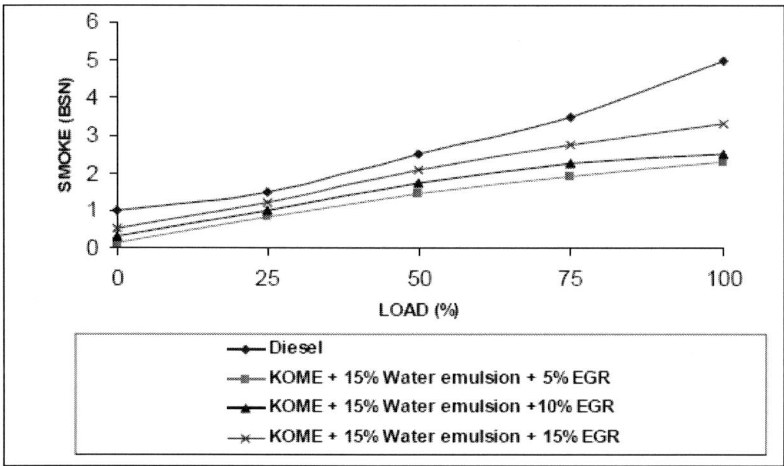

Figure 21. Variation of smoke with load.

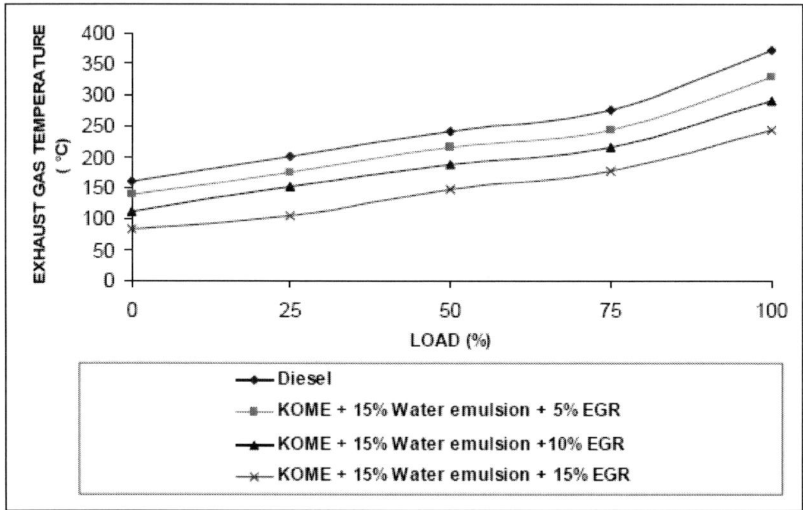

Figure 22. Variation of exhaust gas temperature with load.

Figure 22 portrays the variation of exhaust gas temperature with load at various EGR rate. It is observed from the figure that the EGT increases as the load increases. At full load the exhaust gas temperature in the case of diesel fuel is 372°C where as for the 15% water emulsion of KOME it is 244°C. This is due to the lower calorific value of the fuel. Since the specific heat capacity of the exhaust gas is more and it absorbs the more heat thereby reducing the high temperature and pressure available at the time of combustion.

CONCLUSIONS

Significant conclusions based on experiments conducted using water-KOME emulsions with and without EGR are presented below:

- The use of water-KOME emulsions without EGR decreases the brake thermal efficiency of 1.5 % at full load as compared to that of diesel fuel.
- The use of water-KOME emulsions without EGR increases the brake specific energy consumption, hydrocarbon and carbon monoxide emissions marginally.
- The use of 15 % water emulsion of KOME without EGR decreases the oxides of nitrogen from 10.11 g/kWh to 6.73 g/kWh as compared to that of diesel fuel
- The use of 15 % water emulsion of KOME without EGR decreases the smoke intensity from 5 BSN to 2.3 BSN as compared to that of diesel fuel
- The use of 15 % water emulsion of KOME at 10 % EGR decreases the oxides of nitrogen from 10.11 g/kWh to 3.47 g/kWh as compared to that of diesel fuel
- The use of 15 % water emulsion of KOME at 10 % EGR decreases the smoke intensity from 5 BSN to 2.5 BSN as compared to that of diesel fuel

On the whole it is concluded that using 15 % water emulsion of KOME at 10 % EGR seems to be optimal and it reduces the NO_x and smoke emissions drastically. An improvement in performance and further reduction in emission can be expected by the usage of oxygen enriched air and by altering the injection timing.

ABBREVIATIONS

BTDC	– BEFORE TOP DEAD CENTRE
BSEC	– BRAKE SPECIFIC ENERGY CONSUMPTION
BSN	– BOSCH SMOKE NUMBER
CO	– CARBON MONOXIDE
CO_2	– CARBONDIOXIDE
CRO	– CATHODE RAY OSCILLOGRAPH
EGR	– EXHAUST GAS RECIRCULATION
EGT	– EXHAUST GAS TEMPERATURE
HC	– HYDROCARBON
H_2O	– WATER
KOME	– KARANJI OIL METHYL ESTER
NO_x	– OXIDES OF NITROGEN
O_2	– OXYGEN
PM	– PARTICULATE MATTER
PPM	– PARTS PER MILLION
WBF	– WATER BLEND FUEL

REFERENCES

[1] M. Alam, J. Song, R. Acharya and A. Boehman, K. Miller, "Combustion and Emissions Performance of Low Sulfur, Ultra Low Sulfur and Biodiesel Blends in a DI Diesel Engine" SAE No- 2004-01-3024.

[2] A.K. Babu and G. Devaradjane, "Vegetable oils and Their Derivatives As Fuels for CI Engines: An Overview" SAE No- 2003-01-0767.

[3] Diane L. Lance and Jon D. Anderson, "Emissions Performance of Pure Vegetable Oil in Two European Light Duty Vehicles" SAE No- 2004-01-1881.

[4] Joel Hemanth, " Performance Testing of Diesel Engine using Honge Oil (Pinnata Pongama) an Alternate Fuel for Diesel" SAE No-2005-01-3132.

[5] Kazuyuki Narusawa, Matsuo Odaka, Noriyuki Koike, Yujiro Tsukamoto and Koichi Yoshida, "An EGR Control method for Heavy duty Diesel Engines under Transient Operations" SAE No- 900444.

[6] Kevin F. Brown and John Chadderton, Daniel T. Daly and Deborah A. Langer, David Duncan, "Opportunity for Diesel Emission Reductions Using Advanced Catalysts and Water Blend Fuel" SAE No-2000-01-0182.

[7] M. Leenus Jesu Martin, D. Prithviraj and K.C. Velappan, "Performance and Emission Characteristics of a CI Engine Fueled with Esterified Cottonseed Oil. SAE No- 2005-26-355.

[8] Manuel A. Gonzalez D., Hercilio Rivas, Xiomara Gutierrez and Aymara Leon, "Performance and Emissions using Water in Diesel Fuel Microemulsion" SAE No-2001-01-3525.

[9] M. Muralidharan, Mathew. P. Thariyan, Sumit Roy, J.P. Subramanyam, P.M.V Subbarao, "Use of Pongamia Biodiesel in CI Engines for Rural Applications". SAE No-2004-28-0030.

[10] M.A. A Nazha, H. Rajakaruna and S.A. Wagstaff, "The use of Emulsion, Water Induction and EGR for Controlling Diesel Engine Emissions" SAE No-2001-01-1941.

[11] Norbert Hemmerlein, Volker Korte and Herwig Ritcher and Gunter Schroder, "Performance, Exhaust Emissions and Durability of Modern Diesel Engines Running on Rapeseed Oil" SAE No-910848.

[12] K.A. Subramanian and A. Ramesh, "Experimental Investigation On The Use of Water Diesel Emulsion With Oxygen Enriched Air in a DI Diesel Engine" SAE No-2001-01-0205.

[13] J.G. Suryawanshi and N.V. Deshpande, "Experimental Investigation On a Pongamia Oil Methyl Ester Fuelled Diesel Engine". SAE No-2004-28-0018.

[14] J.G. Suryawanshi and N.V. Deshpande, "Overview of EGR, Injection Timing and Pressure on Emissions and Performance of CI engine with Pongamia Methyl Ester". SAE No- 2005-26-346.

[15] Yasufumi Yoshimoto, Masayuki Onodera and Hiroya Tamaki, "Reduction of NOx, Smoke and BSFC in a Diesel Engine Fueled by Biodiesel Emulsion With Used Frying Oil". SAE. No – 1999-01-3598.

In: Biodiesel: Blends, Properties and Applications
Editors: Jorge M. Marchetti and Zhen Fang

ISBN: 978-1-63117-024-9
© 2014 Nova Science Publishers, Inc.

Chapter 13

LIFE CYCLE ANALYSIS OF BIODIESEL AND OTHER BIOFUELS: A REVIEW

Laura C. Yu[1], Dennis Y.C. Leung[1] and Michael K.H. Leung[2]
[1]Department of Mechanical Engineering, the University of Hong Kong,
Hong Kong, China
[2]School of Energy & Environment, City University of Hong Kong,
Hong Kong, China

ABSTRACT

At a time when the use of fossil fuel means increasing energy scarcity and an environmental crisis in the world in which we live, we need green innovations now more than ever. Growing attention has been drawn to the use of biofuels, such as bioethanol and biodiesel, which have gradually come to make up part of the total energy supply. Despite its rapid development, uncertainties about the environmental and ecological aspects of the production and consumption of biofuel still exist. A life cycle analysis (LCA), with deliberate system boundaries and life cycle inventories, has been applied to evaluate two principal functional parameters, the energy efficiency and GHG balance, from the cradle to the grave of different biofuel feedstocks. However, the results of the entire lifetime estimates varied dramatically in production chains, which make it difficult to take a holistic view about energy intake and yields, economic costs and values, environmental impacts and their benefits. Apart from the diversity in system boundaries and life cycle inventories, a variance in terminologies and the limitations of interdisciplinary communication are the main factors that affect the quality of the results. In this paper, various life cycle analysis models are reviewed and evaluated with a special emphasis on two important biofuels, i.e., biodiesel and ethanol.

Keywords: LCA modeling, greenhouse gases, GHG displacement, GWPs

1. INTRODUCTION

According to the World Energy Outlook, the demand for fossil fuels is poised to peak by 2020, and low-carbon fuels, including solar, wind, geothermal, hydropower and biofuels will possibly increase to about one third of the world's primary sources of energy by 2030 [1]. At present about 40% of the total energy use of the world is supplied by liquid petroleum fuels. Biofuels have been proposed as a suitable substitute for traditional liquid fossil fuels, because other renewable energy sources, like solar energy or wind energy, are used to produce electricity and heat instead of using liquid fuel [1]. Thus, biofuels have drawn a great deal of public attention in recent years, and are contenders to be one of the main low-carbon resources in the near future. Biofuels can be grouped into bio-alcohol (e.g., methanol, ethanol), bio-diesel (e.g., FAME), bio-ether (e.g., Dimethyl ether) and bio-gas (e.g., landfill gas) categories [1]. In order to accurately calculate the energy budgets and GHG impacts of different biofuel feedstocks and biofuel pathways, we review studies that apply life cycle analysis (LCA), which takes into account all the relevant upstream and downstream processes for assessing the comprehensive energy efficiency and GHG balance.

Life cycle analysis is conducted under various system boundaries, life cycle inventories and functional units. "System boundary" refers to the gathering of specific temporal and special boundaries, such as the accounted growing seasons of the plant operation and the area used to grow the biofuels. "Life cycle inventory" is defined as the parameters to be evaluated within the whole life cycle analysis, such as the energy input and output in plant farming, and GHG emission in biofuels conversion [2]. The functional units, as the most important metric of the life cycle analysis framework, refer to the particular units to be assessed, such as energy efficiency and GHG balance [3].

Within all the three of the defined principal frameworks, the results of the entire lifetime estimates still varied dramatically for the biofuel production chains, which make it difficult to take a holistic view of energy intake and yields, economic costs and values, and their environmental harms and benefits. Thus, several commonly used terminologies were developed to more intuitively evaluate the different outcomes of energy efficiency and GHG balance in different biofuel life cycle chains. The most widely used terms for energy efficiency include the Life Cycle Energy Efficiency (LCEE), Net Energy Value (NEV) and Fossil Energy Ratio (FER) [4-8]. The calculation of Life Cycle Energy Efficiency (LCEE) uses the ratio between fuel product energy and total primary energy. It investigates the feedstock energy-losing portion and the process energy added into the production chain; Net Energy Value (NEV) is another widely used term that estimates the difference between fuel product energy and primary energy. Negative NEV values suggest that less energy output can be used when compared with the energy input needed to produce it, and positive values mean the total energy obtained (a net energy gain) in the whole process. It calculates the additional process energy required to make the fuel and energy loss for feedstock energy in the whole life cycle [2, 9, 10]; Fossil Energy Ratio (FER) accounts for the ratio between the fuel product energy produced by the fossil energy inputs. A FER value less than 1 suggests a net energy loss while a value of more than 1 means a net energy gain. If the value tends out to infinity, then the fuel can be considered totally renewable. Thus this ratio can tell us about the renewable extent of fuel [11-13].

Table 1. Major features of various LCA models

LCA Model	GREET	LEM	GHGenius	SimaPro
Developing Organization	Transportation Research Argonne National Laboratory	University of California at Davis	Natural Resources Canada(NRCan)	PRé Consultants
Fuel Cycle Pathways	NG to CNG, LNG, LPG; coal, uranium, renewable energy to electricity; corn, woody biomass, herbaceous biomass to ethanol; soybeans to biodiesel; flared gas to methanol, dimethyl ether, Fischer-Tropsch diesel; landfill gases to methanol	Corn to ethanol; wood to ethanol; soybean to biodiesel; wood to methanol; wood to SNG, methanol, ethanol; switchgrass to ethanol, hydrogen; electricity generation; steam generation.	Corn, wood, grass, wheat to ethanol; wood to methanol, SNG; soybean to biodiesel; wood , corn, wheat to hydrogen	2500 processes in agricultural product, energy (fuels, electricity production, heat production), transportation
Spatial Boundary	Mainly US	More than 30 countries	Specific regions (east, central or west) of Canada, the US, Mexico and India	Based on "Ecoinvent" database
Temporal Boundary	1990- 2020	1970- 2050	1970- 2050	Based on "Ecoinvent" database
Functional Parameters	Total energy consumption; fossil energy consumption; six criteria pollutants	CO_2 -equivalent GHGs by stages; individual pollutant at different stages; ratio of process energy input to energy output for stages	Primary GHGs; criteria pollutants from combustion sources; total energy and fossil energy use	Cumulative energy demand (CED) by fuel types, GHG emissions for fuels
GHGs and CO_2-equivalency	IPCC GWPs for CO_2, CH_4 and N_2O; non-GHGs emissions	LEM CEFs for CO_2, CH_4 and N_2O; non-GHG emissions	LEM CEFs or IPCC GWPs	LEM CEFs or IPCC GWPs
Byproduct Treatment	Distillers' dried grains with soluble (DDGS) displaces corn; excess electricity displaces grid power	DDGS displaces corn grain for corn to ethanol; excess electricity displaces grid power for wood to ethanol and soy meal displace glycerin	DDGS displaces corn grain for corn to ethanol; excess electricity displaces grid power for wood to ethanol; soy meal displace glycerin	User defined
Land Use Changes	USDA analysis of land-use changes along with 1998 version of LEM's C sequestration rate estimation	Overall accounting of land use changes associated with impacts on soil and plant carbon	2003 version of LEM's carbon sequestration rate estimation	User defined
Nitrogen Cycle	NA	Comprehensive accounting of N flows, rates and impacts	Partial accounting of NOx emissions	User defined
Reference	[20-25]	[26-33]	[30, 32, 34, 35]	[40, 41]

GHG emissions can be classified into three different scopes, including direct emissions, energy indirect emissions and other indirect emissions in the biofuel energy chain. The typical sources are carbon dioxide (CO_2), Methane (CH_4), Nitrous Oxide (N_2O), Sulfur Hexafluoride (SF_6), Hydrochlorofluorocarbons (HFC) and Perfluorocarbons (PFC). There are several different expressions for the GHG balance estimates, such as GHG displacement, CO_2-equivalency factors (CEFs) which convert the mass emissions of the non-CO_2 gases into the mass of CO_2 with an equivalent effect on global climate change, and the Global Warming Potentials (GWPs) used by the Intergovernmental Panel on Climate Change (IPCC), and the Mg CO_{2eq} [14-16]. The GHG displacement indicates the amount of GHG emissions that are substituted by the use of biofuels in the replacement of fossil fuels. A negative value suggests a decrease in GHG emissions into the atmosphere and a positive value indicates an increase in GHG emissions [2]. The GWPs is a measure of the global warming effect of a GHG compared with CO_2 of the same mass. The overall environmental burdens raised by multiple GHGs can be estimated by the sum expressed in terms of the CO_2-equivalent, namely the Mg CO_{2eq}. It refers to the GHGs expressed in terms of million gram (= ton) of CO_2 emission, with the equivalent global warming potential of emitted greenhouse gases. Mg $CO_{2eq} = \Sigma_i\ mGHG_i$ *GWP_i, where GWP_i and $mGHG_i$ represent the GWP of a selected GHG_i and its mass, respectively [17-19].

2. LIFE CYCLE ANALYSIS MODELS

As a result of the various lifetime estimates, a number of life cycle analysis models have been developed to assess specific biofuel production chains in particular regions. These models have drawn a great deal of attention; their main features are discussed below, and detailed descriptions of them are given and summarized in Table 1.

2.1 GREET Model

The **G**reenhouse gases, **R**egulated **E**missions, and **E**nergy use in **T**ransportation (GREET) Model was developed by the Systems Assessment Section Center for Transportation Research Argonne National Laboratory in the U.S.. The GREET Model grew out of an original model that examines the energy and environmental impacts of renewable transportation fuels and advanced automotive technologies by the US Department of Energy (DOE) in the middle of the 1980s [25]. The formal version of the GREET model was released in 1996 for the first time, and updated versions were published on a continuous basis [21]. The new GREET Model is developed based on a Microsoft Excel spreadsheet format, which investigates the energy consumption and GHG emission of both well-to-pump and well-to-wheel (entire fuel and vehicles use) results [24]. The results, focused on the whole life cycle for a given vehicle and fuel system, rests on the following aspects: total energy consumption (including the energy in non-renewable and renewable sources), fossil energy consumption (petroleum, natural gas, and coal together), CO_2-equivalent greenhouse gas (GHG) emission (including CO_2, CH_4, N_2O) and six criteria pollutants (VOC_S, CO, NO_X, SO_X, PM_{10}, $PM_{2.5}$) [36]. The GREET has three series of life cycles: The first series (GREET 1.0, 1.1, 1.2, etc.)

calculates fuel-cycle energy use and emissions of light duty vehicles (LDVs); series two (GREET 2.0, 2.1, etc.) calculates vehicle-cycle energy use and emissions of light duty vehicles, especially for HEVs, in a combination of fuel-cycle results in series one and vehicle-cycle results. Series three (GREET 3.0, 3.1, etc) was designed to investigate the fuel-cycle energy use and emissions of heavy-duty vehicles. It takes relevant data for upstream activities from the series one component [20]. The most recent version of series one is the GREET1.8c.0, after some temporary versions, including GREET 1.1, 1.2, 1.3 and 1.4, that were developed before 1996. The advanced version, GREET 1.8, makes some changes in parametric assumptions and adds new fuel subjects and vehicle technologies. It contains more than 30 fuel cycles, 13 types of fuel feedstocks and 14 fuels [37].

In the GREET model, the user can select simulation years between 1990 and 2020, and vehicle types from passenger cars to light duty trucks. Fuels include petroleum, natural gas, bio-ethanol, hydrogen, biodiesel, electricity etc. Fuel pathways may have various compounding or blending rate options. Electricity pathway options, like marginal generation mix for transportation use and average generation mix for stationary use, include several default energy mix allocations of residual oil, natural gas and coal etc. Biofuel pathways include specific options for each biofuel. Vehicle Technology options include the engine types and blending level of gasoline and diesel. It is possible to model different proportions of alternatives fuels blended with conventional petroleum fuels [22, 23].

The GREET provides a session-in and a session-out spreadsheet per research objective. The session-in spreadsheet lists out biofuel pathways, fuel blends options, production assumptions and vehicle assumptions entered. The output spreadsheet lists out the well-to-pump energy consumption and emissions of fuel available at fuel station pumps, the well-to-wheel feedstock, fuel and vehicle operation energy consumption and emissions, and the relative well-to-wheel energy and emission change results for the target biofuel relative to conventional fuels. [20, 22, 23].

2.2. LEM Model

The Lifecycle Emissions Model (LEM) was developed in the early 1990s by the University of California at Davis, which investigates the energy consumption, criteria pollutants and GHG emissions for the complete lifecycle of fuels, materials, vehicles and infrastructures [28, 29]. The system boundaries accounted for in the life cycle analysis include the temporal boundary and spatial boundary. The life cycle research input data come from more than 30 countries, but most details are available for the spatial boundary of the US. The time frame covers any year from 1970 to an expected 2050 [32]. For the original purpose of evaluating electric vehicles and alternative fuel vehicles, it was improved several times in succession, and the most notable part of this new model were changes to data assumptions and model structure, additions of emission sources, and the additions of transportation modes, fuels, and vehicles to the Life Cycle Analysis system boundary [26]. Biomass pathways calculated in LEM include corn to ethanol, soybean to biodiesel, switchgrass to ethanol, wood to ethanol, wood to methanol, wood to synthetic natural gas, switchgrass to hydrogen etc. [33]. For the treatments of co-products, distillers' dried grains with soluble (DDGS) displace corn grain and soybeans for corn to ethanol, excess electricity displaces grid power for wood to ethanol and soy meal displaces glycerin. The whole life cycles are constructed as the following processes: feedstock cultivation, feedstock transport, fuel production, distribution and storage, dispensing of fuels, and fuel end use. They can be classified into three different levels: the farm level (e.g., land-use and cultivation emissions per hectare); the whole upstream fuel lifecycle ("well-to-tank"), and the entire fuel and vehicle use lifecycle ("well-to-wheel"). In the organization of the life cycle analysis in LEM, a general description of basic technical processes could be presented first, then the estimation of CO_2 -equivalent GHGs by stages (the level of farming, well-to-pump lifecycle and the whole well-to-wheel lifecycle), emissions of individual pollutants at different stages and fuel cycles and ratio of process energy input to energy output for selected stages [26, 27].

Life cycles investigated in LEM include the passenger transport mode, freight transport mode, electricity generation, steam generation, heating, cooling, etc. In transportation mode, the LEM calculates life cycle emissions for both passenger transportation and freight transportation. The former includes light-duty passenger cars, electric vehicles, buses, heavy and light-rail transit, and bicycles. The latter includes medium and heavy-duty trucks, diesel trains, tankers, cargos ships and pipelines etc. For motor vehicles, the fuel and feedstock combination used to calculate the lifecycle emissions of different end-use fuel, fuel feedstock and vehicle type combinations varied a great deal.

For the fuel, material, vehicle and infrastructure cycles in the LEM, the use of energy, the emissions of greenhouse gases and urban air pollutants are calculated for the whole life cycle. In the fuel and electricity lifecycle, feedstock production and transportation, fuel production, distribution and storage, fuel dispensing and its end-use are calculated in the process. As for materials, crude ore extraction, final material manufacture and its transportation to end-users are investigated in their life cycle. For vehicle cycles, the material use, vehicle assembly, operation and maintenance, secondary fuel cycle for transport modes (like serving, and administrative support) are taken into account in the life cycle. The lifecycle for infrastructure includes the energy use and emissions associated with materials production, infrastructure construction etc [26].

The GHG sources calculated in LEM include: fuel combustion that supplies process energy, leakage and evaporation of energy feedstocks or biofuels, fugitive dust emissions, chemical transformation that is not allied with burning process fuels, changes of carbon content in soils or biomass, and GHG emissions related to land use changes [26, 33]. The pollutants tracked in LEM are CO_2, CH_4, N_2O, CO, NO_X, non-methane organic compounds (NMOCs), sulfur dioxide (SO_2), total particulate matter (TPM), PM_{10}, chlorofluorocarbons (CFC-12), hydrofluorocarbons (HFC-134a), and the CO_2-equivalent GHGs from a wide range of energy and material life cycles [27]. In LEM, pollutants are first calculated individually and then converted into CO_2-equivalent GHGs, CEFs and GWPs.

2.3. GHGenius Model

The GHGenius Model was initially developed for Natural Resources Canada (NRCan), originated from an LEM version in 1999, and has been updated in numerous ways since then [31]. Delucchi Canadianized LEM in late 1998 by including information about the production and distribution of traditional and renewable fuels provided by Natural Resources Canada (NRCan), Statistics Canada and several other Canadian government agencies. The partially Canadianized version prepared by Delucchi is the basis for the development of the

GHGenius. The models GHGenius and LEM share the same basic approach but have different fuel end-use pathways over the whole life cycle [30, 35].

The GHGenuis model investigates life cycle steps from raw material extraction to energy end-use. The factors considered in the life cycle include: feedstock recovery and production, fertilizer and pesticides manufacture, land use changes and cultivation associated with biomass-derived fuels, leaks and flaring associated with the production of oil and gas, feedstock transport, fuel production (as in production from raw materials), emissions displaced by co-products of alternative fuels, fuel storage and distribution at all stages, fuel dispensing at the retail level, vehicle operation, carbon in air, vehicle assembly and transport, materials used in the vehicles etc [32].

The model focuses on three life cycle results: the primary GHGs, the criteria pollutants from combustion sources and the total energy and fossil energy use. In the case of the greenhouse gases, GHGenius can assess the cost effectiveness of various strategies using the results from the lifecycle assessment together with the fuel and vehicle costs input by the users. This tool is particularly helpful in comparing vehicle-related GHG reduction strategies with fuel-related strategies on a common basis.

The GHG emissions investigated in the model include: CO_2, CH_4, N_2O, CFC-12 and HFC-134a. Like the LEM model, the GHG emissions in GHGenuis are also calculated for the CO_2-equivalent emission like the GWPs used by the IPCC or other values that the user may like to use. Other air contaminants calculated in GHGenius include CO, NOx, NMOCs, SO_2, and TPM [34].

Energy consumption was investigated mainly in the following categories [34]:

- Total energy/fossil energy used per unit of energy produced for each stage of the fuel production steps;
- Energy used per kilo mileage driven, for the fuel used in light/heavy duty internal combustion engines and light/heavy duty fuel cell vehicles;
- The proportions of various types of energy used for each stage of the fuel production cycle.

2.4. SimaPro LCA Model

Apart from the software designed especially for LCA analysis of biofuels, LCA tools like SimaPro have also been used to analyze and compare different biofuels. SimaPro is a useful LCA analysis model that provides a functional platform for people to input the assumption data and build the life cycle process network tree step by step. The availability of all types of useful data can be conveniently added in the modeling process to get the LCA results and some other impact assessment results.

SimaPro has built a systematic and transparent life cycle analysis model that can be used for comprehensive research applications, including carbon footprint calculation, environmental reporting and product and eco-design. The SimaPro model has three different software versions for professional LCA use: SimaPro Compact for quick results, SimaPro Analyst for detailed LCA studies and SimaPro Developer for developing dedicated LCA tools. SimaPro Analyst can be used to import data in the SimaPro database format, create models with powerful LCA Wizards in SimaPro, define and edit parameters, determine uncertainty by using Monte Carlo, export in the SimaPro database format and do the advanced calculation-results and analysis.

The important contribution made by SimaPro to life cycle analyses of biofuel is the method of calculating co-product credits in SimaPro or other LCA models. Two primary methods used to calculate the co-product credits for biofuel life cycle analysis are displacement methods and allocation methods, respectively. The displacement method replaces the energy use and emission burden of byproducts to the primary product, assumes that the co-products are displaced by the conventional products, e.g., replacing soy meal by soybeans on a protein replacement basis, or glycerin by petroleum-based glycerin on a mass replacement basis, or fuel gas by natural gas on an energy value replacement basis. The energy and emissions that are generated or used during the displacement of product are viewed as co-product credits. The energy use and GHG emissions of the displaced products can thus be calculated more clearly and comprehensively [38-40].

The allocation method can be classified into energy-based, market-value-based and hybrid approaches. The formula used in energy allocation method is: Allocation ratio to oil = $(A* LHV_A)/ (A* LHV_A + B* LHV_B)$, where A and B are the mass flow rates of oil and seed in the extraction process, respectively. LHV_A and LHV_B are the lower heating values of oil and seed, respectively [41]. Energy-value based or market-value based allocation ratios for the energy use and emission burdens between primary products and byproducts are used for different biofuel pathways involved in the study [42].

3. MAJOR LCA OF BIOFUEL STUDIES

A number of LCA studies have been conducted on various biofuels (such as biodiesel and ethanol) over the past two decades. Major LCA studies were described below.

3.1. Fuel Cycle GHG Emission of Transportation Fuels Studies

Providing the most comprehensive and extensive study of life cycle analysis for GHG emission and energy balance, Delucchi estimates the fuel cycle GHG emissions of different kinds of transportation fuels and electricity generations in the early years of the 1990s (1991 and 1993). [20] The fuel cycle refers to the production and use of the motor fuel stage, and the GHG here includes CO_2, CH_4 , CO, N_2O, NOx and NMOCs. Apart from the GHG emission and energy efficiency in a specific fuel cycle stage, it also investigates the GHG emission and energy use for motor vehicle manufacturing, transportation systems maintenance, materials manufacturing and land use over changing stages. In the calculation approach, the first step was to estimate the energy use at a specific stage, and then to allocate the total amount of energy use to different fuel types, say residual oil, NG, electricity and coal. The next step was the calculation of CO_2 emissions using a carbon balance approach. The carbon that oxidized into CO_2 is calculated by counting the difference of all available carbon and the carbon contained in VOC, CO and CH_4. The total emissions in the study include both the portion from the combustion of process fuel and the portion from the non-combustion of process fuel, like chemical reactions and leakage and evaporation. The results of the selected fuel cycle and vehicle life cycle GHG emissions are given in the form of CO_2-equivalent emissions per mile by the combination of all GHG emissions and the GWPs. The process energy efficiencies and energy source shares of total energy use data are primarily derived from the Energy Information Administration (EIA), while the emission factors of various energy combustion processes are derived from the US Environmental Protection Agency (EPA) AP-42 document. In 1997, a new report was published based on its earlier study in 1991 and 1993. In the study, many parametric assumptions and methodologies have been revised to account for the energy use and emissions associate with various fuel cycles [20, 28, 29, 43].

As the most comprehensive and extensive study of biofuel LCA, Delucchi's study was widely cited and laid the foundation for LCA studies. It also provides a significant amount of data to the development of the GREET and LEM models.

3.2. GHG Emissions and Energy Balance Study on Soybean Biodiesel

In 1990s, a LCA study was initiated in the US by the National Renewable Energy Laboratory (NREL) and sponsored by the Department of Agriculture and Department of Energy to evaluate the GHG emission and energy balance between soybean biodiesel and conventional petroleum diesel [4, 36]. Although biodiesel can be produced from various feedstocks in various places, the study chose to investigate soybean biodiesels planted in the US. The whole picture of the LCA for soybean biodiesel was explored, from the soybean farming stage to biodiesel combustion on diesel buses, and from the extraction of all raw materials to the end use in automotives for petroleum. The study estimated results of energy balance (primary energy requirements, process energy requirements, fossil energy requirements, petroleum use), the emissions of CO_2, TPM, CO, NO_x and total hydrocarbons (THC), the amount of water and solid waste, and water consumption in the whole production chain [44].

Incomplete data support limits the life cycle analysis findings, and the most reliable results are for overall energy balance and CO2 emissions. As for the energy balance part, the consumption of fossil energy use determines the renewable extent of a fuel. Thus, the lower the amount of fossil energy required in the whole life cycle of a fuel, the more renewable it is.

The result of the study indicated a number of useful indicators as follows:

i.   The consumption of primary energy, namely the cumulative energy content of all unrefined materials from the environment, are about 1.24 MJ and 1.20 MJ input for 1 MJ of biodiesel and petrol-diesel output, which provides a life cycle energy efficiency (LCEE) of 80.6% and 83.3%, respectively.

ii.   The use of fossil energy is about 3.2 units and 1.2 units input per unit of biodiesel and petrol-diesel output, which puts the fossil energy ratio (FER) at 3.2 and 1.2, respectively.

iii.   The CO_2 emission related to petrol-diesel is not surprisingly higher than that related to biodiesel. B100 (100% biodiesel) will reduce about 78% of net CO_2 emissions, and B20 (20% biodiesel with 80% petro-diesel by volume) will offer a 16% reduction over petrol- diesel per unit of power generated by an engine.

iv.   The B100 could save about 32% TPM and 35% CO over the petrol-diesel life cycle. For PM10, the reduction of emissions from an urban bus using B100 is 68% less than that of petrol-diesel. The biggest reduction occurs at the tailpipe of an urban bus.

v.   The biodiesel life cycle had a higher NO_x emission value than petrol-diesel. The NO_x emissions of B100 and B20 are 13% and 2.7% higher than petro-diesel respectively, which also occur primarily at the tailpipe.

vi.   Life cycle emissions of THC for B100 that were 35% higher than those of petrol-diesel were reported. However, the tailpipe THC emissions for B100 are 37% lower than those of petro-diesel the other way round because of the hexane released during soybean processing and the volatilization of agrochemicals applied on the farm.

vii.   The water and solid waste generated by biodiesel are about 80% lower than those generated by petro-diesel. It was also reported that the amount of hazardous waste produced by biodiesel is only 5% of the amount produced from petrol-diesel. However, due to the different composition and disposition methods between biodiesel and petrol-diesel waste, the comparison of consequence data was less reliable than it supposed to be.

Sheehan's study above was designed to identify and quantify the benefits of biodiesel when compared with traditional petroleum diesel [44]. Despite the obvious advantages in energy saving and GHG control for biodiesel as shown above, it also induces problems in NO_X and THC emissions. The following would be focused on some regional life cycle models and feedstock source evaluation as a supplement to Sheehan's work on biodiesel.

3.3. LCA of Ethanol and Reformulated Gasoline

The NREL initiated a comparative life cycle emission analysis of biomass-based ethanol and reformulated gasoline (RFG) together with the Oak Ridge National Laboratory and Pacific Northwest National Laboratory in the early 1990s. The fuels evaluated in the study include RFG, E10 (10% ethanol with 90% gasoline by volume) and E95 (95% ethanol with 5% gasoline by volume). The researcher assumed that RFG was produced using methyl tertiary butyl ether (MTBE) that met the standards of the Clean Air Act Amendment (CAAA). In the study, not all the ethanol is produced from corn; for instance the E10 used in the year 2000 was assumed to be produced from municipal solid waste, while the E95 used in the year 2010 was assumed to be produced from biomass such as switchgrass and cellulosic crop species [45].

The study created an inventory of inputs and outputs of the three selected fuels. It estimated the solid waste, water pollutant and air pollutant emissions. The air pollutants include VOCs, CO, NO_x, SO_x, CO_2 and particulate matters (PM). Petroleum displacement of using E10 and E95 were also calculated in the study.

The results showed that using E10 causes very little change in life cycle emissions when compared with RFG. While E95 would reduce CO_2 emission by 90%, and NO_x, SO_x and PM would also be reduced considerably. On the other hand, the emission of VOC and CO emission of E95 would increase when compared with RFG. The fossil fuel consumption for E10 and E95 would be decreased by 6% and 85% respectively when compared with RFG [46].

NREL assumed favored energy efficiency and low GHG emission in ethanol life cycles, such as a high allocation rate of upstream ethanol life cycle emissions to by-products, a large electricity credit earned in an ethanol plant so that the estimation of GHG emission results are lower than those of Delucchi's study [47, 48].

3.4. LCA of Soybean-Derived Biodiesel and Renewable Fuels

Huo et al. [42], using the GREET model, carried out a well-to-wheel (WTW) analysis of the energy balance and GHG emissions of five fuels for the year 2010. The fuels considered include conventional gasoline, soybean-based biodiesel produced by transesterification (BD), soybean-based renewable diesels I (produced by SuperCetane, RD I) and II (produced by hydrogenation, RD II), and catalytic cracking renewable gasoline (RG). The consumption of total energy, fossil energy, petroleum energy and GHG emissions are the main factors tracked in the research in order to examine the environmental and energy demand benefits of renewable fuels over conventional gasoline. Five different allocation and displacement methods were used in the study to deal with the multiple co-product problems.

The well-to wheel system boundary assumption, including the well-to-pump stage and pump-to-wheel stage, was adopted. The former stage started with the fuel feedstock recovery and ended with the fuels available at the pump station. The latter was carried to the next step until the end use of the fuel. The data of soybean farming, including fertilizer and pesticide, were provided by the US Department of Agriculture. The data of N_2O emission factors, used to determine the amount of N_2O emitted from the nitrogen in soil and water, were a corrected value of both the previous study and the IPCC document. The input and output data for the production of mass and energy balance for soybean-based fuels were obtained according to the ASPEN process simulation model developed by NREL.

The results showed that there was not a big difference between total energy use of renewable fuels and petroleum fuel under various co-product calculation methods. However, fossil energy use showed a two-third reduction between renewable fuels and petroleum fuel. The petroleum energy use of renewable fuels was just about a tenth of that of petroleum fuel. The CO_2-equivalents GHG emissions were calculated under the standards of GWPs from IPCC. The overall reduction of the GHG emissions of the four renewable fuels can achieve a reduction of 64%- 174% versus petroleum fuel in the study [42, 49].

3.5. Other LCA Studies

Kalnes et al. [40] conducted an LCA analysis for synthesized diesel, biodiesel, green diesel (produced via the UOP/Eni Ecofining^{TM} process) and conventional petroleum diesel using SimaPro Analyst. The system boundaries include feedstock farming, oil extraction, transport and processing to the final product, but did not consider the land use change issue. The functional unit in the study is 1 MJ of selected research oils that lay the foundation for the accumulation of energy and the material input inventories of the study. The LCA tools used in SimaPro for this study include the cumulative energy demand (CED) and the Eco-indicator. The CED refers to all the energy consumed throughout the life cycle process, while Eco-indicator was used to calculate the GHG emissions in units of CO_2-equivalents for all

GHGs. The results showed that the total energy demand for synthesized diesel from wood, biodiesel and green diesel are greater than those of petroleum diesel. Green diesel had the lowest GHG emissions, while petroleum counterparts showed the highest GHG emissions.

In another study, Shonnard et al. [41], using the same model, determined the GHG emissions and savings, CED by fuel type and fossil energy demand of camelina-derived jet fuel and green diesel. Impacts on fertilizer rates and model parameter uncertainty can be evaluated as an impact assessment result of the study. With the functional unit of energy content defined as 1 MJ, detailed life cycle inventory data, such as cultivation and processing inputs, were entered along with the material, chemical and energy input from the Ecoinvent database of SimaPro. The allocation methods used in these LCA studies include mass allocation and displacement to deal with the co-product problem in the life cycle analysis. The results of the study showed a total life cycle saving of 75% and 80% GHG emissions for camelina-derived jet fuel and green diesel over petroleum diesel.

4. LCA LIMITATIONS

Over the years, LCA studies were conducted in various forms and with different objectives. Although the LCA format can provide valuable information about the net benefit of the target substance to our environment, it still has some limitations in application, which are summarized as follows [2, 23, 50-53]:

i.   Incomplete Data input: a need for intensive time and resources led to the simplified forms of data collection of life cycle input, which include data inaccuracy, unrepresentative data and model uncertainty; another problematic aspect is the non-transparent assumptions that decrease the LCA credibility.

ii.   Diversity in system boundaries and life cycle inventories: different boundary definitions and variation across life-cycle inventories produces different LCA estimations and increases the uncertainty of the results; the variability between feedstock source and inventoried system also obscured the life cycle results for comparison.

iii.   Variance in terminologies and functional units: the difference of terms and units used to describe the energy efficiency and GHG balance not only obscured the LCA results and destroyed the language consistency among LCA studies, but also limited people in other fields from understanding the outcome.

iv.   Limitation of interdisciplinary communication is another factor that limits the development of LCA. Widely acknowledged standards and integration of related industries would be necessary for drawing the appropriate conclusions.

To reduce the disadvantage of the impact factors mentioned above and increase reliability, the standardization of LCA methodology, including validation of data, estimation of parameter techniques, and critical review, should be applied by practitioners to provide quantitative and qualitative data quality management systems for higher resolution models. Other solutions, like sensitivity analysis and uncertainty analysis, should also be used to overlook uncertainties and improve the affiliation between LCA results and the practical situation [54, 55].

5. CONCLUSION

The LCA of biofuel is a complete calculation of all the inputs and outputs of a production chain, which will often be used to estimate the energy requirements and yields, environmental costs and values, and even economic costs and benefits of the selected biofuels. Growing attention is being paid to LCA studies, and various models and databases have been built to investigate the GHG emissions and energy efficiency under different life cycle system boundaries and functional units. LCA brings about an all-inclusive platform for evaluating biofuel, but it should not be isolated from considerations in other aspects. In order to provide standardized and equally holistic LCA for biofuels, we should lay more emphasis on improving data collection and management, take into account ecological factors like interactive climate and nutrient cycling, and maintain LCA terminology consistency as a whole.

ACKNOWLEDGMENT

The authors would like to acknowledge the ICEE of the University of Hong Kong for providing funding to this study.

REFERENCES

[1]   IEA, *Energy Technology Essentials, Biofuel Production IEA*, Editor 2007.

[2]   Davis, S.C., K.J.A.-. Teixeira, and E.H. DeLucia, Life-cycle analysis and the ecology of biofuels. *Trends in Plant Science,* 2009. 14(3): p. 140-146.

[3]   Malca, J. and F. Freire, Renewability and life-cycle energy efficiency of bioethanol and bio-ethyl tertiary butyl ether (bioETBE): assessing the implications of allocation. *Energy,* 2006. 31(15): p. 3362-3380.

[4]   Sheehan, J., Camobreco, V., Duffield, J., Shapouri, H., Graboski, M.,Tyson, K. S., *Overview of Biodiesel and Petroleum Diesel Life Cycles,* 1998, National Renewable Energy Laboratory: Golden, CO (US).

[5]   Shapouri, H., J. Duffield, and M. Wang, *The energy balance of corn ethanol revisited.*

[6]   Gomez, L.D., C.G. Steele-King, and S.J. McQueen-Mason, *Sustainable liquid biofuels from biomass: the writing's on the walls. New Phytologist,* 2008. 178(3): p. 473-485.

[7]   Von Blottnitz, H. and M.A. Curran, A review of assessments conducted on bio-ethanol as a transportation fuel from a net energy, greenhouse gas, and environmental life cycle perspective. *Journal of cleaner production,* 2007. 15(7): p. 607-619.

[8]   Ragauskas, A.J., Charlotte K. Williams, Brian H. Davison, George Britovsek, The path forward for biofuels and biomaterials. *Science,* 2006. 311(5760).

[9]   Dale, B.E., Thinking clearly about biofuels: ending the irrelevant 'net energy' debate and developing better performance metrics for alternative fuels. *Biofuels, Bioproducts and Biorefining*, 2007. 1(1): p. 14-17.

[10]   Farrell, A.E., Plevin, R.J., Turner, B.T., Jones, A.D., Ethanol can contribute to energy and environmental goals. *Science*, 2006. 311(5760): p. 506.

[11]   Gnansounou, E. and A. Dauriat. *Energy balance of bioethanol: A synthesis*. 2005.

[12]   Hill, J., Nelson, E., Tilman, D., Polasky, S., Tiffany, D., Environmental, economic, and energetic costs and benefits of biodiesel and ethanol biofuels. *Proceedings of the National Academy of Sciences*, 2006. 103(30): p. 11206.

[13]   Halleux, H., Lassaux, S., Renzoni, R., Germain, A., Comparative life cycle assessment of two biofuels ethanol from sugar beet and rapeseed methyl ester. *The International Journal of Life Cycle Assessment*, 2008. 13(3): p. 184-190.

[14]   Armstrong, A., Baro, J., Dartoy, J., Groves, A., Nikkonen, J., Rickeard, D., *Energy and greenhouse gas balance of biofuels for Europe: an update2002:* Concawe.

[15]   Soussana, J., Allard, V., Pilegaard, K., Ambus, P., Amman, C., Full accounting of the greenhouse gas (CO2, N2O, CH4) budget of nine European grassland sites. *Agriculture, Ecosystems & Environment*, 2007. 121(1-2): p. 121-134.

[16]   Harvey, D., A guide to global warming potentials (GWPs). *Energy Policy*, 1993. 21(1): p. 24-34.

[17]   Fulton, L., *Biofuels for Transport: An International Perspective*, 2005, IEA.

[18]   Hill, J., From the Cover: Environmental, economic, and energetic costs and benefits of biodiesel and ethanol biofuels. *Proceedings of the National Academy of Sciences, 2006.* 103(30): p. 11206-11210.

[19]   Lettens, S., Muys, B., Ceulemans, R., Moons, E., Garcia, J., Coppin, P., Energy budget and greenhouse gas balance evaluation of sustainable coppice systems for electricity production. *Biomass and bioenergy*, 2003. 24(3): p. 179-197.

[20]   M.Q.Wang, *GREET 1.5 - transportation fuel-cycle model - Vol. 1 : methodology, development, use, and results.*, 1999, Argonne National Lab., IL (US).

[21]   Wang, M., *GREET 1.0-: Transportation Fuel Cycles Model: Methodology and Use1996:* Argonne National Laboratory.

[22]   Wang, M., *Greenhouse Gases, Regulated Emissions, and Energy Use in Transportation (GREET) Model, Version 1.8 b. Center for Transportation Research,* Argonne National Laboratory, 2008.

[23]   Wang, M., Life-Cycle Analysis of Biofuels, Plant Biotechnology for Sustainable Production of Energy and Co-products. *Biotechnology in Agriculture and Forestry*, 2010. 66: p. 385-408.

[24]   Wang, M., Christopher, M., *Fuel-cycle fossil energy use and greenhouse gas emissions of fuel ethanol produced from US Midwest corn1997: Citeseer.*

[25]   Wang, Q. and M.A. DeLuchi, Impacts of electric vehicles on primary energy consumption and petroleum displacement. *Energy*, 1992. 17(4): p. 351-366.

[26]   Delucchi, M., *A lifecycle emissions model (LEM): lifecycle emissions from transportation fuels, motor vehicles, transportation modes, electricity use, heating and cooking fuels, and materials.* 2003.

[27]   Delucchi, M., *Conceptual and methodological issues in lifecycle analyses of transportation fuels.* 2004.

[28] Deluchi, M.A., *Emissions of greenhouse gases from the use of transportation fuels and electricity*, 1991. p. Medium: ED; Size: Pages: (142 p).

[29] Deluchi, M.A., Greenhouse-gas emissions from the use of new fuels for transportation and electricity. *Transportation Research Part A: Policy and Practice,* 1993. 27(3): p. 187-191.

[30] Fleming, J., *Using life cycle analysis as a tool for transportation fuels policy.* 2007.

[31] Inc., S.T.C., *Documentation for Natural Resources Canada's GHGenius Model 3.0.*

[32] Kammen, D.M., et al., *Energy and greenhouse impacts of biofuels: A framework for analysis.* 2008.

[33] Wang, G. and M. Delucchi, Lifecycle Emissions Model (LEM): Lifecycle Emissions from Transportation Fuels, Motor Vehicles, Transportation Modes, Electricity Use, Heating and Cooking Fuels, and Materials, *Appendix X: Pathways Diagrams. Publication No, 2005*, UCD-ITS-RR-03-17X.

[34] Inc., S.T.C., *Analysis of Biofuel Pathways Using GHGenius* 2010. 2010.

[35] Report, G.; Available from: *http://www.ghgenius.ca/about.php.*

[36] Sheehan, J., Camobreco, V., Duffield, J., Graboski, M., Shapouri, H., Life cycle inventory of biodiesel and petroleum diesel for use in an urban bus, 1998, *National Renewable Energy Lab: Golden*, CO (US).

[37] Wang, M. *Overview of GREET model development at Argonne.*

[38] Wang, M., H. Huo, and S. Arora, Methods of dealing with co-products of biofuels in life-cycle analysis and consequent results within the US context. *Energy Policy*, 2010.

[39] Curran, M.A., Co-product and input allocation approaches for creating life cycle inventory data: a literature review. *ecoinvent Impressum*, 2007: p. 65.

[40] Kalnes, T.N., Koers, K., Marker, T., Shonnard, D., A technoeconomic and environmental life cycle comparison of green diesel to biodiesel and syndiesel. *Environmental Progress & Sustainable Energy*, 2009. 28(1): p. 111-120.

[41] Shonnard, D.R., L. Williams, and T.N. Kalnes, Camelina-derived jet fuel and diesel: Sustainable advanced biofuels. *Environmental Progress & Sustainable Energy*.

[42] Huo, H., Wang, M, Bloyd, C., Putsche, V., *Life-cycle Assessment of Energy Use and GHG emissions of Soybean-derived biodiesel and renewable fuels.*

[43] Delucchi, M.A. and D.I.o.T.S. University of California, A revised model of emissions of greenhouse gases from the use of transportation fuels and electricity1997: *Institute of Transportation Studies*, University of California.

[44] Sheehan, J., Camobreco, V., Duffield, J., Graboski, M., Shapouri, H., Life-cycle assessment of biodiesel versus petroleum diesel fuel. 2002. *IEEE.*

[45] Tyson, K.S., *Fuel cycle evaluations of biomass-ethanol and reformulated gasoline.* 1993.

[46] Lynd, L.R., Overview and evaluation of fuel ethanol from cellulosic biomass: technology, economics, the environment, and policy. *Annual review of energy and the environment,* 1996. 21(1): p. 403-465.

[47] Laboratory, N.R.E., *A comparative Analysis of the Environmental Outputs of Future Biomass-Ethanol Production Cycles and Crude Oil/ Reformulated Gasolone Production Cycles.* 1991.

[48] Laboratory, N.R.E., *Hydrogen Program Plan.* 1992.

[49] Wang, M., *Well-to-wheel energy use and greenhouse gas emissions of advanced fuel/vehicle systems North American analysis,* 2001, Argonne National Lab., IL (US).

[50] Ross, S., D. Evans, and M. Webber, How LCA studies deal with uncertainty. *The International Journal of Life Cycle Assessment*, 2002. 7(1): p. 47-52.

[51] Svoboda, S., *Note on life cycle analysis*. National Pollution Prevention Center for Higher Education. University of Michigan. LCA Note. Ann Arbor, Michigan, 1995.

[52] Finnveden, G., On the limitations of life cycle assessment and environmental systems analysis tools in general. *The International Journal of Life Cycle Assessment,* 2000. 5(4): p. 229-238.

[53] Fredga, K. and K.-G. Mäler, Life Cycle Analyses and Resource Assessments. *Ambio: A Journal of the Human Environment,* 2010. 39, Supplement 1: p. 36-41.

[54] Steen, B., On uncertainty and sensitivity of LCA-based priority setting. *Journal of cleaner production*, 1997. 5(4): p. 255-262.

[55] Bj rklund, A.E., Survey of approaches to improve reliability in LCA. *The International Journal of Life Cycle Assessment*, 2002. 7(2): p. 64-72.

In: Biodiesel: Blends, Properties and Applications   ISBN: 978-1-63117-024-9
Editors: Jorge M. Marchetti and Zhen Fang   © 2014 Nova Science Publishers, Inc.

Chapter 14

QUALITY IMPROVEMENT AND COMPARISON OF THE MAIN BIODIESEL PROPERTIES (ANALYTICAL AND PERFORMANCE) ORIGINATING FROM DIFFERENT SOURCES

Sándor Kovács, Ferenc Kovács, György Pölczmann and Jenő Hancsók[*]
University of Pannonia, Institute of Chemical- and Process Engineering,
Department of MOL Hydrocarbon- and Coal Processing, Veszprém, Hungary

ABSTRACT

In the last one and a half decades efforts for applying biofuels have increased all over the world. The main reasons of this are the rising demands for gas oil and the bio regulations in the European Union. Nowadays approx. 95% of the used bio blending components in diesel fuels is FAME (biodiesel). The use of bio gas oils (mixture of n-and i-paraffins) is gradually increasing. The biodiesels are produced by catalytic transesterification of different triglyceride-containing feedstocks (e.g., vegetable oils, used frying oils, animal fats, algae oils, brown grease, etc.) mainly with methanol. Many factors affect the quality and performance properties of these first generation biofuel; e.g., the type of the feedstock, and the alcohol; the transesterification technology and the catalytic system. The quality of the feedstock has the strongest effect on final FAME quality.

The aim of our systematic research and development activity was to produce high quality biodiesels. The feedstocks from various sources were transesterified in the presence of *Candida antarctica* immobilized lipase in a solvent free medium. The effect of the purity and the fatty acid composition of different feedstocks on the quality and performance properties of FAME were investigated.

Based on the experimental results we concluded that the physical and/or chemical refining of the raw materials is definitely necessary to produce high quality biodiesel. The

---

[*]University of Pannonia, Institute of Chemical- and Process Engineering, Department of MOL Hydrocarbon- and Coal Processing, H-8201 Veszprém, P.O Box: 158, Hungary, E-mail: hancsokj@almos.uni-pannon.hu

significantly different fatty acid composition of the feedstock does not affect the enzymatic transesterification. It can be stated that biodiesels with adequate oxidation stability and CFPP value can be produced only from vegetable oils with high oleic acid content.

INTRODUCTION

The production and application of fuels from agricultural origin have emerged into focus in the last couple of years. A number of environmental, political and economical factors has confirmed and strengthened this process [1]. The main reason of this tendency is the energy policy of the European Union, namely to reduce the green house gas emission of fuels, to decrease the significant dependence of EU on import energy and crude oil and to support rural development. To achieve these objectives, the European Union created the 2003/30/EC and 2009/28/EC directives, which regulate the application of biomass derived fuels. The main purpose is to promote the use of biofuels in transportation by recommending and specifying the share of the bio-components [2,3].

This proposed value (10 energy % share of biofuels in the transport sector by 2020 in the EU) can be reached by the conversion of different triglyceride-containing biofeedstocks (e.g., vegetable oils, used frying oils, animal fats, algae oils, brown grease, etc.) to different biofuels or blending components. Nowadays FAME (Fatty Acid Methyl Esters), called as first generation biofuel, is mostly used as diesel bio blending component. These alternative fuels are produced by catalytic transesterification of different triglyceride-containing feeds with methanol [4-7].

The triglycerides consisting mainly of C_{14} and C_{20} saturated and unsaturated fatty acids, but there is a difference between the ratios of these. This determines the physical and chemical properties of the feedstock (e.g., vegetable oils, used frying oils, animal fats, algae oils, brown grease, etc.) and the finished biodiesel. The applied feeds contain phosphatides, free fatty acids, coloring agents and their possible metabolites, oxidation products of fatty acids, flavors and odorants, metals which catalyzing the oxidation, etc. Therefore the physical and/or chemical refining of the raw materials is definitely necessary before the transesterification [8-11, 12].

Of course not only the properties of applied feeds have a significant effect on the quality of biodiesels (Figure 1) [13-17], but the type of the used alcohol (Table 1.) [18-22] and the transesterification technology (Table 2.), too [18, 23-26].

Based on the abovementioned it can be seen that the the quality and performace properties of the biodiesels are affected by many factors. Among them the type (origin, quality and fatty acid composition) of the applied triglyceride-containing feed has a significant effect. Due to the expanding range of the raw materials their qualities have high importance, too. The „Biofuels action plan" [COM(2006) 34] of the EU was a milestone, because the EU declared the need to use a new kind of feedstocks, too. [27]. Because of this the low-value trygliceride containing feedstocks (non-edible vegetable oils, used frying oils) appeared among the raw materials, which are to replace the food-grade vegetable oils [28, 29].

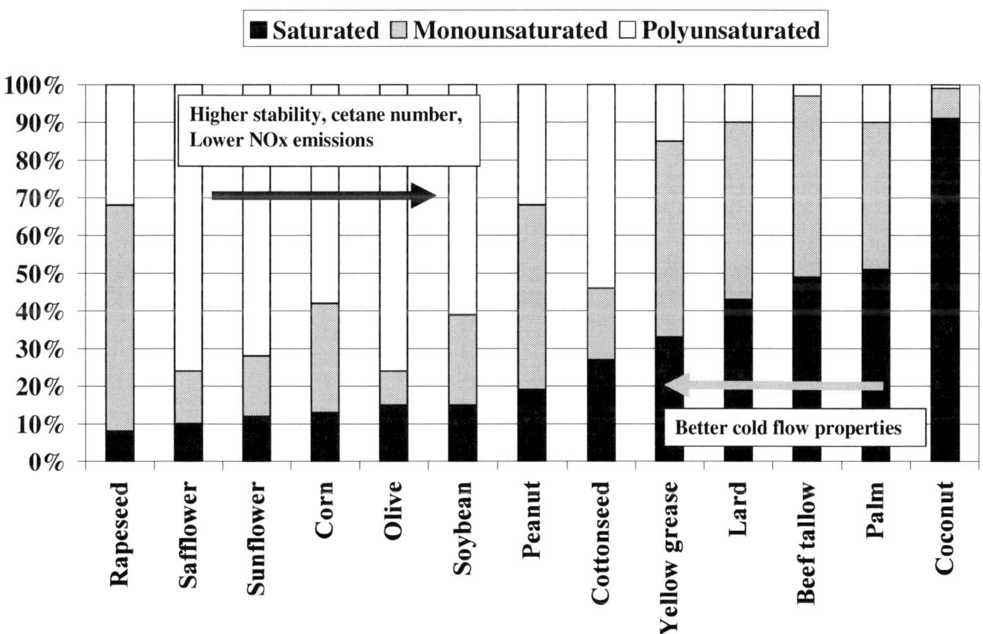

Figure 1. Properties of the biodiesel produced from different feeds.

Table 1. The effect of double bonds and ester type on the cetane number

Fatty acid	Cetane number				
	fatty acid	methyl ester	ethyl ester	propyl ester	butyl ester
C18:0	61.7	75.7	76.8	69.9	80.1
C18:1	46.1	55.0	53.8	55.7	59.8
C18:2	31.4	42.2	37.1	40.6	41.6
C18:3	20.4	22.7	26.7	26.8	28.6

Based on the previously mentioned it is important to enhance the quality of the properties of the biodiesels produced from triglycerides with different origin and fatty acid composition. A lot of papers investigate the quality improvement of biodiesels, and there are several opportunities. The basic difference between these is that they leave unchanged or modify the composition of the fatty acid alkyl ester mixtures. In the first case the application properties are improved with additives (cetane improvers [30], flow improvers [31] and antioxidants [32-35]). The application of additives entails the questions about additive compatibility with each other and with the diesel additives. In the second case the composition of the fatty acid alkyl ester mixtures are modified. One of the solutions is to modify the type of the alcohol used in transesterification. As it was presented before (Table 1.) the FAME application properties can be improved with using alcohols having higher carbon numbers [21, 22]. In this case the extra cost of the other alcohols compared to the methanol has to be mentioned [36]. The other option is the modification of the fatty acid composition with physical methods. In this case components having higher pour point are removed from the biodiesels with winterisation (cycles with multiple cooling and filtering). The disadvantages of this option are the high operation cost and the decrease of oxidation stability and cetane number because of the the saturated fatty acid alkyl esters removal [36-41]. The modification of the biodiesel quality and application properties can be done with applying feedstocks with unconventional fatty acid composition. This could be achieved with hybridization, Target is the increase of the oleic acid amount in the fatty acid composition. One question rises with this development, whether the vegetable oils from this genetically modified plants are applicable in other areas e.g., nutrition [36, 40, 42].

Table 2. The effect of vegetable oil quality and transeterification process on the properties of biodiesels

Properties	Feedstock	Technology	Causes
Acid value		X	conversion, dissolved catalyst
Iodine value	X		fatty acid composition
Methanol content		X	distillation
CFPP	X		fatty acid composition
Viscosity	X	(X)	fatty acid composition (conversion)
Density	X		fatty acid composition
Glycerine content		X	triglyceride conversion
Total contamination	X	X	quality of feedstock /phase separation
Flash point		X	metanoltartalom
Carbon residue	X	(X)	fatty acid composition (glycerine content)
Sulfated ash content		X	quality of feedstock
Phosphorus content	X		quality of feedstock
Cetane number	X		fatty acid composition
Oxidation sability	X		fatty acid composition

The quality of the biodiesels is determined basically by the triglyceride mixtures. The different climate, features of the ground and other factors in the different areas of the Earth significantly determine feedstocks for biodiesel production. So these have to be kept in mind when choosing a quality improvement option for biodiesels.

EXPERIMENTAL

The aim of our experiments was to produce high quality biodiesels by enzyme catalyzed transesterification from various feedstocks which are available in Hungary. The feedstocks from various sources were transesterified in the presence of *Candida antarctica* immobilized lipase in a solvent free medium. The effect of the refining processes (physical and chemical) and the different fatty acid composition on the quality of biodiesels were investigated.

EXPERIMENTAL APPARATUS

The scheme of the experimental apparatus used during the enzyme catalytic experiments is shown in Figure 2. The operational parameters during our experimental work were based on our previous results [43]. Methanol was added to the reaction mixture in 8 parts by applying a methanol-to-triglyceride molar ratio of 4:1 instead of the stochiometric ratio of 3:1, considering that the excess methanol is forcing the reaction. The stepwise addition is necessary to prevent the catalyst poising effect of the methanol. The glycerol was separated by dialysis to prevent the product inhibition. The applied experimental apparatus contained an asymmetric hydrophilic ultrafiltration membrane (made from cellulose acetate, cut off 40 kDa), which was applied in a jacketed, flat sheet membrane module (surface area 100 cm^2). These could also provide the highest possible yield of fatty acid methyl esters.

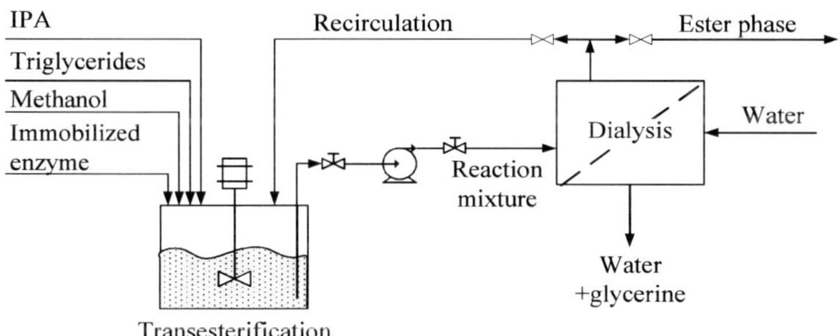

Figure 2. Scheme of the experimental apparatus.

MATERIALS

Based on our previous experiments the applied enzyme catalyst was the macroporous resin immobilized lipase *Candida antarctica* (Novozym 435) (activity: 7000 PLU/g, density: 430 kg/m3, diameter: 0.3-0.9 mm). Analytical grade methanol was used during the transesterification [43].

The main characteristics of the feeds are summarized in Table 3., and their fatty acid compositions in Table 4. Raw materials were selected to fit the iodine value of the finished biodiesels. The main criterion was that the iodine value shoul be below 120 g I$_2$/100g.

The sunflower oil marked 'A' and 'A1' was the same, but 'A1'was raw, untreated oil, while 'A' was a refined one [9]. Sunflower oil marked with 'B' was a conventional, while 'C' was an improved one with high olec acid content [44]. The oil marked'D' was an erucic acid free rapeseed oil and 'E' was a rapeseed oil with low erucic acid content. The 'E' and 'E1'marked rapeseed oil was the same, but 'E1'was raw, untreated, while 'E' was a refined one [9, 44].

Table 3. The main properties of the applied vegetable oils

Properties	Sunflower oil				Rapeseed oil		
	A1	A	B	C	D	E	E1
Density, 15°C, g/cm^3	0.9194	0.9189	0.9156	0.9144	0.9178	0.9200	0.9204
Viscosity, 40°C, mm^2/s	32.5	31.1	39.9	39.2	39.9	34.4	35.0
Flash point, °C	over 300						
Iodine value, g I$_2$/100g	122	108	114	95	105	102	110
Acid value, mg KOH/g	1.3	0.4	2.1	1.9	2.4	2.5	4.0
Carbon residue (on 10 % distillation residue), %	0.40	0.20	0.15	0.15	0.20	0.20	0.5
Free fatty acid content, %	0.65	0.20	1.05	0.85	1.20	1.25	2.0

Table 4. Fatty acid composition of the applied feeds

Fatty acid composition, %*	Sunflower oil			Rapeseed oil	
	A	B	C	D	E
C16:0	7.6	6.3	2.4	4.0	4.7
C16:1	0.1	0.1	0.1	0.2	0.2
C18:0	3.2	3.6	2.5	1.7	2.0
C18:1	20.4	28.5	91.8	75.5	61.0
C18:2	67.6	60.3	1.2	12.3	20.5
C18:3	0.1	0.1	0.6	5.2	9.1
C20:0	0.2	0.2	0.8	0.3	0.6
C20:1	0.1	0.2	0.5	0.7	1.4
C22:0	0.5	0.7	0.1	0.1	0.3
C22:1	0.0	0.0	0.0	0.0	0.2

*The first number represents the number of carbon atoms and the second means the number of double bonds in the molecule

TEST METHODS

Properties of the feed and products were determined by standard test methods specified by EN 14214:2008+A1:2009 (Automotive fuels. Fatty acid methyl esters (FAME) for diesel engines. Requirements and test methods), complying with the specified precision data (Table 5.).

Table 5. The applied test methods

Properties	Test method
FAME content, %	EN 14103
Density, 15°C, g/cm^3	EN ISO 12185
Viscosity, 40°C, mm^2/s	EN ISO 3104
Flash point, °C	EN ISO 2719
Sulfur content, mg/kg	EN ISO 20846
Carbon residue (on 10% distillation residue), %	EN ISO 10370
Cetane number	EN ISO 5165
CFPP, °C	EN 116
Water content, mg/kg	EN ISO 12937
Copper strip corrosion (3h at 50°C)	EN ISO 2160
Oxidation stability, 110 °C	EN 14112
Acid value, mg KOH/g	EN 14104
Iodine value, g I$_2$/100g	EN 14111
Methanol content, %	EN 14110
Monoglyceride content, %	EN 14105
Diglyceride content, %	EN 14105
Triglyceride content, %	EN 14105
Free glycerol, %	EN 14106
Total glycerol,%	EN 14105
Group I metals (Na+K), mg/kg	EN 14538
Group II metals (Ca+Mg), mg/kg	EN 14538
Phosphorus content, mg/kg	EN 14107

THE EFFECT OF THE VEGETABLE OIL REFINING ON THE QUALITY OF FAME

In the refining process, the contaminations – phosphatides being present in the raw vegetable oils in low concentration beside triglycerides, free fatty acids, coloring agents and their possible metabolites, oxidation products of fatty acids, flavors and odorants, metals which are catalyzing oxidation etc. – are to eliminate. A number of published results show that FAME produced from pretreated vegetable oil has not the same the quality compared to FAME reached from raw (untreated) vegetable oil. Therefore it is important to know the physical and chemical vegetable oil refining processes and their effect on the quality of finished biodiesels in terms of manufacturing technologies. During the experiment, after each refining step, the samples of sunflower and rapeseed oils were transesterified. The basis of the comparison was the quality properties of the obtained biodiesels [9].

For this experiments, cold pressed, crude sunflower oil ('A1') and rapeseed oil ('E1') were used (Hungarian origin). After each refining stage enzyme catalyzed transesterification were carried out the (temperature: 50±1°C, methanol to vegetable oil mol ratio was 4:1, the enzyme catalyst was 12% based on total reaction mixture, methanol was added into the reaction mixture in 8 parts, the reaction time was 16 hours) and the quality properties of the products were investigated.

Refining Processes

During the chemical (alkaline) neutralization and physical refining of the vegetable oils the following steps were applied (Table 6.):

Table 6. The applied refining steps

Chemical refining		Physical refining
Crude vegetable oil		
1.	Water degumming, steaming	Water degumming, steaming
2.	Caustic neutralization	Ultra fine neutralization
3.	Bleaching	Bleaching (<100°C)
4.	Deodorization	Winterization
5.	Winterization	Deodorization neutralization

Thereafter, the changes of the properties of the obtained sunflower oil methyl esters after each physical and chemical refinement process are presented.

The properties of biodiesels, produced from sunflower oil and rapeseed oil passing through different refining stages are summarized in (Tables 7 and 8).

Table 7. Properties of the suflower oil methyl ester after different refining treatments

Property	Chemical refining					Physical refining				
	Degumming	Chemical neutralization	Bleaching	Deodorization	Winterization	Degumming	Ultra fine degumming	Bleaching	Winterization	Deodorization neutralization
FAME content, %	90.3	93.2	97.2	97.5	97.9	89.7	92.5	96.8	97.2	97.6
Phosphorus content, mg/kg	55	4	1	-	-	50	2	-	-	-
Polyunsaturated (≥4 double bonds) methyl esters	-	-	0.2	0.8	0.8	-	-	0.4	0.4	0.8
Acid value, mg KOH/g	4	0.1	0.1	0.1	0.1	4	3.9	4	4	0.2
CFPP, °C	+3	+3	0	-1	-6	+3	+2	-1	-5	-5
Oxidation stability, 110°C, h	6.1	6.6	6.0	9.1	9.0	4.9	5.3	4.0	4.4	8.7

Table 8. Properties of the rapseed oil methyl ester after different refining treatments

Property	Chemical refining					Physical refining				
	Degumming	Chemical neutralization	Bleaching	Deodorization	Winterization	Degumming	Ultra fine degumming	Bleaching	Winterization	Deodorization neutralization
FAME content, %	90.9	93.0	97.6	97.8	97.9	89.6	92.6	96.8	97.0	97.4
Phosphorus content, mg/kg	48	2	1	-	-	49	1	-	-	-
Polyunsaturated (≥4 double bonds) methyl esters	-	-	0.2	0.8	0.8	-	-	0.4	0.4	0.8
Acid value, mg KOH/g	4	0.1	0.1	0.1	0.1	4	3.9	4	4	0.2
CFPP, °C	+1	+1	0	-1	-9	+3	+2	+1	-8	-8
Oxidation stability, 110°C, h	6.9	6.9	6.0	10.3	9.9	6.5	6.6	5.0	5.2	9.7

The results obtained from sunflower oil are only presented in the following.

Methyl Ester Content

The methyl ester content of the biodiesel sample obtained by the enzyme catalytic transesterification of the raw sunflower oil was low (76.2 %). With the increasing of the refining steps (effectiveness of contamination removal), higher methyl ester yield could be reached (Figures 3 and 4). After the degumming and bleaching steps, the methyl ester content was higher than 96.5% specified in the EN 14214 standard. Therefore, some refining steps of the raw sunflower oil are needed to provide the minimum methyl ester content in the FAME; otherwise the contaminations inhibit the formation of monoesters.

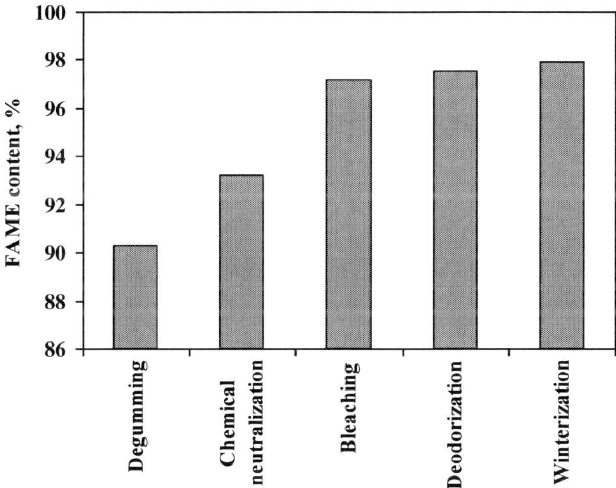

Figure 3. Change of methyl ester content of biodiesels through chemical refining of sunflower oil.

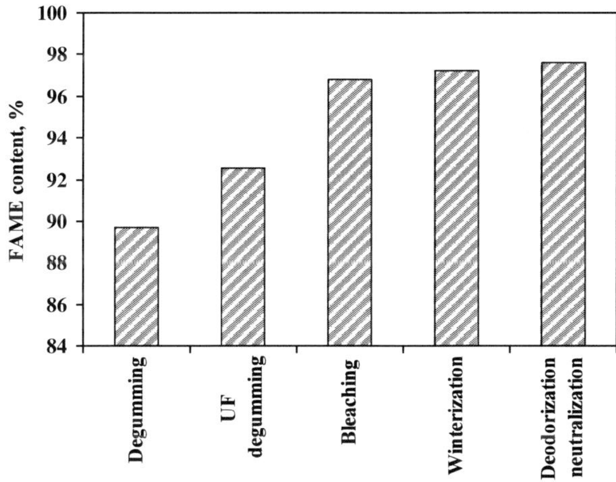

Figure 4. Change of methyl ester content of biodiesels through physical refining of sunflower oil.

Phosphorus Content

The phospholipids being naturally present in vegetable oils are the main origin of the phosphorus content. The phospholipids consist of a glycerol molecule esterified by two molecule of fatty acids and by a phosphate. Phospholipids are polar compounds, so it explains some of their features: solubility in polar solvents, bonding with cations (Ca^{2+}, Mg^{2+}, Ni^{2+}, Fe^{3+}, Cu^{2+}, etc.). Their emulsifying and surfactant properties interfere with the refining process; so the first step is the removal of them. Incomplete elimination of phosphorus containing compounds can cause problem in the downstream refining processes a (e.g., emulsion formation during washing, foaming in the drying step etc.). Due to the polar nature of phospholipids, they react with metals. The metal content of vegetable oils is mainly caused by these compounds. Particularly, iron and copper ions catalyze the autooxidation. With the removal of the phospholipid the concentration of these metals in the finished biodiesel drops below the detection limit, and at the same time improve the oxidation and thermal stability of FAME.

In both refining processes the phosphorus content of sunflower oil methyl esters was ca. 50 ppm after degumming (Figures 5 and 6). In the chemical refining, the phosphorus content of methyl esters was 4 ppm after caustic neutralization and 1 ppm after bleaching, while after the ultra fine degumming and bleaching step of physical refining it was <1 ppm.

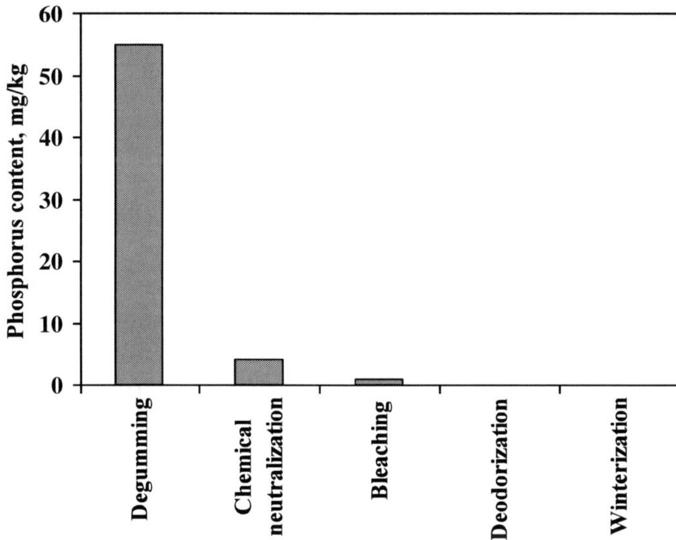

Figure 5. Change of phosphorus content of biodiesels through chemical refining of sunflower oil.

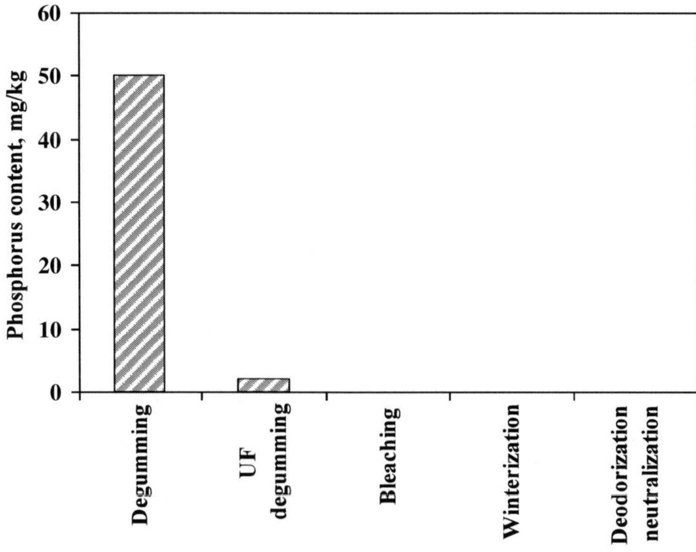

Figure 6. Change of phosphorus content of biodiesels through physical refining of sunflower oil.

Polyunsaturated Fatty Acid Content

Vegetable oils, including sunflower oils naturally do not contain polyunsaturated (≥ 4 double bonds) fatty acids; these are only formed along high temperature treatments and refining steps (Figures 7 and 8). In the new specifications of the currently valid FAME standard the toleration of the multiple unsaturated fatty acids content is maximum 1%. Formation of these compounds could be lowered by the optimization of process parameters of the refining processes. This is essential, since polyunsaturated compounds worsen the storage, thermal and oxidation stability of biodiesel, and they can cause undesired deposits and fouling in the fuel injector or in the combustion chamber.

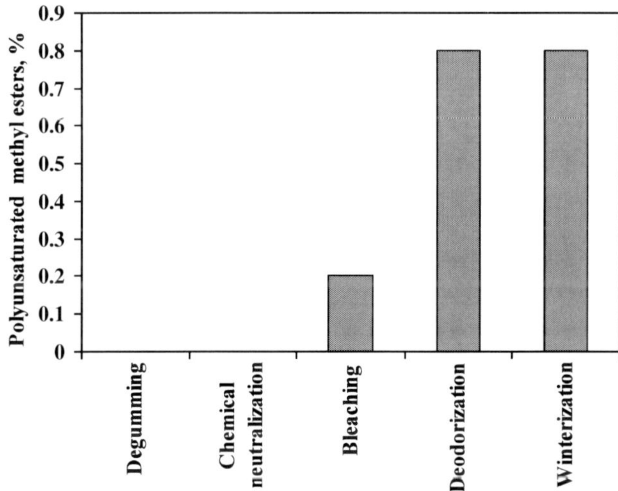

Figure 7. Change of polyunsaturated methyl ester content of biodiesels through chemical refining of sunflower oil.

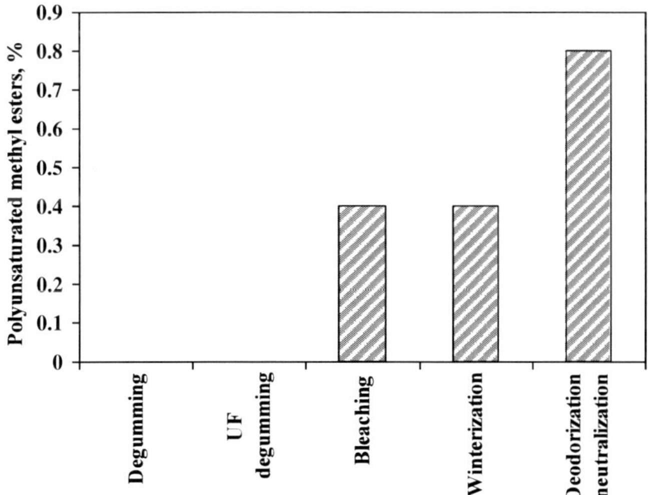

Figure 8. Change of polyunsaturated methyl ester content of biodiesels through physical refining of sunflower oil.

Acid Number

The acid number characterizes the free fatty acid content of vegetable oils. During the refining processes, the acid number of biodiesels is decreased below 0.5 mg KOH/g, only after the neutralization step (Figures 9 and 10). This has a major significance when alkaline catalyst is used for the transesterification, because free fatty acids are forming soaps with the catalyst. In our case, it has no significant consequence, since it did not affect the activity of the enzyme.

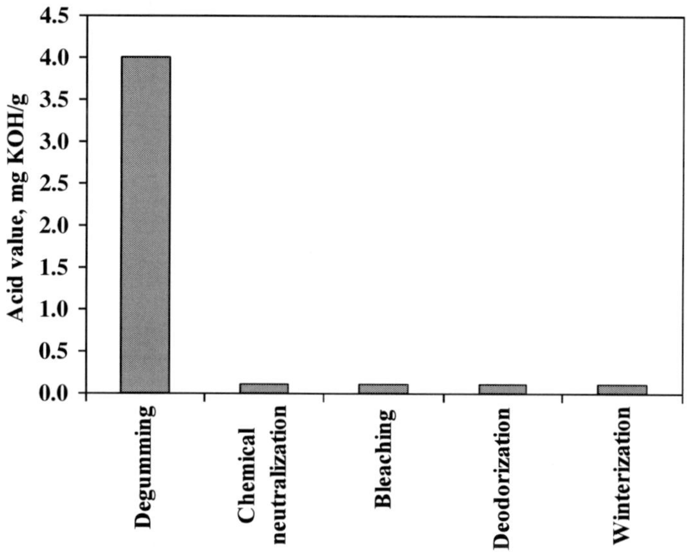

Figure 9. Change of acid number of biodiesel through chemical refining of sunflower oil.

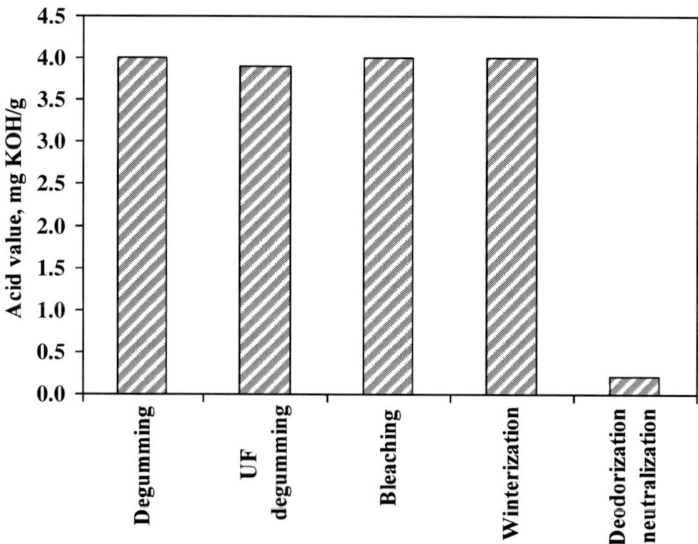

Figure 10. Change of acid number of biodiesel through physical refining of sunflower oil.

Cold Filter Plugging Point (CFPP)

The cold filter plugging point of the crude sunflower oil was +16°C. This value dropped substantially (to ca. +5°C) after transesterification. The average molecular weight of the product mixture decreased significantly during the transesterification, which can be observed through the favorable change of the viscosity and CFPP value. During the refining process the CFPP values are slightly decreased in every step. After the winterization step the CFPP value is significantly decreased because the saturated compounds were eliminated (Figures 11 and 12).

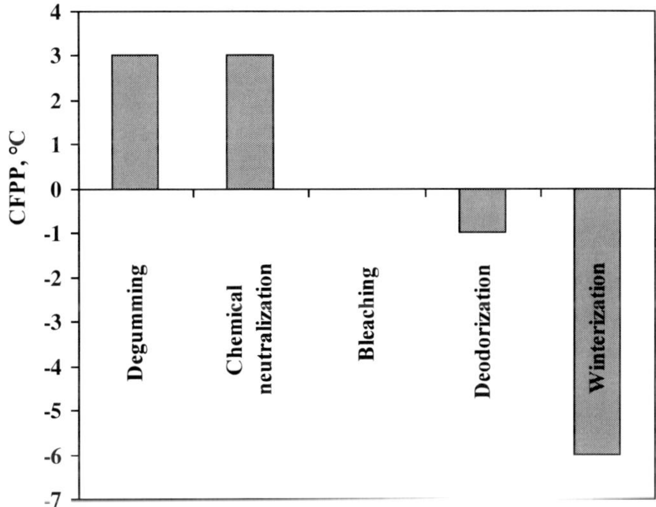

Figure 11. Change of CFPP of biodiesels through chemical refining of sunflower oil.

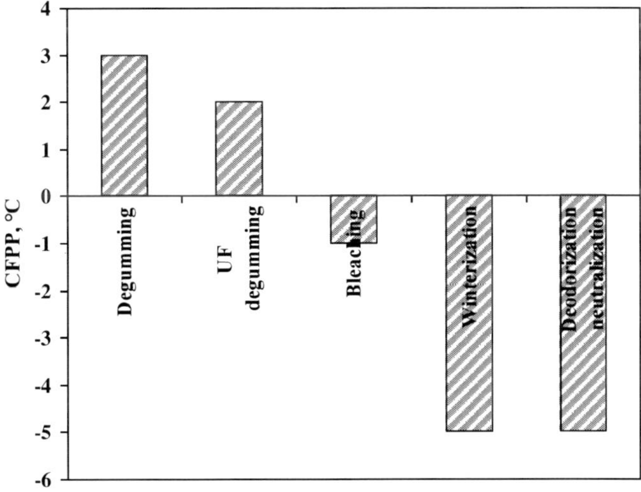

Figure 12. Change of CFPP of biodiesels through physical refining of sunflower oil.

Oxidation Stability

From the aspect storage stability it is an important factor how long can biodiesel preserves its quality. Unwanted, highly corrosive compounds are formed from the esters during oxidation, which can cause serious damage in the fuel system, as well as in the engine. In the refining phase oxidation stability can be improved by eliminating the precursors of oxidation. This stage is known as deodorization. Figures 13 and 14 well illustrate that the oxidation stability of biodiesel which was improved by 50% with deodorization.

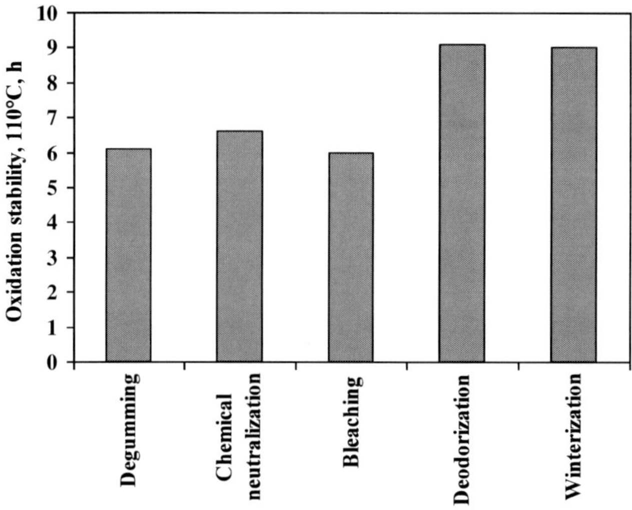

Figure 13. Change of oxidation stability of biodiesels through chemical refining of sunflower oil.

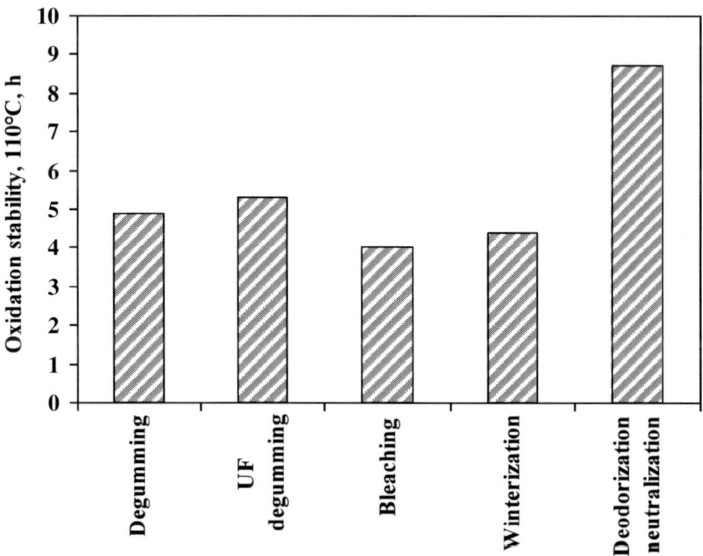

Figure 14. Change of oxidation stability of biodiesels through chemical refining of sunflower oil.

The quality and performance properties of biodiesels produced from chemically and physically refined sunflower oils is compared in Table 9. High quality biodiesel can be produce from both types of sunflower oils (chemically / physically refined) with the selected enzyme catalyst and adjusted process conditions. Similar results were obtained in case of experiments when we applied rapeseed oil (Table 8.).

Table 9. The main properties of sunflower oil methyl esters

Properties	Refining		EN 14214 Standard
	chemical	physical	
FAME content, %	97.8	97.6	>96.5
Density, 15°C, kg/m^3	885	887	860-900
Viscosity, 40°C, mm^2/sec	4.56	4.49	3.50-5.00
Flash point, closed cup, °C	139	132	>101
Sulphur content, mg/kg	3	5	<10
Carbon residue (on 10% distillation residue), %	0.17	0.21	<0.30
Water content, mg/kg	410	370	<500
Copper strip test (3 hours, 50°C), class	Class 1	Class 1	Class 1
Oxidation stability, 110°C, h	9.0	8.7	>6.0
Acid number, mg KOH/g	0.1	0.2	<0.50
Iodine number, g/100g	107	108	<120
Polyunsaturated (≥4 double bonds) methyl esters	0.8	0.8	<1
Methanol content, %	0.05	0.06	<0.20
Monoglycerides content, %	0.25	0.2	<0.80
Diglycerides content, %	0.05	0.1	<0.20
Triglycerides content, %	<0.05	<0.05	<0.20
Free glycerol , %	<0.05	<0.05	<0.02
Total glycerol, %	0.05	0.1	<0.25
Phosphorus content, mg/kg	<1	<1	<4
Cold filter plugging point, °C	-6	-5	to be measured

To summarize the results of the experiments it can be stated that the raw sunflower and rapeseed oils are unsuitable for direct transesterification, as they contain many contamination, which reduce the activity of the catalyst; therefore the produced biodiesel could not meet the requirements of the standard.

THE EFFECTS OF VEGETABLE OILS WITH NEW FATTY ACID COMPOSITION ON THE ENZYME-CATALYZED TRANSESTERIFICATION AND THE FAME QUALITY

The applicability of FAME from sunflower oil and rapeseed oil, expediently chosen and cultivated on our request in Hungary, was studied with enzyme-catalyzed transesterification, furthermore to determine the effect of their different fatty acid compositions on the finished FAME properties reached at same transesterification conditions [9, 44-48].

Experiments were carried out with feedstocks having unconventional fatty acid components. From biodiesel quality point of view the oleic acid methyl ester was found as the best component - based on our theoretical considerations (Tables 10 and 11) [23, 24] – (cetane number, CFPP, oxidation stability) [9, 47, 48].

Table 10. Melting point of several fatty acid methyl esters contained in biodiesels

Name	Symbol	Melting point, °C
Lauric acid methyl ester	C12:0	+ 5.2
Myristic acid methyl ester	C14:0	+19.0
Palmitic acid methyl ester	C16:0	+30.0
Stearic acid methyl ester	C18:0	+39.1
Oleic acid methyl ester	C18:1	-19.9
Linoleic acid methyl ester	C18:2	-35.0
Linolenic acid methyl ester	C18:3	-46.0
Eurcic acid methyl ester	C22:1	-1.2

*The first number represents the number of carbon atoms and the second means the number of double bonds in the molecule

Table 11. The induction periods and relative oxidation rates of several fatty acid methyl esters (25 °C)

Fatty acid methyl ester		Induction period, h	Relative oxidation rate
Name	symbol		
Stearic acid methyl ester	C18:0	-	1
Oleic acid methyl ester	C18:1	82	100
Linoleic acid methyl ester	C18:2	19	1200
Linolenic acid methyl ester	C18:3	1.34	2500

The enzyme-catalyzed transesterifications of the 'A', 'B', 'C' sunflower oils and the 'D' and 'E' rapeseed oils (see composition and properties in Table 3 and 4) were carried out at the advantageous 50°C temperature; 4:1 metanol:triglyceride molar ratio; with the presence of 12% enzyme catalyst and with methanol addition in 8 portions through 16 hours. The yield of the ester phase relating to the reaction time and its methyl ester content was nearly the same; the difference was only 1-2 abs% (Table 12). After the 16 hours of transesterification of two-two types of rapeseed and sunflower oil in every case the amount of ester phase approached the theoretical yield to 99.5-99.8% and their methyl ester content was ≥96.5% [9, 47].

Test for reproducibility were made also, and the fatty acid methyl ester content of the products getting from the same feedstocks resulted differed maximum in only ±0.4 abs % from the average (Table 12). This difference was approx. the same as the allowed deviation of the analytical method (EN 14103). So the technology can be reproduced at laboratory conditions and it was not affected by the different fatty acid compositions of the feedstocks.

The properties of the fatty acid methyl esters made from the 4 different feedstocks with enzyme-catalyzed transesterification (Table 13) were in line with the current standard. The cold filter plugging point value could be enhanced with cold flow improver additive.

Based on the CFPP results the „D ME" fatty acid methyl ester was the best. The main reason for this was the relatively high unsaturated methyl ester content (C18:1; C18:2; C18:3) and the lowest saturated (C18:0, C20:0) methyl ester content (earlier Table 4). Small concentration increase of the saturated compounds increased the CFPP significantly (Table 10).

Table 12. The ester content of the products obtained during reproducibility experiments

Number of experiment	Ester content, %			
	B ME	C ME	D ME	E ME
1.	97.6	99.5	98.6	97.8
2.	97.4	99.7	98.9	97.9
3.	97.8	99.1	98.7	98.0
4.	97.1	99.4	98.6	97.4
5.	97.5	99.6	98.4	97.7
Average	±0.4	±0.4	±0.3	±0.4

Table 13. Main properties of the sunfower oil and rapeseed fatty acid methyl esters

Properties	Fatty acid methyl esters*				EN 14214 standard
	B ME	C ME	D ME	E ME	
Ester content, %	96.9	99.5	98.6	97.2	≥96.5
Density, 15°C, g/cm^3	0.8855	0.886	0.888	0.887	0.860-0.900
Viscosity, 40°C, mm^2/sec	4.30	4.20	4.32	4.40	3.5-5.0
Flash point, closed cup, °C	120	122	132	122	≥101
Cold filter plugging point, °C	-5	-11	-14	-6	to be measured
Sulphur content, mg/kg	5	5	7	6	≤10
Carbon residue (on 10% distillation residue), %	0.15	0.12	0.20	0.19	≤0.3
Water content, mg/kg	257	280	274	285	≤500
Copper strip test (3 hours, 50°C), class	Class 1	Class 1	Class 1	Class 1	Class 1
Oxidation stability, 110°C, h	12.8	14.1	12.0	9.2	≥6.0
Acid value, mg KOH/g	0.10	0.08	0.15	0.25	≤0.5
Methanol content, %	0.1	0.1	0.1	0.1	≤0.2
Monoglycerides content, %	0.28	0.20	0.32	0.41	≤0.8
Diglycerides content, %	0.11	0.10	0.13	0.14	≤0.2
Triglycerides content, %	0.12	<0.05	0.11	0.1	≤0.2
Free glycerol , %	0.01	<0.005	0.015	0.015	≤0.02
Total glycerol, %	0.11	0.14	0.13	0.12	≤0.25
Iodine value, g I$_2$/100g	113	95	103	102	≤120
Phosphorus content, mg/kg	<1	<1	<1	<1	≤4
K content, mg/kg	<1	<1	<1	<1	≤5
Cetane number	51	56	53	50	≥50

* „B ME": fatty acid methyl ester produced from sunflower oil „B"

Based on the oxidation stability results the „C ME" product was the best, because it contained the lowest amount of double bounds (which are sensitive for oxidation) and it contained the highest amount of oleic acid methyl esters (approx. 91.8%) (Table 4).

A good compromise has to be found among the oxidation stability, the CFPP and the cetane number. One solution could be the harmonization of the saturated fatty acid, and the unsaturated (single, double triple bonds) fatty acid content and of course the carbon number of the feedstocks (vegetable oils). Based on the previous results we concluded that the best solution is the application of feedstocks with high oleic acid content Its cetane number (59) is higher than the standard minimum 51, the CFPP is adequate low (-19°C) and the oxidation stability is also good. The ideal biodiesel would be produced from vegetable oil containing only oleic acid. The mentioned results supported our theoretical indications to make the experiments.

Furthermore we concluded that the different fatty acid structure of the trygliceride feedstocks did not affect the catalytic activity of *Candida antarctica* enzyme.

INVESTIGATION OF THE ENZYME CATALYZED TRANSESTERIFICATION OF USED FRYING OILS

Production possibility of biodiesels from non edible raw materials, e.g., properly pretreated used frying oils was studied. The results of these experiments are presented in more publications [49-55]. The quality and the fatty acid composition of the applied used frying oils (UFO) are summarized in Tables 14. and 15. The results of the transesterification (see experimental part, with extender temperature intervall) reactions are summarized in Table 16.

Table 14. The main properties of the applied feedstocks

Properties	Feeds*			
	UFO	HOSO	UFO50	UFO20
Density, 15°C, g/cm^3	0.9216	0.9133	0.920	0.918
Viscosity, 40°C, mm^2/sec	39.8	33.7	35.6	36.2
Sulphur content, mg/kg	10	5	8	6
Nitrogen content, mg/kg	12	6	8	7
CFPP, °C	42	36	39	37
Acid value. mg KOH/g	2.3	0.5	1.5	1.2
Iodine value, gI$_2$/100g	132	89	103	92

*UFO: used frying oil, HOSO: high oleic sunflower oil, UFO50: 50-50% mixture of UFO and HOSO, UFO20: 20-80% mixture of UFO and HOSO

Table 15. The fatty acid composition of the applied feeds

Fatty acid composition, %*	UFO	HOSO	UFO50	UFO20
C16:0	7.9	2.3	0.1	3.4
C16:1	0.2	0.1	4.8	0.1
C18:0	3.8	2.2	0.1	2.5
C18:1	26.3	91.8	3.5	78.7
C18:2	60.3	2.8	59.7	14.3
C18:3	0.2	0.1	30.2	0.1
C20:0	0.2	0.3	0.1	0.4
C20:1	0.2	0.2	0.3	0.2
C22:0	0.6	0.1	0.2	0.2
C22:1	0.0	0.1	0.8	0.1
C24:0	0.2	0.0	0.0	0.0
C24:1	0.1	0.0	0.2	0.0

*The first number represents the number of carbon atoms and the second means the number of double bonds in the molecule

Based on the results of the experiment in case of UFO marked feedstock we were able to produce FAME with high methyl ester content. However, the temperature was higher approx. with 4°C than in case of sunflower oil. The iodine value of the product mixture was high (>120 gI$_2$/100g). The ester content was lower with 3% than the requirement of the standard, but the other properties are fulfilled. Table 14 and 15. contains the important quality properties and the fatty acid composition of high oleic acid content sunflower oil; moreover 50:50% (UFO50) and 80:20% (UFO20) mixtures of used frying oil and the mentioned sunflower oil, too. These two raw materials are important because of the iodine value of the biodiesel produced from the used frying oil alone exceedes the requirements of the valid standard (≤ 120 gI$_2$/100 g). Note that during the transesterification the iodine value of the products do not change compared to the feedstock [48-50].

Of course, the standard can be fulfilled with blending the separately produced fatty acid methyl esters from lower and higher iodine number feedstocks in a suitable ratio. We do not suggest this solution because the reaction temperature of transesterification of the used frying oil was higher with 3°C compared to the sunflower and/or rapeseed oil. Another point is *Candida antarctica* lipase enzyme activity loss (Table 17.).

Table 16. The properties of the produced biodiesels

Properties	Fatty acid methyl ester prodcts*			
	UFOME	HOSOME	UFO50ME	UFO20ME
FAME content,%	94.1	99.0	96.9	98.1
Density, 15°C, g/cm^3	0.883	0.879	0.881	0.880
Viscosity, 40°C, mm^2/sec	4.8	4.6	4.9	4.7
Sulphur content, mg/kg	6	3	4	3
Nitrogen content, mg/kg	8	4	7	5
CFPP, °C	-10	-4	-7	-8
Acid value, mg KOH/g	0.37	0.15	0.23	0.18
Iodine value, gI$_2$/100g	129	86	101	90

* „UFOME": fatty acid methyl ester produced from used frying oil „UFO"

Table 17. Ester content of the products after the reuses of immobilized *Candida antarctica* lipase

Number of uses	Ester content, %				
	Feed				
	SOME	HOSOME	UFOME	UFO50ME	UFO20ME
1	97.6	99.5	94.1	96.9	98.4
2	97.3	99.3	92.8	96.0	97.7
3	96.5	99.0	90.0	93.5	96.6
4	95.0	98.5	84.1	88.3	94.1
5	91.1	97.7	73.2	79.9	90.2
6	83.8	96.4	-	-	-
7	-	94.0	-	-	-

The reason is that the higher polar molecules of the used frying oil adsorb on the surface of the immobilized enzyme catalyst, and they stayed on it partially, covered, and decreased the number of the catalytic active sites. On the other hand this the multiple applications at ca.

3°C higher temperature decreases the activity of the catalyst, too. Another important advantage of the application of the mixed feedstock is that mix has lower viscosity, and the fresh vegetable oil part assists the renewal of the catalyst surface, pushing downward the non-transesterable higher molecules of the used frying oils.

CONCLUSION

Based on the results of our experiments the the main conclusions are the following:

- Investigation of different vegetable oil refining processes:

  - o Raw sunflower oils are unsuitable for direct transesterification because they contain many contaminations, which reduce activity of the catalyst; therefore the produced biodiesel does not meet the requirements of the standard.
  - o Comparing the products reached from chemically or physically refined sunflower oils we concluded that there are no significant differences between the quality and performance characteristics. Using the selected enzyme catalyst and process conditions, it is able to produce sunflower oil fatty acid methyl esters in line with the EN 12414 standard.
  - o It is definitely necessary to use one of the refining processes to be able to produce high quality biodiesel which is preserving its quality for a long time.

- Effect of vegetable oils with new fatty acid compositions:

  - o Based on our experimental results and by applying the favourable process parameters we were able to produce fatty acid methyl esters in laboratory scale. All finished products, reached from the 4 types of vegetable oils, fit the valid EN 14214 standard. The significantly different fatty acid composition does not affect the effectiveness of enzyme catalyzed transesterification.
  - o The quality characteristics of the obtained products satisfied the requirements of the valid standard.
  - o The quality and performance properties of the fatty acid methyl ester obtained from the sunflower oil with high oleic acid content (approx. 92%) were the best. The CFPP value was favourable (-11°C), good oxidation stability (14 h) and relatively low iodine value (95 g I_2/100g) were obtained.
  - o It can be stated that biodiesels with adequate oxidation stability and CFPP value can only be produced from vegetable oils with high oleic acid content. The ideal biodiesel should contain only oleic acid methyl ester, which could be reached by continuing of the plant improvement.

- Investigation of the enzyme catalyzed transesterification of used frying oils:

  - o In case of used frying oil (UFO) feedstock we were able to produce FAME with high methyl ester content. However, the temperature was higher approx. with

4°C than in the case of sunflower oil. The ester content was lower with 3% than the requirement of the standard, but the other properties fulfilled it except for the CFPP.

- o The properties of the products obtained from the sunflower oil with high oleic acid content and their mixtures (UFO50 and UFO20) with the used frying oil fulfilled the requirements of the valid standard. These two raw materials are important because of the iodine value of the biodiesel produced from the used frying oil alone exceedes the requirements of the valid standard (≤ 120 gI$_2$/100 g) during the transesterification the iodine value of the products do not change compared to the feedstock.

- o The maxium reuse of the immobilized *Candida antarctica* lipase was possible only for two times without significant activity loss after the separation of the product mixture from the catalyst when the used frying oil was the feedstock. In case of the other feedstocks the number of possible reuses was 5-7 times. The reason is that the higher polar molecules of the used frying oil adsorb on the surface of the immobilized enzyme catalyst, and they stayed on it partially, covered, and decreased the number of the catalytic active sites.

REFERENCES

[1] Demirbas, A. Biorefineries: Current activities and future developments. *Energy Conversion and Management*, 2009, *50*, 2782–2801.

[2] DIRECTIVE 2003/30/EC OF THE EUROPEAN PARLIAMENT AND OF THE COUNCIL, (2003)

[3] DIRECTIVE 2009/28/EC OF THE EUROPEAN PARLIAMENT AND OF THE COUNCIL, (2009)

[4] Leung, D.Y.C., Wu, X., Leung, M.K.H. A review on biodiesel production using catalyzed transesterification. *Applied Energy*, 2010, *87*, 1083-1095.

[5] Lam, M. K., Lee, K. T., Mohamed, A. R. Homogeneous, heterogeneous and enzymatic catalysis for transesterification of high free fatty acid oil (waste cooking oil) to biodiesel: A review. *Biotechnology Advances*, 2010, *28*, 500–518.

[6] Lee, J.-S., Saka, S., Biodiesel production by heterogeneous catalysts and supercritical technologies. *Bioresource Technology*, 2010, *101*, 7191-7200.

[7] Helwani, Z., Othman, M.R., Aziz, N., Fernando, W.J.N., Kim, J. Technologies for production of biodiesel focusing on green catalytic techniques: A review. *Fuel Processing Technology*, 2009, *90*, 1502–1514.

[8] Karleskind, A., Wolff, J.-P. Oils and Fats manual Volume 2. (ISBN: 2-7430-0087-2), Intercept Ltd., 1996.

[9] Kovács, F., Hancsók, J., Holló, A., Tolvaj, K. *Factors affecting the quality of biodiesel, Motor Fuels* 2004, Slovak Republik, Vyhne, 14-17 June 2004. In Proceedings (ISBN 80-968011-3-9), MF-2231, 11 pp

[10] Canakci, M., Van Gerpen, J. Biodiesel production from oils and fats with high free fatty acids. *Transactions of the ASAE*, 2001, 44(6), 1429-1436.

[11] Masatosi, M., Murakami, S. *Method and equipment of refining plant oil and waste vegetable oil into diesel engine fuel.* US 6364917 B1, April 2, 2002.

[12] Montefrio, M. J., Xinwen, T., Obbard, J. P. Recovery and pre-treatment of fats, oil and grease from grease interceptors for biodiesel production. *Applied Energy*, 2010, *87*, 3155-3166.

[13] Baker, D. BioDiesel – A role for SAE. SAE Motor Vehicle Council, March 31, 2006.

[14] Canakci, M., Sanli, H. Biodiesel production from various feedstocks and their effects on the fuel properties. *Journal of Industrial Microbiology and Biotechnology*, 2008, *35*, 431-441.

[15] Refaat, A.A. Correlation between the chemical structure of biodiesel and its physical properties. *International Journal of Environmental Science and Technology*, 2009, 6, 677-694.

[16] Moser, B. R. Biodiesel production, properties and feedstocks. *In Vitro Cellular & Developmental Biology – Plant*, 2009, *45*, 229-266.

[17] Karmakar, A., Karmakar, S., Mukherjee, S. Properties of various plants and animals feedstocks for biodiesel production. *Bioresource Technology*, 2010, *101*, 7201-7210.

[18] Mittelbach, M., Remschmidt, C.: "Biodiesel the comprehensive handbook", Martin Mittelbach Publisher, Graz, 2004, (ISBN: 3 200 00249 2)

[19] Singh, M., Singh, S., Singh, R.S., Chisti, Y., Banerjee, U.C. Transesterification of primary and secondary alcohols using *Pseudomonas aeruginosa lipase. Bioresource Technology*, 2008, *99*, 2116-2120.

[20] Deng, L., Xua, X., Haraldsson, G.G., Tan, T., Wang, F. Enzymatic Production of Alkyl Esters Through Alcoholysis: A Critical Evaluation of Lipases and Alcohols. *JAOCS*, 2005, *82*, 341-347.

[21] Knothe, G. Dependence of biodiesel fuel properties on the structure of fatty acid alkyl esters. *Fuel Processing Technology*, 2005, *86*, 1059-1070.

[22] Lang, X., Dalai, A.K., Bakhshi, N.N., Reaney, M.J., Hertz, P.B. Preparation and characterization of bio-diesels from various bio-oils. *Bioresource Technology*, 2001, *80*, 53-62.

[23] Schmiedel, H.P. FAME im Mineralölprodukten insbesondere HEL. Deutsche Wissenschaftliche Gesellschaft für Erdöl, Ergas und Kohle e.V., 2001.

[24] Dimming, Th., Haupt, J., Radig, W. Biodiesel aus Pflanzenölen und nativen Recyclingprodukten. *Freiberger. Forschungshefte.A*, 1999, *A852*, 34-60.

[25] Schuchardt, U., Sercheli, R., Vargas, R.M. Transesterification of Vegetable Oils: a Review. *Journal Braz. Chem. Soc.* 1998, *9*, 199-210.

[26] Ma, F., Hanna, M.A. Biodiesel production: a review. *Bioresource Technology*, 1999, *70*, 1-15.

[27] "Biofuels action plan", COM(2006) 34, (2006), 31pp

[28] Mata, T.M., Martins, A.A., Caetano, N.S. Microalgae for biodiesel production and other applications: A review. *Renewable and Sustainable Energy Reviews*, 2010, *14*, 217–232.

[29] Balat, M. Potential alternatives to edible oils for biodiesel production – A review of current work. *Energy Conversion and Management*, 2011, *52*, 1479-1492.

[30] Knothe, G., Bagby, M. O., Ryan, III, T. W. Cetane numbers of fatty compounds: influence of compound structure and of various potential cetane improvers. *SAE Techncal Paper Series*, No. 971681, 1997

[31] Knothe, G., Dunn, R.O., Bagby, M.O. Biodiesel: The Use of Vegetable Oils and Their Derivatives as Alternative Diesel Fuels. in *ACS Symp. Ser. 666 (Fuels and Chemicals from Biomass)*, American Chemical Society, Washington, DC, 1997, pp. 172-208.

[32] Dunn, R.O. Antioxidants for improving storage stability of biodiesel. *Biofuels, Bioproducts and Biorefining*, 2008, *2*, 304–318.

[33] Sarin, A., Singh, N.P., Sarin, R., Malhotra, R.K. Natural and synthetic antioxidants: Influence on the oxidative stability of biodiesel synthesized from non-edible oil. *Energy*, 2010, *35*, 4645-4648.

[34] Karavalakis, G., Hilari, D., Givalou, L., Karonis, D., Stournas, S. Storage stability and ageing effect of biodiesel blends treated with different antioxidants. *Energy*, 2011, *36*, 369-374.

[35] Kivevele, T.T., Mbarawa, M.M., Bereczky, A., Laza, T., Madarasz, J. Impact of antioxidant additives on the oxidation stability of biodiesel produced from Croton *Megalocarpus oil. Fuel Processing Technology*, 2011, *92*, 1244-1248.

[36] Knothe, G. Designer Biodiesel: Optimizing Fatty Ester Composition to Improve Fuel Properties. *Energy and Fuels*, 2008, *22*, 1358-1364.

[37] Lee, I., Johnson, L. A., Hammond, E. G. Reducing the crystallization temperature of biodiesel by winterizing methyl soyate. *Journal of the American Oil Chemists' Society*, 1996, *73*, 631–636.

[38] Gonzalez Gomez, M. E., Howard-Hildige, R., Leahy, J. J., Rice, B. Winterisation of Waste Cooking Oil Methyl Ester to Improve Cold Temperature Fuel Properties. *Fuel*, 2002, *81*, 33–39.

[39] Sahlabji, T., Wichmann, H., Dieckmann, H., Jopke, P., Bahadir, M. Chemical Aspects of Fractional Crystallization of Fatty Acid Methyl Esters Produced from Waste Fats. *Clean*, 2007, *35*, 323–328.

[40] Knothe, G. Improving Biodiesel Fuel Properties by Modifying Fatty Ester Composition. *Energy and Environmental Science*, 2009, *2*, 759-766.

[41] Pérez, Á., Casas, A., C. M. Fernández, Ramos, M. J., Rodríguez, L. Winterization of peanut biodiesel to improve the cold flow properties. *Bioresource Technology*, 2010, *101*, 7375-7381.

[42] Pinzi, S., Garcia, I. L., Lopez-Gimenez, F. J., Luque de Castro, M. D., Dorado, G., Dorado, M. P. The Ideal Vegetable Oil-based Biodiesel Composition: A Review of Social, Economical and Technical Implications. Energy and Fuels, 2009, *23*, 2325–2341.

[43] Bélafi-Bakó, K., Kovács, F., Gubicza, L., Hancsók, J. Enzimatic biodiesel production from sunflower oil by *Candida antarctica* lipase in solvent-free system. *Biocatalysis and Biotransformation*, 2002, *20*, 437-439.

[44] Kovács, F., Hancsók, J., Jánosi, L.: *"Effect of Fatty Acid Composition on the Quality of Biodiesels"*, 2nd World Conference and Technology Exhibition on Biomass for Energy, Industry and Climate Protection, Rome, 10-14 May 2004, Proceedings (Editors: van Swaaij, W. P. M., Fjällström, F. Helm, P. Grassi A.) (ISBN 88 89407 04 2), Volume II., 1577-1580

[45] Hancsók, J., Kovács, F., Krár, M., Magyar, Sz., Neményi, M. Effect of fatty acid composition on the quality of fatty acid methyl esters in enzymatic transesterification. *Chemical Engineering Transactions*, 2005, *7*, 291-295.

[46] Krár, M., Hancsók, J., Kovács, F.: *"Effect of fatty acid composition on performance characteristics of biodiesels"*, 42th IPC Conference, Slovakia, Bratislava, 11-12 October 2005

[47] Krár, M., Hancsók, J., Kovács, S., Neményi, M.: *"Enzymatic Transesterification of Hungarian Rapseed and Sunflower Oils"*, European Congress of Chemical Engineering – 6, Denmark, Copenhagen, 16-20 September 2007, ECCE-6 Book of Abstracts Vol(1), 193-194. , Conference CD, 9 pages

[48] Kovács, F., Hancsók, J., Krár, M., Nagy, G., Neményi, M.: *"Enzymatic transesterification of high quality sunflower oils"*, 14th European Biomass Conference & Exhibition. Biomass for Energy, Industry and Climate Protection, Paris, 17-21 October 2005. Proceeding of the International Conference held in Paris (ISBN 88 89407 07 7), 1415-1418.

[49] Kovács, S., Krár, M., Beck, Á., Hancsók, J.: *"Enzymatic transesterification of used frying oils"*, 15th European Biomass Conference & Exhibition. Biomass for Energy, Industry and Climate Protection, Germany, Berlin, 7-11 May 2007, Proceedings (ISBN 978 88 89407 59 X) 1747-1750.

[50] Kovács, S., Krár, M., Hancsók, J.: *"Comparison of Conventional (combined acidic/alkali-catalyzed) and enzyme-catalyzed transesterification of used frying oil"*, 43rd International Petroleum Conference, Slovakia, Bratislava, 24-26 September 2007, Conference CD, 10 pages

[51] Kovács, S., Krár, M., Hancsók, J.: *"Investigation of Enzyme-Catalysed Transesterification of Vegetable Oils"*, 8th International Symposium MOTOR FUELS 2008, Tatranské Matliare, 23-26 June 2008. Proceedings Part 2 (ISBN 978 80 969710 2 2), 552-561.

[52] Krár, M., Kovács, S., Hancsók, J.: *"Enzymatic transesterification of used frying oils"*, 5th International Conference on Environmental Catalysis, Belfast, 31 August – 3 September 2008, In Proceedings 433.

[53] Kovács, S., Krár, M., Hancsók, J.: *"Investigation of enzyme-catalyzed transesterification of used frying oils"*, Interfaces'08, Sopron, 24-26 September 2008, In Proceedings (ISBN 978-963-9319-87-5), 11 pages

[54] Kovács, S., Krár, M., Hancsók, J. Investigation of enzyme-catalyzed transesterification of used frying oils. *Hungarian Journal of Industrial Chemistry*, 2008, *36*, 59-63.

[55] Kovács, S., Hancsók, J.: *"Enzymatic transesterification of used frying oils"*, 4th International Bioenergy Conference, Finland, Jyväskylä, 31 August – 4 September 2009, In Proceedings, Part II. (ISBN 978-952-5135-44-2), 613-618.

In: Biodiesel: Blends, Properties and Applications   ISBN: 978-1-63117-024-9
Editors: Jorge M. Marchetti and Zhen Fang   © 2014 Nova Science Publishers, Inc.

Chapter 15

PILOT PLANT STUDIES OF BIODIESEL PRODUCTION USING UFO (USED FRYING OIL) AS RAW MATERIAL

José Aracil[*]*, Mercedes Martínez, Marta Arriaga,*
Abderrahim Bouaid, Rubén Oliveros, Marta Serrano,
Marcos Sánchez, Noureddin El-Boulifi and Elisa Peña
Laboratory of Development, Optimization and Scale-up of Integrated Chemical and
Biochemical Processes, Chemical Engineering Department, Faculty of Chemistry,
Universidad Complutense de Madri, Ciudad Universitaria s/n, Madrid, Spain

ABSTRACT

In recent years, the development of alternative fuels from renewable resources, like biomass, has received considerable attention. Biodiesel is defined as fatty acid methyl or ethyl esters from vegetable oils, when it is used as fuel in diesel engines and heating systems.

In this context, the use of used frying oil (UFO) as a raw material for biodiesel production has gained special interest since it has provided a potential added value to what has traditionally been treated as a residue.

Methyl or ethyl esters are the product of transesterification of vegetable oils with alcohol (methanol/ethanol) using an alkaline catalyst. In addition, the process yields glycerol, which has large applications in the pharmaceutical, food and plastics industries.

In the present work, the process of biodiesel production for pilot plant using sunflower oil as raw material with methanol and using potassium hydroxide as catalyst has been studied providing a mass and energy balance of the process.

The biodiesel quality meets European specifications defined by pr EN 14214:2003(E). The obtained results have been used for industrial scale up of the process.

---

[*] To whom correspondence should be addressed, e-mail: jam1@quim.ucm.es

1. INTRODUCTION

1.1. Biodiesel Definition

American Standards for Testing and Materials (ASTM) defines biodiesel as Mono-alkyl Esters of long chain fatty acids from renewable lipids as vegetable oils used in compression ignition engines or boilers.

Table 1. Acceptance properties and limits of FAME according to EN 14214 standard

PROPERTY	Units	LIMITS Minimum	LIMITS Maximum
SPECIFICATIONS AS DIESEL			
Ester content	%(w/w)	96,5	
Density (15ºC)	kg/m^3	860	900
Viscosity (40ºC)	mm^2/s	3,5	5
Inflammation Point	ºC	120	-
CFFP	ºC		10$^{(1)}$ 0$^{(2)}$
Sulfur content	mg/kg		10,0
Carbonaceous residue	%(w/w)	-	0,30
Sulfur ash contain	%(w/w)	-	0,02
Water	mg/kg		500
Total contamination	mg/kg	-	24
Copper Strip Corrosion		Class 1	
SPECIFICATIONS AS FAME			
Acid Value	mg KOH/g	-	0,5
Iodine Value		-	140
Methanol Content	%(w/w)	-	140
Monoglyceride Content	%(w/w)	-	0,80
Diglyceride Content	%(w/w)	-	0,20
Monoglyceride Content	%(w/w)	-	0,20
Free Glycerin	%(w/w)	-	0,02
Total Glycerin	%(w/w)	-	0,25
Group I metals (Na+K)	mg/kg	-	5,00
Group II metals (Ca+Mg)	mg/kg	-	5,00
Phosphorous Content	mg/kg	-	10,00

[1] Winter (October 1st-March 31st)
[2] Summer (April 1st-30th September)
[3] Spain Specification

This definition includes, apart from methyl and ethyl esters, other monoalcohol esters as isopropyl and butyl esters for example.

Nevertheless, in Europe biodiesel is considered the mixture of this kind of esters while it is assured a performance in some fuel quality test according to EN 1421 standard (proved for the Standard European Committee: Fuels. Methyl esters of fatty acids (FAME) for diesel engines. Requirements and tests) whose specifications are showed in Table 1.

Biodiesel can be used with different mixture percentages according to the supplier, and the current legislation in each country, however the most used percentages are:

B100 (100% biodiesel)
B70 (70% biodiesel)
B50 (50% biodiesel)
B30 (30% biodiesel)
B10 (10% biodiesel)
B5 (5% biodiesel)

For all of them, the applicable standard regarding quality is given by the EN 14214 and 1700/2003 Spanish decree-law.

1.2. Biodiesel Production Process

1.2.1. Main Reaction

The most common process in order to obtain biodiesel is transesterification reaction. In this reaction, triglycerides, which are present in oils, react with a low molecular weight alcohol (methanol usually) in order to get methyl esters of fatty acids that can be used as diesel fuels. Transesterification is a reaction equilibrium and it does not involve big energy changes. Furthermore, methanol can split the ester through methanolysis reaction, with the adequate catalyst.

According to the scheme reaction [1] each mol of oil reacts with three mol of methanol in order to obtain three moles of methyl ester and one of glycerol. This glycerol must be removed from the reaction stream.

$$
\begin{array}{ccccc}
\text{CH}_2\text{-O-C-R}_1 & & \text{CH}_3\text{O-C-R}_1 & & \text{CH}_2\text{-OH} \\
\text{CH-O-C-R}_2 & + \ 3\text{CH}_3\text{OH} \longrightarrow & \text{CH}_3\text{O-C-R}_2 & + & \text{CH-OH} \\
\text{CH}_2\text{-O-C-R}_3 & & \text{CH}_3\text{O-C-R}_3 & & \text{CH}_2\text{-OH} \\
\text{Triglyceride} & \text{Methanol} & \text{Methyl Ester} & & \text{Glycerol}
\end{array}
\tag{1}
$$

As catalyst, it is usually used alkaline homogeneous catalysts like alkaline hydroxides (KOH or NaOH) and alkaline alkoxides (sodium and potassium methoxides).

Chemically, transesterification consists of three reversible and consecutive reactions where triglycerides are turned in diglyceride, monoglyceride and glycerine, consecutively. In each stage aforementioned, it is released a methyl ester mol. Equations [2], [3] and [4] represent these reactions, where k_1, k_3 and k_5 are the direct kinetic constants while k_2, k_4 and k_6 are the inverse kinetic constants for each of the three consecutive and reversible reactions.

$$
\begin{array}{llll}
CH_2\text{-}O\text{-}CO\text{-}R_1 & & & CH_2\text{-}O\text{-}CO\text{-}R_1 \\
| & & & | \\
CH\text{-}O\text{-}CO\text{-}R_2 & + CH_3OH & \underset{k_2}{\overset{k_1}{\rightleftarrows}} \; CH_3\text{-}O\text{-}CO\text{-}R_3 \; + & CH\text{-}O\text{-}CO\text{-}R_2 \\
| & & & | \\
CH_2\text{-}O\text{-}CO\text{-}R_3 & & & CH_2\text{-}OH \\
\\
\text{Triglyceride} & \text{Methanol} & \text{Methyl Ester} & \text{Diglyceride}
\end{array}
\tag{2}
$$

$$
\begin{array}{llll}
CH_2\text{-}O\text{-}CO\text{-}R_1 & & & CH_2\text{-}O\text{-}CO\text{-}R_1 \\
| & & & | \\
CH\text{-}O\text{-}CO\text{-}R_2 & + CH_3OH & \underset{k_4}{\overset{k_3}{\rightleftarrows}} \; CH_3\text{-}O\text{-}CO\text{-}R_2 \; + & CH\text{-}OH \\
| & & & | \\
CH_2\text{-}OH & & & CH_2\text{-}OH \\
\\
\text{Diglyceride} & \text{Methanol} & \text{Methyl Ester} & \text{Monoglyceride}
\end{array}
\tag{3}
$$

$$
\begin{array}{llll}
CH_2\text{-}O\text{-}CO\text{-}R_1 & & & CH_2\text{-}OH \\
| & & & | \\
CH\text{-}OH & + CH_3OH & \underset{k_6}{\overset{k_5}{\rightleftarrows}} \; CH_3\text{-}O\text{-}CO\text{-}R_1 \; + & CH\text{-}OH \\
| & & & | \\
CH_2\text{-}OH & & & CH_2\text{-}OH
\end{array}
\tag{4}
$$

Reactants from transesterification, oil and methanol, are two immiscible liquids. For that reason, at the beginning of the reaction, the reaction system is formed by two phases and the kinetic process is controlled by mass transfer. As soon as methyl esters are formed, those act as cosolvent in the reaction system, because they are miscible oil and methanol both. Therefore, the system is formed by one phase and chemical stage can control the reaction.

The mathematical model, which describes the kinetics of tranesterification, ("Kinetic of sunflower oil Methanolisis", Vicente et al. Ind. Eng. Chem. Res. 2005, 44, 5447-5454), considers a pseudohomogeneous kinetic model of second order for the direct reactions as for the inverse reactions ([2],[3] and [4]). That is to say, first order respect to all the reactants and products. In addition, the reaction is first order respect to the catalyst and, on the other hand, reactions do not take place in catalyst absence. In conclusion, the kinetic equations for all the components which are presented in transesterification are:

$$\frac{d[TG]}{dt} = -k_1'[TG][MeOH] + k_2'[ME][DG]$$

$$\frac{d[DG]}{dt} = -k_1'[TG][MeOH] - k_2'[ME][DG] + k_3'[DG][MeOH] + k_4'[ME][MG]$$

$$\frac{d[MG]}{dt} = k_3'[DG][MeOH] - k_4'[ME][MG] - k_5'[MG][MeOH] + k_6'[ME][GL]$$

$$\frac{d[GL]}{dt} = -k_5'[MG][MeOH] + k_6'[ME][GL]$$

$$\frac{d[ME]}{dt} = k_1'[TG][MeOH] + k_3'[DG][ME] + k_5'[MG][MeOH] - k_2'[ME][DG] - k_4'[ME][MG] - k_6'[ME][GL]$$

TG: Triglyceride　　　　　　　　　　MeOH: Methanol
DG: Diglyceride　　　　　　　　　　 ME: Methyl ester
MG: Monoglyceride　　　　　　　　　GL: Glycerol

1.2.2. Secondary Reactions

When the catalyst of transesterification is an alkaline catalyst (KOH, NaOH, methoxides), it exists a secondary reaction in water presence which is called saponification. Triglyceride saponification produces soaps and glycerol ([6]).

CH_2-O-CO-R　　　　　　　　　　　　　　　　　　　　　CH_2-OH

|

CH-O-CO-R　　+ 3KOH　　$\xleftarrow{\ H_2O\ }$　3R-COO-K^+　+　　CH-OH　　　[5]

|

CH_2-O-CO-R　　　　　　　　　　　　　　　　　　　　　CH_2-OH

Moreover, if the catalyst is a hydroxide, it exists another secondary reaction which is the neutralization of the free fatty acids of the oil ([6]).

$$R\text{-}COOH \quad + \quad KOH \quad \xleftrightarrow{\ H_2O\ } \quad R\text{-}COO\text{-}K^+ \quad + \quad H_2O$$

[6]

Both reactions are undesirables (and faster than the main reaction from a kinetic point of view) because they cause a yield reduction of methyl ester phase as well as process catalyst consumption. For this reason, water from high-humidity oil must be eliminated by evaporation and reaction conditions must be taken in account (especially temperature and catalyst amount) to reduce the most the saponification.

Influence of the Reaction Parameters.

Different parameters have influence on transesterification reaction: molar ratio between methanol and oil, temperature, pressure, reaction time, catalyst amount and agitation speed. These variables have influence on yield and selectivity of methyl esters.

Agitation speed must be high enough to avoid possible mass transfer limitations in the reaction and therefore, to obtain chemical reaction as the unique control stage.

The reaction takes place at atmospheric pressure, because higher pressures might decrease process profitability.

Time reaction depends on the amount of product that is wanted to obtain, reaction conditions and energy requirements. In order to obtain maximum conversions, reaction time is usually fixed in one hour.

Temperature reaction is determined for methanol boiling temperature (65ºC), for oil fusion temperature and for temperatures which make easier secondary soap reactions. Therefore, work temperature is fixed between 25 and 65ºC.

Catalyst concentration is between 0,5-1,5 % w/w respect to oil amount. When these catalysts are used, especial care must be taken in order to reduce the saponification. An excessive increase in this amount can produce emulsions which increase viscosity and gel formation.

According to stoichiometry, methanolisis requires three moles of alcohol for one of triglyceride but, in addition, it is a reversible reaction and it is necessary to use alcohol in excess in order to shift the equilibrium to products formation. For that reason, it must be used an alcohol excess which allows to obtain high conversions without making difficult product separation and purification. According to this, molar ratio between methanol and oil can swing between 4,5:1 and 7,5:1 which correspond to 75% and 125% methanol excess respect to the stoichiometry amount.

2. GENERAL DESCRIPTION OF THE PROCESS

The target of this process is to obtain biodiesel from used frying oil (UFO) as raw material. The process is based on the technology developed by the Chemical Engineering Department of the Universidad Complutense de Madrid "Method for the trans-esterification of triglycerides with monoalcohols having a low molecular weight in order to obtain light alcohol esters using mixed catalysts" (WO/2003/062358) where low molecular weight esters are obtained which –after a purification process- can be used as fuel, combustible or dissolvent.

Figure 1 shows a general scheme of the Biodiesel production process.

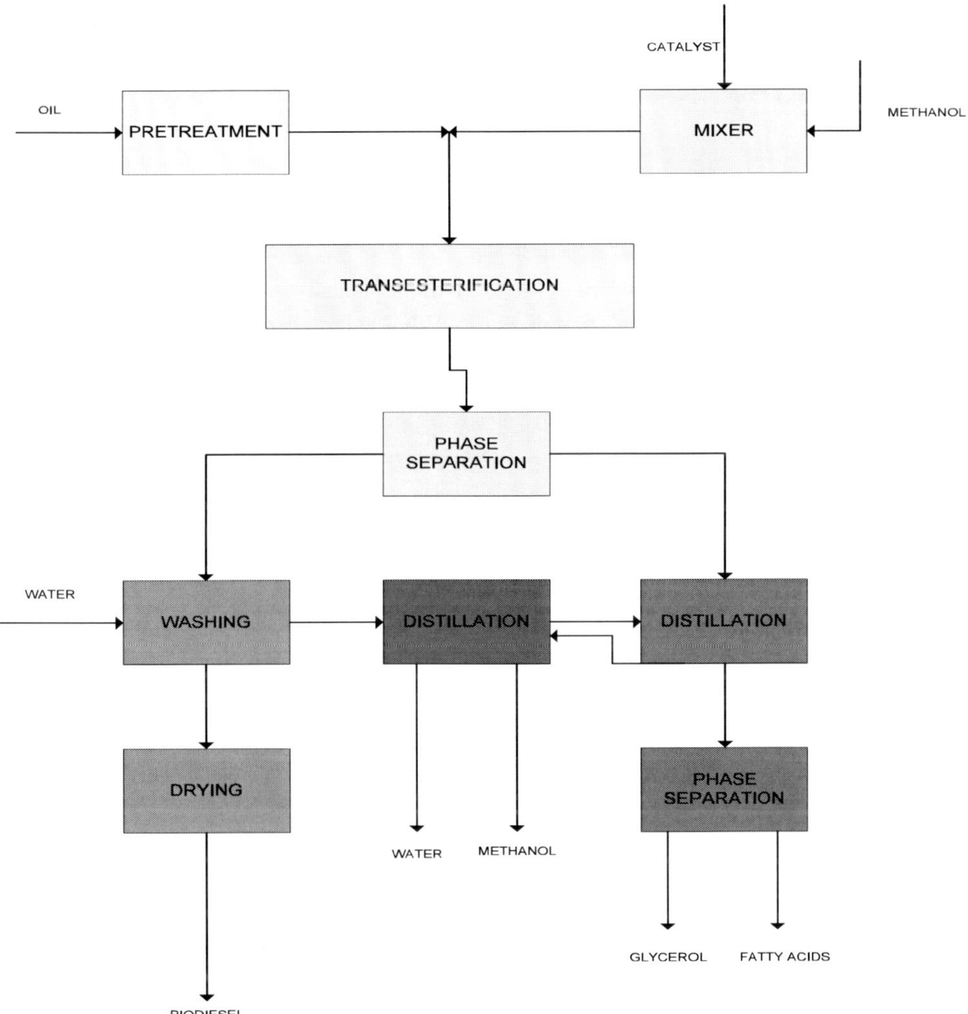

Figure 1. Scheme of the biodiesel production process from used frying oil.

In order to describe the process, it can be divided into five different sections:

- Storage of raw materials and products.
- Transesterification.
- Methyl ester purification.
- Glycerol purification.
- Methanol recovery.

2.1. Storage Section

The plant can use different types of vegetable oils that can be stored and classified in six storage tanks.

Methanol, which is used as a reactant for the transesterification is stored in two tanks, as well as the glycerol produced.

There are also eight tanks in this section for the storage of the Biodisel produced from the vegetable oils.

2.2. Transesterification Section

The grounds for the production of biodiesel are placed in the transesterification reaction in which the vegetable oils become methilic esters, using mixed catalytic system formed by an alkaline catalyst (potassium methoxide) and an alkaline salt of the fatty acids found in the oil (soaps), under mild temperature conditions (25-60°C) and at atmospheric pressure.

The oil, filtered and later stored in a buffer tank is introduced into a first reactor where the reaction takes place. In that reactor is also introduced the mixture methanol-potassium methoxide previously prepared in a mixer tank before the reactor.

This reactor reaches an oil conversion above 98%. The reactor is designed to allow certain variations on the reaction parameters. Thus, the reaction time can vary from 20 minutes until 2 hours, reaction temperature can vary from ambient temperature to 60°C in order to increase the reaction rate and stirring speed can also be modified to assure a homogenous mixture. At the outlet of each reactor is placed a decanter which separates glycerol from the methyl ester phase.

2.3. Methyl Ester Purification Section

The transesterification reaction of the oil with the mixture methanol-catalyst leads to a mixture of the two main products (methyl ester and glycerol) with the excess of methanol and also a limited amount of soaps (coming from the saponification of methyl esters).

The separation of the reaction mixture at the outlet of the reactor generates a heavy phase (glycerol that contains most of the soaps) which is leaded to the glycerol purification section.

The methyl ester phase (light phase also called crude methyl ester) contains impurities such as methanol, traces of glycerol and soaps. This phase is sent to a liquid-liquid extraction column using water as extraction fluid making soaps and methanol dissolve in the water that is then separated in a centrifuge to finally obtain purified methyl ester (light phase) and soapy water with methanol and glycerol traces that is sent to the methanol recovery section.

Purified methyl ester obtained by centrifugation is sent to a flash distiller in which the traces of water and methanol that still remains in this phase are removed.

This drying section contains a previous heating in a heat exchanger, an evaporator and a recirculation pump. The evaporation is made using a liquid ring vacuum pump which generates a slight depression in the drying tank that guarantees the methanol and water removal below the limits required by the European Standard. To avoid the crystal formation that –due to its high cloud point- can appear on the biodiesel when used in a motor, it is necessary to introduce an additive to low the Cold Filter Plug Point (CFPP) following the evaporator.

The water from the vacuum pump polluted with some methanol traces is recovered and sent to a methanol-water tank and chilled in a heat exchanger in order to reduce the effluents.

Dried and methanol free methyl ester is sent to the storage tanks.

2.4. Glycerol Purification Section

This stage of the process is only used when fatty acids from the transesterification reaction are to be recovered.

The heavy phase from the methyl ester centrifugation, that contains glycerol and soaps, is treated with sulphuric acid and water (dilution is needed to achieve a suitable density and viscosity for the operation) in a neutralization reactor where crystalized potassium sulphate, water and fatty acids from soap are obtained.

Potassium sulphate crystals are separated in a basket centrifuge where the crystals remain inside the mesh. Finally they are stored in a hopper in order to use them for fertilizers.

Once the glycerol phase has the soaps removed, still contains fatty acids, methanol and water. In order to recover methanol this phase is introduce into an evaporation circuit that operates at 100°C and 0.2 bar. This circuit is made up by a tubular heat exchanger and an evaporator that operates under vacuum thanks to a pump connected to the methanol recovery section in order to treat the gas phase from the evaporator.

Purified glycerol still contains fatty acids which are separated by cooling the stream to 40°C in a heat exchanger and introducing it into a centrifuge where fatty acids are removed, obtaining as well technologic quality glycerol with a concentration from 70% to 80% that is stored in tanks in order to be sold to different industries.

2.5. Methanol Recovery

Biodiesel wash water coming from the centrifuge carries a big amount of methanol and some traces of soap. The pH of this water is slightly alkaline due to the soap that it contains. On the other hand, it is necessary to remove these soaps before the stream is sent to the distillation column since they can cause packing fouling and accumulation which will cause an efficiency drop.

For that reason, washing water is sent to a neutralization tank where is treated with sulphuric acid until pH becomes neutral.

In the distillation column for methanol purification, there is an inlet of the washing water which comes from a buffer tank that has already collected the water from the sections of methyl ester and glycerol purification. Distillation is carried out in a packed column at atmospheric pressure where methanol is obtained on top with a purity of 99.5% to be condensed and partly sent to the column as reflux while the other part will be recycled to the process.

A cool compressor is also provided in order to condensate the possible gas outcome from the methanol condenser venting, reducing its temperature to -15°C, guaranteeing that way the minimum content of methanol in the gases generated by the process.

3. DESIGN BASIS

In this chapter the mass and energy balance for the biodiesel production using used frying oil (UFO) as raw material is shown. Data showed in this section are obtained from a production plant placed in Alcalá de Henares (Spain) for biodiesel production designed, set up and operated under UCM technical direction. The allowed acidity for this oil is 1% - otherwise, a pretreatment stage will be needed- and glycerol phase will be purified in order to recover the fatty acids from the soaps formed due to this acidity, to obtain fertilizer and finally, glycerol with purity around 80%.

3.1. Production Capacity

The biodiesel plant is designed for a capacity of 50.000 Tm/year based on a 8.000 hours/year basis.

3.2. Plant Availability

The plant operates 24 hours a day, seven days a week within a total of 334 days a year. Plant availability, referred to the percentage of operation time during a year (excluding the time the plant is off due to maintenance or breakdown) is 91.5%.

Biodiesel consumption is not seasonal, but it remains almost constant all over the year, so production capacity of the plant will be spread uniformly along the year. It is forecast that all biodiesel will be sold at a constant price so there is no need to vary the production due to business purposes.

Every raw material needed for the biodiesel production is supposed to be commercial products available all over the year and can be acquired through different suppliers.

3.3. Raw Materials Specifications

The plant can be used to process every type of oil seeds, in this case used frying oil is proposed as long as the basic requirements of constant quality and fixed production are met.

Table 2. Oil properties

Property	Units	Used Frying Oil
Density	(g/cm^3)	0.8906
Viscosity	(mm^2/s)	83
Acid value	(mg KOH/kg)	2,03
Free Fatty acids Content	(% oleic acid w/w)	1
Water	(% w/w)	0,1
Unsaponificables	(mg KOH/kg)	187,26
Iodine Value		105,94
Peroxide Value	(meq/kg)	12,56

Tables 2 and 3 show the composition and physical properties of the raw materials
Tables 4 and 5 show the composition and physical properties of the products.

Table 3. Properties of other raw materials

Property	Units	Methanol	Potassium Methoxide	Sulphuric Acid	Additive
Composition	(% weight)	>99,9	32	10-50	
Density	(g/cm^3)	0,792 (20°C)	0,99 (20°C)	1,8	0,9
Viscosity		0,74 (mm2/s) (20°C)			500-1000 cP
Solubility a 25°C		Miscible in water		110 g/ 100 ml water	
Molecular Weight	(g/mol)	32	70,12	98	
Boiling Point	(°C)	65	93	340	
Fusion Point	(°C)	-94		10	
Inflammation Point	(°C)	12			
Autoignition Temperature	(°C)	385			
Vapor Pressure	(kPa)	0,79 (20°C)		0,13 (146°C)	
Realtive vapor Density	(air=1)	1,1			
Explosion Limit	(% vol in air)	6 - 35,6			
Color		No color	No color	No color	
Aparience		Liquid	Liquid	Transparent Liquid	
Odor		Alcohol		Characteristical	
Humidity	(% peso)	< 0,1	< 0,1		

Table 4. Composition and physical properties of the products

	Glycerin		Washing Water	Potassium sulphate	Fatty Acids
Composition (% weight)	Composition			> 98 potassium sulphate <2 alkaline water	> 95 fatty acids <2 salts and glycerol
	GLYCERIN	80	2,75		
	WATER	17	88		
	SALTS		2,1		
	MONOGLYCERIDES	0,5	3,6		
	METHANOL	0,01	traces		
	IMPURITIES				
Density (g/cm^3)	1,112 (25°C)		0,9101 (25°C)		
Kinematic viscosity (mm^2/s)	362 (25°C)		1,375 (25°C)		

Table 5. Properties of biodiesel obtained from used frying oil

Property	Units	FAME UFO
Ester Content	% (w/w)	98
Density (15°C)	kg/m^3	873
Viscosity (40°C)	mm^2/s	3,8
Inflammation Point	°C	120
Sulfur Content	mg/kg	
Methyl Linoleate	% (w/w)	
Polyunsaturated ($\geq$ 4 double bonds)	% (w/w)	
Sulfated Ash Content	% (w/w)	
Water Content	mg/kg	traces
Oxidation Stability 110°C	h	6
Acid Value	mg KOH/ kg	0,013
Iodine Value	g I$_2$/100g	128
Methanol Content	% (w/w)	traces
Monoglycerides Content	% (w/w)	0.08
Diglicerides Content	% (w/w)	0.02
Triglycerides Content	% (w/w)	0.01
Free Glycerin	% (w/w)	traces
Total Gycerin	% (w/w)	traces

4. MASS AND ENERGY BALANCES

Table 5 shows the physical properties that have been considered to carry out the mass and energy balances.

Table 6. Physical properties of raw materials and products

	Molecular weight	Density	Viscosity	Cp (liq)	Cp (vap)	ΔH vap	Boiling Temperature
	g/mol	kg/m3	cP	kcal/kg°C	kcal/kg°C	kcal/°C	°C
Vegetable Oil	882	880		0,47			
Methylester	293			0,58			
Glycerol	92	1.260		0,58			290
Methanol	32	792	0,59	0,59	0,33	280	65
Sulphuric Acid	98	1.800					337
Water	18	1.000	1,00	1,00	0,50	500	100

4.1. Global Yield

In the described biodiesel production process from UFO is obtained for every 1000 kg of oil, 981 kg of biodiesel, 122 kg of glycerol and 8.5 kg of fertilizer.

Figure 2 represents the block diagram of the process which includes all sections.

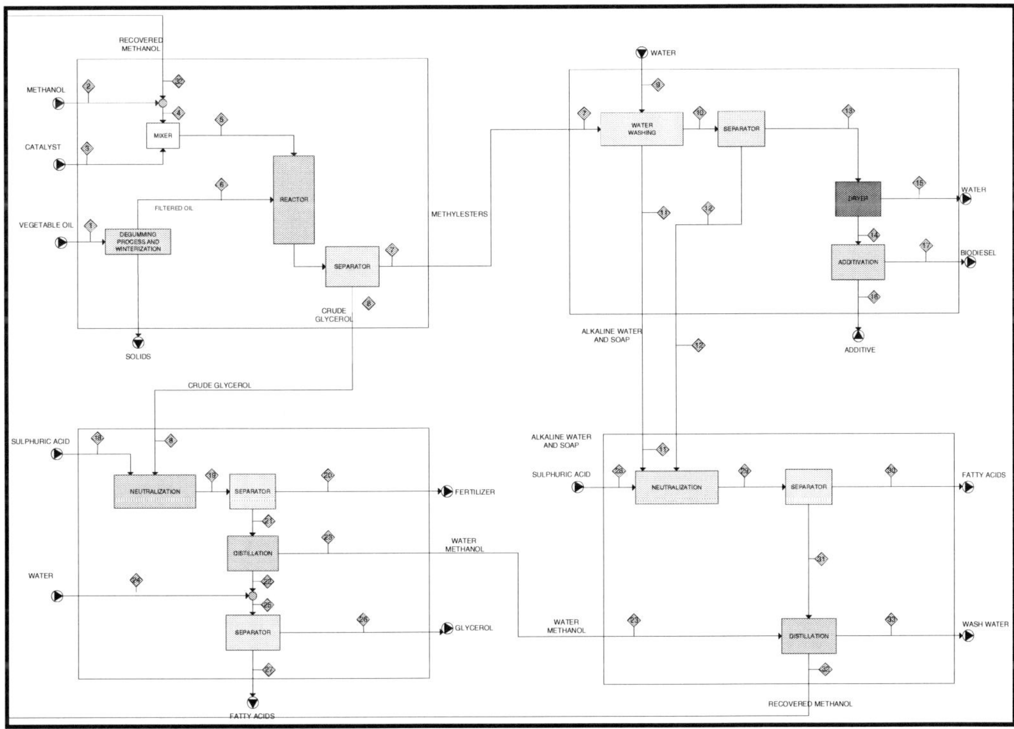

Figure 2. Block diagram for the biodiesel production process using oil with 1% of acidity.

4.2. Mass Balance

4.2.1. Global Mass Balance

In the next tables the amount of raw materials and products generated by the plant in one year are shown.

Table 7. Raw material consumption in a year

RAW MATERIAL	QUANTITY (Ton/year)
Vegetable Oil	50.954
Methanol	4.294
Potassium Methoxide 32% weight	1.578
Sulphuric Acid 50% weight	7.056
Water	9.082
Additive	to be determined

Table 8. Products and residues obtained in a year

PRODUCTS AND RESIDUES	QUANTITY (Ton/year)
Biodiesel	50.000
Glycerin	6.212
Water	9.559
Potassium Sulfate (fertilizer)	430
Free Fatty Acids	504

Table 9. Raw material consumption per hour

INPUTS (kg/h)	
Oil	6.369,2
Methanol	536,7
Potassium Methoxide (32% w/w)	197,2
Sulfuric Acid (50% w/w)	88,2
Water	1.135,3
Additive	to be determined
TOTAL kg/h	8.326

Table 10. Products and residues obtained per hour

OUTPUTS (kg/h)	
Biodiesel	6.237,9
Glycerin	776,5
Water	1.194,0
Fatty Acids	63,0
Potassium Sulfate (fertilizer)	53,8
TOTAL kg/h	8.326

4.2.2. Transesterification Section

This section includes oil filtering, mixing of catalyst and methanol and transesterification reaction.

Table 11. Mass balance for transesterification section

COMPOSITION (kg/h)	1	2	3	4	5	6	7	8	32
Methylester							6.230,6	12,6	
Glycerin							32,8	621,2	
Oil	6.299,9					6.299,9	34,5		
Monoglycerides							48,8	4,3	
Diglycerides							9,3	0,8	
Free Fatty acids	63,0					63,0			
Methanol		536,1	134,2	1.253,8	1.388,0		278,5	417,7	717,7
Potassium Methoxide			63,0		63,0		18,9	28,3	
Soaps							3,8	67,8	
Water	6,3	0,6		1,2	1,2	6,3		11,6	0,6
Additive									
Sulphuric Acid									
Salts									
Suspended Solids									
TOTAL (kg/h)	6.369,2	536,7	197,2	1.255,0	1.452,2	6.369,2	6.657,1	1.164,3	718,3

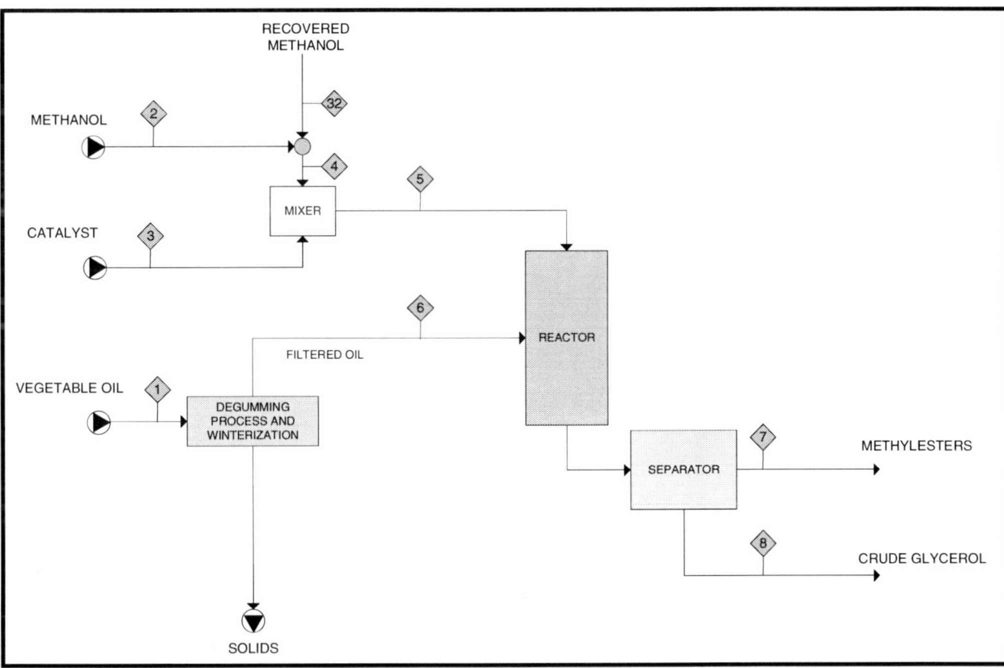

Figure 3. Transesterification section.

4.2.3. Methyl Ester Purification Section

This section includes the separation of the ester phase from the glycerol, water wash, methyl ester/water phase separation, methyl ester drying and additivation.

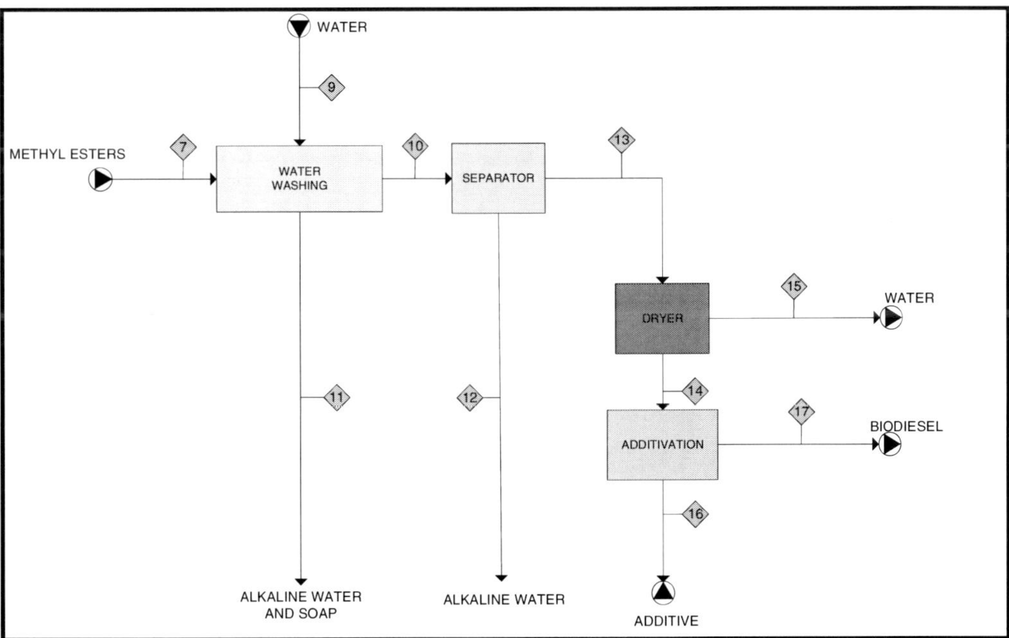

Figure 4. Methyl ester purification section.

Table 12. Mass balance for methyl ester purification section

COMPOSITION (kg/h)	7	9	10	11	12	13	14	15	16	17
Methylester	6.230,6		6.230,6			6.230,6	6.230,6			6.230,6
Glycerin	32,8			32,8						
Oil	34,5		0,6	33,9		0,6	0,6			0,6
Monoglycerides	48,8		5,0	43,5		5,0	5,0			5,0
Diglycerides	9,3		1,3	8,0		1,3	1,3			1,3
Free Fatty acids										
Methanol	278,5			278,5						
Potassium Methoxide	18,9			18,9						
Soaps	3,8			3,1	0,7					
Water		997,0	127,9	869,1	124,8	3,1	<500 ppm	3,1		<500 ppm
Additive									*	*
Sulphuric Acid										
Salts										
Suspended Solids										
TOTAL (kg/h)	6.657,1	997,0	6.365,4	1.287,8	125,5	6.240,6	6.237,5	3,1		6.237,9

* to be determined

4.2.4. Glycerol Purification Section

This section includes glycerol neutralization, fertilizer separation, neutralized glycerol distillation and fatty acid separation.

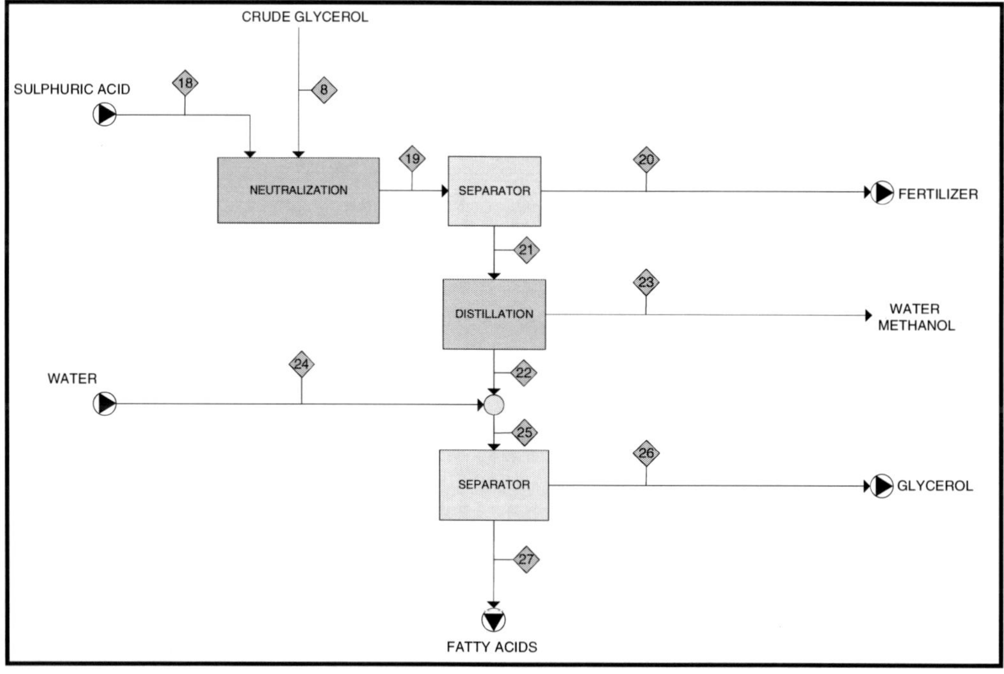

Figure 5. Glycerol purification section.

Table 13. Mass balance for glycerol purification section

COMPOSITION (kg/h)	8	18	19	20	21	22	23	24	25	26	27
Methylester	12,6		12,6		12,6	12,6			12,6	12,6	
Glycerin	621,2		621,2		621,2	621,2			621,2	621,2	
Oil											
Monoglycerides	4,3		3,8		3,8	3,8			3,8	3,8	
Diglycerides	0,8		0,6		0,6	0,6			0,6	0,6	
Free Fatty acids			59,7		59,7	59,7			59,7		59,7
Methanol	417,7		430,6		430,6		430,6				
Potassium Methoxide	28,3										
Soaps	67,8										
Water	11,6	30,3	41,9		41,9		41,9	138,3	138,3	138,3	
Additive											
Sulphuric Acid		30,3									
Salts			53,8	53,8							
Suspended Solids											
TOTAL (kg/h)	1.164,3	60,6	1.224,2	53,8	1.170,4	697,9	472,5	138,3	836,2	776,5	59,7

4.2.5. Methanol Recovery Section

This section includes soapy water neutralization and water/methanol distillation.

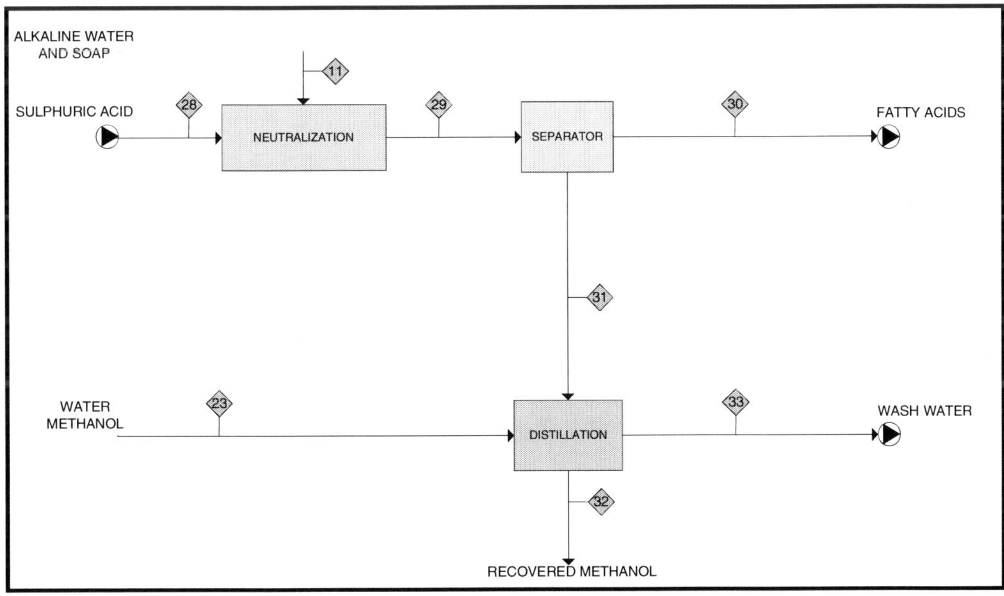

Figure 6. Methanol recovery section.

Table 14. Mass balance for methanol recovery section

COMPOSITION (kg/h)	11	12	28	29	31	30	31	23	33	32
Methylester										
Glycerin	32,8			32,8	32,8		32,8		32,8	
Oil	33,9			33,9	33,9		33,9		33,9	
Monoglycerides	43,5			43,5	43,5		43,5		43,5	
Diglycerides	8,0			8,0	8,0		8,0		8,0	
Free Fatty acids				3,3		3,3				
Methanol	278,5			287,1	287,1		287,1	430,6		717,7
Potassium Methoxide	18,9									
Soaps	3,1	0,7								
Water	869,1	124,8	13,8	1007,7	1007,7		1007,7	41,9	1049,0	0,6
Additive										
Sulphuric Acid			13,8							
Salts				24,5	24,5		24,5		24,5	
Suspended Solids										
TOTAL (kg/h)	1287,8	125,5	27,6	1440,9	1437,6	3,3	1437,6	472,5	1191,7	718,3

4.3. Energy Balance

4.3.1. Global Energy Balance

For the energy balance the calculation basis is 1 hour, and the reference state is: liquid phase, temperature 25°C and atmospheric pressure.

Table 15. Global energy balance of the biodiesel production process from UFO

INPUT HEAT	kcal/h
Reaction Heat	115.380
Wet Biodiesel Heat	162.884
To dry biodiesel	271.069
Glycerin-Methanol Flash	251.538
Methanol-.Water Column	858.118
TOTAL	1.658.989
REFRIGERATION	kcal/h
To cool biodiesel	67.903
To cool glycerin after flash	109.109
Methanol/water Condensated after flash	140.952
To cool methanol up the column	285.593
To cool water at bottom of the column	623.259
TOTAL	1.226.816

4.3.2. Transesterification Section

Enthalpy of the transesterification reaction is negligible. In this stage is just necessary to supply the heat for rising reactants temperature to reaction temperature.

Table 16. Energy balance and operation conditions for transesterification section

STREAM	3	4	5	6	7	8	32
Temperature °C	25	25	60	60	40	40	30
Boiling Temperature °C	-	65	-	-	-	-	65
Cp_l (kcal/kg°C)	-	0,594	-	-	≈0,5	-	0,594
Cp_v (kcal/kg°C)	-	0,329	-	-	-	-	0,329
Enthalpy (kcal/h)	0	0	25.574	89.805	57.829	151	2.128

Operation Conditions	
Temperature (°C)	REACTION
Inlet Temperature (°C)	25
Operation Temperature (°C)	60
Outlet Temperature (°C)	40
Total Input Heat(kcal/h)	115.380
Total Refrigeration (kcal/h)	67.903

4.3.3. Methyl Ester Purification Section

Heat supplied in this stage is used in the drying operation at 95°C and then, heat is released when cooling the biodiesel to 60°C in order to be stored.

Table 17. Energy balance and operation conditions for methyl ester purification section

STREAM	7	9	10	11	12	13	14	15	16	17
Temperature °C	40	25	25	25	25	70	95	40	25	40
Boiling Temperature °C	-	100	-	-	-	-	-	-	-	-
Cp_l (kcal/kg°C)	≈0,5	1	-	-	-	-	-	-	-	-
Cp_v (kcal/kg°C)	-	0,5	-	-	-	-	-	-	-	-
Enthalpy (kcal/h)	57.829	0	0	0	0	162.883	271.336	54.267	0	54.267

Operation Conditions	
Temperature (°C)	Dry Process
Input Temperature (°C)	70
Operation Temperature (°C)	95
Output Temperature (°C)	Water:25 Biodiesel:60
Total Input Heat(kcal/h)	Pre-heat: 162.884 Dry Process: 271.336
Total Refrigeration (kcal/h)	Biodiesel: 217.069

4.3.4. Glycerol Purification Section

Heat supplied in this stage is the needed amount to perform the flash evaporation of the traces of methanol and water that remain in the glycerol once it has been neutralized and sulphate has been separated.

Table 18. Energy balance and operation conditions for glycerol purification section

STREAM	8	18	19	20	21	22	23	24	25	26
Temperature °C	40	25	25	25	115	35	30	25	25	25
Boiling Temperature °C	-	-	-	-		-	-	100	-	-
Cp_l (kcal/kg°C)	-	-	-	-	0,58	-	-	1	-	-
Cp_v (kcal/kg°C)	-	-	-	-	-	-	-	0,5	-	-
Enthalpy (kcal/h)	151	0	0	0	251538	4832	1279	0	0	0

Operation Conditions	
Temperature (°C)	Flash Distillation
Input Temperature (°C)	25
Operation Temperature (°C)	115
Output Temperature (°C)	Glycerin: 35
	Methanol/Water: 35
Total Input Heat(kcal/h)	Flash: 251.538
Total Refrigeration (kcal/h)	Glycerin cooling: 109.109
	Metanol/water cooling: 140.952

4.3.5. Methanol Recovery

In this stage, the heat exchanged corresponds to the methanol/water distillation.

Table 19. Energy balance and operation conditions for methanol recovery section

STREAM	11	12	23	28	29	30	31	32	33
Temperature °C	25	25	30	25	25	25	25	30	30
Boiling Temperature °C	-	-	65	-	-	-	-	65	100
Cp_l (kcal/kg°C)	-	-	0,594	-	-	-	-	0,8	1
Cp_v (kcal/kg°C)	-	-	0,329	-	-	-	-	-	-
Enthalpy (kcal/h)	0	0	1.279,0	0	0	0	0	2.127,2	16.564,7

Operation Conditions	
Temperature (°C)	Distillation
Input Temperature (°C)	25
Operation Temperature (°C)	-
Output Temperature (°C)	Water: 40
	Methanol: 30
Total Input Heat(kcal/h)	Distillation: 858.118
Total Refrigeration (kcal/h)	Methanol (Head): 285.593
	Water (bottom): 623.259

5. CONCLUSION

In this chapter, it has been shown the fundaments, process and balances to produce biodiesel from used frying oil developed by the Universidad Complutense de Madrid. To conclude this chapter, the overall consumption of raw materials and utilities needed to successfully carry out this process is presented together with the yield of the process itself.

Table 20. Raw material consumption, utilities and yield of the process.

INPUTS	
Raw Materials	
Oil	1.000 kg
Methanol	84 kg
Potassium Methoxide	31 kg
Sulphuric Acid (50%)	14 kg
Utilities	
Heat vapour (540 kcal/kg)	480 kg
Refrigeration water	21 m^3
Process water	178 m^3
YIELD	
Products	
Biodiesel	979 kg
Glycerol (80%)	122 kg
Fatty Acids	10 kg
Fertilizer (salts)	9 kg

REFERENCES

[1] "Application of the Factorial Design of Experiments y Response Surface Methodology to Optimize Biodiesel Production".G. Vicente, A. Coterón, M. Martínez and J. Aracil. *Industrial Crops and Products*. 8 (1). 29-35 (1998).

[2] "Ésteres metílicos como combustibles. Materias primas y porpiedades". G. Vicente, M. Martínez and J. Aracil. *Tecno-Ambiente* 85 (10), 9-12 (1998).

[3] *"Method for the trans-esterification of triglycerides with monoalcohols having a low molecular weight in order to obtain light alcohol esters using mixed catalysts"* J. Aracil, M. Martínez, L. Fernando Bautista, M. Isabel Guijarro (WO/2003/062358)

[4] "Integrated Biodiesel production: a comparison of different homogeneous catalyst systems". G. Vicente, M. Martínez and J. Aracil. *Bioresource Technol*. 92 (3), 297-305 (2004).

[5] "Kinetics of Sunflower Oil methanolysis". G. Vicente, M. Martínez and J. Aracil.*Ind. Eng. Chem. Res*. 44(15), 5447-5454 (2005).

[6] "Pilot Plants Studies of Biodiesel Production". A. Bouaid, Y. Diaz, M. Martínez and J. Aracil. *Catalysis Today*. 106, 193-196 (2005).

[7] "Optimization of brassica carinata oil methanolysis for biodiesel production" G. Vicente, M. Martínez and J. Aracil. *Journal of the American Oil Chemists Society* 82, 899-904 (2005).

[8] "A comparative study of vegetables oils for Biodiesel Production in Spain". G. Vicente, M. Martínez and J. Aracil. Energy and Fuels. 20, 394-398 (2006).

[9] "Optimisation of integrated biodiesel production. Parte I: A study of the biodiesel purity and yield". G. Vicente, M. Martínez and J. Aracil. *Bioresource Technology*. 98, 1724-1733 (2006).

[10] "Optimisation of integrated biodiesel production. Parte II: A study of material balance". G. Vicente, M. Martínez and J. Aracil. *Bioresource Technology* 98, 1754-1761 (2006).

[11] "Kinetics of Brassica carinata oil methanolysis". G. Vicente, M. Martínez and J. Aracil. *Energy & Fuels* 20, 1722-1726 (2006).

[12] "Long storage stability of biodiesel from vegetables and used frying oils". A. Bouaid, M. Martínez and J. *Aracil. Fuel*, 86, 2596-2602 (2007).

[13] "A comparative study of the production of ethyl esters from vegetableoils as a biodiesel fuel optimization by factorial design". A. Bouaid, M. Martínez and J. Aracil, *Chemical engineering Journal* 134, 93-99 (2007).

[14] "Optimization of biodiesel production from joroba oil". A. Bouaid, L. Bajo M. Martínez and J. Aracil. *Process safety and environmental Protection* 85, 378-382 (2007).

[15] "Production of biodiesel from bioethanol and Brassica carinata oil: Oxidation stability study". A. Bouaid, M. Martínez and J. Aracil, *Bioresource Technology*, 100, 2234-2239 (2009).

[16] "Biorefinery approach for coconut oil valorisation: A statistical study". A. Bouaid, M. Martínez and J. Aracil. *Bioresource Tech.* 101, 4006-4012 (2010)

In: Biodiesel: Blends, Properties and Applications
Editors: Jorge M. Marchetti and Zhen Fang

ISBN: 978-1-63117-024-9
© 2014 Nova Science Publishers, Inc.

Chapter 16

BIODIESEL PROPERTIES AND EMISSIONS BASED ON THE TYPE OF BLEND AND RAW MATERIAL

Jorge Mario Marchetti[*]

Departamento de Física. Instituto de Física del Sur (IFISUR),
Universidad Nacional del Sur – CONICET.), Bahía Blanca, Argentina

ABSTRACT

Biodiesel production is gaining more and more relevance due to its environmental benefits such as its biodegradability, it has no CO or sulfur emissions, its produce a reduction in the CO_2 emissions, prolongs the engines lives, could be blend with regular diesel and it is produce from renewable and sustainable sources.

The name "biodiesel" is normally associated to any blend of regular diesel and biodiesel; however, the ASTM defines as biodiesel to the B100 (100% pure biodiesel) while any other blend is label in relation to the percentage of biodiesel in the mixture. The most common blend is B20, which has a 20% of biodiesel.

Based on the percentage of the biodiesel, the amount of regular diesel and the raw material from where the biofuel is produce, different properties, emissions and engine performance can be seen. A comparison of the biofuel based on these three main parameters will be presented with the aim of pointing out the differences of each blend, the influence on the emissions based on the quantity of biodiesel in the fuel and how this percentage affects the combustion of the internal engine of a car.

Even more, it will be shown international policies regarding the type of blend that should be used and when this should be reached.

1. INTRODUCTION

Biodiesel, a renewable fuel produce from liquid feedstock, such as vegetable oil or animal fat, is gaining more and more relevance over the years due to its environmental friendly characteristic such as having less particulate and less CO_2 and sulfur emissions, it is

[*] Corresponding author: jmarchetti@plapiqui.edu.ar, Tel: 54-291-4595100 ext: 2818, Fax:54-291-4595142

biodegradable, produce almost no emission of CO due to a better combustion because of the oxygen in its molecule, and many other.

This fuel is produce by the interaction of a triglyceride with an alcohol in the presence of a catalyst, normally a base homogeneous catalyst (sodium hydroxide or potassium hydroxide) [1-3]. The reaction takes place in three steps, where from triglycerides, diglycerides are produce; from this last one, monoglycerides are generated and from the latest glycerol is produce. In all the steps, fatty acid alkyl ester (biodiesel) is form. Figure 1 shows a condense scheme of the reaction.

$$
\begin{array}{llllll}
CH_2\text{-}COO\text{-}R_1 & & & CH_2\text{-}O\,H & & R_1\text{-}COO\text{-}R' \\
| & & Catalyst & | & & \\
CH\text{-}COO\text{-}R_2 & +\quad 3R'OH & \leftrightarrow & CH\text{-}OH & + & R_2\text{-}COO\text{-}R' \\
| & & & | & & \\
CH_2\text{-}COO\text{-}R_3 & & & CH_2\text{-}OH & & R_3\text{-}COO\text{-}R' \\
Triglycerides & Alcohol & & Glycerol & & Esters
\end{array}
$$

Figure 1. Typical transesterification reaction: a triglyceride with an alcohol to produce biodiesel. (R_n is carbon chain of different length)

As presented in Figure 1, different fatty acid could be found in a triglyceride molecule, giving this different properties and characteristic to each raw material used for biodiesel production. Even more if there is a presence of free fatty acid in the raw material due to a less pure vegetable oil.

If a more impure oil is used, with some free fatty acids in it, and sodium hydroxide is employ, the saponification reaction of the fatty acid will take place consuming the catalyst. The soap produce will also be a problem for the downstreaming separation of the biodiesel and the glycerol. The saponification reaction could be seen in Figure 2 [1-6]

$$
\begin{array}{llllll}
R\text{-}COOH & + & NaOH & \leftrightarrow & H_2O & + & R\text{-}COO\text{-}Na \\
Fatty\ acid & & Base & & Water & & Soap
\end{array}
$$

Figure 2. Saponification reaction of free fatty acid and sodium hydroxide

Due to environmental concern, such as the green house effect and global warming, the use of alternative fuel is increasing, having a key role in today´s environmental friendly energy [1-7]. However, it is important to assure that the compounds that are produce by combustion or use of alternative energies are less harmful than those produce from petroleum fuel. Is important to reduce the amount of CO_2, but also important not to increase the amount of nitrogen or sulfuric compounds in the process.

Since biodiesel composition changes accordingly to the raw material as well as the purity and the final quality, associated with the production process used, in this chapter will be presented a short summary of the different technologies as well as raw materials, and will be presented the result of different emissions tests and how the mayor pollutants concentration changes according to the raw material and fuel properties and composition. It will be presented a comparison with the result from petroleum in order to check which pollutant concentration decreases.

2. BIODIESEL COMPOSITION BASED ON THE RAW MATERIAL

Biodiesel is normally produce based on vegetable oils, it could also be done on animal fats but this is a source that is being under consideration nowadays [1-7]. The type of oil used by different countries to produce their oil is mainly associated with climate considerations, market and oil nutrients, considering that not all the oils could grow in all places. O´Brien et al. [8] have done a very good work, showing where some of the main seeds are being cultivate around the world, showing that for the mayor types of seed different countries could produce different amounts of oil

Vegetable oil composition is based in hydrocarbon chain, mainly from C16 to C24; however, each different vegetable oil has a different composition on its fatty acid percentage and type. O'Brien [8] presented a table with some composition for the most common vegetable oils, such as sunflower, soybean, rapeseed, olive, coconut, palm, palm kernel, corn, and many other. Another important point is that the vegetable oil main composition is unsaturated fatty acid chains, being those with one double bond the most common ones

Table 2, extracted from reference [8] shows these percentages.

Table 1. %wt. of the different fatty acid in the main types of oil [8]

Vegetable oil	Fatty acid composition % by weight								
	16:1	18:0	20:0	22:0	24:0	18:1	22:1	18:2	18:3
Corn	11.67	1.85	0.24	0.00	0.00	25.16	0.00	60.60	0.48
Cottonseed	28.33	0.89	0.00	0.00	0.00	13.27	0.00	57.51	0.00
Crambe	20.7	0.70	2.09	0.80	1.12	18.86	58.51	9.00	6.85
Peanut	11.38	2.39	1.32	2.52	1.23	48.28	0.00	31.95	0.93
Rapeseed	3.49	0.85	0.00	0.00	0.00	64.4	0.00	22.30	8.23
Soybean	11.75	3.15	0.00	0.00	0.00	23.26	0.00	55.53	6.31
Sunflower	6.08	3.26	0.00	0.00	0.00	16.93	0.00	73.73	0.00

In order to produce the biofuel, refine oil has been mainly use as raw material, due to its properties and also due to the possibility of carry on the reaction while using a basic homogeneous catalyst such as sodium hydroxide that has been used with great results for quite some time already [9-15].

Among the most use refine oil, each of them has different properties and different characteristic and it is required to know each of them in order to choose the best alternative to what it is desire to produce. Also, in the selection of the type of vegetable oil is related to the production location as established by O'Brien [8].

Table 3 presents some major properties for some vegetable oils focusing in some of the most important properties for different oils. It could be seen that in some cases the values are quite similar; this is the case of the specific gravity and the smoke point. On the other hand the flash point is quite diverse; changing from 225 °C to 327°C, another properties that has a wider range is the acid value, going from 0.6 to 6.6 accordingly.

Table 2. Physical Chemical Properties of different vegetable oil [16-28]

Parameters	Vegetable Oils						
	Soybean	Sunflower	Rapeseed	Olive	Canola	Palm	Corn
Viscosity	50*	50*	38*	84+	78.2+	107+	28.6*
Specific gravity	0.920	0.917	0.915	0.917	0.917	0.92	0.918
Flash point [°C]	280	274	246	225	327	300	255
Smoke point [°C]	241	246	204	242	230	235	236
Saponifcation value [mg KOH/]	193	190	186	190	193	195	
Cetane number	37.9	37.1	37.6	----	45	50	37.6
Acid value [mg KOH/g]	2.0	0.6	2.0-	6.6	----	1	0.15
Iodine value	123	130	107	85	115	48	122
HG [kJ/kg]	39623	39575	39709	40037	25110	39300	39500

+ these values are of dynamic viscosity [cP]

* these values are of kinematic viscosity [cSt]

The vegetable oil compared up supra are refine oil, however, due to an increase concern of oil for food vs. oil for fuel, less pure sources of vegetable oil, mainly waste, cooking and frying oil or soapstocks are being considered as new candidates of raw material. However, as much impure the source is, the more different its physical-chemistry properties are due to the amount of different pollutants in within.

Table 4 presents a comparison of some of the main properties for the refine, crude and waste oil, showing how the impurities affect some of the main physical-chemical variables. As it can be seen, the sulfur content increases considerable when comparing a refine with a waste oil, similar situation is for the acid value and the main pollutants that are present in a waste cooking oil. On the other hand, the viscosity does not change that much, it does increase due to impurities but not in orders of magnitude as other variables studied.

Table 3. Properties of the mayor types of oil

Property	Refine [29-30]	Crude [31]	Waste [32-34]
Kinematics viscosity [mm^2/s]	30.2	36	40.2
Carbon residue [wt%.]	0.24	0.278	0.18
Cetane number	38.01	39	---
Higher heating value [MJ/kg]	39.41	39.2	24.67
Ash content [wt%]	0.012	0.01	---
Sulfur content [wt%]	0.013	0.075	5
Iodine value [centigram I/g Oil]	112.86	125	13.2
Acid value [mg KOH/g Oil]	<0.2	variable	5.96

There are two alternatives sources of oil for biofuel that does not compete with food oil, the first one is non-edible oils, that could be cultivate with no food purposes, however, and its production will for sure use land fields that could have been use to produce other grains that are more suitable for human consumption. And therefore it could still produce some controversy. The second option is oil form microalgae, it has several advantages such as that they required CO_2 to grow, consuming a pollutant normally produce from internal combustion engines, ii) they grow in non drinkable water, not competing with drinkable sources, iii) algae could grow in land fields where no other vegetable oil could grow, iv) algae could be use as raw materials for several other chemical compounds easily produce.

Different types of microalgae has variable amount of oil content, mainly changing from 15 to 75% in so called low, middle and high oil content algae [35]. Some of the most common one are presented by Chisti [35], such as: Botryococcus braunii, Chlorella sp., Crypthecodinium cohnii, Cylindrotheca sp., Dunaliella primolecta, Isochrysis sp., Monallanthus salina, Nannochloris sp., Nannochloropsis sp., Neochloris oleoabundans, Nitzschia sp., Phaeodactylum tricornutum, Schizochytrium sp., Tetraselmis sueica,

Based on the type of oil, and therefore the characteristic fatty acid in within, the final biodiesel will have different properties due to the length of the carbon chain of the fatty acid; also due to the type of alcohol employed to carry on the reaction. As it could be seen from Table 5 [36], when comparing the methyl esters from fatty acid with no double bonds, the increase in the chain length produce a higher viscosity. However, when double bonds are being considered, the viscosity tend to decrease as long as the amount of double bonds increases, this could be seen for C18:0 ME, C18:1 ME, C18:2 ME and 18:3 ME. When looking into the type of alcohol used, the longer the alcohol chain is the higher viscosity of the biodiesel.

A similar scenario could be seen when comparing the cetane number, when the carbon chain or the alcohol chain increases, the cetane number increases as well, however, the presence of double bonds will decrease this property considerably

Table 4. Properties of different biodiesel. Extracted from reference [36]

FAEE	Melting point (°C)	Δ_cH (MJ/mol)	Kinematics viscosity (mm²/s)	Oil stability index (h)	Cetane Number	Lub (μm)
C12:0 ME	5	8.14	2.43	>40	67	416
C12:0 EE	-2		2.63	>40		
C14:0 ME	19	10.67	3.30	>40		353
C14:0 EE	12		3.52	>40		
C16:0 ME	31	10.67	4.38	>40	86	357
C16:0 EE	19		4.57		93	
C16:1 ME	-34	10.55	3.67	2.1	51	246
C16:1 EE	-37					
C18:0 ME	39	11.96	5.85	>40	101	322
C18:0 EE	32		5.92	>40	97	
C18:0 BE	28		7.59		92	
C18:1 ME	-20	11.89	4.51	2.5	59	290
C18:1 EE	-20		4.78	3.5	68	
C18:1 BE	-26		5.69		62	303
C18:2 ME	-35	11.69	3.65	1.0	38	236
C18:2 EE			4.25	1.1	40	
C18:3 ME	-52	11.51	3.14	0.2	23	183
C18:3 EE			3.42	0.2	27	

3. COMPARISON OF PRODUCTION TECHNOLOGIES

Before analyzing the emissions from biodiesel combustion and its use in vehicles, we would like to introduce some comments on how biodiesel is produced. This will be a short summary and it is important to notice that there are many other alternatives in the literature; this is just a compilation of some of them.

Figure 3 shows a summary of the production/separation/purification process for biodiesel. It could use different types of vegetable oil as well as different catalyst and several types of alcohol. The most common process involved the use of refine oil (any type) methanol and sodium hydroxide, being a very fast reaction but with several separation and purifications units due to the homogeneous nature of the catalyst

A more completed scenario, production from seed, could be seen in the work done by Balat [37]. For a more detail process when algae is used it is recommended to see the work done by Lin et al. [38].

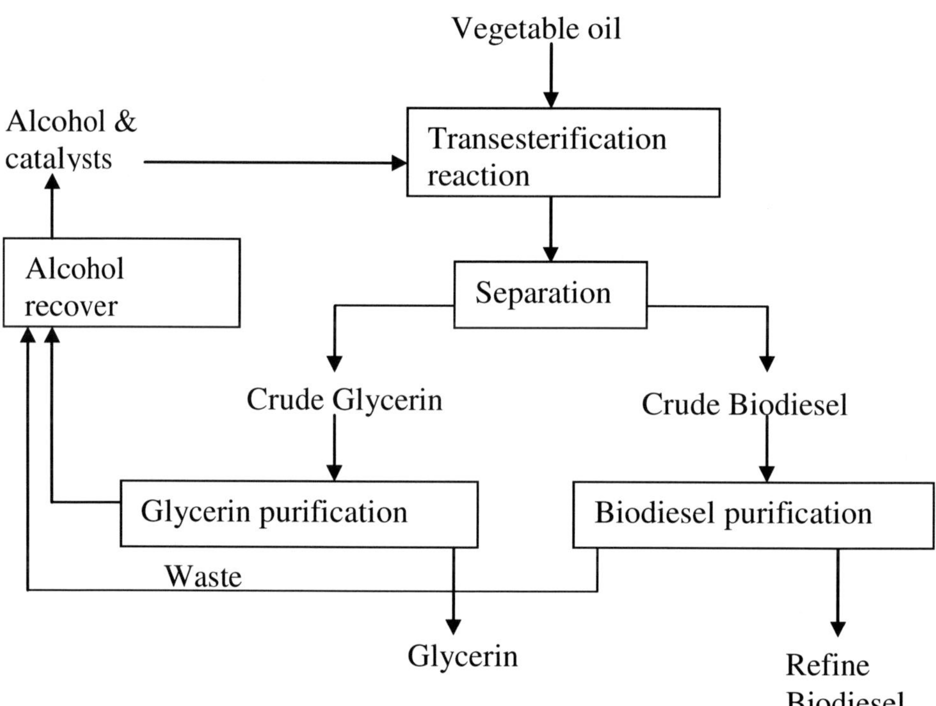

Figure 1. Flow diagram of a conventional process [1-4].

Other options include the use of acid homogeneous catalyst, normally sulfuric acid. In this case the reaction is 4000 times slower compared with the previous scenario [39-44]. However, the presence of free fatty acid could be treated as part of the raw material and not as an impurity. Similar to the base alternative, separation and purification are required due to the homogeneous state of the catalyst.

In order to avoid some of the neutralization and purification equipments, the use of heterogeneous catalyst shows up as a good new alternative. In this case, solid resin, acid and basic, could be used showing good results [45-53]. Some of the advantages of this catalyst are the less amount of equipment, ergo the lower investment, the more pure biodiesel and glycerin (allowing glycerin to be sold in a pharmaceutical grade). The easy the separation of the products and raw materials. However, there are some disadvantages to take into account, in some cases it is possible to carry on the reaction in the presence of water as well as other impurities such as fatty acid. Some of these solid catalysts get deactivated with water and therefore the raw material must be refine oil, getting back on the fuel vs. food debate. In some cases, the catalyst could treat some amount of water and fatty acid, however, in those scenarios; the reaction time is so large that it is not of industrial interest [54].

Enzymes seem as the best natural and environmental friendly option, it is produce in nature, it required not drastic conditions to work and it could treat some impurities in the raw material. Reaction temperature could not been too high, otherwise the organisms will die, some water is need it to start the reaction, but too much water will deactivated the catalyst, it will allow the transesterification as well as the esterification reaction to take place simultaneously [55-59]. One of the fastest technologies available nowadays considered the use of supercritical alcohols, either methanol or ethanol mainly. In this case the reaction temperature and pressure produce that the alcohol is in a supercritical state and therefore not catalyst is required. With this technology, a full conversion of quite non high quality oil could be reached in less than 5 minutes [60-67]. The non-presence of a catalyst allows the system to treat triglycerides as well as fatty acid and the presence of water with no concerns. Even more, in order to reduce the amount of alcohol, some works, such as those of Sawangkeaw et al. [68], Tan et al. [69], Yin et al. [70] and Han et al. [71] have shown that it is possible to reduce the alcohol molar ratio by increasing a secondary supercritical fluids such as CO_2, hexane, heptanes, propane or tetrahydrofuran, being those much cheaper. Other new options are being considered, such as the use of membrane reactors as well as monolithic catalyst and monolithic reactors. Each of them has mayor advantages over the conventional process but, unfortunately, there are some disadvantages as well [72-79].

In order to compare the mayor variables previously described, Table 9 has been developed where it can be seen the different technologies and the difference variables in each of them, some process variables have been considered, some other might have been chosen but these ones are the ones considered the most relevant.

Based on the variables presented, it could be seen than all the cases, except for the base technology, will produce ester from the presence o fatty acid, however, not all of them have the same effect when water is presented, in the case of solid resin and enzyme, water could have a negative effect. In the case of monolithic, the leaching is due to the contact with a liquid phase, not necessarily water and therefore the effects of water itself are quite not know yet.

The reaction temperature is quite low for the base, acid and enzymatic process but could be quite high for the other three options, this is one of the biggest reasons for the cost of the technology to be from medium to quite expensive. In the case of enzymes, even thought the reaction temperature is quite low, the cost of the catalyst make that the general investment is quite big, in some cases, making the technology not viable.

Table 5. Comparison of different technologies for Biodiesel production [1,6]

Variable	Base	Enzyme	Supercritical	Monolithic	Resin	Acid
Temp. [°C]	60-70	30-50	200-350	50-180	60-180	50-80
Products from FFA	Soaps	Esters	Esters	Esters	Esters	Esters
*Effect of Water**	↓	↓	——	——	— ↓	——
Yield to ester	Normal	High	High	Normal	Good	Normal
Purification of glycerol	Difficult	Simple	Simple	Simple	Simple	Difficult
Reaction time+	1-2 h	8-70 h	4-10 min	6 h	variable	4-70 h
Ester purification	Difficult	Simple	Simple	Simple	Simple	Difficult
Cost	Cheapest	Expensive	Expensive	Medium	Medium	Cheaper
Amount of equipment	High	Low	Low	Low	Low	High

* in this case the down arrow mean that water is a draw back while the line means that the is not effect and the system will be able to treat a raw material with some amount of water. For the Enzyme case, a down arrow has been supply, in this case is important to say that is believe that some water is require for enzyme activation; however, a lot of water will produce a deactivation of the catalyst. In the case of the resin, it could be seen a down arrow as well as a line, this is due to the fact that water has different effect over different solid catalyst. In the case of the monolithic scenario, a line has been selected due to the fact that leaching it is not causing by water per se but for a non stability of the catalyst.

+the reaction time set in this table is what it is most likely, however, it is important to point out that other times for the same technology could be found in the open literature

The purification of ester as well as the quality of it and the quality of glycerin is quite related to the type of technology used, especially when the catalyst is in a homogeneous phase. In the last case, the catalyst needs to be separated and the process needs neutralization, separation, purification and in some cases the need of effluent treatment. For all the heterogeneous as well as the non catalytic alternatives, the need of purification is simpler and the needs of equipment are also lower. In the case of the monolithic reactor, the leaching problem could make this technology less suitable due to the possible need of separation and purification, so far this problem have been presented [79] but not much more has been done in order to solve it, and therefore, it has been considered that this disadvantage might be easier to overcome.

In all scenarios, the yield for biodiesel is good, even thought in some cases this could be much higher, however, due to industrial scenarios, the final product will have a high yield and high conversion, leaving the considerations of conversion to only the reaction step and the need of recycle and other treatments.

4. EMISSIONS CHARACTERISTICS

Emissions from any type of combustion engine are important and relevant in order to understand how the green house gasses are reduce due to the use of alternative energies such as biodiesel, bioalcohols, etc. It is a major concern that all the alternative energies produce less emissions that the regular petroleum fuels, to not only shows that these energies are renewable and sustainable but also since it will decreases the gas emissions from combustion engines.

The mayor concern gases, according to the Kyoto protocol [80] are methane (CH_4), Carbon dioxide (CO_2), nitrous oxide (N_2O), sulphur hexafluoride (SF_6), hydroflourcarbons (HCF) and perflourcarbons (PFC). However these are the major GHG (green house gases) the use of biodiesel will also have an effect on the amount of other gases such as NOx, sulphur compounds and particulate matter for example.

Even more, the properties of petroleum diesel fuel could be improve if a blend is used (meaning that a certain amount of biodiesel is added into regular diesel), the most typical blend is either B5 or B20, meaning that in the regular diesel fuel tank 5 or 20 %, respectively, is biodiesel. These blends have shown several advantages over the regular diesel, however, the most important goal is related to the possibility of changing the entire diesel market for biodiesel, something that is still a challenge due to the millions of liter of fuel that are being required every day.

Accordingly to Xue et al. [81], in the literature there are several works that address this concern, the authors have done a comparison of several papers and arrange the results in different categories such as outcome of NOx emissions, CO emissions, HC emissions, CO_2 emissions, PM emissions (particulate matter), and some others. The authors have compared the emissions from a pure biodiesel combustion exhaust emissions engine with a regular diesel one. They have found that for CO, HC and PM, the reduction in the percentage of the emissions is over 85%, while for NOx and CO_2 there are more works that claims that these emissions will increase in a higher percentage, being it 65% for NOx and 46% for CO_2 (See reference 81 and reference in within). The case of aromatic compounds present no works that claim that this percentage could increase, it will remains as the same or it will decrease (85% decrease), however, for carbonyl compounds 80% is the augmentation in the emissions. Even more, it is reported by Song and Zhang [82] that the break power and torque increased with the increase in biodiesel percentage in the blend. The higher amount of biodiesel gives more oxygen to the system, reducing the ignition time considerable. However, an opposite result is presented by Carraretto et al. [83] whom had found that the engine power and torque is reduced due to an increase of the amount of biodiesel in the blends, similar results was found by Ayding et al. [84] and by Murillo et al. [85]. A middle situation was found by Gumus & Kasifoglu [86], they found that the power increase until the blend is B20 and from there to B100 the power decreases, this was done in a single 4stroke,DI air cooled diesel engine. Usta et al. [87] have reported a similar result to that from Gumus & Kasigoflu [86].

According to the review done by Xue et al. [81] the reduction in the engine power could be due to the high viscosity that is associated to biodiesel, this could have produce a mayor atomization problem in the combustion chamber and therefore a reduction in the engine power. Similar work was done by Fazal et al. [88], the authors have compared several works and how the emissions increases or not when comparing different blends, the majority of the works they have reported, 10 out of 15, using different raw materials, claim that the NOx emissions increases in an average of 8.6% while the average of decreasing is around 11.7% showing that despite the fact that most works claims a increase on the NOx, the reduction average is higher. All the referenced work shows a net decrease in the CO and CO_2 emissions, as well as on particulate matter, hydrocarbons, smoke density and SO_2. Showing that in general terms the use of biodiesel has a net positive effect on the mayor emissions as requested by the Kyoto protocol [80]. However, there are some drawbacks when using biodiesel such as the fuel pump failure, filter plugging sticking, etc.

The study of the different types of emissions that could provide from biodiesel, diesel or blends combustions has been studied extensively. Here, it will be presented some works and what have they studied and the results they have achieved, comparing them in order to find tendency on which compositions will have the better chance to less pollute the environment.

Qi et al. [89] compared the effect of biodiesel, regular diesel and blends with ethanol and diethyl ether in a single cylinder 4 –stroke ID engine. They have found that the presence of oxygen in the molecule produce a reduction in the smoke emissions, especially at high engine loads; however, there is not net modification on the NOx emissions between diesel and blends, but the presence of ethanol. The hydrocarbons emissions also increase in the presence of additives. Similar work was done by Kegl [90] where they showed that the torque is reduce in 10% when using biodiesel due to the lower heating value of the pure rapeseed oil. They have results agreeing with Qi et al. [89] when comparing the effect on the smoke and with Fazal et al. [88] in the effect of biodiesel over NOx emissions. However, in lower engines speed, the different from biodiesel and regular diesel NOx emissions became smaller. They results on the amount of Hydrocarbons emissions is in opposite direction to the expected results, showing a small increase in these compounds emissions, due maybe to unburned compounds in the engine cylinders. The higher viscosity, density and elasticity bulk modulus produce that when biodiesel is combusted there is a higher injection pressure and higher injection timing required, these produce a higher temperature and pressure in the cylinders. Park et al. [91] studied similar exhaust emissions but using a different injection configurations, they proposed a multi injection system, they have found that the emissions from the most commons compounds decreases when using multi injections compared to simple injection. Other blends could be found in the literature in order to see if the properties of the biofuel could be improve, that is the case of the work done by Zhu et al. [92]. Hey have compared different biodiesel blends with methanol or ethanol with Euro V diesel fuel. The effect over the NOx is very good due to a decrease of them. However, the presence of alcohols in the mixture produces an increase over the hydrocarbons and the CO emissions. Even more, the presence of alcohols produce an increase on the break specific fuel consumption due to the heating values of methanol and ethanol. The best combinations is achieved when using 5% blends of ethanol in the mixture, in this case, the HC and CO are reduced in comparison with the regular diesel and all the biodiesel blends. Banapurmath & Tewari [93] studied the addition of ethanol to biodiesel, the smoke opacity it is reduce by adding ethanol to jatropha biodiesel, similar tendency it is obtained for the unborn hydrocarbons due to the complete combustion of ethanol in the engine. Due to a better combustion of the molecule, the amount of CO is reduced when ethanol is blend with biodiesel if compared to regular diesel and pure biodiesel. Opposite to the previous results is the concentration of NOx, this pollutant increase its emission when using high loads in the combustion engine.

Barabás et al. [94] have concluded that, when comparing a diesel, biodiesel and blends (with no extra additives as done before), the amount of CO emissions is always higher for diesel when different loads of fuel are given to the engine. Not the same happens for the blends, where some of the blends have less emission for lower loads to the engine, while at higher loads the mix D70B25E5 has the highest concentration. Due to the reduction of the CO emissions, the authors believe that this is the key factor to the increase of the CO2 emission when comparing the blends with pure diesel. Increasing the emissions as the load of fuel increases. In the case of the NOx, for higher load of fuel, the NOx from diesel the smallest value or close to being the smallest. While for a low loads of fuel, the concentration of NOx is quite similar for all types of fuels. The smoke pollutants becomes higher for the high loads of fuel in the engine, while for low and medium loads the contributions are 5, or more, times smaller. Moon et al. [95] have done a similar study but using biodiesel blends with GTL diesel. Unfortunately they have found that the NOx increases due to the presence of oxygen in the molecule especially in the biodiesel diesel blends. However, a reduction in the amount of particulate matter could be achieved due to the mix of biodiesel with GTL fuel.

The study of the emissions of polycyclic aromatic hydrocarbons (PAHs), nitro PAHs and oxy PAHs in a passenger car were studied by Karavalakis et al. [96], when running it with biodiesel. They have found that the emission of these compounds does not follow any type of tendency but biodiesel might increase the toxicity of the emissions due to the presence of toxically compounds. Comparing different biodiesel emissions, Valente et al. [97] compared the emissions from castor oil and soybean oil biodiesel. The authors have studied blends from 5 to 85% of biodiesel. For high loads on the engine the authors have found that the concentration of CO2, CO and HC. The concentration of CO and HC was higher in all the range when biodiesel was use over diesel while the concentration of CO2 was smaller when using small loads. The only exception was when using soybean oil and in high concentration on the blend. In those cases, the HC emission where less when biodiesel was employed.

A study on the emissions of NOx, from biodiesel combustion engines, was done by Sun et al. [98]. A quite complete work was done analyzing a large amount of literature information. It has been concluded that there are several causes for the changes in the NOx concentration from fuels, present them in a qualitative way. However, it is inconclusive that NOx emissions increase if compared biodiesel with regular petroleum diesel. There not a net tendency when considering different variables such as engine type, or technology. Even more, there is not a net tendency when comparing different operational conditions. The most possible reasons for difference in the NOx concentration are: the system response issue, advance in start combustion, adiabatic flame temperature, decrease radiation heat transfer with biodiesel,

Other alcohol additives that have been tested is butanol, alcohol that could also be employed to produce biodiesel, it is a new alternative in order to improve the properties of the final diesel and to reduce the emissions from engines. Lebedevas et al. [99] have done a research where biodiesel, butanol, and biodiesel and butanol are compared. The biodiesel have been produce by using two different alcohols, methanol as well as butanol. The authors have found that the mixture of rapeseed methyl and rapeseed butyl ester with butanol and regular diesel are extremely promising in being used due to the high reduction of pollutants from their combustion, such as 4 times for hydrocarbons and for CO. It has also been found that having a 30% of biocomponents is the optimum amount, a higher amount of biocomponents will produce an increase in the CO emissions by 25%. A reduction of 60% in smoke capacity, 11% for break specific fuel consumption, and up to 32 % for CO are the mayor results obtained by Buyukkaya [100] when comparing rapeseed oil and its blends (5, 20 and 70%) with regular diesel. This was done in a six cylinder, four stroke, turbocharged injection diesel engine. From their results it is concluded that blends with less than 20% of biodiesel are quite promising for the full scale up

A comparison between karanja biodiesel and petroleum diesel have shown that the injection timing is smaller when using biodiesel and also helps the ignition time, making it shorter. It does produce a reduction in HC and smoke, showing no changes for the CO concentration but showing an increase in the NOx emissions [101]. Biodiesel could also be compared with Fischer-Tropsch reaction as well as blends. Nabi & Hustad [102] have presented that when using FT, there is less contamination due to CO, HC, CO2 Smoke, NOx. However, jathropa biodiesel presented lower amounts of CO, THC, but higher concentrations of NOx. However, as long as the percentage of the blends increases, the thermal efficiency of the engines decreases. In the case of particulate matter, the presence of biodiesel in the FT fuel reduced considerable the emissions as well as the size of the particles emitted. The authors claim that a B25% blend is environmentally attractive. A similar work was done by Lapuerta et al. [103], where biodiesel was mix with Fischer-Tropsch fuels. The authors have found similar results for the smoke, particulate matter, and hydrocarbons.

So far we have seen the effects over the mayor pollutants based on vegetable oils biodiesels vs. petroleum diesel engines. Batan et al. [104] have done a similar work but using algae oil to produce biodiesel and compared this fuel with others. They have compared three majors' compounds, CO2, CH4 N2O, it has been found that it has a smaller reduction on the CO2 emission when compared with soybean diesel, but it does have a reduction on the N2O emissions when compared with soybean as well as regular diesel. Showing that the biodiesel from microalgae is a suitable and potential source of clean raw material. Other alternatives are the use of antioxidants, in this case, the bio part of the fuel could be compromised due to the presence of petroleum derivates but the properties of the fuel might change. Tert-butylhydroquinone was found by Ryu [105] as the best antioxidant showing a great stability and no effects on the emissions form a combustion engine if compared with biodiesel.

As it can be seen, from all the previous works, which are some but not all you can find on the literature, there is a not net answer to what biodiesel, biodiesel blends, and other fuel does produce to the exhaust emissions, it does depend in many variables, however, most of them, do present positive results towards it use and a reduction on CO, CO_2, and HC compounds, showing that alternative fuels could be the key answer to achieved the desired emissions as established in international agreements.

5. CONCLUSIONS

Different works have been presented in this chapter with the aim of showing how different technologies, different raw materials and therefore different final biodiesel have different performance in exhaust emissions measurements. It has been point out that most of the results presented shown that carbon monoxide, carbon dioxide and hydrocarbons pollutants are reduced in different amounts depending on the type of biodiesel but also in the engine used. However, the most controversial pollutant, NOx, has a divided opinion on if this pollutant concentration increases or decreases due to the use of alternative fuels. All these results shown that the use of biodiesel, or biodiesel blends with regular diesel or other additives such as alcohols, appear as a net possibility for greenhouses gas reduction. The use of biodiesel and its blends could be part of the answer to achieved international standards of allow gas emissions form combustion engines.

REFERENCES

[1]   Marchetti. J.M.. Miguel. V.U.. Errazu. A.E. *Possible methods for biodiesel production.* Renewable and Sustainable Energy Reviews. 11. (2007), 1300-1311.

[2]   Srivastava A. Prasad R.. *Triglycerides-based diesel fuels.* Renewable Sustainable Energy Reviews. 4. (2000), 111–133.

[3]   Ma F. Hanna MA.. *Biodiesel production: A review.* Bioresource Technology. 70. (1999), 1–15.

[4]   Fukuda. H., Kondo. A., Noda, H. *Biodiesel fuel production by transesterification of oils.* Journal of Bioscience & Bioengineering. 92(5). (2001), 405–416.

[5]   Knothe, G., Van Gerpen, J., Krahl, J. (Eds.), *The biodiesel handbook*, AOCS Press, Champaign, Illinois, 2005, 302 pp.

[6]   Marchetti, J.M. *Biodiesel Production Technologies.* Nova Publishers. (2010). New York, New York.

[7]   Schuchardt, U., Sercheli, R., Vargas, R.M. *Transesterification of vegetable oils: a review.* JBCS, 9(1). (1998), 199–210.

[8]   O'Brien, R., Farr, W., Wan, P. Editors. *Introduction to Fats and Oils Technology*: Second Edition, 2000. Editorial AOCS.

[9]   Noureddini, H., Zhu, D. *Kinetics of transesterification of soybean oil.* Journal of American Oil Chemists Society. 74(11). (1997), 1457-1463.

[10]   Freedman, B., Pryde, E.H., Mounts, T.L. *Variables affecting the yields of fatty esters from transesterified vegetable oils.* Journal of American Oil Chemists Society. 61(10). (1984), 1638-1643

[11]   Vicente, G., Martínes, M., Aracil, J. *Integrated biodiesel production: a comparison of different homogeneous catalysts systems.* Bioresource Technology. 92, (2004), 297-305.

[12]   Meng, X., Chen, G., Wang, Y. *Biodiesel production from waste cooking oil via alkali catalyst and its engine test.* Fuel Processing Technology. 89, (2008), 851-857.

[13]   Alamu, O.J., Waheed, M.A., Jekayinfa, S.O. *Biodiesel production from Nigenerian palm kernel oil: effect of KOH concentration on yield.* Energy for Sustainable Development. XI(3). (2007), 77-82.

[14]   Alamu, O.J., Waheed, M.A., Jekayinfa, S.O. *Effect of ethanol-palm kernel oil ratio on alkali-catalyzed biodiesel yield.* Fuel. 87. (2008), 1529-1533.

[15]   Colucci, J.A., Borrero, E.E., Alape, F. *Biodiesel from an alkaline transesterification reaction of soybean oil using ultrasoni mixing.* Journal of the American Oil Chemists Society. 82(7). (2005), 525-530.

[16]   http://www.elsbett.com/engl/oil.htm

[17]   http://journeytoforever.org/biodiesel_yield2.html

[18]   http://www.indiamart.com/company/2466820/products.html

[19]   Fuse, T., Kusu, F., Takamura, K. *Determination of acid values of fats and oils by flow injection analysis with electrochemical detection.* Journal of Pharmaceutical and Biomedical Analysis. 15, (1997), 1515-1519.

[20]   http://vegburner.co.uk/oils.htm

[21]   http://journeytoforever.org/biofuel_library/fatsoils/fatsoils2.html

[22]   http://journeytoforever.org/biodiesel_SVO-palm.html

[23] Dalai, A,K. *Applications of vegetable oil derived esters as a diesel additive.* Energeia. 15(6), (2004), 1-6.

[24] http://www.diracdelta.co.uk/science/source/f/l/flash%20point/source.html

[25] http://www.canolacouncil.org/uploads/Chemical1-6.pdf

[26] Purdy, R.H. *High oleic sunflower: Physical and chemical characteristics.* Journal of American Chemists Society. 63(8), (1986), 1062-1066.

[27] http://www.sunflowernsa.com/oil/product-specifications/nusun-refined/

[28] http://hypertextbook.com/facts/2000/IngaDorfman.shtml

[29] Chhetri, A.B., Watts, K.C., Islam, M.R. *Waste Cooking Oil as an Alternate Feedstock for Biodiesel Production.* Energies. 1. (2008), 3-18.

[30] Demirbas, A. *Biodiesel. A Realistic Fuel Alternative for Diesel Engines.* Editorial Springer 2008.

[31] http://globalsmartinvestment.com/SoyOil.aspx

[32] Zhang, Y., Dubé, M.A., McLean, D.D., Kates, M. *Biodiesel production from waste cooking oil: 2 Economic assessment and sensitivity analysis.* Bioresource Technology. 90. (2003), 229-240.

[33] Dmytryshyn, S.L., Dalai, A.K., Chaudhari, S.T., Mishra, H.K., Reaney, M.J. *Synthesis and Characterization of Vegetable Oil Derived Esters: Evaluation for their Diesel Additive Properties.* Bioresource Technology. 92. (2004), 55.64.

[34] Phan, A.N., Phan, T.M. *Biodiesel Production from Waste Cooking Oils.* Fuel, 87. (2008), 3490-3496.

[35] Chisti, Y. *Biodiesel from microalgae.* Biotechnology Advances. 25. (2007), 294-306.

[36] Moser, B.R. *Biodiesel production, properties and feedstocks.* In Vitro Cellular and Developmental Biology – Plants. 45. (2009), 229-266.

[37] Balat, M. *Potential alternatives to edible oils for biodiesel production – a review of current work.* Energy Conversion & Managament. 52, (2011), 1479-1492.

[38] Lin, L., Cunshan, Z., Vittayapadung, S., Xiangqian, S., Mingdong, D. *Opportunities and challenges for biodiesel fuel.* Applied Energy. 88, (2011), 1020-1031.

[39] Zheng, S., Kates, M., Dubé, M.A., McLean, D.D. *Acid-catalyzed production of biodiesel form waste frying oil.* Biomass & Bioenergy. 30(3). (2006), 267–272.

[40] Canakci M, Van Gerpen J. *Biodiesel production via acid catalysis.* Transactions of the ASAE. 42(5). (2003), 1203–1210.

[41] Canakci M, Van Gerpen J. *Biodiesel production from oils and fats with high free fatty acids.* Transactions of the ASAE. 44(6). (2003), 1429–1436.

[42] Zhang, Y., Dubé, M.A., McLean, D.D., Kates, M. *Biodiesel production from waste cooking oil: 2 Economic assessment and sensitivity analysis.* Bioresource Technology. 90. (2003), 229-240.

[43] Van Gerpen, J. *Biodiesel processing and production.* Fuel Processing Technology. 86. (2005), 1097-1107.

[44] Marchetti, J.M., Errazu, A.F. *Esterification of free fatty acids using sulfuric acid as catalysts in the presence of triglycerides.* Biomass & Bioenergy. 32. (2008), 892-895.

[45] Bournay, L., Casanave, D., Delfort, B., Hillion, G., Chodorge, J.A. *New heterogeneous process for biodiesel production: A way to improve the quality and the value of the crude glycerin produced by biodiesel plants.* Catalysis Today. 105, (2005), 190-192.

[46] Di Serio, M., Tesser, R., Dimiccoli, M., Cammarota, F., Nastasi, M., Santacesaria, E. *Synthesis of biodiesel via homogeneous Lewis acid catalyst.* Journal of Molecular Catalysis A: Chemical. 239, (2005) 111-115.

[47] Soriano Jr, N.U., Venditti, R., Argyropoulos, D.S. *Biodiesel synthesis via homogeneous Lewis acid-catalyzed transesterification.* Fuel. 88, (2009) 560-565.

[48] Hamad, B., Lopes de Souza, R.O., Sapaly, G., Carneiro Rocha, M.G., Pries de Oliveira, P.G., Gonzalez, W.A., Andrade Sales, E., Essayem, N. *Transesterification of rapeseed oil with ethanol over heterogeneous heteropolyacids.* Catalysis Communications. 10, (2008), 92-97.

[49] Suppes, G.J., Dasari, M.A., Doskocil, E.J., Mankidy, P.J., Goff, M.J. *Transesterification of soybean oil with zeolite and metal catalysts.* Applied Catalysis A: General 257, (2004), 213-223.

[50] Di Serio, M. Ledda, M., Cozzolino, M., Minutillo, G., Tesser, R., Santacesaria, E. *Transesterification of Soybean Oil to Biodiesel Using Heterogeneous Basic Catalysts.* Ind. Eng. Chem. Res. 45, (2006,) 3009-3014.

[51] Kulkarni, M.G., Gopinath, R., Meher, L.C., Dalai, A.K. *Solid acid catalyzed biodiesel production by simultaneous esterification and transesterification.* Green Chemistry. 8, 1056-1062, 2006.

[52] López, D.E., Goodwin Jr., J.G., Bruce, D.A., Furuta, S. *Esterification and transesterification using modified zirconia catalysts.* Applied Catalysis A: General. 339, (2008), 76-83.

[53] Cao, F., Chen, Y., Zhai, F., Li, J., Wang, J., Wang, X., Wang, S., Zhu, W. *Biodiesel production from high acid value waste frying oil catalyzed by superacid heteroployacid.* Biotechnology and Bioengineering. 101(1), (2008), 93-100.

[54] Marchetti, J.M., Errazu, A.F. *Biodiesel production from acid oils and ethanol using a solid basic resin as catalyst.* Biomass & Bioenergy. 34(3), (2010), 272-277.

[55] Antczak, M.S., Kubiak, A., Antczak, T., Bielecki, S. *Enzymatic biodiesel synthesis - key factors affecting efficiency of the process.* Renewable Energy. 34. (2009), 1185-1194.

[56] Rodrigues, R.C., Volpato, G., Wada, K., Ayub, M.A.Z. *Enzymatic synthesis of biodiesel from transesterification reactions of vegetble oils and short chaing alcohols.* Journal of American oil Chemists Society. 85, (2008) 925-930.

[57] Dalla Rosa, C., Morandim, M.B., Ninow, J.L., Oliveira, D., Trichel, H., Vladimir Oliveira, J. *Lipase-catalyzed production of fatty acid ethyl esters from soybean oil in compressed propane.* The Journal of Supercritical Fluids. 47, (2008), 49-53.

[58] Matassoli, A.L.F., Corrêa, I.N.S, Portillo, M.F., Veloso, C.O., Langone, M.A.P. *Enzymatic synthesis of biodiesel via alcoholysis of palm oil.* Applied Biochemistry & Biotechnology. 155, (2008), 44-52.

[59] Dizge, N., Aydiner, C., Imer, D.Y., Bayramoglu, M., Tanriseven, A., Keskinler, B. *Biodiesel production from sunflower, soybean and waste cooking oils by transesterification using lipase immobilized onto a novel microporous polymer.* Bioresource Technology. 100, (2009), 1983-1991.

[60] Demirbaş, A. *Biodiesel fuels from vegetable oils via catalytic and non catalytic supercritical alcohol transesterification and other methods: a survey.* Energy Conversion and Management. 44, (2003), 2093-2109.

[61] Saka, S., Kusdiana, D. *Biodiesel fuel from rapeseed oil as prepared in supercritical methanol.* Fuel. 80, (2001), 225-231.

[62] Kusdiana, D., Saka, S. *Kinetics of transesterification in rapeseed oil to biodiesel fuel as treated in supercritical methanol.* Fuel. 80, (2001), 693-698.

[63] Kusdiana, D., Saka, S. *Effects of water on biodiesel fuel production by supercritical methanol treatment.* Bioresource Technology. 91, (2004), 289-295.

[64] Demirbaş, A. *Biodiesel fuels from vegetable oils via transesterification in supercritical methanol.* Energy Conversion and Management. 43, (2002), 2349-2356.

[65] Hawash, S., Kamal, N., Zaher, F., Kenawi, O., El Diwani, G. *Biodiesel fuel from Jatropha oil via non-catalytic supercritical methanol transesterification.* Fuel. 88, (2009), 579-582.

[66] Gui, M.M., Lee, K.T., Bhatia, S. *Supercritical ethanol technology for the production of biodiesel: process optimization studies.* The Journal of Supercritical Fluids. Doi:10.1016/j.supflu.2008.12.014.

[67] Kasim, N.S., Tsai, T.H., Gunawan, S., Ju, Y.H. *Biodiesel production from rice bran oil and supercritical methanol.* Bioresource Technology. 100, (2009), 2399-2403.

[68] Sawangkeaw, R., Bunyakiat, K., Ngamprasertsith, S. *Effect of co-solvents on production of biodiesel via transesterification in supercritical methanol.* Green Chemistry. 9, (2007), 679-685.

[69] Tan, K.T., Lee, K.T., Mohamed, A.R. *Effects of free fatty acids, water content and co-solvent on biodiesel production by supercritical methanol reaction.* The Journal of Supercritical Fluids. 53, (2010), 88-91.

[70] Yin, J.Z., Xiao, M., Song, J.B. *Biodiesel from soybean oil in supercritical methanol with co-solvent.* Energy Conversion & Management. 49, (2008), 908-912.

[71] Han, H., Cao, W., Zhang, J. *Preparation of biodiesel from soybean oil using supercritical methanol and CO_2 as co-solvent.* Process Biochemistry. 40, (2005), 3148-3151.

[72] Dubé, M.A., Tremblay, A.Y., Liu, J. *Biodiesel production using a membrane reactor.* Bioresource Technology. 98, (2007), 639–647.

[73] Baroutian, S., Aroua, M., Raman, A.A., Sulaimna, N.M.N. *A packed bed membrane reactor for production of biodiesel using activated carbon supported catalyst.* Bioresource Technology. 102(2), (2011), 1095-1102.

[74] Zhu, M., He, B., Shi, W., Feng, Y., Ding, J., Li, J., Zeng, F. *Preparation and characterization of PSSA/PVA catalytic membrane for biodiesel production.* Fuel. 89(9), (2010), 2299-2304.

[75] Cheng, L,H., yen, S.Y., Su, L.S., Chen, J. *Study on membrane reactors for biodiesel production by phase behaviors of canola oil methanolysis in batch reactors.* Bioresource Technology. 101(17), (2010), 6663-6668.

[76] Cao, P., Dubé, M.A., Tremblay, A. *Methanol recycling in the production of biodiesel in a membrane reactor.* Fuel. 87(6), (2008), 825-833.

[77] Kolaczkowski, S.T., Asli, U.A., Davidson, M.G. *A new heterogeneous ZnL_2 catalyst on a structured support for biodiesel production.* Catalysis Today. 147(1), (2009), S220-S224.

[78] Dizge, N., Aydinet, C., Imer, D.Y., Bayramoglu, M., Tanriseven, A., Keskinler, B. *Biodiesel production from sunflower, soybean, and waste cooking oils by*

transesterification using lipase immobilized onto a novel microporous polymer. Bioresource Technology. 100(6), (2009), 1983-1991.

[79] Tonetto, G.M., Marchetti, J.M. *Transesterification of soybean oil over Me/Al$_2$O$_3$ (Me = Na, Ba, Ca, and K) catalysts and monolith K/Al$_2$O$_3$-cordierite.* Topics in Catalysis. 53(11-12), (2010), 755-762.

[80] http://unfccc.int/resource/docs/convkp/kpeng.pdf

[81] Xue, J., Grift, T.E., Hansen, A.C. *Effect of biodiesel on engine performances and emissions.* Renewable and Sustainable Energy Reviews. 15, (2011), 1098-1116.

[82] Song, J.T., Zhang, C.H. *An experimental study on the performance and exhaust emission fo a diesel engine fuelled with soybean oil methyl esters.* P I Mech. Eng. D-J. 222, (2008), 2487-2496.

[83] Carraretto, C., Macor, A., Mirandola, A., Stoppato, A., Tonon, S. *Biodiesel as alternative fuel: experimental analysis and energetic evaluations.* Energy. 29, (2004), 2195-2211.

[84] Ayding, H., Bayindir, H. *Performance and emission analysis of cottonseed oil methyl ester in a diesel engine.* Renewable Energy. 35, (2010), 588-592.

[85] Murillo, S., Miguez, J.L., Porteiro, J., Granada, E., Moran J.C. *Performance and exhaust emissions in the use of biodiesel in outboard diesel engines.* Fuel. 86, (2007), 1765-1771.

[86] Gumus, M., Kasifoglu, S. *Performance and emission evaluation of a compression ignition engine using a biodiesel (apricot seed kernel oil methyl ester) and its blends with diesel fuel.* Biomass & Bioenergy. 34, (2010), 134-139.

[87] Usta, N., Ozturk, E., Can, O., Conkur, E.S., Nas, S., Çon, A.H., Can, A.Ç., Topcu, M. *Combustion of biodiesel fuel produced from hazelnut soapstock/waste sunflower oil mixture in a diesel engine.* Energy Conversion & Management. 46, (2005), 741-755.

[88] Fazal, M.A., Haseeb, A.S.M.A., Masjuki, H.H. *Biodiesel feasibility study: An evaluation of material compatibility; performance; emission and engine durability.* Renewable and Sustainable Energy Reviews. 15, (2011), 1314-1324.

[89] Qi, D.H., Chen, H., Geng, L.M., Bian, Y.Z. *Effect of diethyl ether and ethanol additives on the combustion and emission characteristics of biodiesel-diesel blended fuel engine.* Renewable Energy. 36, (2011), 1252-1258.

[90] Kegl, B. *Influence of biodiesel on engine combustion and emission characteristics.* Applied Energy. 88(5), (2011), 1803-1812.

[91] Park, S.H., Yoon, S.H., Lee, C.S. *Effects of multiple-injection strategies on overall spray behavior, combustion, and emissions reduction characteristics of biodiesel fuel.* Applied Energy. 88, (2011), 88-98.

[92] Zhu, L., Cheung, C.S., Zhang, W.G., Huang, Z. *Emissions characteristics of a diesel engine operating on biodiesel and biodiesel blended with ethanol and methanol.* Science of the Total Environment. 408, (2010), 914-921.

[93] Banapurmath, N.R., Tewari, P.G. *Performance, combustion, and emissions characteristics of a single-cylinder compression ignition engine operated on ethanol-biodiesel blended fuels.* Proc. IMECHE. JAUTO. 224, (2009), 533-543.

[94] Barabás, I., Todorut, A., Băldean, D. *Performance and emission characteristics of an CI engine fueled with diesel-biodiesel-bioethanol blends.* Fuel. 89, (2010), 3827-3832.

[95]   Moon, G., Lee, Y., Choi, K., Jeong, D. *Emission characteristics of diesel, gas to liquid, and biodiesel-blended fuels in a diesel engine for passengers cars.* Fuel. 89, (2010), 3840-3846.

[96]   Karavalakis, G., Boutsika, V., Stournas, S., Bakeas, E. *Biodiesel emissions profile in modern diesel vehicles. Part 2: effects of biodiesel origin on carbonyl, PAH, nitro-PAH and oxy-PAH emissions.* Science of the Total Environment. 409, (2011), 738-747.

[97]   Valente, A.S., da Silva, M.J., Pasa, V.M.D., Belchoir, C.R.P., Sodré, J.R. *Fuel consumption and emissions from a diesel power generator fuelled with castor oil and soybean biodiesel.* Fuel. 89, (2010), 3637-3642.

[98]   Sun, J., Caton, J.A., Jacobs, T.J. *Oxides of nitrogen emissions from biodiesel-fuelled diesel engines.* Progress in Energy and Combustion Science. 36, (2010), 677-695.

[99]   Lebedevas, S., Lebedeva, G., Sendzikiene, E., Makareviciene, V. *Investigation of the performance and emission characteristics of biodiesel fuel containing butanol under the conditions of diesel engine operation.* Energy & Fuels. 24, (2010), 4503-4509.

[100] Buyukkaya, E. *Effects of biodiesel on a DI diesel engine performance, emission and combustion characteristics.* Fuel. 89, (2010), 3099-3105.

[101] Anand, K., Sharma, R.P., Metha, P.S. *Experimental investigations on combustion, performance, and emissions characteristics of a neat biodiesel-fuelled, turbocharged, direct injection diesel engine.* Proc. IMECHE. JAUTO. 224, (2009), 661-679.

[102] Nabi, M.N., Hustad, J.E. *Influence of Biodiesel adittion ot Fischer-Tropsch fuel on diesel engine performance and exhaust emissions.* Energy & Fuels. 24, (2010), 2868-2874.

[103] Lapuerta, M., Armas, O., Hernández, J.J., Tsolakis, A. *Potential for reducing emissions in a diesel engine by fuelling with conventional biodiesel and Fischer-Tropsch diesel.* Fuel. 89, (2010), 3106-3113.

[104] Batan, L., Quinn, J., Willson, B., Bradley, T. *Net energy and greenhouse gas emission evaluation of biodiesel derived from microalgae.* Environmental Science & Technology. 44, (2010), 7975-7980.

[105] Ryu, K. *The characteristics of performance and exhaust emissions of a diesel engine using a biodiesel with antioxidants.* Bioresource Technology. 101, (2010), 578-582.

INDEX

D

E

F

G

J

K

L

S